Reading Essentials

An Interactive Student Workbook

life.msscience.com

New York, New York Columbus, Ohio Chicago, Illinois Peoria, Illinois Woodland Hills, California

To the Student

In today's world, knowing science is important for thinking critically, solving problems, and making decisions. But understanding science sometimes can be a challenge.

Reading Essentials takes the stress out of reading, learning, and understanding science. This book covers important concepts in science, offers ideas for how to learn the information, and helps you review what you have learned.

In each chapter:

- **Before You Read** sparks your interest in what you'll learn and relates it to your world.
- **Read to Learn** describes important science concepts with words and graphics. Next to the text you can find a variety of study tips and ideas for organizing and learning information:
  - The **Study Coach** offers tips for getting the main ideas out of the text.
  - **Foldables™ Study Organizers** help you divide the information into smaller, easier-to-remember concepts.
  - **Reading Checks** ask questions about key concepts. The questions are placed so you know whether you understand the material.
  - **Think It Over** elements help you consider the material in-depth, giving you an opportunity to use your critical-thinking skills.
  - **Picture This** questions specifically relate to the art and graphics used with the text. You'll find questions to get you actively involved in illustrating the concepts you read about.
  - **Applying Math** reinforces the connection between math and science.
- Use **After You Read** to review key terms and answer questions about what you have learned. The **Mini Glossary** can assist you with science vocabulary. Review questions focus on the key concepts to help you evaluate your learning.

See for yourself. ***Reading Essentials*** makes science easy to understand and enjoyable.

The McGraw·Hill Companies

Send all inquiries to:
Glencoe/McGraw-Hill
8787 Orion Place
Columbus, OH 43240

ISBN 0-07-867125-6
Printed in the United States of America
5 6 7 8 9 10 024 09 08 07 06

Table of Contents

Exploring and Classifying Life

section 1 What is science?

Before You Read

Look at the title of Section 1. On the lines below, write what you think science is.

____

____

What You'll Learn

- how to apply scientific methods to problem solving
- how to measure using scientific units

Read to Learn

Study Coach

Identify the Main Idea As you read, identify the important ideas. For each question subhead, write the main idea of the subhead. Use the main ideas as a study guide for the section.

The Work of Science

One way scientists find out about the world is by asking questions. Science is an organized way of studying things and finding answers to questions.

There are many types of science. The names of the sciences describe what is being studied. For example, a life scientist might study the millions of different animals, plants, and other living things on Earth. Life scientists who study plants are botanists. Those who study animals are zoologists.

Critical Thinking

You solve problems every day. To figure things out, you have to think about them. You have to think about what will work and what will not. Suppose your portable CD player stops working. To figure out the problem, you have to think about it. You know that the CD player runs on batteries, so your first thought would be to replace the batteries. If replacing the batteries does not work, you have to think of other possible solutions. Thinking this way is called critical thinking. It is the way you use skills to solve problems. Separating important information from information that is not important is a skill. Identifying the problem is another skill you may have. ☑

☑ Reading Check

1. **Identify** two critical-thinking skills that might be used to solve problems.

____

____

____

FOLDABLES

A Classify Make a three-tab Foldable, as shown below. Under each tab, write facts about scientific methods, units of measurement, and science safety.

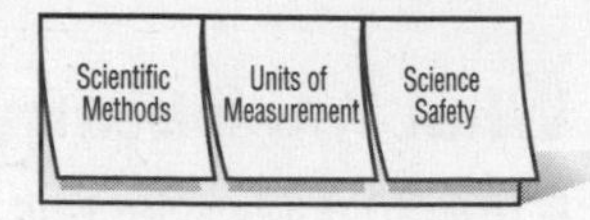

Picture This

2. **Explain** to a partner the steps in solving a problem scientifically when the hypothesis is supported. Then have your partner explain the steps when the hypothesis is not supported.

Solving Problems

Scientists use critical-thinking skills to try to solve problems and answer questions. Solving problems requires organization. In science, this organization often takes the form of a series of procedures called **scientific methods**. The procedures shown in the figure below are one way that scientific methods might be used to solve a problem.

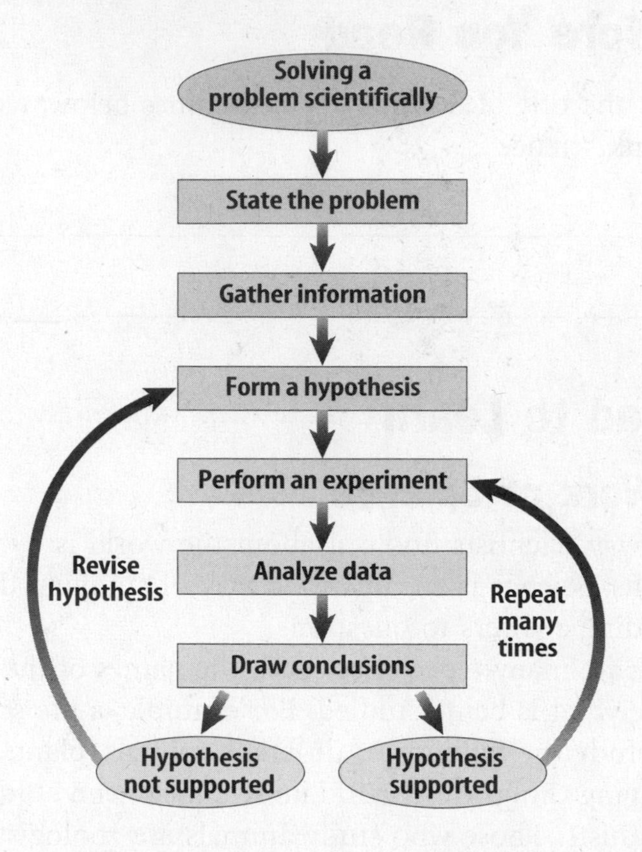

How are scientific problems stated?

The first step in solving a problem scientifically is to state the problem. For example, four cats were being boarded at a veterinarian's office. The veterinary technician noticed that two of the cats were scratching and had patches of skin with red sores. They had both been fine when they first arrived. The technician identified this as a problem. Now she must decide how to gather information about the problem.

How is scientific information gathered?

Scientists collect information through laboratory observations and experiments. Sometimes information is collected from fieldwork done outside the laboratory. For example, scientists might find out how a bird builds a nest by going outside and watching it.

Observation The technician gathers information about the problem with the cats by watching them closely. She watches to see if the behavior of the cats with the sores is different than the behavior of the other two cats. She observes that other than the scratching, the behavior of the four cats is the same. The technician finds out that the brand of cat food used at the clinic is the same as the one the cats get at home. She decides that the two cats are reacting to something in their environment. She notices that they seem to scratch most after using their litter boxes.

How do scientists form a hypothesis?

After scientists collect the information, they form a hypothesis. A **hypothesis** is a prediction that can be tested. After collecting the information, the technician hypothesizes that something in the cat litter is irritating the cats' skin. ☑

How do scientists test a hypothesis?

Scientists test a hypothesis by performing an experiment. In an experiment, the hypothesis is tested using controlled conditions.

The technician gets permission from the owner to test her hypothesis by running an experiment. The technician reads the label on two brands of cat litter. She finds that the ingredients of each brand are the same except that one has a deodorant. The cat litter used in the clinic has a deodorant. The technician finds out that the litter the cats have at home does not have a deodorant.

How do scientists use controls in an experiment?

The technician separates the cats with sores from the other two cats. She puts each of the cats with sores in a cage by itself. One cat is called the experimental cat. This cat gets a litter box with litter that does not have deodorant. The other cat gets a litter box that has cat litter with deodorant. The cat with deodorant cat litter is the control.

A **control** is the standard to which the outcome of a test is compared. The control cat will be compared with the experimental cat at the end of the experiment. Whether or not the cat litter has deodorant is the variable. A **variable** is something in an experiment that can change. An experiment should have only one variable. Other than the difference in the cat litter, the technician treats the cats the same.

Reading Check

3. **Define** What is a hypothesis?

____

____

____

Think it Over

4. **Analyze** Why is it important to have only one variable in an experiment?

____

____

____

How do scientists analyze data?

During the week, the technician observes both cats. She collects data on how often and when the cats scratch or chew. She records the data in a journal. The data show that the control cat scratches more often than the experimental one does. The sores on the experimental cat begin to heal. The sores on the control cat do not.

How do scientists draw conclusions?

The technician draws the conclusion that the deodorant in the cat litter probably irritated the skin of the two cats. To draw a conclusion is to get a logical answer to a question based on data and observation. The next step is to accept or reject the hypothesis. In this case, the technician accepts the hypothesis. If she had rejected it, then she would need to conduct new experiments.

The technician realizes that even though she accepted the hypothesis, she should continue her experiment to be surer of her results. She should switch the cats to see if she gets the same results again. However, if she did this, the healed cat might get new sores. She makes an ethical decision and does not continue the experiment. Ethical decisions are important in deciding what experiments should be done.

Think it Over

5. **Draw Conclusions** If the hypothesis is rejected, the scientist has failed. Is this statement true or false? Explain your reasoning.

____

____

____

____

What do scientists do with results of experiments?

It is important to share the information when using scientific methods. The veterinary technician shares her results with the cats' owner. She tells him she has stopped using the cat litter with the deodorant.

Developing Theories

After a scientist reports the results of an experiment that supports the hypothesis, many scientists repeat the experiment. If the results always support the hypothesis, the hypothesis can be called a theory. A scientific **theory** is an explanation of things or events based on scientific knowledge that is the result of many observations and experiments. It is not a guess. A theory usually explains many hypotheses. For example, scientists made observations of cells and experimented for more than 100 years before enough information was collected to propose a theory. A theory raises many new questions. Data or information from new experiments might change conclusions and theories can change.

Think it Over

6. **Compare** How is a theory different than a hypothesis?

____

____

____

What is scientific law?

A scientific **law** is a statement about how things work in nature that seems to be true all the time. Although laws can be modified as more information becomes known, they are less likely to change than theories. ☑

Laws tell you what will happen under certain conditions but do not necessarily explain why it happened. For example, in life science you might learn about laws of heredity. These laws explain how genes are inherited, but do not explain how genes work. There is a great variety of living things, but few laws to explain them. A law that describes how all cells work may never be developed.

How do scientific methods help answer questions?

You can use scientific methods to answer all kinds of questions. Using these methods does not guarantee that you will get an answer. Often they lead to more questions and more experiments. Science is about looking for the best answers to your questions.

Measuring with Scientific Units

An important part of scientific investigations is making accurate measurements. Many things you use every day are measured. In your science classes, you will use the same standard system of measurement that scientists use in their work. This system is called the International System of Units, or SI. For example, you may need to calculate the distance a bird flies in kilometers. Perhaps you will be asked to measure the amount of air your lungs can hold in liters. Some of the SI measurements are shown in this table.

Common SI Measurements

Measurement	Unit	Symbol	Equal to
Length	1 millimeter	mm	0.001 (1/1,000) m
	1 centimeter	cm	0.01 (1/100) m
	1 meter	m	100 cm
	1 kilometer	km	1,000 m
Volume	1 milliliter	mL	0.001 (1/1,000) L
	1 liter	L	1,000 mL
Mass	1 gram	g	1,000 mg
	1 kilogram	kg	1,000 g
	1 tonne	t	1,000 kg = 1 metric ton

Reading Check

7. **Explain** what scientific laws can tell you about nature.

______

______

______

Applying Math

8. **Calculate** How many centimeters are there in 2.5 meters? (Show your work.)

______

______

______

Safety First

Some of the scientific equipment that you use in the classroom or laboratory is the same as the equipment scientists use. Laboratory safety is important. It is important to wear proper eye protection. Make sure you wash your hands after handling materials. Following safety rules, as shown in the figure, will protect you and others from injury.

Picture This

9. Apply What part of your body will be protected by wearing safety goggles?

____

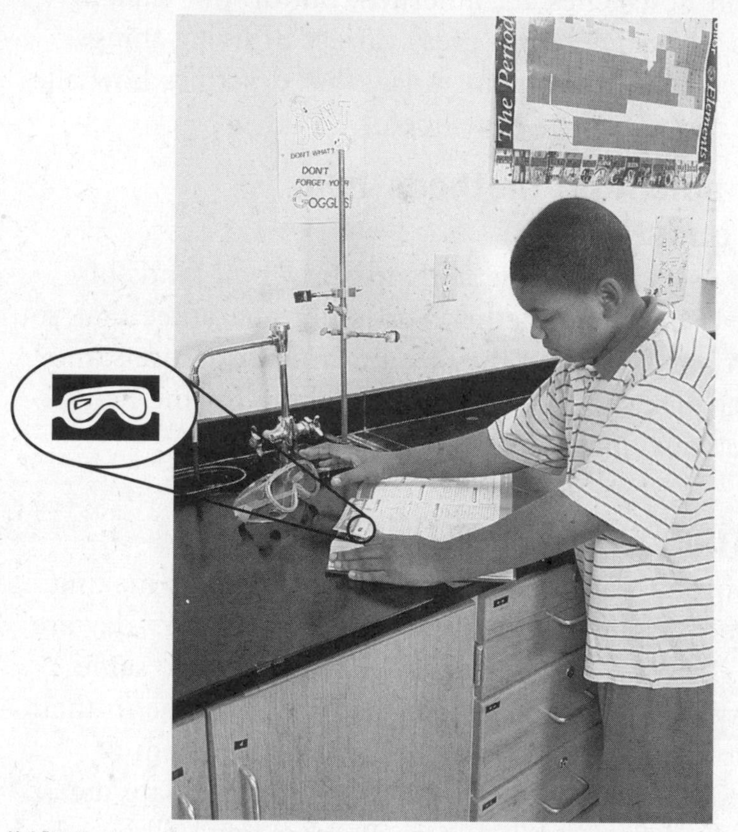

Mark Burnett

Symbols are often used in texts and laboratories to alert you to situations that require special attention. Some of these symbols are shown below.

Picture This

10. Identify Several symbols indicate safety equipment that you might wear when working in the lab. Circle two of the symbols.

After You Read

Mini Glossary

control: the standard to which the outcome of a test is compared

hypothesis: a prediction that can be tested

law: a statement about how things work in nature

scientific methods: an organized series of procedures used to solve a problem

theory: an explanation of things or events based on scientific knowledge that is the result of many observations and experiments

variable: something in an experiment that can change

1. Review the terms and their definitions in the Mini Glossary. Write a sentence that explains the difference between a control and a variable.

2. Decide on a problem you would like to solve. Use the steps in the scientific method to explain how you would solve the problem.

Visit **life.msscience.com** to access your textbook, interactive games, and projects to help you learn more about what science is.

chapter 1

Exploring and Classifying Life

section 2 Living Things

What You'll Learn

- the difference between living and nonliving things
- what living things need to survive

Before You Read

List three living things in your environment. What do these things need to live?

____

____

____

Study Coach

Create a Quiz As you study the information in this section, create questions about the information you read. Be sure to answer your questions.

FOLDABLES™

B Explain Make a half sheet Foldable, as shown below, to list the traits of living organisms.

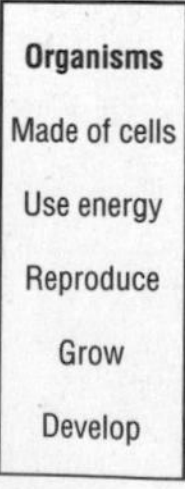

Read to Learn

What are living things like?

Any living thing is called an **organism**. Organisms vary in size from microscopic bacteria to giant trees. They are found just about everywhere. They have different behaviors and food needs. However, all organisms have similar traits. These traits determine what it means to be alive.

How are living things organized?

Living things are made up of small units called cells. A **cell** is the smallest unit of an organism that carries on the functions of life. Some organisms are made up of just one cell. Others are made up of many cells. Cells take in materials from their surroundings. They use the materials in complex ways. Each cell has an orderly structure and has hereditary material. The hereditary material has instructions for cell organization and function. All the things organisms can do are possible because of what their cells can do.

How do living things respond?

Living things interact with their surroundings. Anything that causes some change in an organism is a stimulus (plural, *stimuli*). The reaction to a stimulus is a response. Often that response results in movment.

Response to Stimuli To carry on its daily activity and to survive, an organism must respond to stimuli. Organisms respond to external stimuli such as movement and light.

Living things also respond to stimuli that occur inside them. For example, water or food levels in organisms' cells can increase or decrease. The organisms then make internal changes to keep the right amounts of water and food in their cells. An organism's ability to keep the proper conditions inside no matter what is going on outside the organism is called **homeostasis**.

How do living things get energy?

The energy that most organisms use to perform life activities comes from the Sun. Plants and some other organisms get energy directly from the Sun. They do this by combining sunlight with carbon dioxide and water to make food. People and most other organisms cannot use the energy of sunlight directly. Instead, they take in and use food as a source of energy. People get food by eating plants or other organisms that eat plants. Most organisms, including plants, must take in oxygen in order to release the energy of foods.

Some organisms, such as bacteria that live at the bottom of the oceans where sunlight cannot reach, cannot use the Sun's energy to make food. These organisms use chemical compounds and carbon dioxide to make food. They do not need oxygen to release the energy found in their food.

How do living things grow and develop?

Organisms grow by taking in raw materials. One-celled organisms grow by increasing in size. Most growth in many-celled organisms is due to an increase in the number of cells.

Organisms change as they grow. All of the changes that take place during an organism's life are called development. Complete development can take a few days for the butterfly shown below, or several years for a dog. The length of time an organism is expected to live is its life span. Some organisms have a short life span. Some have long life spans.

FOLDABLES

C Identify Make a four-tab book, as shown below. On each flap identify things organisms need to live.

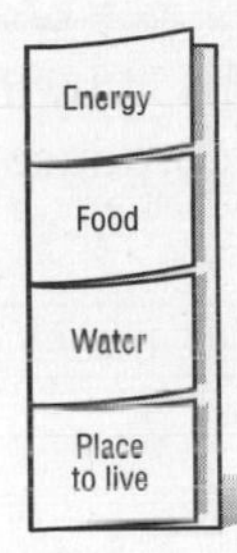

Picture This

1. **Circle** the animal that completes its development cycle in a few days.

Why do living things reproduce?

Organisms eventually reproduce. They make more of their own kind. Some bacteria reproduce every 20 minutes. A pine tree might take two years to produce seeds. Without reproduction, living things would not exist to replace those individuals that die. ☑

Reading Check

2. Explain Why do living things reproduce?

____

____

____

What do living things need?

To survive, all living things need a place to live and raw materials. The places where they live and the raw materials they use can vary.

In what places do organisms live?

The environment limits where organisms can live. Not many organisms can live in extremely hot or extremely cold environments. Most cannot live at the bottom of the ocean or on the tops of mountains. All organisms need living space in their environment. For example, thousands of penguins build their nests on an island. The island becomes too crowded for all the penguins. They fight for space and some may not find space to build nests. An organism's surroundings must provide for all its needs.

What raw materials do organisms need?

Water is important for all living things. Most organisms are made of more than 50 percent water. Humans are made of 60 to 70 percent water. Plants and animals take in and give off large amounts of water each day. Organisms use homeostasis to balance the amount of water taken in and lost.

Organisms use water for many things. Blood is about 90 percent water. Blood transports food and wastes in animals. Plants use water to transport materials between roots and leaves.

Living things are made up of substances such as sugars, proteins, and fats. Animals get these substances from the food they eat. Plants and some bacteria make the substances using raw materials from their surroundings. These important substances are used over and over again. When organisms die, substances from their bodies are broken down and released into the soil or air. The substances can then be used again by other organisms. ☑

Reading Check

3. List three substances found in living things.

____

____

____

After You Read

Mini Glossary

cell: the smallest unit of an organism that carries on the functions of life

homeostasis: an organism's ability to regulate internal, life-maintaining conditions

organism: any living thing

1. Review the terms and their definitions in the Mini Glossary. Write a sentence explaining how homeostasis works in humans.

____

____

____

2. Complete the web diagram below by describing traits that tell what living things are like.

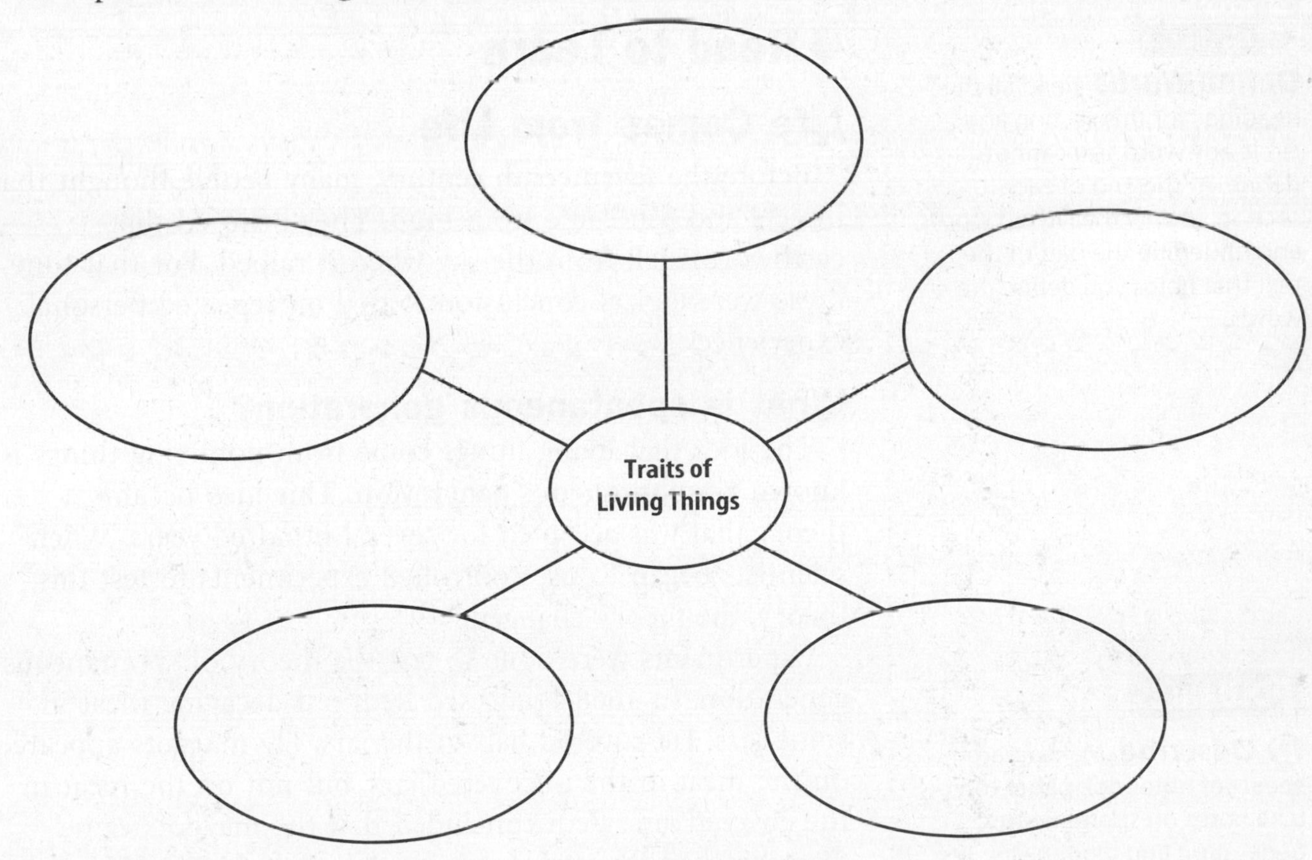

Science Online Visit **life.msscience.com** to access your textbook, interactive games, and projects to help you learn more about living things.

End of Section

Exploring and Classifying Life

section 3 Where does life come from?

What You'll Learn
- about spontaneous generation experiments
- how scientific methods led to the idea of biogenesis

Before You Read

On the lines below, describe what you think the universe was like before Earth and the solar system developed.

____

____

____

Mark the Text

Define Words Read all the headings for this section and circle any word you cannot define. At the end of each section, review the circled words and underline the part of the text that helps you define the words.

FOLDABLES™

D Describe Make quarter sheets of notebook paper to take notes on spontaneous generation and biogenesis.

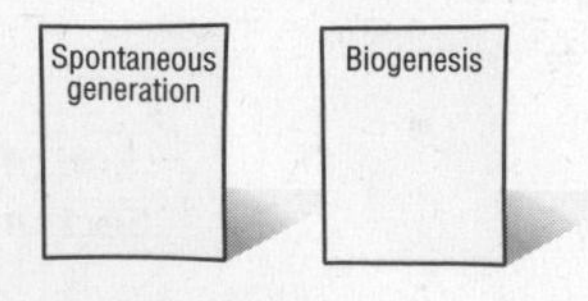

Read to Learn

Life Comes from Life

Before the seventeenth century, many people thought that insects and fish came from mud. They believed that earthworms fell from the sky when it rained. For that time, these were logical conclusions based on repeated personal experiences.

What is spontaneous generation?

The idea that living things come from nonliving things is known as **spontaneous generation**. This idea became a theory that was accepted for several hundred years. When scientists began to use controlled experiments to test this theory, the theory changed.

Experiments were done to test the theory of spontaneous generation. In 1668 Francesco Redi put decaying meat in some jars. He covered half of the jars. Fly maggots appeared on the meat in the uncovered jars, but not on the meat in the covered jars. Redi concluded that the maggots came from hatched fly eggs, not from the meat. In the 1700s, John Needham and Lazzaro Spallanzani conducted more experiments to test the theory of spontaneous generation. These early experiments did not disprove the theory entirely.

What is biogenesis?

In the mid-1800s, the work of Louis Pasteur, a French chemist, provided enough evidence to disprove the theory of spontaneous generation. The theory was replaced with biogenesis (bi oh JE nuh suss). **Biogenesis** is the theory that living things come only from other living things.

Life's Origins

If living things come only from other living things, then how did life on Earth begin? Some scientists hypothesize that about 5 billion years ago, Earth's solar system was a whirling mass of gas and dust. They hypothesize that the Sun and planets formed from this mass.

It is estimated that Earth is about 4.6 billions years old. Fossils of once-living organisms more than 3.5 billion years old have been found. Where did these living organisms come from?

What is Oparin's hypothesis?

In 1924, Alexander I. Oparin, a Russian scientist, suggested that Earth's early atmosphere had no oxygen. He said that it was made up of the gases ammonia, hydrogen, methane, and water vapor. Oparin hypothesized that these gases could have combined to form the more complex compounds found in living things.

American scientists Stanley L. Miller and Harold Urey set up an experiment to test Oparin's hypothesis in 1953. They used the gases and conditions that Oparin described. The experiment sent electricity through a mixture of Earth's earliest gases. When the gases cooled, they condensed to form an oceanlike liquid. The liquid contained materials such as amino acids that are found in present-day cells. Their experiment is summarized below. However, it did not prove that life began in this way. Today, some scientists continue to investigate ideas about life's origins. ☑

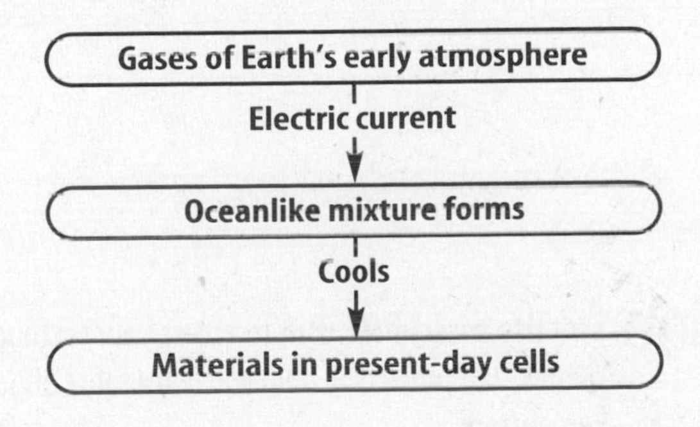

Think it Over

1. **Analyze** How would the age of fossils help suggest Earth's age?

____

____

____

Reading Check

2. **Describe** What did the experiments run by American scientists tell about Oparin's hypothesis?

____

____

____

After You Read

Mini Glossary

biogenesis (bi oh JE nuh suss): the theory that living things come only from other living things

spontaneous generation: the idea that living things come from nonliving things

1. Review the terms and their definitions in the Mini Glossary. Write a sentence that explains the difference between biogenesis and spontaneous generation.

____

____

____

2. Choose one of the question headings in the Read to Learn section. Write the question in the space below. Then write your answer to that question on the lines that follow.

Write your question here.

____

____

____

3. How did circling and underlining words and definitions help you understand where life comes from?

____

____

____

Science Online Visit **life.msscience.com** to access your textbook, interactive games, and projects to help you learn more about where life comes from.

Exploring and Classifying Life

section 4 How are living things classified?

Before You Read

On the lines below, list the different ways you could classify the place where you live (examples: city, suburb, state).

____

____

____

____

What You'll Learn

- how early scientists classified living things
- how similarities are used to classify organisms
- the system of binomial nomenclature
- how to use a dichotomous key

Read to Learn

Classification

When similar items are placed together, they are being classified. Organisms also are classified into groups. Early classifications of organisms included grouping plants that were used in medicines. Animals were often classified by human traits. For example, lions were classified as courageous animals and owls were classified as wise.

More than two thousand years ago, Aristotle, a Greek, decided that any organism could be classified as either a plant or an animal. Then he broke these two groups into smaller groups. For example, his groups included animals that had hair and animals that did not have hair, and animals with and without blood.

Who was Carolus Linnaeus?

In the late 1700s, Carolus Linnaeus, a Swedish naturalist, developed a new system of grouping organisms. His system was based on organisms with similar structures. For example, plants that had a similar flower structure were grouped together. His system was accepted and used by most other scientists. ☑

Study Coach

Ask and Answer Questions Read each subhead. Then work with a partner to write questions about the information in each subhead. Take turns asking and answering the questions.

Reading Check

1. **Describe** the system used by Linnaeus to group organisms.

____

____

____

What classification do modern scientists use?

Modern scientists also use similarities in structure to classify organisms. They also use similarities in both external and internal features. For example, scientists use the number of chromosomes in cells to understand which organisms may be genetically related to each other.

In addition, scientists study fossils, hereditary information, and early stages of development. Scientists use the information to determine an organism's phylogeny. **Phylogeny** (fi LAH juh nee) is the organism's evolutionary history. This tells how the organism has changed over time. It is the basis for the classification of many organisms. ☑

Reading Check

2. **Explain** What can you learn from the phylogeny of an organism?

____

____

____

How are organisms grouped?

A classification system commonly used today groups organisms into six kingdoms. A **kingdom** is the first and largest category. Kingdoms are divided into smaller groups. The smallest classification is a species. Organisms in the same species can mate and produce fertile offspring. The figure below shows how a bottle-nosed dolphin can be classified.

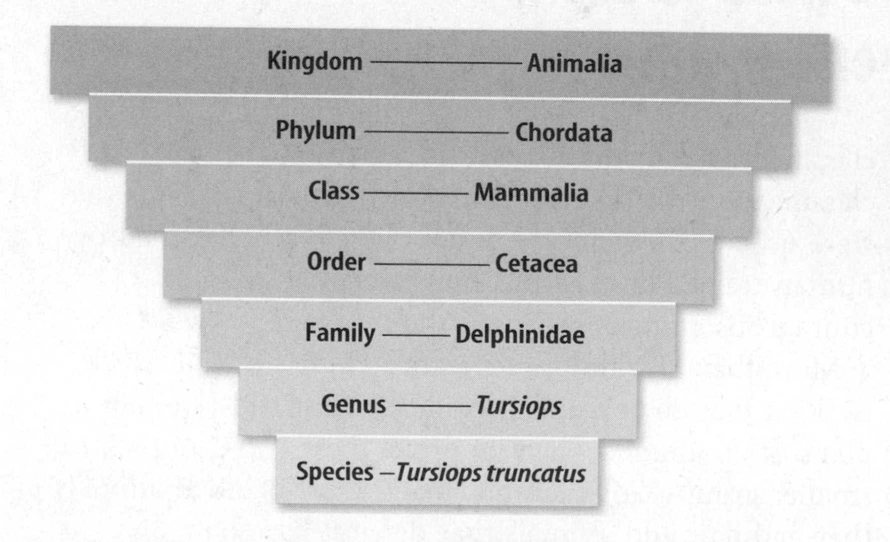

Picture This

3. **Classify** Circle the names of the kingdom and species of the bottle-nosed dolphin.

Scientific Names

If scientists used only common names of organisms, it would be confusing. For example, a jellyfish is neither a fish nor jelly. A sea lion is more closely related to a seal than a lion. To avoid confusion, scientists use a naming system developed by Linnaeus when referring to a particular species. Each species has a unique, two-word scientific name.

What is binomial nomenclature?

The two-word naming system used to name organisms is called **binomial nomenclature** (bi NOH mee ul · NOH mun klay chur). The first word of the two-word name identifies the genus of the organism. A **genus** is a group of similar species. The second word of the name might tell you something about the organism. It might tell what it looks like or where it is found. ☑

Why are scientific names used?

Two-word scientific names are used for four reasons.

- They help avoid mistakes.
- Animals with similar evolutionary history are classified together.
- Scientific names give descriptive information about the species.
- Scientific names allow information about organisms to be organized easily and efficiently.

Tools for Identifying Organisms

Tools used to identify organisms include field guides and dichotomous (di KAH tuh mus) keys. Field guides include descriptions and pictures of organisms. They give information about where each organism lives. You can use a field guide to identify species from around the world. ☑

What are dichotomous keys?

A dichotomous key is a detailed list of identifying characteristics that includes scientific names. The keys are set up in steps. Each step has two descriptive statements, such as hair or no hair. You can use a dichotomous key, such as the one below, to identify and name a species.

Key to Some Mice of North America	
1. Tail hair	**a.** no hair on tail; scales show plainly; house mouse, *Mus musculus* **b.** hair on tail, go to 2
2. Ear size	**a.** ears small and nearly hidden in fur, go to 3 **b.** ears large and not hidden in fur, go to 4
3. Tail length	**a.** less than 25 mm; woodland vole, *Microtus pinetorum* **b.** more than 25 mm; prairie vole, *Microtus ochrogaster*
4. Tail coloration	**a.** sharply bicolor, white beneath and dark above; deer mouse, *Peromyscus maniculatus* **b.** darker above than below but not sharply bicolor; white-footed mouse, *Peromyscus leucopus*

Reading Check

4. Explain What does the first word in an organism's binomial nomenclature indicate?

____

Reading Check

5. List two tools that can be used to identify organisms.

____

____

____

Picture This

6. Identify the mouse that has a mostly dark, hairy tail and large ears.

____

After You Read

Mini Glossary

binomial nomenclature (bi NOH mee ul · NOH mun klay chur): the two-word naming system used to name organisms

genus: a group of similar species

kingdom: the first and largest category of organisms

phylogeny (fi LAH juh nee): the evolutionary history of an organism

1. Review the terms and their definitions in the Mini Glossary. Choose one of the terms and explain its role in classifying organisms.

___

___

___

2. Complete the diagram below by explaining what binomial nomenclature is and the reasons for using it.

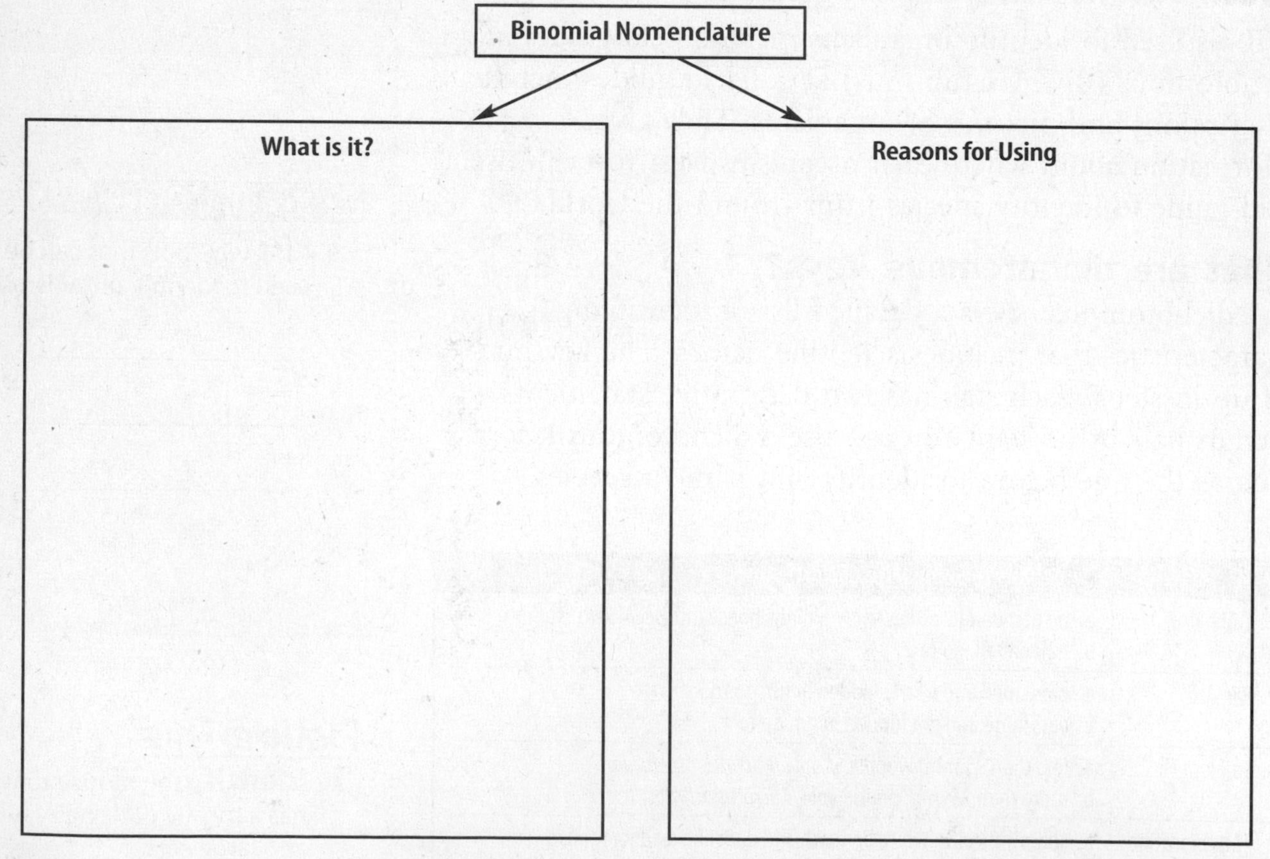

End of Section

Science Online Visit **life.msscience.com** to access your textbook, interactive games, and projects to help you learn more about how living things are classified.

Cells

section 1 Cell Structure

Before You Read

Think about the different jobs people have in a restaurant. List three of those jobs on the lines below. Then explain how these people work together to provide food to customers.

____

____

____

What You'll Learn

- the names and functions of cell parts
- the importance of a nucleus in a cell
- about tissues, organs, and organ systems

Read to Learn

Common Cell Traits

Living cells have many things in common. A cell is the smallest unit that can perform life functions. All cells have an outer covering called a **cell membrane**. Inside every cell is a gelatinlike material called **cytoplasm** (SI tuh pla zum). Cytoplasm contains hereditary material that controls the life of the cell.

Mark the Text

Identify Important Words As you read the section, circle all the words you do not understand. Highlight the part of the text that helps you define those words.

How do cells differ?

Cells come in different sizes and shapes. A cell's shape might tell you something about its function. A nerve cell has many branches that send and receive messages to and from other cells. A nerve cell in your leg could be a meter long. A human egg cell is no bigger than the dot on this *i*. A human blood cell is much smaller than the egg cell. A bacterium is even smaller—8,000 of the smallest bacteria can fit inside one red blood cell.

A nerve cell cannot change its shape. Muscle cells and some blood cells can change shape. Some cells in plant stems are long and hollow and have openings at their ends. These cells carry food and water throughout the plant. ✔

Reading Check

1. **Infer** Why are cells in plant stems hollow with openings at both ends?

____

____

____

Picture This

2. Identify Circle the features that are the same in both types of cells.

What types of cells are there?

Scientists separate cells into two groups, as shown in the figure below. A prokaryotic (proh KAYR ee yah tihk) cell does not have membrane-bound structures inside the cell. A cell with membrane-bound structures inside the cell is called a eukaryotic (yew KAYR ee yah tihk) cell.

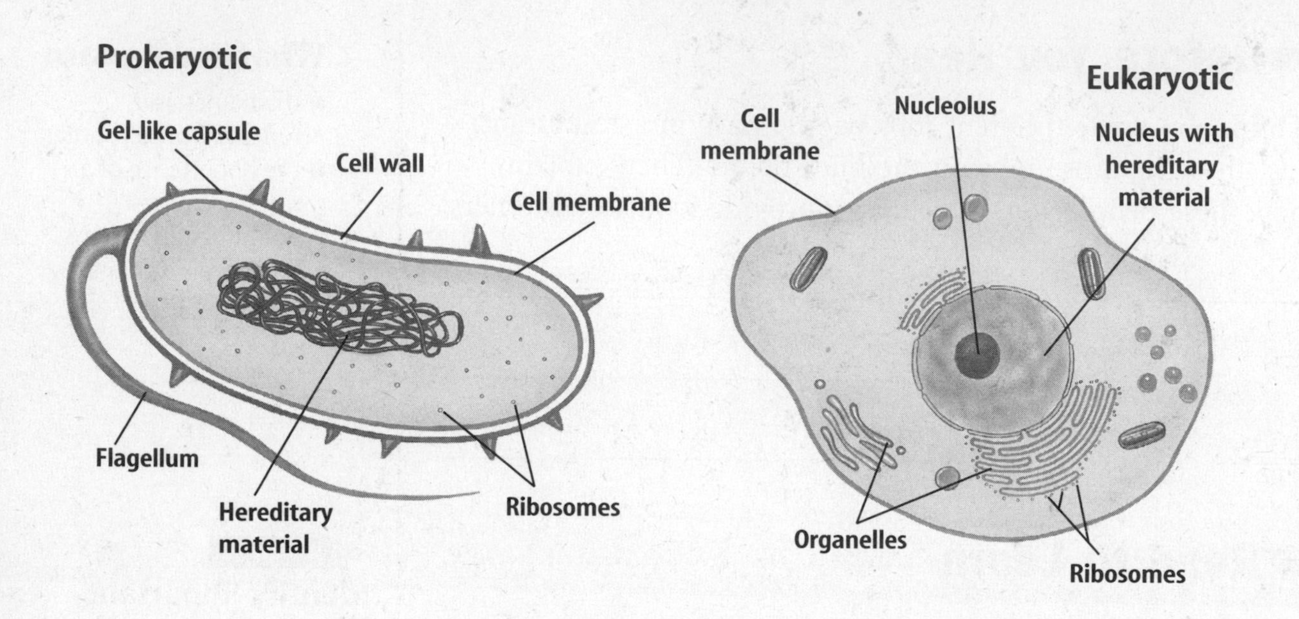

Cell Organization

Just as restaurant workers have specific jobs, each cell in your body has a certain job to do. Cells take in nutrients, release and store chemicals, and break down substances 24 hours a day.

What protects a cell and gives it shape?

A **cell wall** is a tough, rigid outer covering that protects the cell and gives it shape. The cells of plants, algae, fungi, and most bacteria are enclosed in a cell wall.

A plant cell wall is mostly made up of a carbohydrate called cellulose. The long, threadlike fibers of cellulose form a thick mesh. The mesh allows water and dissolved materials to pass through the cell wall.

Cell walls may contain pectin, which is used to thicken jams and jellies. Cell walls also contain lignin. Lignin is a compound that makes cell walls rigid. Plant cells responsible for support have large amounts of lignin in their walls. ☑

Reading Check

3. List three things found in the cell wall of a plant.

____

____

____

What is the function of the cell membrane?

The protective layer around all cells is the cell membrane. If a cell has a cell wall, the cell membrane is inside the cell wall. The cell membrane controls what happens between a cell and its environment. Water and some food particles move freely into and out of a cell through the cell membrane. Waste products leave through the cell membrane.

What is cytoplasm?

Cytoplasm is a gelatinlike material in the cell. Many important chemical reactions occur within the cytoplasm. Cytoplasm has a framework called the cytoskeleton, which helps the cell keep or change its shape. The cytoskeleton helps some cells to move. The cytoskeleton is made up of thin, hollow tubes of protein and thin, solid protein fibers. ☑

What are the functions of organelles?

Most of a cell's life processes happen in the cytoplasm. Within the cytoplasm of eukaryotic cells are structures called **organelles**. Some organelles process energy. Others make materials needed by the cell or other cells. Some organelles move materials. Others store materials. Most organelles are surrounded by membranes.

Why is the nucleus important?

The **nucleus** (NEW klee us) directs all cell activities. The nucleus usually is the largest organelle in a cell. It is separated from the cytoplasm by a membrane. Materials enter and leave the nucleus through openings in the membrane. The nucleus contains DNA. DNA is the chemical that contains the code for the cell's structure and activities. ☑

Which organelles process energy?

Cells need energy to do their work. In plant cells, food is made in green organelles called **chloroplasts** (KLOR uh plasts). Chloroplasts contain chlorophyll (KLOR uh fihl), which captures light energy that is used to make a sugar called glucose. Animal cells and some other cells do not have chloroplasts. Animals must get food from their environment.

The energy in food is stored until it is released by organelles called mitochondria (mi tuh KAHN dree uh). **Mitochondria** release energy by breaking down food into carbon dioxide and water. Some types of cells, such as muscle cells, are more active than other types of cells. These cells have large numbers of mitochondria.

Reading Check

4. **Describe** What is the cytoskeleton?

____

____

____

Reading Check

5. **Explain** the function of the nucleus.

____

____

____

What organelle makes proteins?

Protein takes part is almost every cell activity. Cells make their own proteins on structures called **ribosomes**, which are shown below. Ribosomes are considered organelles, even though they are not membrane bound. Hereditary material in the nucleus tells ribosomes how, when, and in what order to make proteins. Ribosomes are made in the nucleolus (new klee OHL us) and move out into the cytoplasm. Some ribosomes are free-floating in the cytoplasm and some attach to the endoplasmic reticulum. ☑

Reading Check

6. **Identify** Where are ribosomes made?

____

Animal Cell

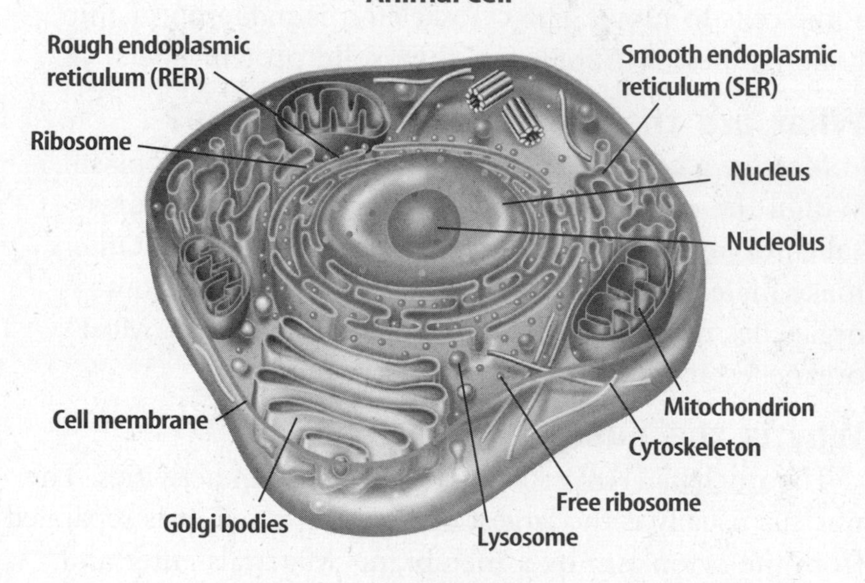

Plant Cell

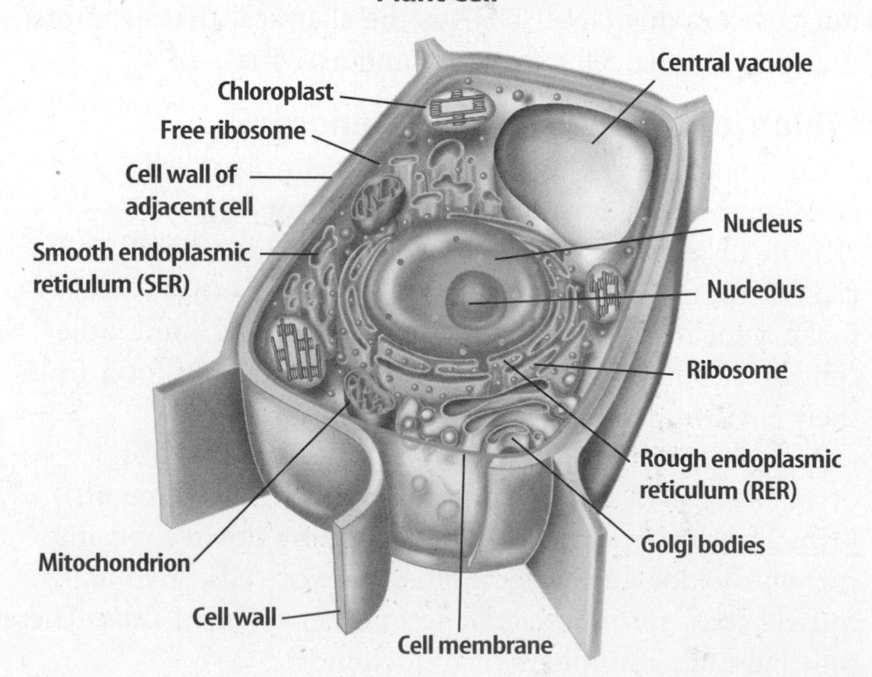

Picture This

7. **Compare** Circle the organelles that direct cell activities in each cell. Highlight the organelle that contains chlorophyll.

What is the endoplasmic reticulum?

The **endoplasmic reticulum** (en duh PLAZ mihk • rih TIHK yuh lum), or ER, is a series of folded membranes in which materials can be processed and moved around inside the cell. Smooth ER processes materials such as lipids that store energy. Rough ER has ribosomes that make proteins. The proteins are used within the cell or moved out of the cell.

What types of organelles transport or store materials?

The **Golgi** (GAWL jee) **bodies** sort proteins and other cellular materials and put them into structures called vesicles. Vesicles deliver the cellular materials to areas inside the cell and to the cell membrane where they are released. Cells have membrane-bound spaces called vacuoles. Vacuoles store cellular materials, such as water, wastes, and food.

How does a cell recycle its materials?

Active cells break down and recycle materials. An organelle called a lysosome (LI suh sohm) contains digestive chemicals that help break down materials in the cell. The lysosome's membrane stops the digestive chemicals from leaking into the cytoplasm and destroying the cell. When a cell dies, a lysosome's membrane breaks down. The released digestive chemicals destroy the cell's contents.

From Cell to Organism

The figure below shows how a many-celled organism is organized. A cell in a many-celled organism performs its own work and depends on other cells in the organism. Similar cells grouped together to do one job form a **tissue**. Each cell works to keep the tissue alive. Tissues are organized into organs. An **organ** is made up of two or more different types of tissue that work together. For example, your heart is an organ that is made up of cardiac tissue, nerve tissue, and blood tissues. An organ system is a group of organs that work together to perform a function. Your cardiovascular system is made up of your heart, arteries, veins, and capillaries. Organ systems work together to keep an organism alive.

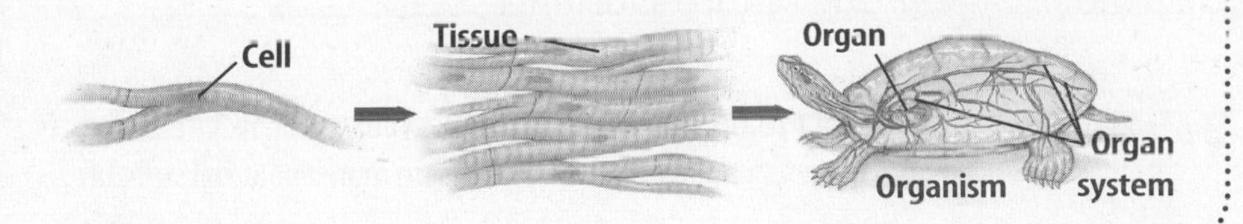

FOLDABLES™

A Describe Make a three-tab Foldable, as shown below. Use the Foldable to describe how cells are organized to work together.

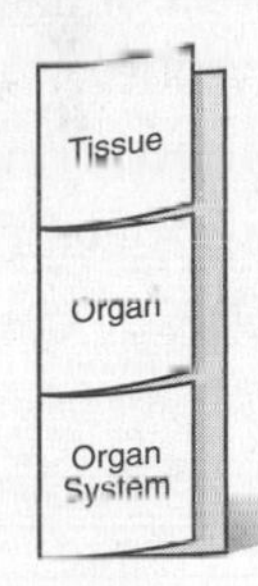

Picture This

8. **Sequence** Write a number from 1 to 5 beside each label on the diagram. A 1 is the simplest level of organization and a 5 is the most complex level of organization.

After You Read

Mini Glossary

cell membrane: the protective layer around a cell, which controls what happens between a cell and its environment

cell wall: a tough, rigid outer covering that protects the cell and gives it shape

chloroplast (KLOR uh plast): a green organelle that makes food in plant cells

cytoplasm (SI tuh pla zum): gelatinlike material inside every cell where hereditary material is contained

endoplasmic reticulum: a series of folded membranes in which materials can be processed and moved around inside the cell

Golgi (GAWL jee) bodies: organelles that sort proteins and other cellular materials and put them into structures called vesicles

mitochondria: organelles that release energy by breaking down food into carbon dioxide and water

nucleus (NEW klee us): directs all cell activities

organ: a structure made up of two or more different types of tissues that work together

organelle: a structure within a eukaryotic cell; some process energy and others make substances needed by the cell or other cells

ribosome: a small structure where a cell makes its own protein

tissue: a group of similar cells that work together to do one job

1. Review the terms and their definitions in the Mini Glossary. Choose one term that describes a cell structure and write a sentence to explain its function.

____

____

2. Complete the diagram below to show the organization of many-celled organism.

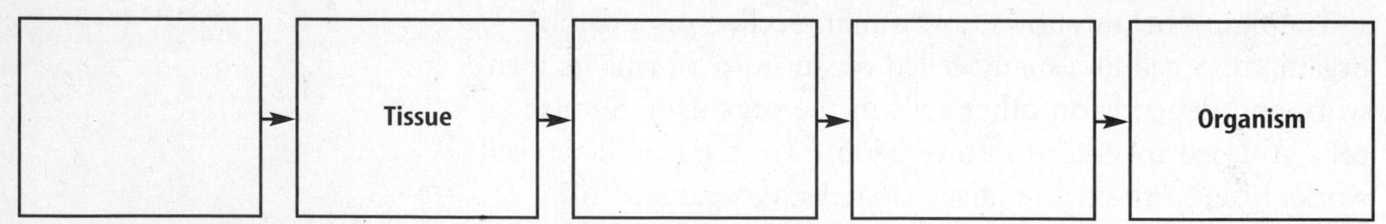

3. Beside each organelle listed below, write *Plant, Animal,* or *Both* to show where the organelle is found.

a. Nucleus ______________

b. Chloroplast ______________

c. Golgi bodies ______________

d. Ribosome ______________

e. Lysosome ______________

f. Mitochondrion ______________

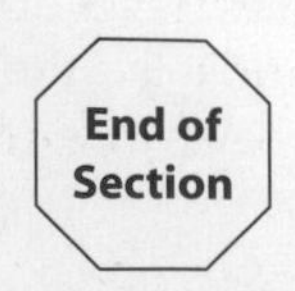

Visit **life.msscience.com** to access your textbook, interactive games, and projects to help you learn more about cell structure.

Cells

section 2 Viewing Cells

Before You Read

Have you ever looked at anything using a magnifying lens or a microscope? On the lines below, describe what you saw.

____

____

____

What You'll Learn
- the differences among microscopes
- the discoveries that led to the cell theory

Read to Learn

Study Coach

K-W-L Fold a sheet of paper into three columns. In the first column, write what you know about microscopes. In the second column, write what you want to know about microscopes. Fill in the third column with facts you learned about microscopes after you have read this section.

Magnifying Cells

The number of living things in your environment that you can't see is greater than the number you can see. Many of the things you can't see are only one cell in size. Most cells are so small that you need a microscope to see them. A microscope has one or more lenses that directs light toward your eye and enlarges the appearance of the cell so you can see its parts.

What were early microscopes like?

In the mid-1600s, Antonie van Leeuwenhoek made a simple microscope with a tiny glass bead for a lens. It could magnify images of things up to 270 times their normal size. The microscope's lens had a power of 270×. Early microscopes made an image larger, but that image was not always sharp or clear.

What are modern microscopes like?

Today there are simple and compound microscopes. A simple microscope has just one lens. It is like a magnifying lens. The compound light microscope has two sets of lenses—eyepiece lenses and objective lenses. The eyepiece lenses are placed in one or two tubelike structures. Compound light microscopes have two to four movable objective lenses.

FOLDABLES™

B Compare Make a three-tab Foldable, as shown below. Compare compound light microscopes and electron microscopes.

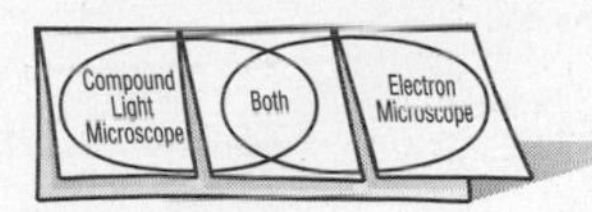

Magnification The larger image produced by a microscope is called magnification. The powers of the eyepiece and objective lenses determine the total magnification of a compound microscope. For example, if the eyepiece lens has a power of 10× and the objective lens has a power of 43×, then the total magnification is 430× (10× times 43×). A 430× microscope can make an image of an object 430 times larger than its actual size.

How does an electron microscope magnify objects?

Things that are too small to be seen with compound light microscopes can be viewed with an electron microscope. An electron microscope uses a magnetic field in a vacuum to direct beams of electrons. Some electron microscopes can magnify up to one million times. Electron microscope images must be photographed or electronically produced.

FOLDABLES™

C Describe Make a four-tab Foldable, as shown below, to list facts about the cell theory.

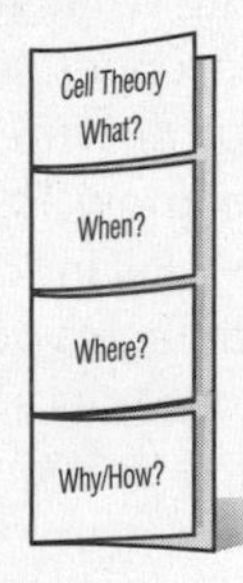

Cell Theory

In 1665, Robert Hooke looked at a thin slice of cork under a microscope. He thought the cork looked like it was made up of empty little boxes, which he named cells.

What discoveries led to the cell theory?

In the 1830s, Matthias Schleiden used a microscope to study plants. He concluded that all plants are made of cells. Theodor Schwann observed different animal cells. He concluded that all animals are made of cells. These two scientists put their ideas together and concluded that all living things are made of cells.

Several years later, Rudolf Virchow hypothesized that cells divide to form new cells. He suggested that every cell came from a cell that already existed. Virchow's ideas about cells and those of other scientists are called the **cell theory,** as described in the table below.

Picture This

1. **Identify** Highlight Rudolf Virchow's contribution to cell theory.

The Cell Theory	
1. All organisms are made up of one or more cells.	An organism can have one or many cells. Most plants and animals have many cells.
2. The cell is the basic unit of organization in organisms.	Even in complex organisms such as humans, the cell is the basic unit of life.
3. All cells come from cells.	Most cells can divide to form two new cells that are exactly the same.

After You Read

Mini Glossary

cell theory: states that all organisms are made up of one or more cells; the cell is the basic unit of organization in organisms; and all cells come from other cells

1. Review the term and its definition in the Mini Glossary. Write a sentence describing one part of the cell theory. Include the name of the scientist connected to that part of the cell theory.

____

____

____

2. Complete the table below to list the discovery made by each scientist that led to the cell theory.

Matthias Schleiden	Theodor Schwann	Rudolf Virchow

3. How did the K-W-L chart help you organize the information about microscopes?

____

____

____

Visit **life.msscience.com** to access your textbook, interactive games, and projects to help you learn more about viewing cells.

Cells

section 3 Viruses

What You'll Learn
- how a virus copies itself
- how vaccines help people
- some uses of viruses

Before You Read

Think about the vaccinations you have had when at the doctor's office or at a health clinic. On the lines below, list the kinds of diseases these shots will help prevent.

____

____

____

Study Coach

Create a Quiz Write a question about the main idea under each heading. Exchange quizzes with another student. Together discuss the answers to the quizzes.

Read to Learn

What are viruses?

Cold sores, measles, chicken pox, colds, the flu, and AIDS are some diseases caused by nonliving particles called viruses. A **virus** is a strand of hereditary material surrounded by a protein coating.

What are characteristics of viruses?

Viruses don't have a nucleus or other organelles. They also lack a cell membrane. Viruses have a variety of shapes.

A virus is so small it can be seen only by an electron microscope. Before the electron microscope was invented, scientists only hypothesized about viruses.

FOLDABLES™

D Identify Make quarter sheets of notebook paper, as shown below. On the first sheet, answer the question "What are viruses?". On the second sheet, answer the question "How do viruses multiply?".

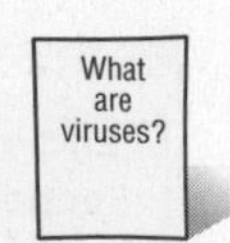

How do viruses multiply?

The only way a virus can reproduce is by making copies of itself. A virus, however, must have the help of a living cell called a **host cell.** Crystalized forms of some viruses can be stored for years. Then, if they enter an organism, they can multiply quickly.

Once a virus enters a host cell, the virus can act in two ways. It can be either active or latent, which is an inactive stage.

What happens when a virus is active?

When a virus enters a cell and is active, it causes the host cell to make new viruses. This process destroys the host cell. Follow the steps in the figure below to see one way that an active virus works inside a cell.

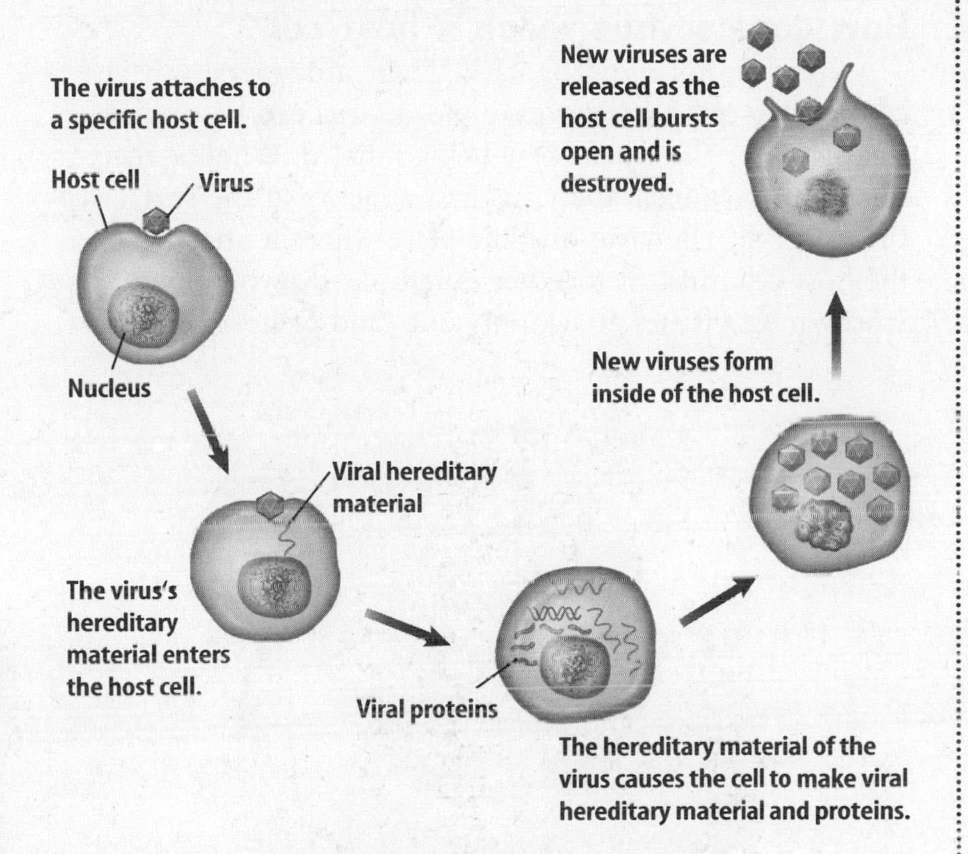

The virus attaches to a specific host cell.
Host cell
Virus
Nucleus
Viral hereditary material
The virus's hereditary material enters the host cell.
Viral proteins
The hereditary material of the virus causes the cell to make viral hereditary material and proteins.
New viruses form inside of the host cell.
New viruses are released as the host cell bursts open and is destroyed.

Picture This

1. **Sequence** Circle the step that shows the virus's hereditary material entering the host cell. Highlight the step that shows new viruses forming inside a host cell.

What happens when a virus is latent?

When a latent, or inactive, virus enters a host cell, its hereditary material can become part of the cell's hereditary material. It does not immediately make new viruses or destroy the cell. As the host cell reproduces, the virus's DNA is copied. A virus can be inactive for many years. Then, at any time, something inside or outside the body can make the virus active. ☑

If you have a cold sore on your lip, a latent virus in your body has become active. The cold sore is a sign that the virus is active and destroying cells in your lip. When the cold sore goes away, the virus has become latent again. The virus is still in your body's cells, but it is hiding and doing no harm.

✓ Reading Check

2. **Describe** what happens to a latent virus's DNA when a host cell reproduces.

____

____

How do viruses affect organisms?

Viruses attack animals, plants, fungi, protists, and all prokaryotes. Some viruses can infect only certain kinds of cells. For example, the potato leafroll virus infects only potato crops. A few viruses can infect many kinds of cells. The rabies virus can infect humans and many other animal hosts.

How does a virus reach a host cell?

A virus cannot move by itself. There are several ways it can reach a cell host. For example, a virus can be carried to a host cell by the wind or by being inhaled. When a virus infects an organism, the virus first attaches to the surface of the host cell. The virus and the place where it attaches on the host cell must fit together exactly, as shown below. This is why most viruses attack only one kind of host cell.

Picture This

3. Identify Circle each of the places that the virus is attached to the cell.

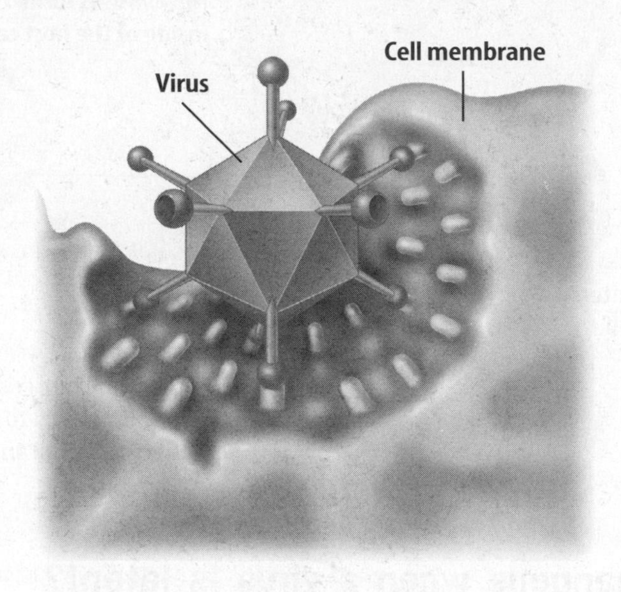

What are bacteriophages?

Viruses that infect bacteria are called bacteriophages (bak TIHR ee uh fay jihz). They differ from other kinds of viruses in the way that they enter bacteria. Bacteriophages attach to a bacterium and inject their hereditary material. The entire cycle takes about 20 minutes. Each virus-infected cell releases an average of 100 viruses.

4. Define the term *vaccine.*

____

____

____

Fighting Viruses

A vaccine is a kind of medicine used to prevent a disease. It is made from weakened virus materials that cannot cause disease anymore. Vaccines have been made to prevent many diseases, including chicken pox, measles, and mumps.

How was the first vaccine developed?

Edward Jenner developed the first vaccine in 1796. The vaccine was for smallpox. Jenner noticed that people who got cowpox did not get smallpox. He made a vaccine from the sores of people who had cowpox. He injected the cowpox vaccine into healthy people. The cowpox vaccine protected them from smallpox. ☑

How are viral diseases treated?

One way your body can fight viral infections is by making interferons. Interferons are proteins that are made quickly by virus-infected cells and move to noninfected cells in the host. Interferons cause the noninfected cells to make protective materials.

Antiviral drugs can be given to an infected patient to help fight a virus. A few drugs are helpful against viruses. Some of these drugs are not used widely because they have harmful side effects.

How can viral diseases be prevented?

There are many ways to prevent viral diseases. People can get vaccinated against diseases. Sanitary conditions can be improved. People who have viral diseases can be kept away from healthy people. Animals, such as mosquitoes, that spread disease can be kept under control. ☑

Research with Viruses

Scientists are discovering helpful uses for some viruses. One use, called gene therapy, substitutes normal hereditary material for a cell's flawed hereditary material. Normal hereditary material is placed into viruses. These altered viruses then are used to infect those cells that contain flawed hereditary material. The normal hereditary material in the altered viruses enters the cells and replaces the flawed hereditary material. Using gene therapy, scientists hope to help people with genetic disorders and find a cure for cancer.

Reading Check

5. **Identify** Who developed the first vaccine?

____

Reading Check

6. **Describe** one way viral diseases can be prevented.

____

____

____

After You Read

Mini Glossary

host cell: a living cell that a virus enters

virus: a strand of hereditary material surrounded by a protein coating that can infect and multiply in a host cell

1. Review the terms and their definitions in the Mini Glossary. Write a sentence that describes the relationship between a virus and a host cell.

2. Choose one of the question headings in the Read to Learn section. Write the question in the space below. Then write your answer to that question on the lines that follow.

Write your question here.

3. Complete the diagram below to identify two ways viruses can act inside host cells.

Viruses enter host cells and become

or

End of Section

ScienceOnline Visit **life.msscience.com** to access your textbook, interactive games, and projects to help you learn more about viruses.

Cell Processes

section 1 Chemistry of Life

Before You Read

On the lines below, list five things that are in the room around you. Then write something that these items have in common.

____

____

____

What You'll Learn

- the differences among atoms, molecules, and compounds
- how chemistry and life science are related
- the differences between organic compounds and inorganic compounds

Read to Learn

Mark the Text

Identify the Main Point Underline the important ideas in this section. This will help you remember what you read.

The Nature of Matter

Look at the things around you. What are they made of? Each item looks different, but all of them are made of matter. Matter is anything that has mass and takes up space. You are made of matter. The chair you sit in is made of matter. The book you read is made of matter.

Energy holds matter together and breaks matter apart. Energy is anything that brings about change. The food you eat is matter. It is held together by chemical energy. When you cook food, energy in the form of heat breaks some of the bonds that hold the matter in food together.

What makes up matter?

Matter exists in three forms—solids, liquids, and gases. All forms of matter are made up of atoms. The oxygen atom shown on the next page will help you understand the parts of the atom. The nucleus is the center of the atom. The nucleus holds the protons and neutrons. Notice that the protons and neutrons are about the same size. They also have about the same masses. Protons have a positive charge, while neutrons have no charge.

FOLDABLES™

A Describe Make a two-tab book, as shown below. Write notes about matter and energy.

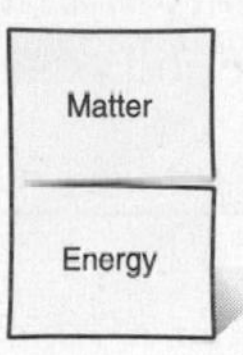

Where are electrons found?

Electrons are outside the atom's nucleus. They have a negative charge. It takes about 1,837 electrons to equal the mass of one proton. Electrons are the part of the atom that is involved in chemical reactions.

Look at the figure of the oxygen atom again. It shows that most of the atom is empty space. Energy holds the parts of an atom together.

Picture This

1. **Identify** Highlight the electrons. Circle the nucleus containing the protons and neutrons.

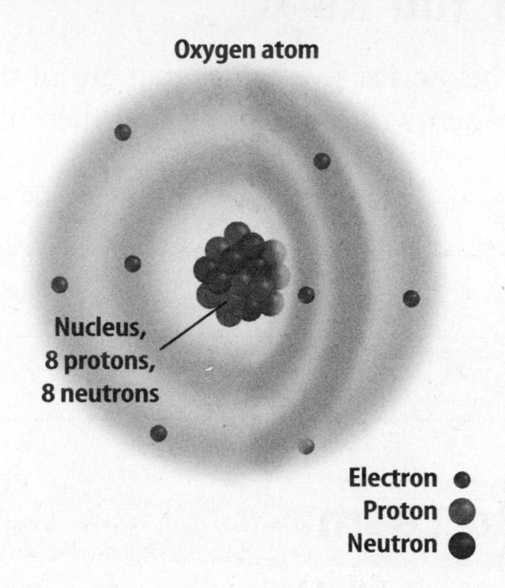

What are elements?

When something is made up of only one kind of atom, it is called an element. An element can't be broken down into a simpler form by chemical reactions. The element oxygen is made up of only oxygen atoms. Hydrogen is made up of only hydrogen atoms. Scientists give each element its own one- or two-letter symbol.

All matter is made up of elements. Most things, including all living things, are made up of a combination of elements.

Six elements make up 99 percent of living matter—oxygen (O), carbon (C), hydrogen (H), nitrogen (N), phosphorus (P), and sulfur (S).

What are compounds?

Water is a compound made up of two elements—oxygen and hydrogen. Compounds are made up of two or more elements in set ratios. For example, the ratio of hydrogen and oxygen in water is always two hydrogen atoms to one oxygen atom. Compounds have properties different from the elements they are made of. There are two types of compounds—molecular compounds and ionic compounds. ☑

2. **Determine** What two elements make up the compound water?

____

____

What is a molecule?

The smallest part of a molecular compound is a molecule. A molecule is a group of atoms held together by the energy of chemical bonds. When chemical reactions occur, chemical bonds break and the atoms move around to form new bonds. The molecules formed after the reaction are different from those that began the reaction.

How do molecular compounds form?

Molecular compounds form when different atoms share their electrons that are farthest from the nucleus. Water is a molecular compound that has two hydrogen atoms and one oxygen atom. The two hydrogen atoms each share one electron on one oxygen atom. This molecular compound is shown in the figure at the right.

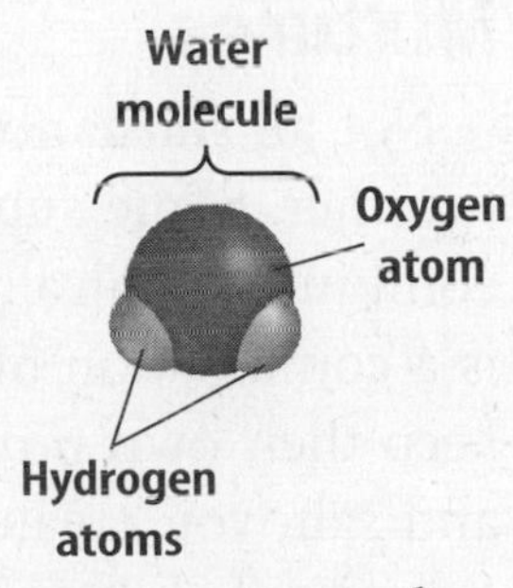

Water does not have the same properties as oxygen and hydrogen. Oxygen and hydrogen are gases. Yet, water can be a liquid, a solid, or a gas. When hydrogen and oxygen combine, a change occurs and a new substance forms.

What are ions?

Atoms also combine when they become negatively or positively charged. Most atoms have no electric charge. They are neutral. When an atom loses an electron, it has more protons than electrons. Protons have a positive charge so the atom becomes positively charged. When an atom gains an electron, it has more electrons than protons. This creates a negatively charged atom. These electrically charged atoms—positive or negative—are called ions.

How do ionic compounds form?

Ions of opposite charges attract each other to form neutral compounds, or compounds with the same number of protons and electrons. These neutral compounds are called ionic compounds. ☑

Table salt is an ionic compound made up of sodium (Na) and chlorine (Cl) ions. When sodium and chlorine atoms combine, the chlorine atom gains an electron from the sodium atom. The chlorine atom becomes a negatively charged ion. The sodium atom becomes a positively charged ion. These opposite charges attract each other. The neutral ionic compound sodium chloride (NaCl) is formed.

Applying Math

3. **Use Numbers** Write a simple addition problem that shows how the compound water is formed.

____

Reading Check

4. **Explain** What are ionic compounds?

____

____

____

Why are ions important?

Ions are important to many life processes that take place in your body and in other organisms. For example, when you touch something hot, a message is sent from your hand to your brain to tell you to move your hand. This message travels along your nerve cells as potassium and sodium ions move in and out of the nerve cells. Ions also help move oxygen throughout your body. Some substances could not move into and out of a cell without ions.

Mixtures

Not all substances form compounds when combined together. Some substances do not change each other or combine chemically when they are put together. A **mixture** is a combination of substances in which individual substances keep their own properties. For example, if you combine sugar and salt, you create a mixture. No chemical reaction occurs. You simply have sugar and salt mixed together. Mixtures can be solids, liquids, gases, or any combination of them.

Most chemical reactions in living organisms take place in mixtures called solutions. A solution is a mixture in which substances are mixed evenly. Sweat is a solution of salt and water.

Living things also contain mixtures called suspensions. A suspension forms when a liquid or a gas has another substance evenly spread throughout it. Unlike solutions, the substances in a suspension eventually sink to the bottom. For example, if a blood sample is left standing, the red blood cells and white blood cells will sink to the bottom of the test tube. In your body, your heart keeps your blood moving, and the red and white blood cells stay suspended.

Think it Over

5. **Apply** Name two other mixtures that you have made.

____

____

____

Organic Compounds

All compounds are classified as organic or inorganic. Rocks and other nonliving things are made up of inorganic compounds. Living things such as humans and plants are made up of organic compounds. **Organic compounds** always have carbon and hydrogen atoms. Some nonliving things also include organic compounds. Coal, for example, is a nonliving thing that was formed from dead and decaying plants. It contains organic compounds because the plants were once living things. ✔

✔ Reading Check

6. **Determine** What do all organic compounds contain?

____

____

How are organic compounds organized?

Organic molecules contain many atoms that can be arranged in many different ways. Organic compounds are organized into four groups—carbohydrates, lipids, proteins, and nucleic acids. The table below describes these groups of organic compounds.

Organic Compounds Found in Living Things

	Carbohydrates	Lipids	Proteins	Nucleic Acids
Elements	carbon hydrogen oxygen	carbon oxygen hydrogen phosphorus	carbon oxygen hydrogen nitrogen sulfur	carbon oxygen hydrogen nitrogen phosphorus
Examples	sugars starch cellulose	fats oils	skin hair	DNA RNA
Function	• supply energy for cell processes • short-term energy storage • form plant structures	• store large amounts of energy long term • form boundaries around cells	• regulate cell processes • build cell structures	• carry hereditary information • used to make proteins

What are carbohydrates?

Carbohydrates supply energy for cell processes. Cells use carbohydrates for energy. Sugars and starches are carbohydrates. Some carbohydrates also are important parts of cell structures. For example, cellulose is a carbohydrate that is an important part of plant cells.

What are lipids?

Lipids are organic compounds that do not mix with water. Fats and oils are lipids that store energy. These lipids release more energy than carbohydrates. One type of lipid, the phospholipid, is a major part of cell membranes.

Why are proteins important?

Proteins have many important jobs in living organisms. They are made up of smaller molecules called amino acids. Proteins are the building blocks of many structures in living organisms. Certain proteins called **enzymes** control most chemical reactions in cells.

FOLDABLES™

B Describe Make a four-door book, as shown below. Write notes about the four groups of organic compounds.

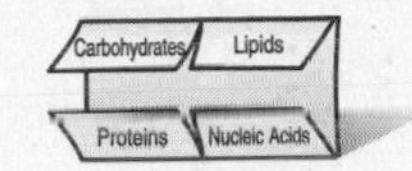

Picture This

7. **Analyze** Use the table to answer the following questions.

a. What organic compound supplies energy for cell processes?

____

b. What elements are found in all four organic compounds?

____

____

____

What are nucleic acids?

Nucleic acids are large molecules that store important coded information in cells. One nucleic acid is deoxyribonucleic acid, or DNA—genetic material. DNA is found in all cells at some point in their life. It carries the information that tells the cell what to do. Ribonucleic acid, or RNA, is another nucleic acid. It makes enzymes and other proteins.

Inorganic Compounds

Reading Check

8. Identify Name an important inorganic compound.

Most **inorganic compounds** are made from elements other than carbon. They usually have fewer atoms than organic molecules. Many foods you eat contain inorganic compounds. Your body needs the elements found in inorganic compounds. Water is an inorganic compound that is important to all living things.

Why is water important?

Living things are made up of more than 50 percent water and depend on water to survive. You can live for weeks without food but only a few days without water.

Some seeds and spores can exist without water. But they, like all organic compounds, need water to grow and reproduce. All chemical reactions in living things happen in water solutions. Most organisms use water to move materials throughout their bodies. For example, plants use water to move minerals and sugars between the roots and leaves. Water also helps cells keep their temperature constant.

About two-thirds of your body's water is located in your body's cells, as the circle graph shows. Water helps the cells keep their shapes and sizes. About one-third of your body's water is outside the cells.

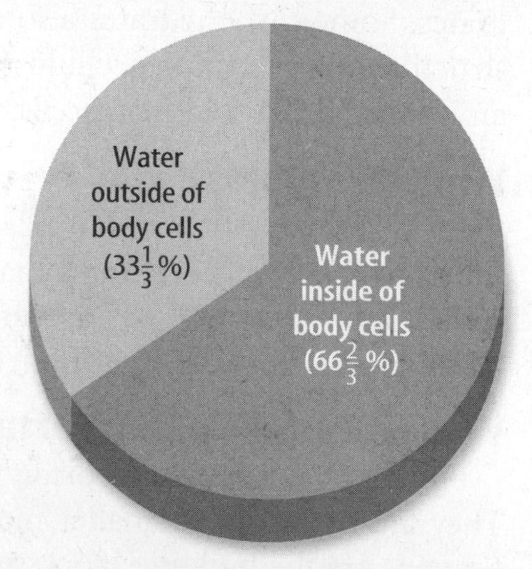

Picture This

9. Conclude Is there more water inside or outside body cells?

After You Read

Mini Glossary

enzymes: proteins that control most chemical reactions in cells

inorganic compound: compound that is made from elements other than carbon; compounds that make up most nonliving things

mixture: a combination of substances in which individual substances keep their own properties

organic compound: compounds that have carbon and hydrogen atoms; compounds that make up all living things

1. Review the terms and their definitions in the Mini Glossary. Write a sentence that explains the difference between inorganic and organic compounds.

____

____

____

2. Choose one of the question headings in the Read to Learn section. Write the question in the space below. Then, write your answer to that question on the lines that follow.

Write your question here.

____

____

____

3. How did underlining the main ideas help you understand what you read in this section?

____

____

____

Science Online Visit **life.msscience.com** to access your textbook, interactive games, and projects to help you learn more about the chemistry of life.

End of Section

chapter 3 Cell Processes

section 2 Moving Cellular Materials

What You'll Learn

- how selectively permeable membranes work
- about diffusion and osmosis
- the differences between passive transport and active transport

Before You Read

On the lines below, describe the purpose of window screens. Think of what they keep out and what they allow to pass through.

____

____

____

Study Coach

Make Flash Cards Think of a quiz question for each paragraph. Write the question on one side of the flash card and the answer on the other side. Keep quizzing yourself until you know all of the answers.

FOLDABLES™

C Describe Make a two-tab book, as shown below. Use the Foldable to take notes about active and passive transport.

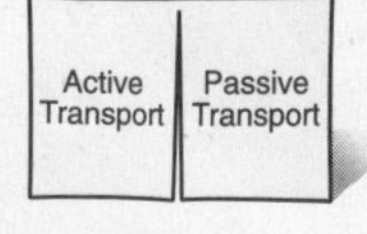

Read to Learn

Passive Transport

Window screens keep unwanted things, such as bugs, leaves, and birds, outside. But screens do let some things, such as air and smoke, pass through.

Cells get food, oxygen, and other substances from their environments. They release waste materials into their environments. The membrane around the cell works like a window screen works for a room. A window screen is selectively permeable (PUR mee uh bul). It lets things like air come into the room and keeps some things like bugs out of the room. A cell's membrane also is selectively permeable. It lets some things come into or leave the cell. It also keeps other things from entering or leaving the cell.

Things move through a cell membrane in several ways. The movement depends on the size of the molecules, the path the molecules take, and whether energy is needed. When substances move through the cell membrane without using energy, this movement is known as **passive transport**. Three types of passive transport are diffusion, osmosis, and facilitated diffusion. The type of transport depends on what is moving through the cell membrane.

How does diffusion create equilibrium?

Molecules move constantly and randomly. You might smell perfume when you walk past someone who is wearing it. The perfume molecules move freely throughout the air. This random movement of molecules from an area where there are more of them into an area where there are fewer of them is called **diffusion**. Diffusion is a type of passive transport. Molecules will keep moving from one area to another until the number of these molecules is equal in the two areas. When this occurs, **equilibrium** is reached and diffusion stops. ☑

All cells in your body use oxygen. Oxygen moves through your body in the red blood cells. When your heart pumps blood to your lungs, your red blood cells contain few oxygen molecules. Your lungs have many oxygen molecules. Oxygen molecules move, or diffuse, from your lungs into your red blood cells. The blood continues its journey through your body. When the blood reaches your big toe, there are more oxygen molecules in your red blood cells than in the cells of your big toe. The oxygen diffuses from your red blood cells to your big toe's cells. The process is shown in the figure below.

Reading Check

1. **Determine** How does diffusion create equilibrium?

____

____

____

Picture This

2. **Explain** Use the figure to explain to a partner how diffusion works.

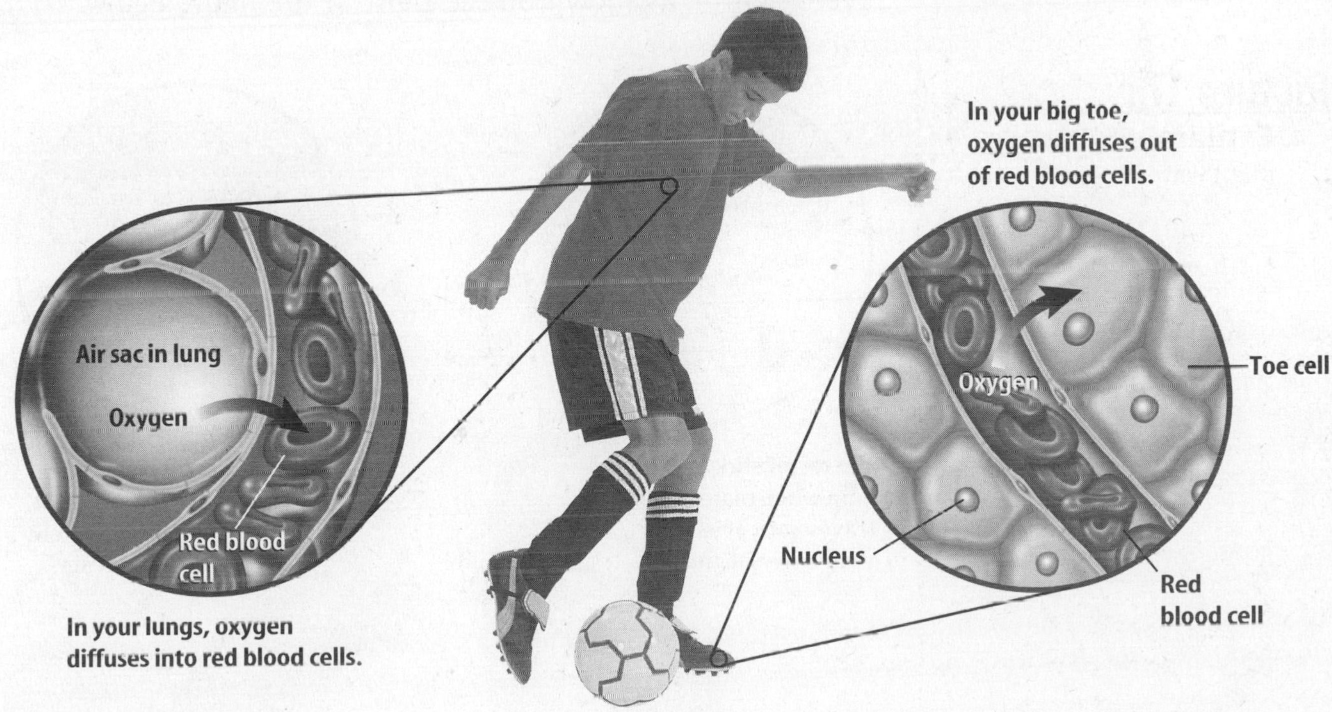
Air sac in lung
Oxygen
Red blood cell
In your lungs, oxygen diffuses into red blood cells.
In your big toe, oxygen diffuses out of red blood cells.
Oxygen
Toe cell
Nucleus
Red blood cell

KS Studios

What is facilitated diffusion?

Some substances pass easily through the cell membrane by diffusion. Larger substances may need help passing through the cell membrane. Transport proteins in the cell membrane help these substances enter the cell. This process is called facilitated diffusion. Transport proteins are similar to the gates at a stadium. Gates are used to move people into and out of the stadium. Similarly, transport proteins are used to move substances into and out of a cell. ☑

Reading Check

3. **Describe** What do transport proteins do?

____

____

____

What is osmosis?

Remember that water makes up a large part of living matter. Water molecules move by diffusion in and out of cells. The diffusion of water through the cell membrane is called **osmosis.**

What happens when you do not water plants? As a plant cell loses water, its cell membrane pulls away from the cell wall. This reduces pressure against the cell wall, and the plant cell becomes limp, as shown on the left in the figure below. The plant wilts because more water leaves the plant's cells than enters them.

When you water the plant, the water moves through the cell membranes and fills the cells with water. The plant's cell membranes push against their cell walls, and the cells become firm, as shown on the right in the figure below.

Picture This

4. **Explain** Why does a plant wilt?

____

____

____

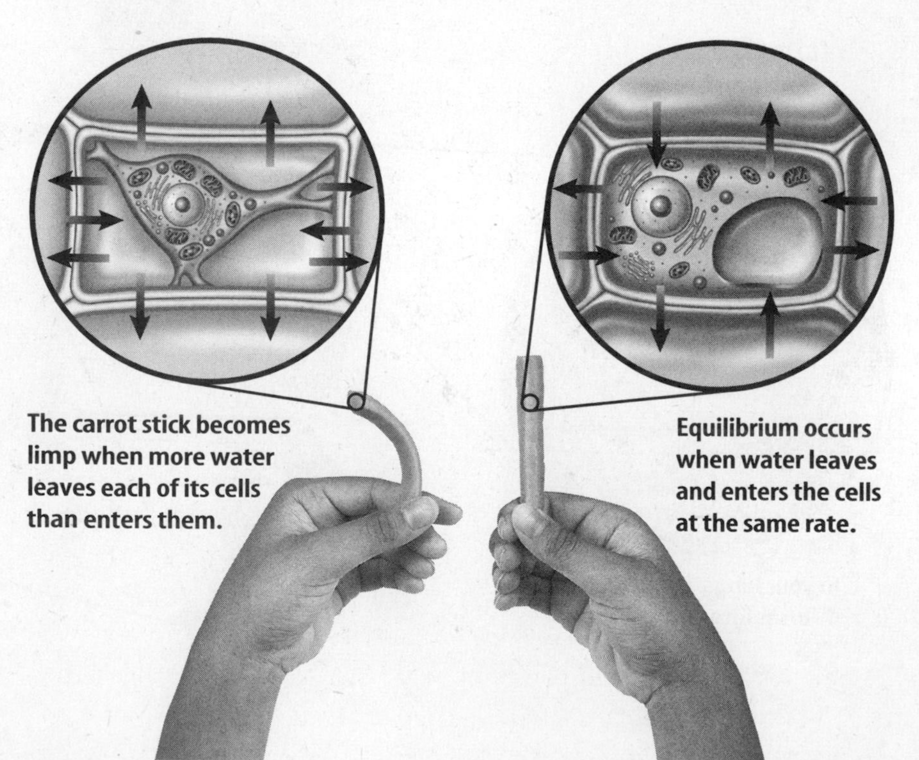

The carrot stick becomes limp when more water leaves each of its cells than enters them.

Equilibrium occurs when water leaves and enters the cells at the same rate.

Aaron Haupt

Osmosis in Animal Cells Osmosis also takes place in animal cells. If animal cells were placed in pure water, they too would swell up. However, animal cells are different from plant cells. Just like an overfilled balloon, animal cells will burst if too much water enters the cell.

Active Transport

Suppose you have just left a theater at the end of a movie when you remember that you left your jacket inside. You have to move against the crowd to enter the theater and get your jacket. Which takes more energy—leaving the theater with the crowd or moving against the crowd to get back into the theater? Something similar to this happens in cells.

Active transport takes place when energy is needed to move substances through a cell membrane. For example, root cells require minerals from the soil. The root cells already have more molecules of the minerals than the surrounding soil. Normally, the mineral molecules would move out of the root into the soil until equilibrium is reached. But the root cells need to take in the minerals from the soil.

Like facilitated diffusion, active transport uses transport proteins. In active transport, transport proteins bind with the needed substance and cellular energy is used to move it through the cell membrane. ☑

Reading Check

5. **Explain** What must be used with transport proteins to move a substance through a cell membrane in active transport?

____

____

Endocytosis and Exocytosis

Some molecules are too large to move through the cell membrane by diffusion or by using transport proteins. Large protein molecules, for example, can enter a cell when they are surrounded by the cell membrane. The cell membrane folds around the molecule, completely surrounding it. The sphere created is called a vesicle. The sphere pinches off and moves the molecule into the cell. The process of taking substances into a cell by surrounding it with the cell membrane is known as **endocytosis** (en duh si TOH sus). Some one-celled organisms take in food this way.

Exocytosis (ek soh si TOH sus) is the process in which the contents of a vesicle are moved outside a cell. A vesicle's membrane joins with a cell's membrane, and the vesicle's contents are released. Exocytosis occurs in the opposite way that endocytosis does. Cells in your stomach use exocytosis to release chemicals that help digest food. ☑

Reading Check

6. **Explain** What happens during exocytosis?

____

____

____

After You Read

Mini Glossary

active transport: takes place when energy is needed to move substances through a cell membrane; uses transport proteins

diffusion: random movement of molecules from an area where there are more of them into an area where there are fewer of them

endocytosis (en duh si TOH sus): process of taking substances into a cell by surrounding them with the cell's membrane

equilibrium: the number of molecules in two areas are the same

exocytosis (ek soh si TOH sus): process in which the contents of a vesicle are moved outside a cell

osmosis: the diffusion of water through the cell membrane

passive transport: movement of substances through the cell membrane without using energy

1. Review the terms and their definitions in the Mini Glossary. Choose one term that explains how substances move into and out of cells and write a sentence explaining how the process works.

____

____

____

2. Complete the Venn diagram below to help you compare active and passive transport.

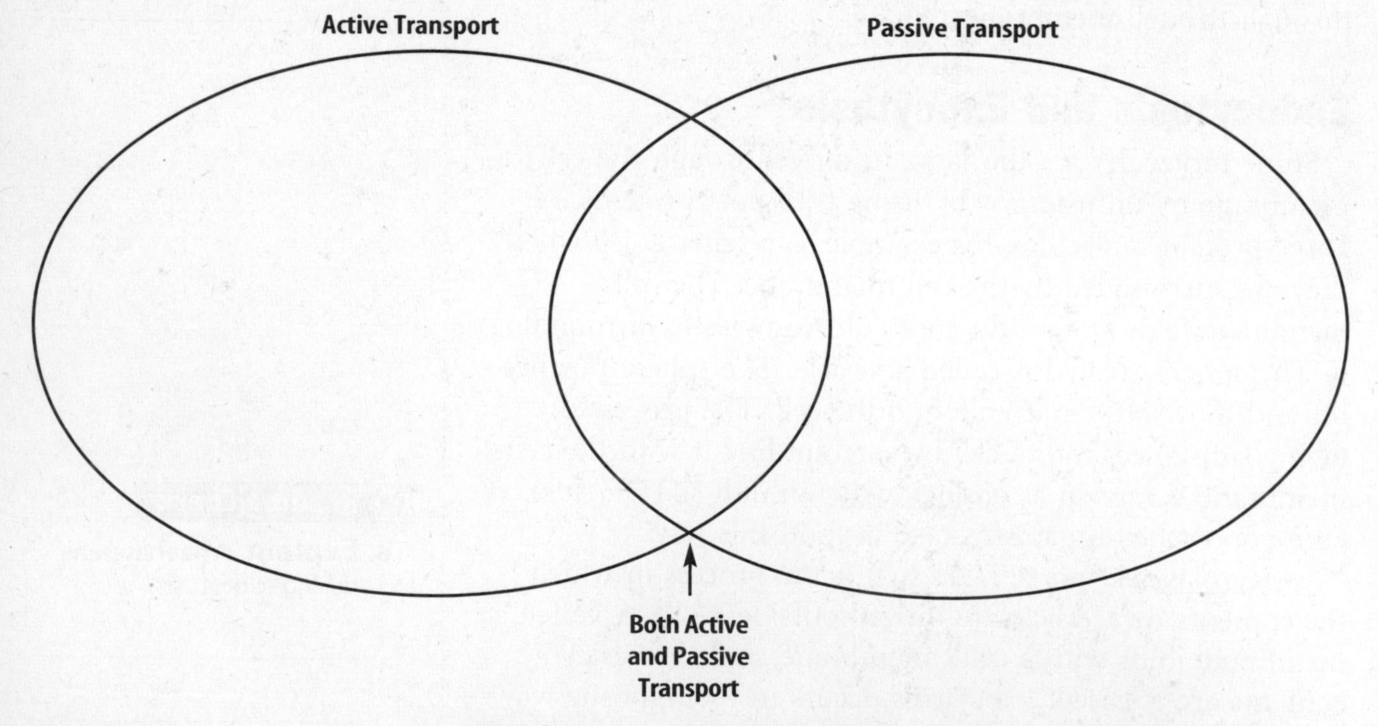

End of Section

Science Online Visit **life.msscience.com** to access your textbook, interactive games, and projects to help you learn more about the movement of cellular materials.

chapter 3 Cell Processes

section 3 Energy for Life

Before You Read

Describe on the lines below why you think your body needs food.

____

____

____

What You'll Learn

- the differences between producers and consumers
- that photosynthesis and respiration store and release energy
- how cells get energy

Read to Learn

Trapping and Using Energy

Chemical energy is stored in food molecules. This chemical energy is changed inside cells into other forms of energy needed for life. In every cell, these changes involve chemical reactions. In fact, all of an organism's activities involve chemical reactions. All the chemical reactions in an organism make up **metabolism**.

The chemical reactions of metabolism need enzymes. Enzymes cause changes, but the enzymes are not changed during the reaction and can be used again. In the figure below, an enzyme attaches to a large molecule and helps it to change. At the end of the chemical reaction, the molecule has changed into two smaller molecules, but the enzyme has not changed.

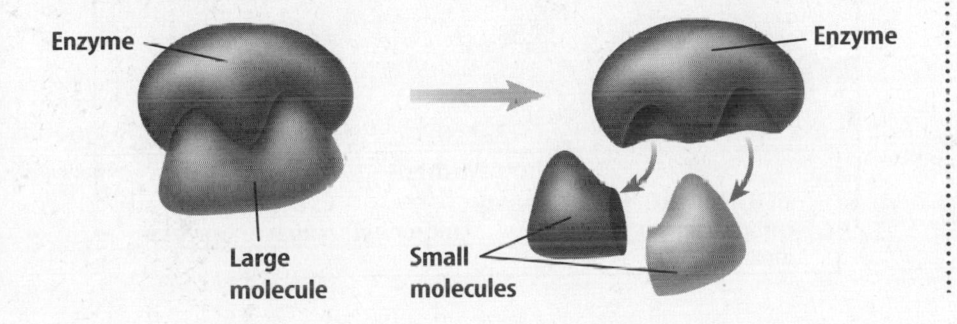

Mark the Text

Locate Information Read all the headings for this section and circle any word you cannot define. At the end of each section, review the circled words and underline the part of the text that helps you define the words.

Picture This

1. **Explain** What happens to the enzyme during the chemical reaction?

____

____

____

FOLDABLES™

D Describe Make a three-tab book, as shown below. Write the names of the following processes on the tabs: photosynthesis, respiration, and fermentation. Inside each tab, describe the process.

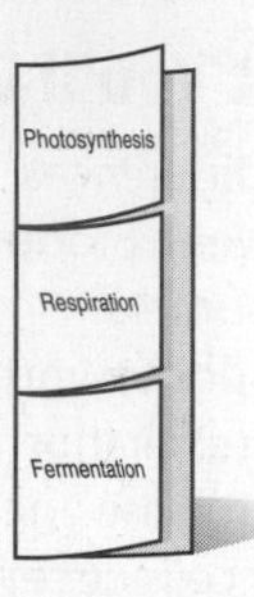

What happens during photosynthesis?

Living things are divided into two groups—producers and consumers—based on how they obtain their food. Organisms that make their own food, such as plants, are producers. Organisms that cannot make their own food are consumers.

Plants and many other producers can convert light energy into chemical energy. Producers use a process called **photosynthesis** to change light energy from the Sun into sugars, which can be used for food. Plants and other producers that use photosynthesis are usually green because they contain a green pigment called chlorophyll (KLOR uh fihl).

In plant cells, these pigments are found in chloroplasts. Chlorophyll is used in photosynthesis to capture light energy. Plants use chlorophyll to make sugar and oxygen (O_2) from the raw materials carbon dioxide (CO_2), water (H_2O), and light energy. Plants get their raw materials from the air, soil, and Sun. Some of the light energy is stored in the chemical bonds that hold the sugar molecules together. Enzymes also are needed for the reactions to occur. The process of photosynthesis is shown in the figure below. Review the chemical equation for photosynthesis to identify the raw materials and the results of the chemical process.

Picture This

2. Identify Circle the three things needed for photosynthesis to take place.

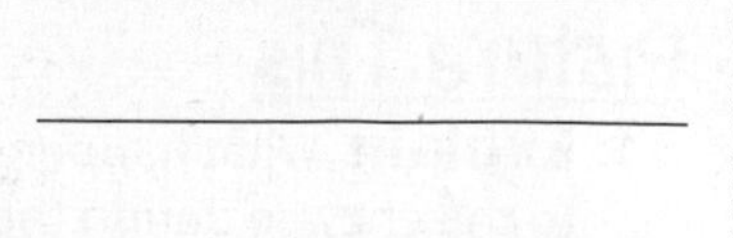

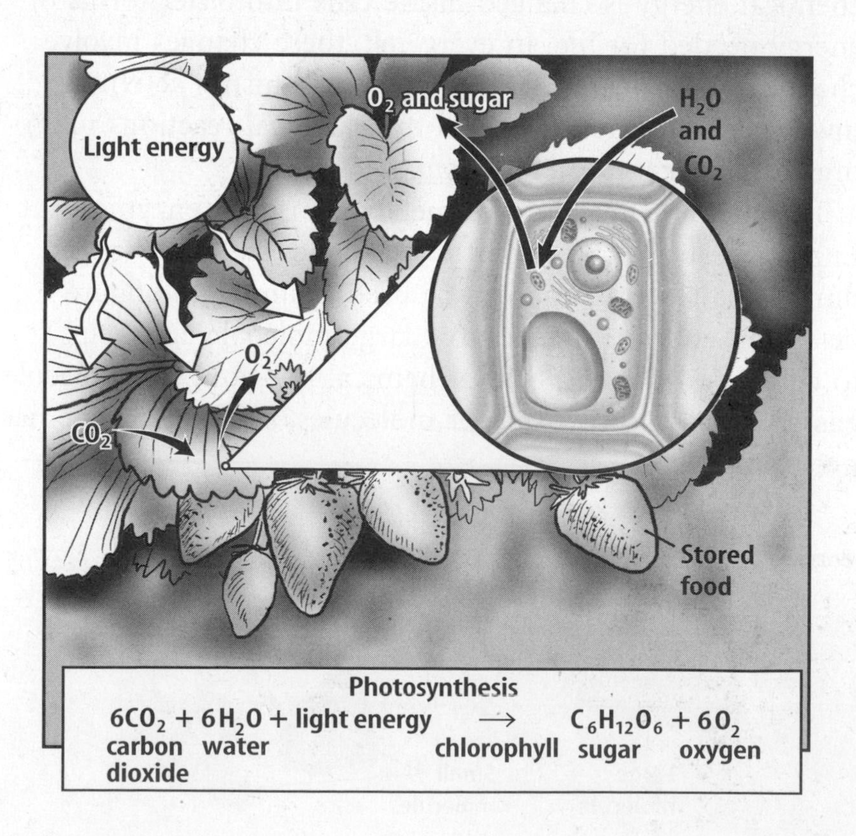

Photosynthesis

$$6CO_2 + 6H_2O + \text{light energy} \xrightarrow{\text{chlorophyll}} C_6H_{12}O_6 + 6O_2$$

carbon dioxide, water, chlorophyll, sugar, oxygen

How do plants store and use carbohydrates?

Plants make more sugar during photosynthesis than they need for survival. The extra sugar is changed and then stored as starches and other carbohydrates. Plants, such as apple trees, use these carbohydrates for growth, for keeping up cells, and for reproduction. ☑

Why is photosynthesis important to consumers?

Consumers get energy by eating producers and other consumers. No matter what food you eat, photosynthesis was involved directly or indirectly in its production. For example, an apple tree uses photosynthesis to make apples. When you eat an apple, the stored sugars help feed your body. Some cheese comes from milk, which is produced by cows that eat plants. The plants the cows eat are producers. The cows and humans are consumers.

How do you use energy?

Imagine that you get up late for school. You dress quickly and run three blocks to school. When you get to school, you feel hot and are breathing fast. Why? Your muscle cells use a lot of energy when you run. To get this energy, muscle cells break down food. Some of the energy in the food is used when you run and some of it becomes thermal energy, which is why you feel warm or hot. Most cells need oxygen to break down food. You are breathing fast because your body was working to get oxygen to your muscles.

What is respiration?

When you ran, your muscle cells were using the oxygen for the process of respiration. During **respiration**, chemical reactions break down food molecules into simpler substances and release stored energy. Just as in photosynthesis, enzymes are needed for the chemical reactions of respiration.

Respiration occurs in the cells of all living things. As you are reading this page, millions of cells in your body are breaking down food molecules and releasing energy. Two waste products, carbon dioxide and water, are produced during respiration. Your body gets rid of the carbon dioxide and some of the water when you breathe out, or exhale.

Reading Check

3. **Explain** How do plants use carbohydrates?

____

____

____

Think it Over

4. **Draw Conclusions** Why does respiration occur only in living things?

____

____

____

What is fermentation?

Even though you breathe harder when you run, your muscle cells might not receive enough oxygen for respiration. When this happens, a process in the muscle cells known as **fermentation** releases some of the energy stored in glucose (sugar) molecules.

Fermentation also releases energy and produces wastes. The type of wastes produced depends on the type of cell. They may be lactic acid, alcohol, and carbon dioxide. Fermentation in your muscle cells changes simple molecules into lactic acid while releasing energy, as shown in the figure below. The presence of lactic acid is why your muscles might feel stiff or sore after you have run to school.

Picture This

5. Identify What are three waste products created during fermentation?

____

____

____

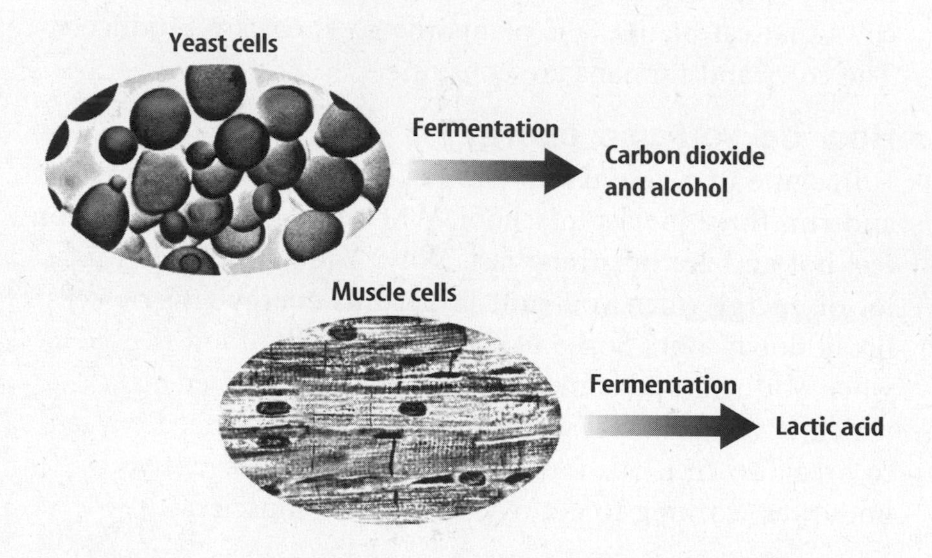

What products come from fermentation?

Some organisms, such as bacteria, carry out fermentation and make lactic acid. Some of these organisms are used to make yogurt and some cheeses. These organisms break down a sugar in milk to release energy. The lactic acid produced causes the milk to become more solid. Some of the flavor in yogurt and cheese comes from this process.

Have you ever used yeast to make bread? Yeasts are one-celled living organisms. Fermentation in yeast cells breaks down the sugar in bread dough. The cells produce alcohol and carbon dioxide as wastes. The carbon dioxide waste is a gas that makes the bread dough rise. The alcohol is lost as the bread bakes. ☑

☑ Reading Check

6. Identify What waste products come from the fermentation of yeast cells?

____

____

How do photosynthesis and respiration work together?

Some producers make food through photosynthesis. All living things release energy stored in food through respiration or fermentation. If you think carefully about photosynthesis and respiration, you will note that what is produced by one process is used by the other process.

Photosynthesis and respiration are almost the opposite of each other. Photosynthesis produces sugars and oxygen, and respiration uses these products. The carbon dioxide and water produced during respiration are used during photosynthesis.

As you fill in the products in the figure below, review how the products of one process are the wastes of the other process. Photosynthesis and respiration cannot take place without each other. And most life would not be possible without these important chemical reactions.

Picture This

7. **Illustrate** In the figure below, fill in the products released by photosynthesis and respiration.

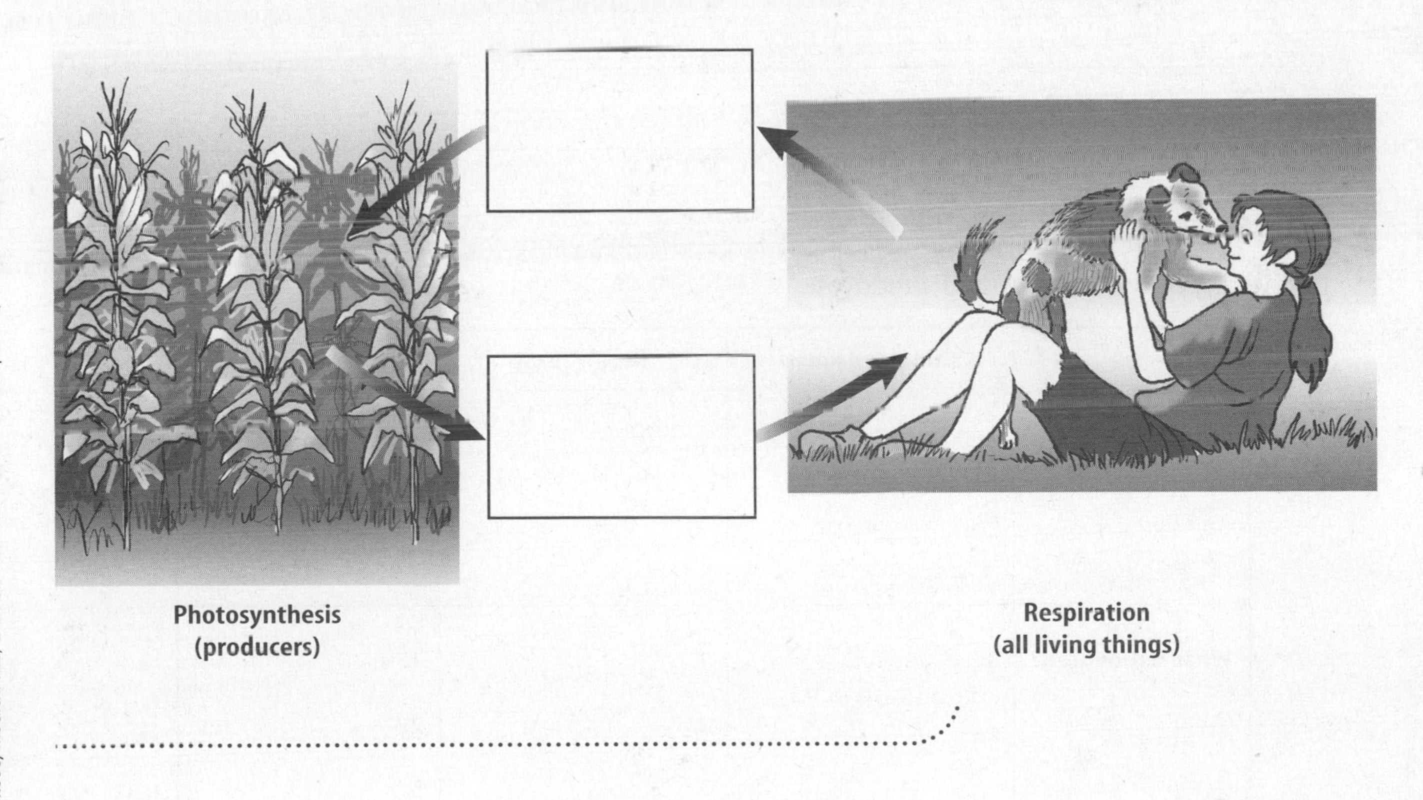

After You Read

Mini Glossary

fermentation: chemical reaction that releases energy stored in glucose (sugar) molecules and produces carbon dioxide, lactic acid, and alcohol as wastes

metabolism: all chemical reactions that take place in an organism

photosynthesis: process that uses light energy, carbon dioxide, and water to produce the sugars and oxygen needed by all living things

respiration: chemical reaction that uses oxygen and breaks down food molecules into simpler substances to release their stored energy

1. Review the terms and their definitions in the Mini Glossary. Write a short paragraph that describes how photosynthesis and respiration are related.

____

____

____

____

____

2. Fill in the table below to identify what is needed by each chemical reaction and what is produced by each chemical reaction.

	Photosynthesis	Respiration	Fermentation
What is needed?	1. 2. 3.	1. 2.	1. glucose molecules
What is produced?	1. 2.	1. 2. 3.	1. 2. 3.

Visit **life.msscience.com** to access your textbook, interactive games, and projects to help you learn more about energy for life.

Cell Reproduction

section 1 Cell Division and Mitosis

Before You Read

List five living things on the lines below. Then write one thing that these items have in common with each other and with you.

____

____

____

What You'll Learn

- why mitosis is important
- the steps of mitosis
- the similarities and differences between mitosis in plant and animal cells
- examples of asexual reproduction

Read to Learn

Why is cell division important?

All living things are made up of cells. Many organisms start as one cell. The cell divides and becomes two cells, two cells become four, four become eight, and so on. Through the process of cell division, the organism grows.

Cell division is still important after an organism stops growing. For example, every day billions of your red blood cells wear out and are replaced through cell division. During the time it takes you to read this sentence, your bone marrow produced about six million red blood cells.

Cell division is the way a one-celled organism makes another organism of its kind. When a one-celled organism reaches a certain size, it reproduces by dividing into two cells.

Mark the Text

Identify Details Highlight each question head. Then use another color to highlight the answer to that question.

The Cell Cycle

Every living organism has a life cycle. A life cycle has three parts. First, the organism forms. Next, it grows and develops. Finally, the life cycle ends when the organism dies. Right now, you are in a part of your life cycle called adolescence (a doh LEH sence), which is a time of active growth and development.

FOLDABLES™

A Describe Use quarter sheets of notebook paper, as shown below, to describe cell growth and development and cell division.

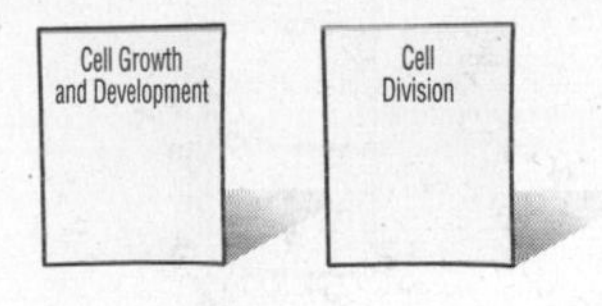

How long is the life cycle of a cell?

Every cell has a life cycle. A cell's life cycle is called a cell cycle, as shown in the figure below. A cell cycle is not completed in the same amount of time in all cells. For example, the cell cycle of some human cells takes about 16 hours. The cell cycle of some plant cells takes about 19 hours. A cell cycle has three parts—interphase, mitosis, and cytoplasm division.

Picture This

1. Identify Draw an outline around the interphase part of the cell cycle to the right. Approximately how much of the cell cycle is interphase?

____

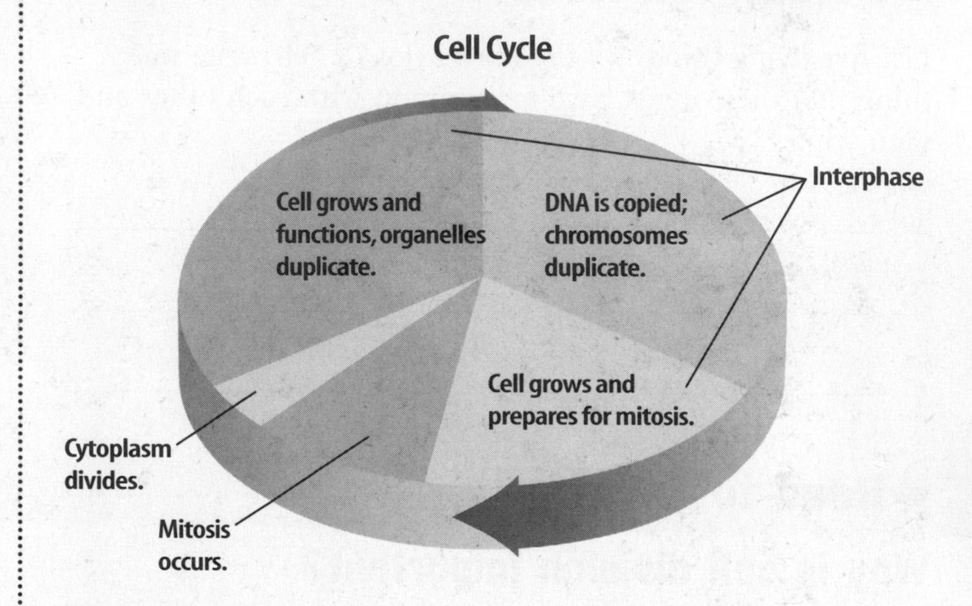

Cell Cycle
Interphase
Cell grows and functions, organelles duplicate.
DNA is copied; chromosomes duplicate.
Cell grows and prepares for mitosis.
Cytoplasm divides.
Mitosis occurs.

What is the longest part of the cell cycle?

For cells that have a nucleus, the longest part of the cell cycle is a period of growth and development called **interphase.** Cells in your body that no longer divide, such as nerve and muscle cells, are always in interphase.

During interphase, an actively dividing cell, such as a skin cell, copies its DNA and prepares for cell division. DNA is the chemical code that controls an organism's growth and operation. A copy of a cell's DNA must be made before dividing so that each of the two new cells will get a complete copy. Each cell needs a complete set of hereditary material to carry out life functions.

FOLDABLES™

B Sequence Make a four-tab book, as shown below. Use the Foldable to identify facts about the four steps of mitosis.

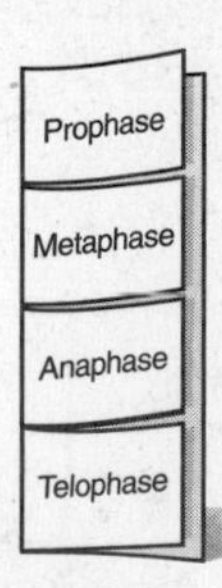

Mitosis

After interphase, cell division begins. Mitosis is the first step in cell division. **Mitosis** (mi TOH sus) is the process in which the cell's nucleus divides to form two nuclei. Each new nucleus is identical to the original nucleus. The steps of mitosis are called prophase, metaphase, anaphase, and telophase.

What happens to chromosomes during cell division?

A chromosome (KROH muh sohm) is a structure in the nucleus that contains DNA. During interphase, each chromosome is copied. When the nucleus is ready to divide, the two copies of each chromosome coil tightly into two thickened, identical DNA strands called chromatids (KROH muh tidz). In the figure to the right, the chromatids are held together at a place called the centromere.

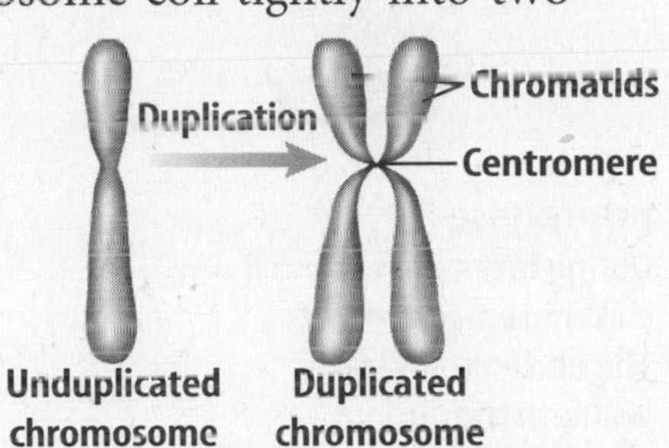

Prophase During prophase, the chromatid pairs can be seen. The nuclear membrane breaks apart. Two small structures called centrioles (SEN tree olz) move to opposite ends of the cell. Between the centrioles, threadlike spindle fibers stretch across the cell. Animal cells have centrioles, but plant cells do not. ☑

Metaphase In metaphase, the chromatid pairs line up across the center of the cell. The centromere of each pair usually becomes attached to two spindle fibers—one from each side of the cell.

Anaphase In anaphase, each centromere divides. The spindle fibers become shorter, and each chromatid separates from its partner. The separated chromatids begin to move to opposite ends of the cell. They are now called chromosomes.

Telophase The final step of mitosis is telophase. During telophase, the spindle fibers start to disappear. The chromosomes start to uncoil, and a new nucleus forms.

How does the cytoplasm divide?

For most cells, after the nucleus divides, the cytoplasm separates and two new cells are formed. Each new cell contains one of the new nuclei. In animal cells, the cell membrane pinches in the middle, like a balloon with a string tightened around it. The cell divides at the pinched area to form two new cells. Each new cell contains half the cytoplasm from the old cell.

After the division of the cytoplasm, most new cells begin interphase again. Use the figure on the next page to review the cell division of an animal cell.

Picture This

2. **Identify** Circle the place where the chromatids are held together.

Reading Check

3. **Explain** what happens to the centrioles during prophase.

____

____

____

Picture This

4. **Describe** Highlight the chromosomes in each phase of mitosis. As you highlight the step, explain to a partner what is happening to the chromosome.

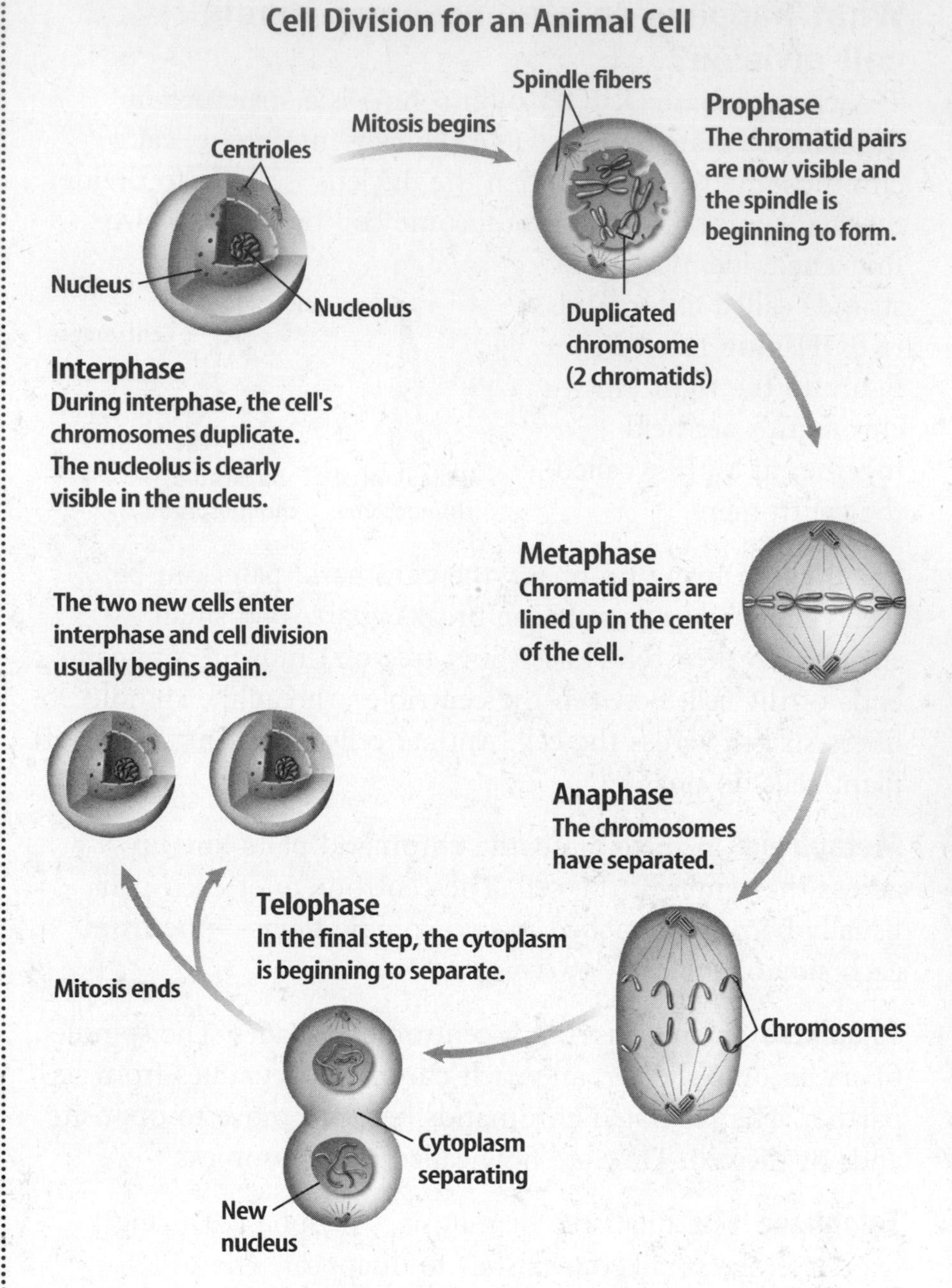

Cell Division for an Animal Cell
Spindle fibers
Mitosis begins
Centrioles
Prophase
The chromatid pairs are now visible and the spindle is beginning to form.
Nucleus
Nucleolus
Duplicated chromosome (2 chromatids)
Interphase
During interphase, the cell's chromosomes duplicate. The nucleolus is clearly visible in the nucleus.
Metaphase
Chromatid pairs are lined up in the center of the cell.
The two new cells enter interphase and cell division usually begins again.
Anaphase
The chromosomes have separated.
Telophase
In the final step, the cytoplasm is beginning to separate.
Mitosis ends
Chromosomes
Cytoplasm separating
New nucleus

Reading Check

5. **Explain** In plant cells, what divides the cytoplasm into two parts?

____

How do plant cells divide after mitosis?

In plant cells, a cell plate forms in the middle of the cell. The cell plate divides the cytoplasm into two parts. New cell walls form along the cell plate, and new cell membranes develop inside the cell walls. ☑

What are the results of mitosis?

You should remember two important things about mitosis. First, mitosis is the division of a cell's nucleus. Second, it produces two new nuclei that are identical to each other and to the original nucleus. Every cell in your body, except sex cells, has a nucleus with 46 chromosomes—23 pairs. This is because you began as one cell with 46 chromosomes in its nucleus. Skin cells, produced to replace or repair your skin, have the same 46 chromosomes as the original single cell you developed from.

The 46 chromosomes of a human cell are shown below. Notice that the last pair is labeled XY. This is the chromosome pair that determines sex. The XY label indicates a male. Females have XX chromosome pairs.

Chromosomes of a human cell

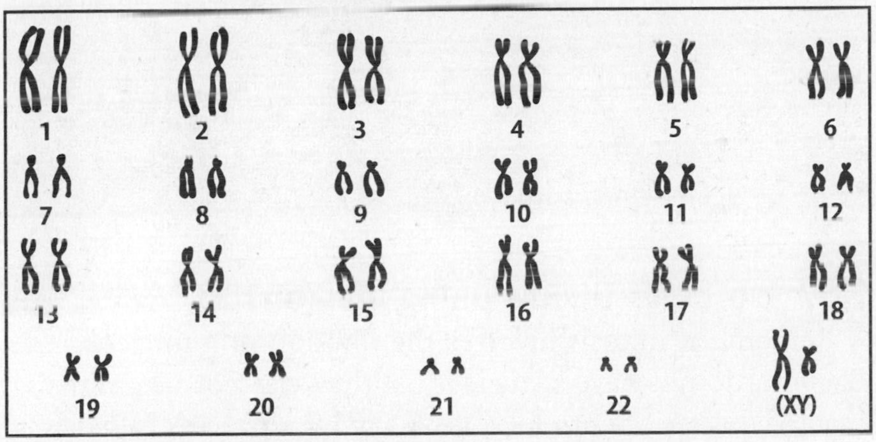

______ (No. of chromosome pairs) × 2 = ______ (No. of chromosomes)

Each of the trillions of cells in your body, except sex cells, has a copy of the same DNA. All of your cells, however, use different parts of the DNA to become different types of cells. Skin cells and blood cells contain a copy of the same DNA. They use different parts of the DNA to perform their different functions.

Cell division allows growth and replaces worn out or damaged cells. You are much larger than you were when you were a baby. This is possible because of cell division. If you cut yourself, the wound heals because cell division replaces damaged cells. ☑

Picture This

6. Solve Complete the equation at the bottom of the figure using the information in the figure.

✔ Reading Check

7. Explain What is the purpose of cell division?

______

______

______

Asexual Reproduction

The way an organism produces others of its kind is called reproduction. Among living organisms, there are two types of reproduction—sexual and asexual. Sexual reproduction usually involves two parent organisms. In **asexual reproduction**, a new organism (sometimes more than one) is produced from only one parent organism. The new organism has the same DNA as the parent. New strawberry plants can be reproduced asexually from horizontal stems called runners. The figure below shows the asexual reproduction of a strawberry plant.

Picture This

8. **Identify** How many organisms were needed to produce the strawberry runner?

____

How do cells divide using fission?

Remember, mitosis involves the division of a nucleus. Bacteria do not have a nucleus, so they can not use mitosis. Instead, bacteria reproduce asexually by a process called fission. During fission, a bacteria cell's DNA is copied. The cell then divides into two identical organisms. Each new organism has a complete copy of the parent organism's DNA.

How do organisms reproduce using budding?

Budding is a type of asexual reproduction in which a new organism grows from the body of the parent. When the bud on the adult becomes large enough, it breaks away to live on its own. ☑

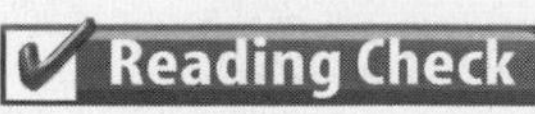

9. **Explain** budding, which is a form of asexual reproduction.

____

____

____

How do some organisms regrow body parts?

Some organisms, such as sponges and sea stars, can regrow damaged or lost body parts. The process that uses cell division to regrow body parts is called regeneration. If a sea star breaks into pieces, a whole new organism can grow from each piece.

After You Read

Mini Glossary

asexual reproduction: the way a new organism is produced from one organism

chromosome (KROH muh sohm): a structure in the nucleus that contains hereditary material

mitosis (mi TOH sus): the process in which the nucleus divides to form two identical nuclei; the four steps include prophase, metaphase, anaphase, telophase

1. Review the terms and their definitions in the Mini Glossary. Write a sentence to explain mitosis using a skin cell as an example.

2. Complete the Venn diagram below to help you compare mitosis in plant and animal cells. Write one similarity at each phase in the overlapping area.

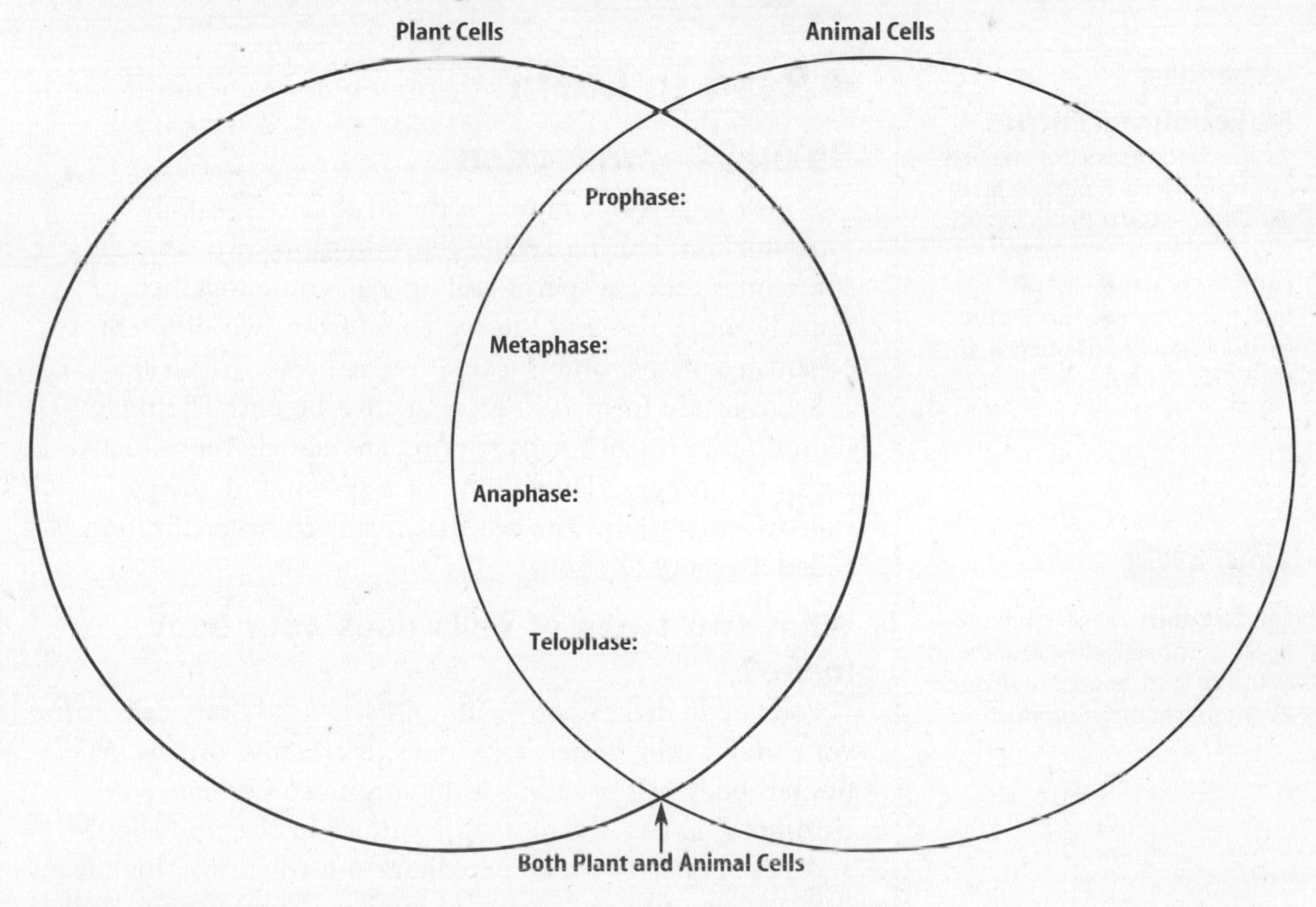

Visit **life.msscience.com** to access your textbook, interactive games, and projects to help you learn more about cell division and mitosis.

Cell Reproduction

section 2 Sexual Reproduction and Meiosis

What You'll Learn

- the stages of meiosis
- how sex cells are produced
- why meiosis is needed for sexual reproduction
- the names of the cells involved in fertilization
- how fertilization occurs in sexual reproduction

Before You Read

On the lines below, explain what makes you different from anyone else in your class.

____

____

____

Study Coach

Make Journal Entries As you read the section, write a question for each paragraph in a journal. Answer the question with information from the paragraph. Make a list of questions you have about the section that are still unclear and then find the answers.

FOLDABLES™

C Explain Make a three-tab book, as shown below. Use the Foldable to make a Venn diagram explaining sexual reproduction.

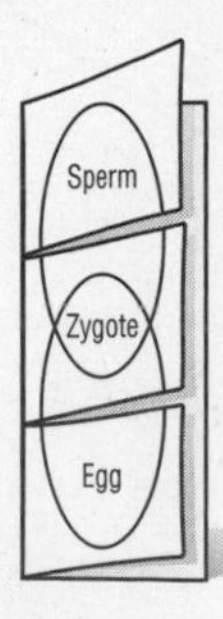

Read to Learn

Sexual Reproduction

A new organism can be produced through sexual reproduction. During **sexual reproduction,** two sex cells, sometimes called a sperm and an egg, come together. Usually the sperm and the egg come from two different organisms of the same species.

Sex cells are formed in reproductive organs. The male reproductive organ forms **sperm.** The female reproductive organ forms **eggs.** The joining of a sperm and an egg is called **fertilization.** The cell that forms from fertilization is called a **zygote** (ZI goht).

What two types of cells does your body make?

Your body makes body cells and sex cells. Body cells form your brain, skin, bones, and other tissues and organs. A human body cell usually has 46 chromosomes. Each chromosome has a mate that is similar in size and shape and has similar DNA, or hereditary information. This means that a body cell has 23 pairs of similar chromosomes. Cells that have pairs of similar chromosomes are called **diploid** (DIH ployd) cells.

What are haploid cells?

A sex cell has half the number of chromosomes found in a body cell, or 23 chromosomes. A sex cell has only one chromosome from each pair. A cell that does not have pairs of chromosomes is called a **haploid** (HA ployd) cell.

Meiosis and Sex Cells

A process called **meiosis** (mi OH sus) produces haploid sex cells. During meiosis, two divisions of the nucleus occur. These divisions are called meiosis I and meiosis II. The steps of each division of meiosis are named like the steps in mitosis—prophase, metaphase, anaphase, and telophase. The figure below shows what happens during meiosis I.

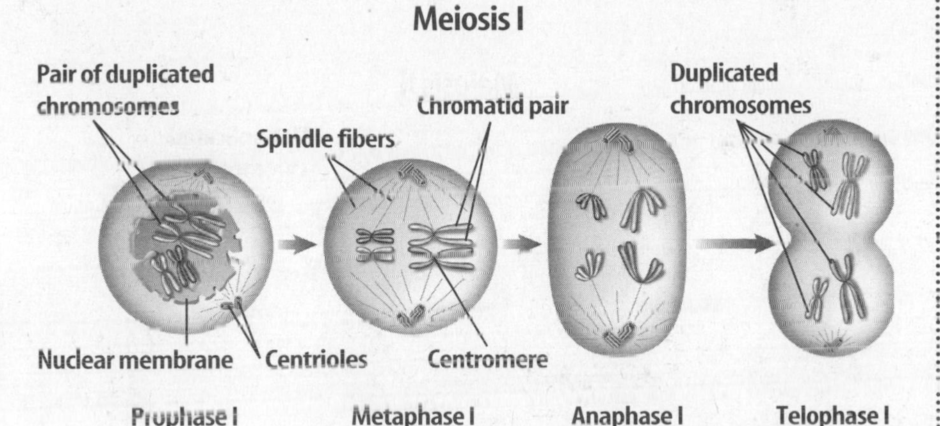

Picture This

1. Identify How many cells form in meiosis I?

What happens to a cell during meiosis I?

Before meiosis begins, each chromosome is copied. When the cell is ready for meiosis, the two copies of each chromosome can be seen under a microscope as two chromatids. Follow the steps in meiosis I in the figure above. Notice that in prophase I, each pair of duplicated chromosomes comes together. ☑

In metaphase I, the pairs of duplicated chromosomes line up in the center of the cell. As you can see, the centromere of each chromatid pair attaches to one spindle fiber.

In anaphase I, the two copies of the same chromosome, the chromatids, move away from each other to opposite ends of the cell. Notice that each duplicated chromosome still has two chromatids.

In telophase I, the cytoplasm divides and two new cells form. Each new cell has one duplicated chromosome from each similar pair.

Reading Check

2. Explain What happens in a cell before meiosis I begins?

What happens in meiosis II?

The two cells that formed in meiosis I now begin meiosis II. Follow the steps in meiosis II in the figure below. As you can see in prophase II, the duplicated chromosomes and spindle fibers reappear in each new cell.

In metaphase II, the duplicated chromosomes move to the center of each cell. The centromere of each chromatid pair attaches to two spindle fibers.

In anaphase II, the centromere in each cell divides. Then the chromatids separate and move to opposite ends of each cell. Each chromatid becomes an individual chromosome.

In telophase II, the spindle fibers disappear, and a nuclear membrane forms around the chromosomes at each end of the cell. When meiosis II is finished, the cytoplasm of each cell divides.

Think it Over

3. **Explain** how metaphase I and metaphase II differ.

____

____

____

____

____

Meiosis II

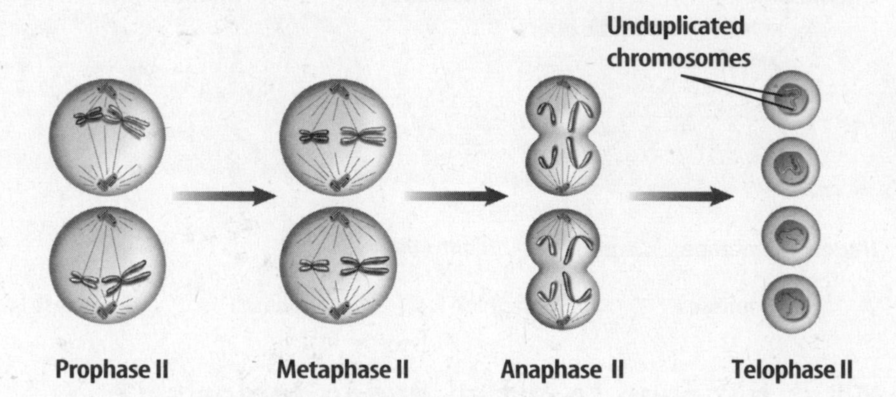

What is the final result of meiosis?

During meiosis I, one cell divides into two cells. During meiosis II, those two cells divide. When meiosis II ends, there are four sex cells. Each sex cell has 23 unpaired chromosomes. This is one-half the number of chromosomes that were in the original nucleus—46 chromosomes.

What can go wrong in meiosis?

Mistakes sometimes occur during meiosis. These mistakes can produce sex cells with too many or too few chromosomes. Zygotes, cells that form from fertilized eggs, produced from these sex cells sometimes die. If the zygote lives, every cell that grows from the zygote will have the wrong number of chromosomes. Organisms with the wrong number of chromosomes usually do not grow normally. ☑

Reading Check

4. **Explain** What is the usual result of too many or too few chromosomes?

____

____

____

After You Read

Mini Glossary

diploid (DIH ployd): cells that have pairs of similar chromosomes
egg: sex cell formed in the female reproductive organs
fertilization: the joining of a sperm and an egg
haploid (HA ployd): cells that do not have pairs of chromosomes, such as sex cells
meiosis (mi OH sis): a process that produces haploid sex cells
sexual reproduction: two sex cells come together to produce a new organism
sperm: sex cell formed in the male reproductive organs
zygote (ZI goht): the cell that forms from fertilization

1. Review the terms and their definitions in the Mini Glossary. Choose the terms that explain the process of sexual reproduction and write one or two sentences explaining how the process works.

2. Complete the graphic organizer below to label the steps that occur during meiosis I and meiosis II.

3. How do your journal entries help you understand sexual reproduction and meiosis?

Visit **life.msscience.com** to access your textbook, interactive games, and projects to help you learn more about sexual reproduction and meiosis.

chapter 4 Cell Reproduction

section 3 DNA

What You'll Learn

- the parts of a DNA molecule and its structure
- how DNA copies itself
- the structure and role of each kind of RNA

Before You Read

Write on the lines below how police departments use DNA to solve crimes.

____

____

____

Study Coach

Discuss Read a paragraph to yourself, then take turns with your partner saying something about what you have learned. Continue your discussion until you and your partner understand the paragraph. Then repeat the process with the remaining paragraphs in the section.

Read to Learn

What is DNA?

Before you could learn to read, you learned the alphabet. The letters of the alphabet are a code you needed to know before you could read. A cell also uses a code. That code contains information for an organism's growth and function. It is stored in a cell's hereditary material. The code is a chemical called deoxyribonucleic (dee AHK sih ri boh noo klay ihk) acid, or **DNA**. The figure to the right shows the spiral-shaped structure of DNA.

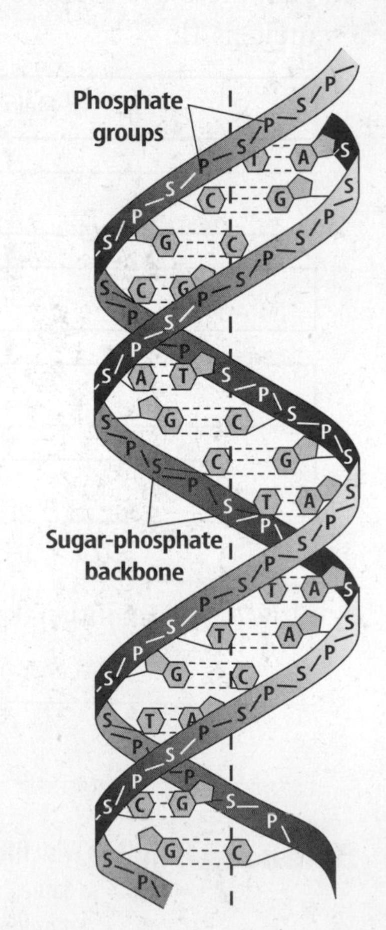

When a cell divides, the DNA code is copied and passed to the new cells. New cells get the same DNA code that was in the original cell. Every cell that has ever been formed in your body or in any organism has DNA.

Picture This

1. **Infer** Examine the DNA strand in the figure. What do the letters "P" and "S" represent?

____

____

What does DNA look like?

In 1952, scientist Rosalind Franklin discovered that DNA is two chains of molecules. As you can see in the figure on the previous page, DNA looks like a twisted ladder. Each side of the ladder is made up of sugar-phosphate molecules. The sugar in each molecule is called deoxyribose (dee AHK sih ri bohs). In 1953, scientists James Watson and Francis Crick made a model of a DNA molecule. ☑

What are the four nitrogen molecules that make up DNA?

The rungs, or steps, of the DNA ladder are made up of molecules called nitrogen bases. The four kinds of nitrogen bases found in DNA are adenine (A duh neen), guanine (GWAH neen), cytosine (SI tuh seen), and thymine (THI meen). In the DNA model on the previous page, the first letters of the name of each base, A, G, C, and T, are used to stand for the bases. Also notice that adenine (A) always pairs with thymine (T), and guanine (G) always pairs with cytosine (C).

How is DNA copied?

When chromosomes are copied before mitosis or meiosis, the amount of DNA in the nucleus is doubled. The figure below shows how the DNA copies itself. The two sides of DNA unwind and separate. Each side then becomes a pattern on which a new side can form. The new DNA pattern is exactly the same as the original DNA pattern.

Reading Check

2. Identify What did Rosalind Franklin discover?

____

____

____

Picture This

3. Determine Write one quiz question in the space below based on one of the steps in this figure.

____

____

____

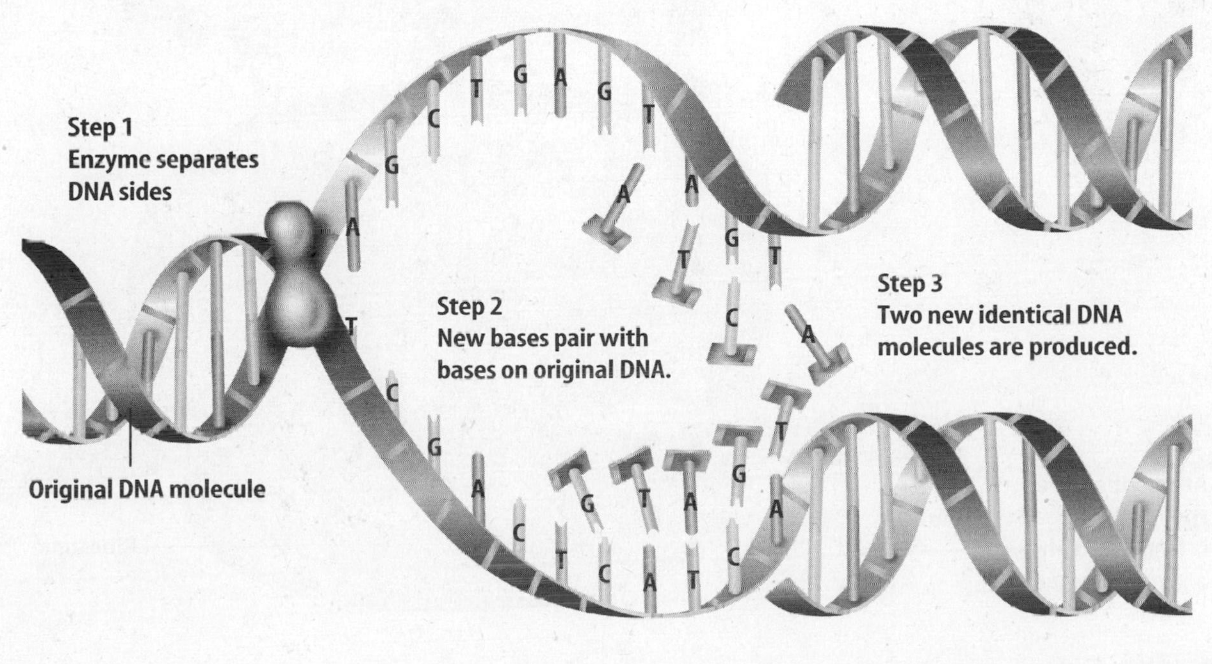
Step 1
Enzyme separates DNA sides
Step 2
New bases pair with bases on original DNA.
Step 3
Two new identical DNA molecules are produced.
Original DNA molecule

Genes

What color are your eyes? How tall are you? The answers to questions like these depend on the kinds of proteins your cells make. Proteins build cells and tissues or work as enzymes. The instructions for making certain proteins are found in genes. A **gene** is a section of DNA on a chromosome. Each chromosome has hundreds of genes. ☑

Reading Check

4. **Explain** where the instructions for making certain proteins are found.

____

____

____

What are proteins?

Proteins build cells and tissues. Proteins are made of chains of many amino acids. The gene decides the order of amino acids in a protein. Changing the order of the amino acids makes a different protein. Genes are found in the nucleus, but proteins are made on ribosomes in cytoplasm.

What is RNA?

The codes for making proteins are carried from the nucleus to the ribosomes by ribonucleic acid, or **RNA.** RNA is made in the nucleus on a DNA pattern, but it is different from DNA. Look at the model of an RNA molecule below. Notice that RNA is like a ladder with its rungs sawed in half. Like DNA, RNA has the bases A, G, and C. But it has the base uracil (U) instead of thymine (T). The sugar-phosphate molecules in RNA contain the sugar ribose.

Picture This

5. **Apply** Fill in the two circles in the figure with the correct letter.

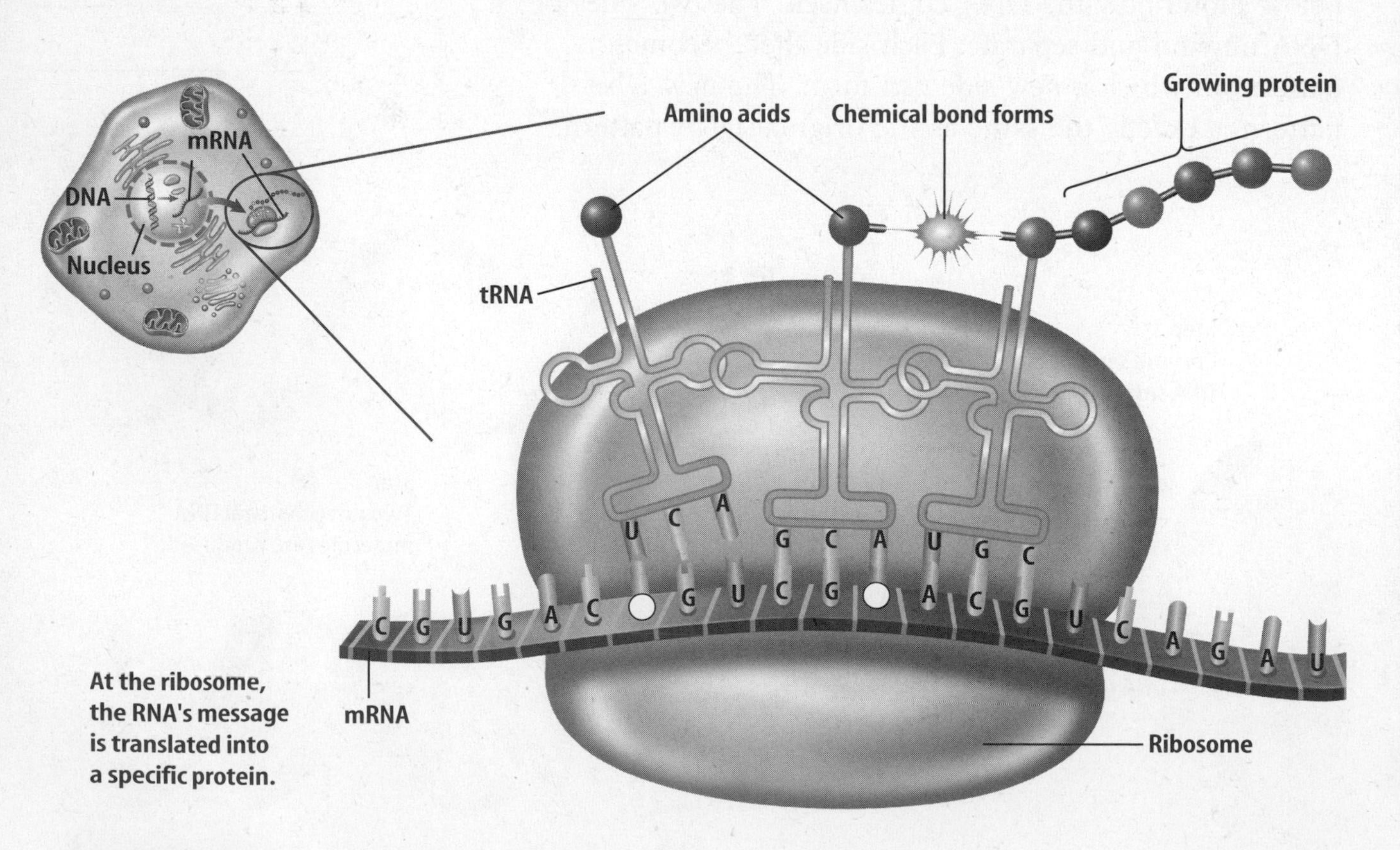

At the ribosome, the RNA's message is translated into a specific protein.

What does RNA do?

There are three main kinds of RNA made from DNA in a cell's nucleus. They are messenger RNA (mRNA), ribosomal RNA (rRNA), and transfer RNA (tRNA). Protein is made when mRNA moves into the cytoplasm. In the cytoplasm, ribosomes, which are made of rRNA, attach to the mRNA. The ribosomes get amino acids from tRNA molecules that are already in the cytoplasm. Inside the ribosomes, three nitrogen bases on the mRNA temporarily match with three nitrogen bases on the tRNA. The same thing happens for the mRNA and another tRNA molecule. The amino acids that are attached to the two tRNA molecules connect. This is the beginning of a protein.

FOLDABLES™

D Identify Make a three-tab book, as shown below. Use the Foldable to write facts about the three types of RNA.

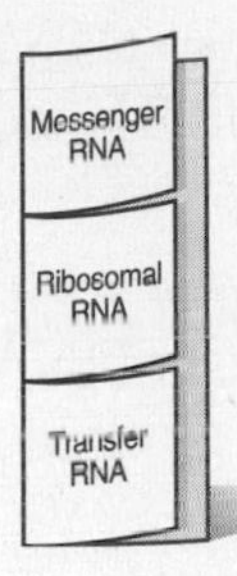

How do cells control genes?

Even though most cells in an organism have exactly the same genes, they do not make the same proteins. Each cell uses only the genes that make the proteins that it needs. For example, muscle proteins are made in muscle cells but not in nerve cells.

Cells control genes by turning some genes off and turning other genes on. Sometimes the DNA is twisted so tightly that no RNA can be made. Other times, chemicals attach to DNA so that it cannot be used.

Mutations

If DNA is not copied exactly, proteins may not be made correctly. These mistakes, called **mutations**, are permanent changes in the DNA sequence of a gene or chromosome. ☑

Reading Check

6. Explain What is a mutation?

____

____

____

What are the results of a mutation?

An organism with a mutation may not be able to grow, repair, or maintain itself. A mutation in a body cell may or may not cause problems for the organism. A mutation in a sex cell, however, makes changes to the species when the organism reproduces. Many mutations are harmful to organisms, often causing their death. Some mutations have no effect on an organism. Other mutations can be helpful to an organism.

After You Read

Mini Glossary

DNA: a chemical in a cell that contains information for an organism's growth and function

gene: a section of DNA on a chromosome that contains the instructions for making a specific protein

mutations: any permanent change in the DNA sequence of a gene or chromosome of a cell

RNA: a nucleic acid that carries the codes for making proteins from the nucleus to the ribosomes

1. Review the terms and their definitions in the Mini Glossary. Write a short paragraph that contrasts DNA and RNA.

____

____

____

____

____

2. Moving from left to right, write the letters (A, T, C, or G) in the empty circles of the bases that will pair with the bases on the top strand to this DNA molecule. The first three pairs have been created for you.

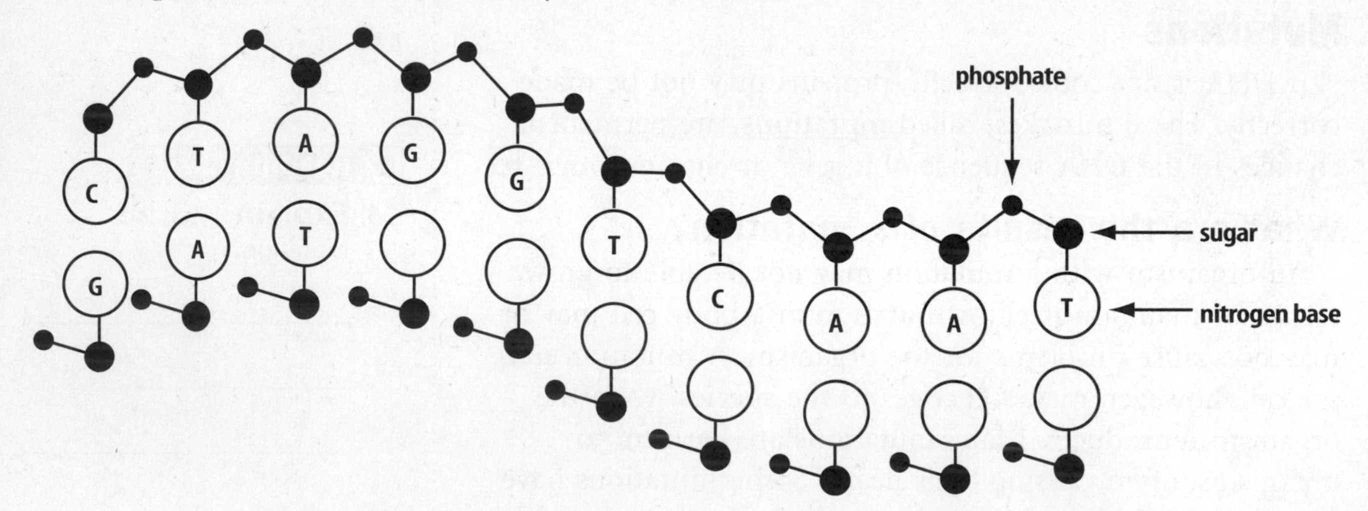

Science Online Visit **life.msscience.com** to access your textbook, interactive games, and projects to help you learn more about DNA.

Heredity

section 1 Genetics

Before You Read

Think of a parent and a child that you know. On the lines below, list four ways the child looks like the parent.

____

____

____

____

What You'll Learn

- how traits are inherited
- Mendel's role in the history of genetics
- how to use a Punnett square
- the difference between genotype and phenotype

Read to Learn

Inheriting Traits

Do you look more like one parent or grandparent? Do you have your father's eyes? Eye color, nose shape, and many other physical features are traits. Traits also include things that cannot be seen, such as your blood type. An organism is a collection of traits, all inherited from its parents. **Heredity** (huh REH duh tee) is the passing of traits from parent to offspring, or children.

Study Coach

Create a Vocabulary Quiz Write a question about each vocabulary word or term in the section. Exchange quizzes with another student. Together discuss the answers to the quizzes.

What is genetics?

Usually, genes on chromosomes control an organism's shape and function. The different forms of a trait that a gene may have are called **alleles** (uh LEELZ). When a pair of chromosomes separates during meiosis (mi OH sus), alleles for each trait also separate into different sex cells. As a result, every sex cell has one allele for each trait, as shown in the figure on the next page. The allele in one sex cell may control one form of the trait, such as dimples. The allele in another sex cell may control a different form of the trait, such as no dimples. The study of how traits are inherited through the interactions of alleles is called **genetics** (juh NE tihks). ☑

Reading Check

1. **Define** the word genetics.

____

____

____

Chromosomes Separate During Meiosis

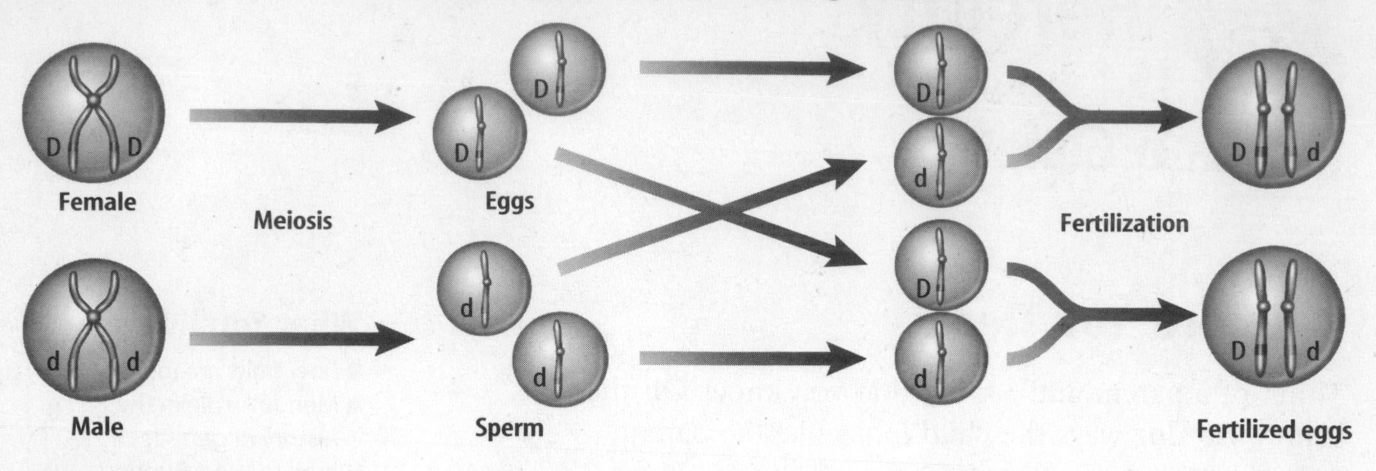

A The alleles that control a trait are located on each duplicated chromosome.

B During meiosis, duplicated chromosomes separate.

C During fertilization, each parent donates one chromosome. This results in two alleles for the trait in the new individual formed.

Mendel—The Father of Genetics

Did you know that an experiment with pea plants helped scientists understand why your eyes are the color they are? Gregor Mendel was an Austrian monk who studied mathematics and science. His job at the monastery where he lived was gardening. His interest in plants began as a boy in his father's orchards. He learned to predict the possible types of flowers and fruits that would result from crossbreeding plants.

In 1856, Mendel began experimenting with garden peas. He wanted to know the connection between the color of a pea flower and the type of seed the plant produced. Before Mendel, scientists relied on observation and description. They often studied many traits at one time. This made it hard to develop good hypotheses about how traits are inherited. Mendel used scientific methods in his study. Mendel was the first person to trace one trait through many generations. He was the first person to record the study of how traits pass from one generation to another. He was also the first person to use the mathematics of probability to explain heredity.

In 1900, three plant scientists repeated Mendel's experiments and reached the same conclusions as Mendel. For this reason, Mendel is known as the father of genetics.

Picture This

2. Identify Circle the sex cells on the diagram.

Think it Over

3. Analyze When Mendel studied traits, how did his methods differ from those of other scientists?

____

____

____

Genetics in a Garden

When Mendel studied a trait, he crossed two plants with different forms of the trait. He found that the new plants all looked like one of the two parents. Mendel called each new plant a **hybrid** (HI brud) because it received different genetic information, or different alleles, for a trait from each parent.

What is a purebred?

Garden peas are easy to breed for pure traits. An organism that always produces the same traits, generation after generation, is called a purebred. For example, plants can be purebred for the trait of tall height. The table below shows the pea plant traits that Mendel studied.

Picture This

4. Identify How many traits did Mendel study?

____

Traits Compared by Mendel

Traits	Shape of Seeds	Color of Seeds	Color of Pods	Shape of Pods	Plant Height	Position of Flowers	Flower Color
Dominant Trait	Round	Yellow	Green	Full	Tall	At leaf junctions	Purple
Recessive Trait	Wrinkled	Green	Yellow	Flat, constricted	Short	At tips of branches	White

What are dominant and recessive factors?

In nature, insects carry pollen as they move from plant to plant. The pollination by insects is random. In his experiments, Mendel pollinated the plants by hand to control the results. He used pollen from the flowers of purebred tall plants to pollinate the flowers of purebred short plants. This process is called cross-pollination. He found that tall plants crossed with short plants produced seeds that produced all tall plants. Mendel called the tall form the **dominant** (DAH muh nunt) factor because it dominated, or covered up, the short form. He called the short form the **recessive** (rih SE sihv) factor because this form seemed to disappear. Today, these factors are called dominant alleles and recessive alleles.

FOLDABLES™

A Describe Make a two-tab Foldable, as shown below. Write notes under the tabs to describe dominant and recessive alleles.

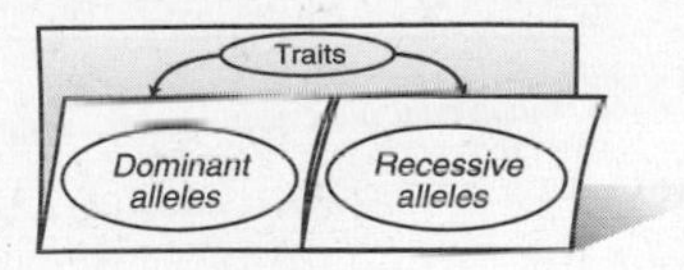

What is probability?

A branch of mathematics that helps you predict the chance that something will happen is called probability. For example, there are two sides to a coin. If you toss the coin in the air, the probability that one side of the coin will land facing up is one out of two, or 50 percent. Mendel used probabilities in his study of genetics. His predictions were very accurate because he studied large numbers of plants over a long period of time. He studied almost 30,000 pea plants over a period of eight years. This increased Mendel's chances of seeing a repeatable pattern. Valid scientific conclusions need to be based on results that can be repeated.

What is a Punnett square?

Scientists use a tool called a Punnett (PUH nut) square to predict results in genetics. A **Punnett square** is used to predict the number of times certain traits will occur. In a Punnett square, letters stand for dominant and recessive alleles. An uppercase letter stands for a dominant allele, and a lowercase letter stands for a recessive allele. The letters are a form of code. They show the **genotype** (JEE nuh tipe), or genetic makeup, of an organism. The way an organism looks and behaves as a result of its genotype is its **phenotype** (FEE nuh tipe). If you have brown hair, the phenotype for your hair color is brown. ☑

Reading Check

5. Identify How is a dominant allele shown in a Punnett square?

____

____

How do alleles determine traits?

Most cells in your body have two alleles for every trait. An organism with two alleles that are the same is called **homozygous** (hoh muh ZI gus). In his experiments, Mendel would have written *TT* (homozygous for the tall-dominant trait) or *tt* (homozygous for the short-recessive trait). An organism that has two different alleles for a trait is called **heterozygous** (he tuh roh ZI gus). Mendel would have written *Tt* for plant hybrids that were heterozygous for height.

Think it Over

6. Contrast What is the difference between a homozygous organism and a heterozygous organism?

How do you make a Punnett square?

The letters representing the two alleles from one parent are written in the top row of the Punnett square. The letters representing the two alleles from the other parent are written down the left column. Each square in the grid is then filled in with one allele from each parent. The combinations of letters in the completed Punnett square are the genotypes of the possible offspring those parents could produce.

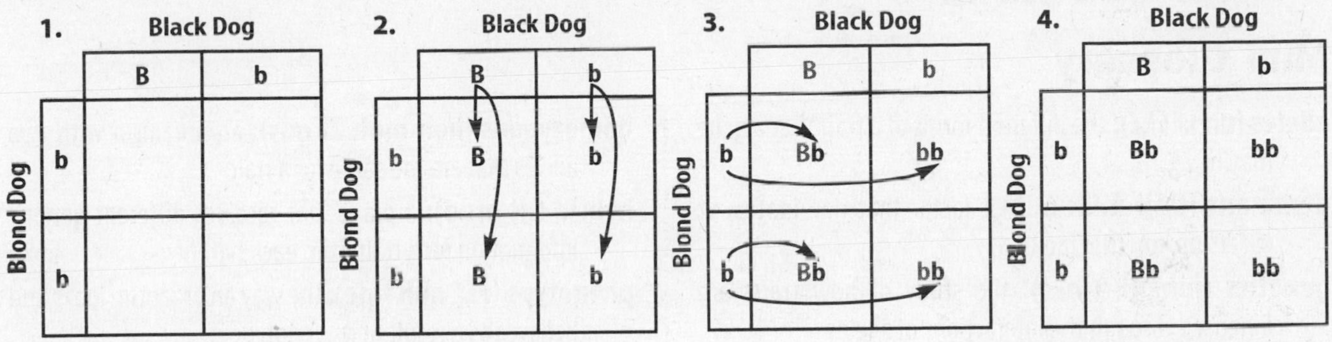

1. Black Dog

Blond Dog	B	b
b		
b		

2. Black Dog

Blond Dog	B	b
b	B	b
b	B	b

3. Black Dog

Blond Dog	B	b
b	Bb	bb
b	Bb	bb

4. Black Dog

Blond Dog	B	b
b	Bb	bb
b	Bb	bb

How do you use a Punnett square?

You want to know the possible offspring of two dogs. One dog carries heterozygous black-fur traits (*Bb*). The other dog carries homogeneous blond-fur traits (*bb*). How do you complete the Punnett square to find the results? Follow the steps in the figure above.

1. Write the letters representing the alleles from the black dog (Bb) in the top row. Write the letters from the blond dog (bb) in the left column.
2. Write the letter in each column (B or b) in the two squares for that column.
3. Add the letter for each row (b or b) to the squares. You then have two letters in each square.
4. The squares show the possible genotypes of the offspring.

An offspring with a *Bb* genotype will have black fur, and an offspring with a *bb* genotype will have blond fur. In this case, there is one chance in two, or a 50 percent chance, that the offspring will have black fur.

What are the main principles of heredity?

Mendel spent many years repeating his experiments and observing the results. He analyzed the results and reached several conclusions. Mendel's principles of heredity are summarized in the table below.

Mendel's Principles of Heredity
Traits are controlled by alleles on chromosomes.
An allele's effect is dominant or recessive.
When a pair of chromosomes separates during meiosis, the different alleles for a trait move into separate sex cells.

Picture This

7. Identify In step 4, shade the two squares that would result in an offspring with blond fur.

Think it Over

8. Explain What controls traits?

______

______

______

After You Read

Mini Glossary

alleles (uh LEELZ): the different forms of a trait that a gene may have

dominant (DAH muh nunt): factor that dominates, or covers up, another factor

genetics (juh NE tihks): the study of how traits are inherited through the interactions of alleles

genotype (JEE nuh tipe): genetic makeup of an organism

heredity (huh REH duh tee): passing of traits from parent to offspring

heterozygous (he tuh roh ZI gus): an organism that has two different alleles for a trait

homozygous (hoh muh ZI gus): an organism with two alleles that are the same for a trait

hybrid (HI brud): a plant that receives different genetic information for a trait from each parent

phenotype (FEE nuh tipe): the way an organism looks and behaves as a result of its genotype

Punnett (PUH nut) square: a tool used to predict the number of times certain traits will occur

recessive (rih SE sihv): factor that disappears if a dominant trait is present

1. Review the terms and their definitions in the Mini Glossary. Write a sentence that explains the difference between a dominant allele and a recessive allele.

____

____

2. Complete the Punnett square below to show the probability of an offspring having the *DD*, *Dd*, and the *dd* genotypes.

	D	d
D		
d		

3. How can taking a quiz that another student wrote help you prepare for a test?

____

____

ScienceOnline Visit **life.msscience.com** to access your textbook, interactive games, and projects to help you learn more about genetics.

Heredity

section 2 Genetics Since Mendel

Before You Read

At dog and cat shows, an animal's owner may be asked to show its pedigree. What do you think a pedigree shows?

____

____

____

What You'll Learn

- how traits are inherited by incomplete dominance
- the difference between multiple alleles and polygenic inheritance
- how sex-linked traits are passed to offspring

Read to Learn

Incomplete Dominance

A scientist crossed purebred red four-o'clock plants with purebred white four-o'clock plants. He thought the new plants would have all red flowers, but they were pink. Neither allele for flower color was dominant. Next, he crossed the pink-flowered plants with each other. The new plants had red, white, and pink flowers.

He discovered that when the allele for red flowers and the allele for white flowers combined, the result included red flowers, white flowers, and an intermediate, or in-between, phenotype—pink flowers. When the offspring of two homozygous parents show an intermediate phenotype, this inheritance is called **incomplete dominance**.

What are multiple alleles?

A trait that is controlled by more than two alleles is said to be controlled by multiple alleles. A trait controlled by multiple alleles will produce more than three phenotypes of that trait.

Mark the Text

Build Vocabulary Skim the section, circling any words you do not know. After you read the section, review the circled words. Write any words you cannot define on a separate sheet of paper and look up the definitions.

FOLDABLES™

B Explain Make a layered-look Foldable, as shown below. Write notes under the flaps to explain inheritance patterns.

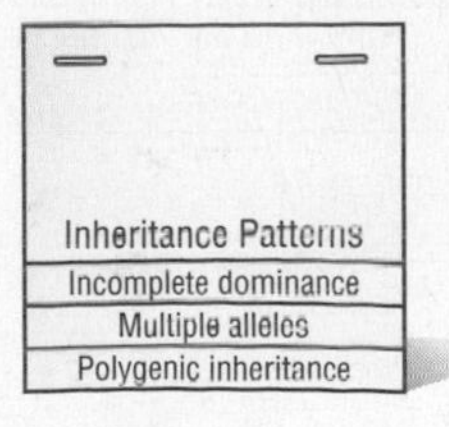

What traits are controlled by multiple alleles?

Blood type in humans is an example of a trait controlled by multiple alleles. The alleles for blood type produce six genotypes but only four phenotypes. The alleles for blood type are called A, B, and O. The O allele is recessive to both the A and B alleles. When a person inherits one A allele and one B allele, his or her phenotype is AB. A person with phenotype A blood has the genotype AA or AO. Someone with the phenotype B blood has the genotype BB or BO. A person with phenotype O blood has the genotype OO. ☑

Reading Check

1. **Identify** What are the six different blood type genotypes?

____

____

____

Polygenic Inheritance

Eye color is an example of a trait that is produced by a combination of many genes, or polygenic (pah lih JEH nihk) inheritance. **Polygenic inheritance** occurs when a group of gene pairs acts together to produce a trait. Polygenic inheritance results in a wide variety of phenotypes. Examine the eye colors of your classmates. You will likely notice many different shades. For example, you may notice several shades of brown, several shades of green, and so on.

How does the environment affect your genes?

Your environment plays a role in how some of your genes are expressed. Genes can be influenced by an organism's internal or external environment. For example, most male birds are more brightly colored than females. Chemicals in their bodies determine whether or not the gene for brightly colored feathers is expressed.

Your environment plays a role in whether your genes are expressed at all. For example, some people have genes that make them at risk for developing skin cancer. Whether or not they get cancer might depend on external environmental factors. If people who are at risk for skin cancer limit their time in the sun and take care of their skin, they may never develop skin cancer.

Think it Over

2. **Draw Conclusions** What environmental factors might affect the size of leaves on a tree?

____

____

____

Human Genes and Mutations

Sometimes genes change. Also, sometimes errors occur in the DNA when it is being copied during cell division. These changes and errors are called mutations. Many mutations are harmful. Some mutations are helpful or have no effect on an organism. Certain chemicals, X rays, and radioactive materials can cause mutations.

What are chromosome disorders?

Problems can happen if the incorrect number of chromosomes is inherited. Mistakes in the process of meiosis can result in an organism with more or fewer chromosomes than normal. Down's syndrome is a disorder in which the person has one more chromosome than normal.

Recessive Genetic Disorders

Many human genetic disorders are caused by recessive genes. Such genetic disorders occur when both parents have a recessive allele responsible for the disorder. Because the parents are heterozygous, they do not show any symptoms of the disorder. However, if each parent passes a recessive allele to the child, the child inherits two recessive alleles and will have the disorder. Cystic fibrosis is a homozygous recessive disorder. It is the most common genetic disorder that can lead to death among Caucasian Americans. People with cystic fibrosis produce thicker mucus than normal. The thick mucus builds up in the lungs and makes it hard to breathe. ☑

Reading Check

3. Explain How is cystic fibrosis inherited?

____

____

____

Sex Determination

Each egg produced by a female normally contains one X chromosome. Males produce sperm that normally have either one X or one Y chromosome. When a sperm with an X chromosome fertilizes an egg, the offspring is a female, XX. When a sperm with a Y chromosome fertilizes an egg, the offspring is a male, XY. Sometimes chromosomes do not separate during meiosis. When this happens, a person can inherit an unusual number of sex chromosomes.

Sex-Linked Disorders

Some inherited conditions are linked with the X and Y chromosomes. An allele inherited on a sex chromosome is called a **sex-linked gene**. Color blindness is a sex-linked disorder in which people cannot tell the difference between certain colors. The color-blind trait is a recessive allele on the X chromosome. Because males have only one X chromosome, a male with this recessive allele on his X chromosome is color-blind. However, a color-blind female occurs only when both of her X chromosomes have the allele for this trait. ☑

Reading Check

4. Identify What is one sex-linked disorder?

____

____

____

Picture This

5. Infer In the pedigree, why are there no color-blind women in this family?

____

____

____

____

____

Think it Over

6. Draw Conclusions Why do you think pedigrees are important for animals bred for show, such as dogs?

____

____

____

Pedigrees Trace Traits

You can trace a trait through a family using a pedigree like the one shown below. Males are represented by squares. Females are represented by circles. A completely filled square or circle shows that the person has the trait. A half-colored square or circle shows that the person carries an allele for the trait, but does not have the trait. The pedigree in the figure below shows how the trait for color blindness is carried through a family. In this pedigree, the grandfather was color blind. He married a woman who did not carry the color-blind allele.

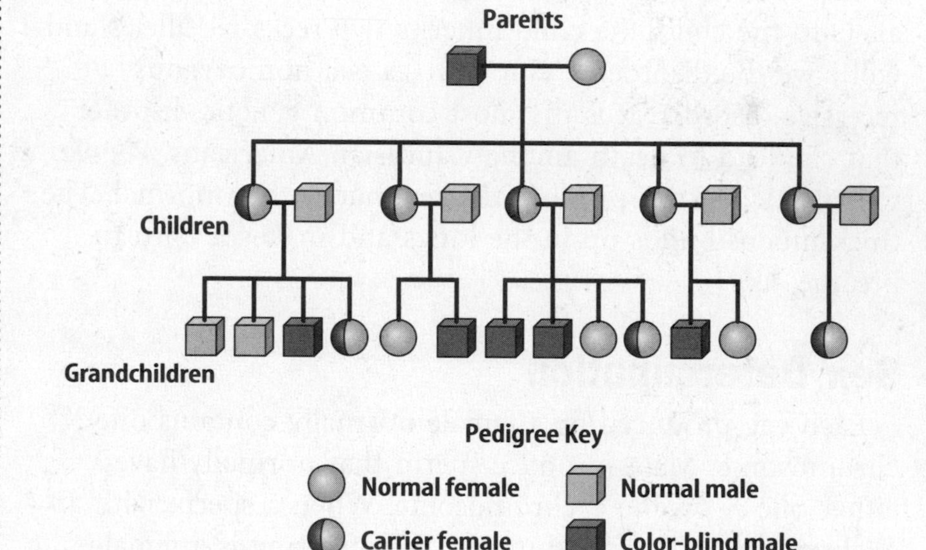

How can pedigrees be helpful?

A pedigree can be used by a geneticist to trace a trait in members of a family over several generations. The pedigree allows the geneticist to determine the trait's pattern of inheritance. The geneticist can identify if the trait is recessive, dominant, sex-linked, or follows some other pattern. Geneticists use this information to predict the probability that a baby will be born with a specific trait.

Pedigrees also are used to breed animals and plants for desirable traits. Livestock and plant crops are food sources for humans. Using pedigrees, these organisms can be bred to increase their yield and nutritional content.

After You Read

Mini Glossary

incomplete dominance: the offspring of two homozygous parents show an intermediate phenotype

polygenic (pah lih JEH nihk) inheritance: a group of gene pairs act together to produce a trait

sex-linked gene: an allele inherited on a sex chromosome

1. Review the terms and their definitions in the Mini Glossary. Choose one term and use it to explain one way that traits can be inherited.

____

____

____

2. Choose one of the question headings in the Read to Learn section. Write the question in the space below. Then write your answer to that question on the lines that follow.

Write your question here.

____

____

____

3. List the words that you circled in the Read to Learn section. Select one of those words and write its definition below.

____

____

____

Visit **life.msscience.com** to access your textbook, interactive games, and projects to help you learn more about genetics since Mendel.

Heredity

section 3 Biotechnology

What You'll Learn
- the importance of advances in genetics
- the steps in making genetically engineered organisms

Before You Read

Describe on the lines below what you have heard or read about recent advances in medical research.

____

____

____

Mark the Text

Identify Main Points Highlight the main idea in each paragraph. Underline the details that support the main idea.

Read to Learn

Why is genetics important?

New developments in genetic research are happening all the time. The principles of heredity are being used to change the world.

Genetic Engineering

Genetic engineering is the use of biological and chemical methods to change the arrangement of DNA that makes up a gene. One use for genetic engineering is to produce large amounts of different medicines. Genes also can be inserted into cells to change how those cells perform their normal functions. Genetic engineering researchers are also looking for new ways to improve crop production and quality.

FOLDABLES™

C Describe Make a three-tab book, as shown below. Use the Foldable to describe genetic engineering, recombinant DNA, and gene therapy.

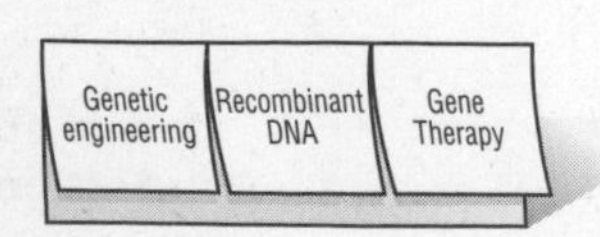

How is recombinant DNA made?

Making recombinant DNA is one method of genetic engineering. Recombinant DNA is made by inserting a useful section of DNA from one organism into a bacterium. This process is used to make large amounts of insulin, which is used to treat diabetes. Other uses include the production of a growth hormone to treat dwarfism and chemicals used to treat cancer.

How does gene therapy work?

Gene therapy is another kind of genetic engineering. It is used to replace abnormal alleles. In gene therapy, a normal allele is placed in a virus, as shown in the figure below. The virus then delivers the normal allele when it infects the target cell. The normal allele replaces the abnormal one. Scientists are conducting experiments that use gene therapy to test ways of controlling cystic fibrosis and some kinds of cancer. With continued research, gene therapy may be used to cure genetic disorders in the future. ☑

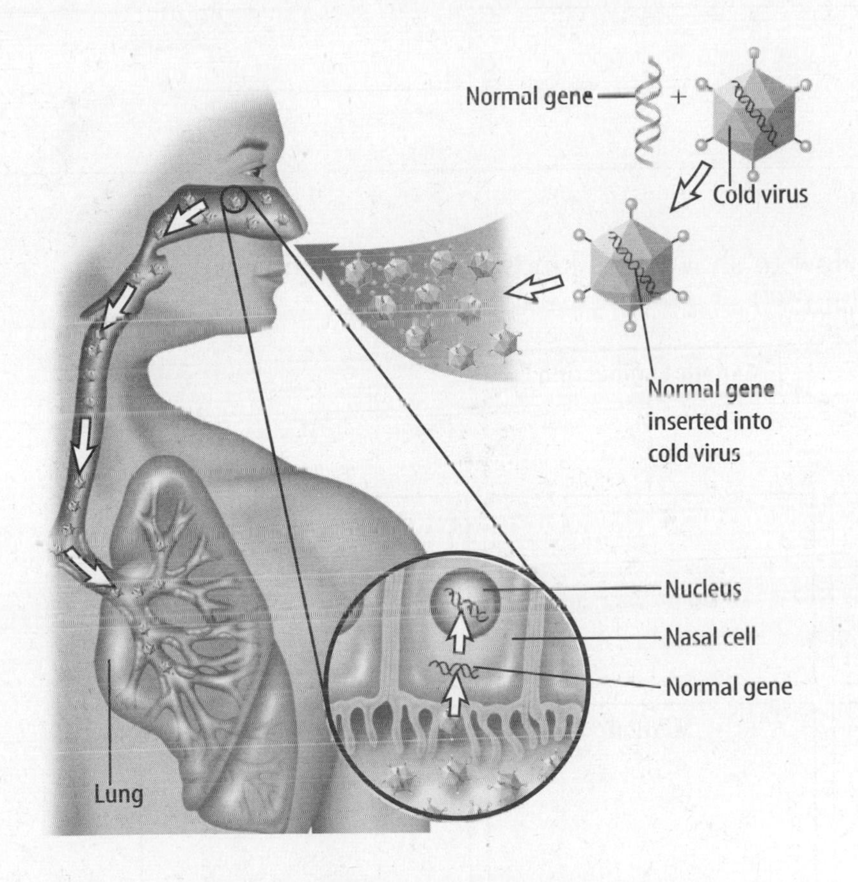

How are plants genetically engineered?

Before people knew about genotypes, they selected plants with the most desired traits to breed for the next generation. This process is called selective breeding. Today people also use genetic engineering to improve crop plants. One method is to find the genes that produce desired traits in one plant and then insert those genes into a different plant. Scientists recently made genetically engineered tomatoes with a gene that allows them to be picked green. As these tomatoes are being sent to stores, they continue to ripen. You can then buy ripe, firm tomatoes in the store. The long-term effects of eating genetically engineered plants are not known.

Reading Check

1. **Identify** What is replaced in gene therapy?

____

____

____

Picture This

2. **Explain** Use the figure to explain to a partner how gene therapy works.

After You Read

Mini Glossary

genetic engineering: biological and chemical methods to change the arrangement of DNA that makes up a gene

1. Review the term and its definition in the Mini Glossary. Write a sentence that explains how genetic engineering can improve crop plants.

____

____

____

2. Complete the concept web below to show three kinds of genetic engineering and the methods used to carry them out.

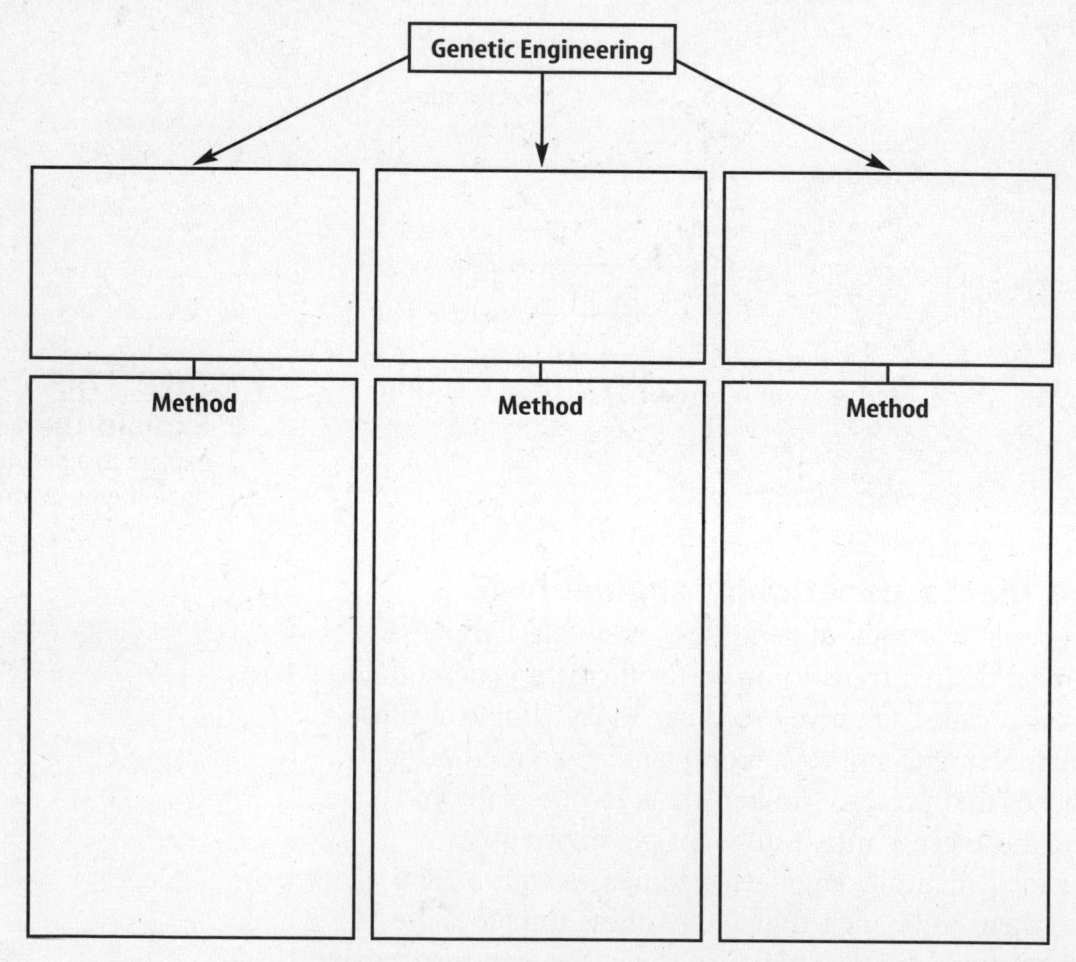

End of Section

Science Online Visit **life.msscience.com** to access your textbook, interactive games, and projects to help you learn more about advances in genetics.

Adaptations over Time

section 1 Ideas About Evolution

Before You Read

In what ways are you like your parents or other relatives?

What You'll Learn

- Lamarck's hypothesis of acquired characteristics
- Darwin's theory of natural selection
- variations in organisms
- the difference between gradualism and punctuated equilibrium

Read to Learn

Early Models of Evolution

There are millions of species of plants, animals, and other organisms living on Earth today. A **species** is a group of organisms that share similar characteristics and can reproduce among themselves to produce fertile offspring. The characteristics of a species that are passed from parent to offspring are called inherited characteristics. Change in these inherited characteristics over time is **evolution**.

Study Coach

Ask Questions Read each question heading. Then work with a partner to write questions about the information related to the heading. Take turns asking and answering the questions. Use the questions as a study guide about evolution.

What was Lamarck's hypothesis?

In 1809, Jean Baptiste de Lamarck proposed a hypothesis to explain how species change over time. He said that characteristics, or traits, that a parent organism develops during its lifetime are inherited by its offspring. Lamarck's hypothesis is called the inheritance of acquired characteristics. According to Lamarck's hypothesis, if a parent develops large muscles through exercise or hard work, the trait of large muscles would be passed on to the offspring. Scientists tested Lamarck's hypothesis by collecting data on traits that are passed from parent to offspring. The data did not support Lamarck's hypothesis.

Think it Over

1. **Conclude** Do scientists today support Lamarck's hypothesis? Explain.

FOLDABLES

A Organize Make a four-door Foldable, as shown below, to organize information of how evolution occurs.

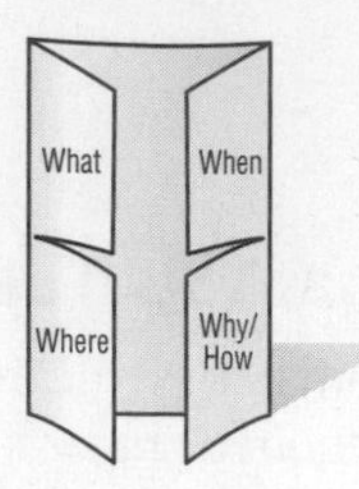

Darwin's Model of Evolution

In 1831, Charles Darwin set out on a journey from England that took him to the Galápagos Islands. The Galápagos Islands, shown on the map below, are off the coast of Ecuador. Darwin was amazed by the variety of life he saw on these islands. He hypothesized that plants and animals living on the Galápagos Islands originally came from Central and South America. He noted that the species on the islands were similar in many ways to the species he had seen on the mainland. However, Darwin observed different traits in many species on the islands as well. Darwin studied several species and developed hypotheses to explain the differences in traits he observed.

Picture This

2. Describe the location of the Galápagos Islands.

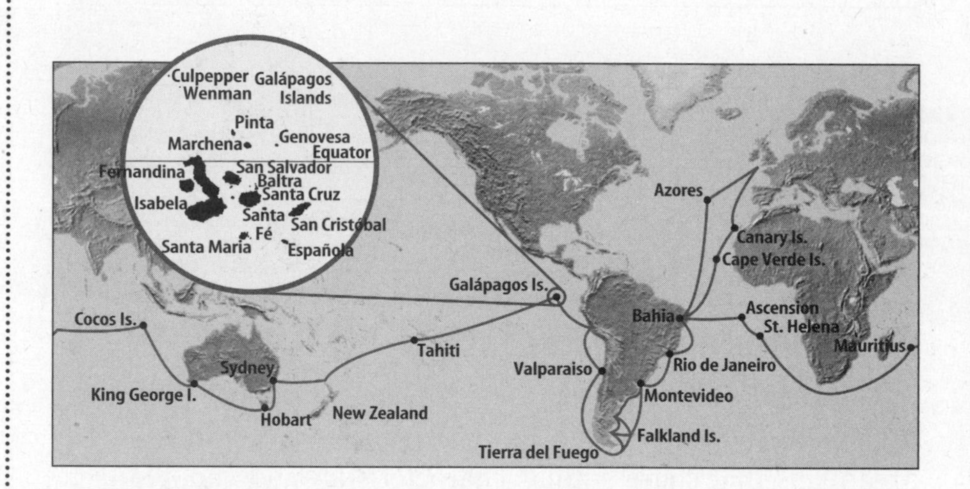

What did Darwin observe?

Darwin observed 13 species of finches on the Galápagos Islands. He noticed that all 13 species were similar except for three characteristics—body size, beak shape, and eating habits. Darwin concluded that the different species of finches must have had to compete with each other for food. Finches that had beak shapes that allowed them to eat available food survived longer and had more offspring than finches without those kinds of beak shapes. After many generations, these groups of finches became separate species.

Darwin observed that the beak shape of each species of Galápagos finch is related to its eating habits. Darwin observed finches that ate nuts and seeds. Their beaks were short and strong for breaking hard shells. He observed finches that fed on insects. They had long, narrow beaks for finding the insects beneath tree bark. ☑

Reading Check

3. Identify How did finches who ate seeds and nuts use their beaks?

Natural Selection

In the mid-1800s, Darwin developed a theory of evolution that is accepted by most scientists today. He described his ideas in a book called *On the Origin of Species.*

What was Darwin's theory?

Darwin's theory became known as the theory of evolution by natural selection. **Natural selection** means that organisms with traits best suited to their environment are more likely to survive and reproduce. Their traits are passed to more offspring. The principles of natural selection are shown in the table below. ☑

The Principles of Natural Selection
1. Organisms produce more offspring than can survive.
2. Differences, or variations, occur among individuals of a species.
3. Some variations are passed to offspring.
4. Some variations are helpful. Individuals with helpful variations are better able to suvive and reproduce.
5. Over time, the offspring of individuals with helpful variations increase and become a larger percentage of the population. Eventually, they may become a separate species.

Reading Check

4. **Explain** According to the theory of evolution by natural selection, which organisms are most likely to survive and reproduce?

______

______

______

Variation and Adaptation

Darwin's theory of evolution by natural selection focuses on the variations of species' members. A **variation** is an inherited trait that makes an individual organism different from other members of its species. Variations happen when there are permanent changes, or mutations, in an organism's genes. Some mutations produce small variations, such as differences in the shape of human hairlines. Other mutations produce large variations, such as fruit without seeds. Over time, more and more members of a species might inherit these variations. If individuals with these variations continue to survive and reproduce over time, a new species can evolve.

Some variations are more helpful than others. An **adaptation** is any variation that makes an organism better suited to its environment. Adaptations can include an organism's color, shape, behavior, or chemical makeup. Camouflage (KA muh flahj) is an adaptation. An organism that is camouflaged can blend into its environment. Camouflage makes it easier for the organism to hide, increasing the chances that it will survive and reproduce. ☑

Reading Check

5. **Determine** How does camouflage help an organism survive?

______

______

______

How do changes in genes affect species?

Over time, changes in the genes of a species might change the appearance of the species. As the inherited traits of a species of seed-eating Galápagos finch changed, so did the size and shape of its beak. Environmental conditions can help bring about these changes. When individuals of the same species move into an area, they bring genes and variations. When they move out of an area, they remove their genes and variations. Suppose a family from a different country moves to your neighborhood. They might bring different foods, customs, and ways of speaking. In a similar way, when new individuals enter an existing population, they can bring different genes and variations.

Reading Check

6. Infer What two factors affected the development of new species?

____

Does geographic isolation affect evolution?

Sometimes geologic features such as mountains or lakes can separate a group of individuals from all the other members of the population. Over time, variations that are not found in the larger population might become common in the smaller, separate population. Also, gene mutations could add variations to the smaller population. After many generations, the two populations can become so different that they can no longer breed with each other. They become two different species. For example, Portuguese sailors brought European rabbits to the Canary Islands. European rabbits feed during the day and grow fairly large. In order to survive the warm temperatures of the Canary Islands, the European rabbits, over many generations, developed large eyes and fed at night. The Canary Island rabbits eventually became a separate species. ☑

The Speed of Evolution

FOLDABLES™

B Compare Make a three-tab Foldable, as shown below, to compare gradualism and punctuated equilibrium.

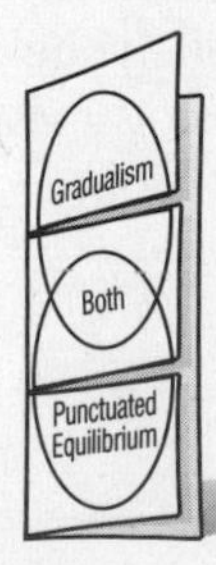

Scientists do not agree on how quickly evolution happens. Some hypothesize that it happens slowly, over hundreds of millions of years. Others hypothesize that it can happen quickly. Most scientists agree that there is evidence to support both hypotheses.

What is gradualism?

Darwin hypothesized that evolution happens slowly. His hypothesis is called gradualism. **Gradualism** is a hypothesis that describes evolution as a slow, continuing process in which one species changes to a new species over millions or hundreds of millions of years.

What is punctuated equilibrium?

Gradualism does not explain the evolution of all species. For some species, fossil records show that one species suddenly changes into another. **Punctuated equilibrium** is a hypothesis that describes evolution as a rapid process in which one species changes suddenly to a new species. Rapid evolution happens when the mutation of a few genes results in a new species over a fairly short period of time. The figure below shows how punctuated equilibrium describes the evolution of the brown bear.

Hypothesized Evolution of the Brown Bear

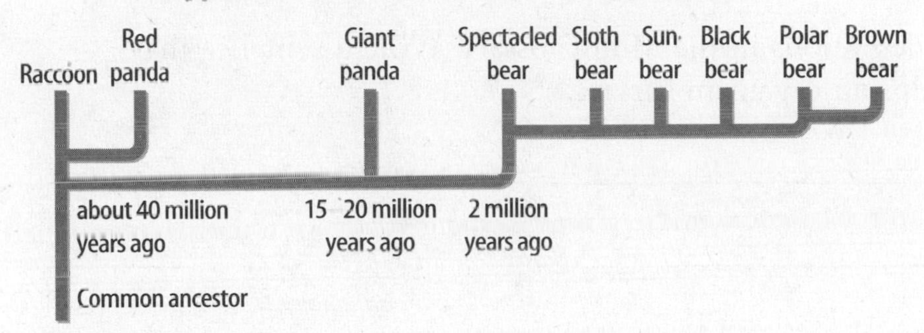

Picture This

7. Identify Circle the name of the common ancestor of the giant panda and the brown bear.

Is punctuated equilibrium happening today?

Evolution by punctuated equilibrium can happen over a few thousand or hundreds of thousands of years. Sometimes, evolution can happen even faster than that. For example, many species of bacteria have changed into new species in only a few decades. Many disease-causing bacteria species were once easily killed by the antibiotic penicillin. Some of these species are no longer harmed by penicillin. These bacteria have become resistant to penicillin.

These penicillin-resistant bacteria evolved quickly. The bacteria changed because some individuals had variations that allowed them to survive even when exposed to penicillin. Other individuals could not survive. The bacteria that had the penicillin-resistant variation survived to reproduce and pass this trait to their offspring. Over a period of time, all of the bacteria in the population had the variation for penicillin resistance.

Think it Over

8. Analyze What allowed some bacteria to survive while other bacteria were killed by penicillin?

____

____

____

After You Read

Mini Glossary

adaptation: any variation that makes an organism better suited to its environment

evolution: change in inherited characteristics over time

gradualism: hypothesis that describes evolution as a slow, ongoing process by which one species changes to a new species

natural selection: theory that states that organisms with traits best suited to their environment are more likely to survive and reproduce

punctuated equilibrium: hypothesis that says rapid evolution comes about when the mutation of a few genes results in the appearance of a new species over a relatively short period of time

species: group of organisms that share similar characteristics and can reproduce among themselves to produce fertile offspring

variation: inherited trait that makes an individual different from other members of its species

1. Review the terms and their definitions in the Mini Glossary. Write a sentence that describes a variation that helps an organism survive.

____

____

____

2. Complete the chart below to explain the models of evolution listed in the chart.

Theory or Model	Description
Hypothesis of acquired characteristics	
Theory of evolution by natural selection	
Gradualism	
Punctuated equilibrium	

Visit **life.msscience.com** to access your textbook, interactive games, and projects to help you learn more about early models of evolution.

Adaptations over Time

section 2 Clues About Evolution

Before You Read

Have you ever seen a fossil? On the lines below, tell what kind of fossil it was and where you saw it.

____

____

____

What You'll Learn

- why fossils provide evidence of evolution
- how relative and radiometric dating are used to estimate the age of fossils
- five types of evidence for evolution

Read to Learn

Clues from Fossils

Paleontologists are scientists who study the past by collecting and examining fossils. A fossil is the remains of an ancient organism or an imprint left behind by the organism.

The Green River Formation is one of the richest fossil deposits in the world. It covers parts of Wyoming, Utah, and Colorado. About 50 million years ago, during the Eocene Epoch, this area was covered by lakes. By studying fossils from the Green River Formation, paleontologists have learned that fish, crocodiles, and lizards lived in the lakes. After the animals died, they were covered with silt and mud. Over millions of years, they became fossils.

Mark the Text

Identify Unfamiliar Words Skim the reading and underline any word that you do not know. At the end of each paragraph review the words you have underlined and see if you can define them. If you cannot, look up the word and write its definition in the margin.

Types of Fossils

Most of the evidence for evolution comes from fossils. Most fossils are found in sedimentary rock. **Sedimentary rock** is formed when layers of sand, silt, clay, or mud are pressed and cemented together or when minerals are deposited from a solution. Fossils are most often found in a sedimentary rock called limestone. ☑

Reading Check

1. **Explain** What is the main source of evidence for evolution?

____

____

FOLDABLES™

C Describe Make a two-tab concept map Foldable, as shown below, to describe the two methods scientists use to date fossils—relative dating and radiometric dating.

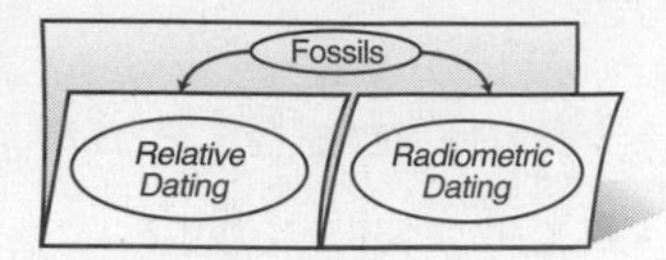

Determining a Fossil's Age

Paleontologists study the rock layers that fossils are found in. The rocks provide clues about the age of the fossils. Some of these clues include information about the geologic time period in which it was formed. Information may include weather, geology, and other organisms that were alive. Paleontologists have two ways of estimating the age of rocks and fossils—relative dating and radiometric dating.

What is relative dating?

Relative dating is based on the fact that younger rock layers usually lie on top of older rock layers. Relative dating gives only an estimate of a fossil's age. Scientists compare the ages of rock layers found above and below the fossil layer. For example, if a 50-million-year-old rock layer lies below a fossil and a 35-million-year-old rock layer is above the fossil, then the fossil is probably between 35 million and 50 million years old.

What is radiometric dating?

Radiometric dating gives an estimate of the age of a rock layer that is more exact. This method of dating fossils uses radioactive elements. A **radioactive element** gives off a steady amount of radiation as it slowly changes to a nonradioactive element. Each radioactive element gives off radiation at a different rate. Scientists estimate the age of the rock by comparing the amount of radioactive element with the amount of nonradioactive element in the rock.

Fossils and Evolution

Fossils provide a record of organisms that lived in the past. However, the fossil record has gaps, much like missing pages in a book. The gaps exist because most organisms do not become fossils. Even though there are gaps, scientists have still been able to draw conclusions from the fossil records. For instance, they have learned that simple organisms were the first forms of life to appear on Earth. More complex forms of life appeared later.

Fossil discoveries are made all over the world. When scientists find fossils, they make models that show what the organisms might have looked like when they were alive. Scientists can use fossils to find out whether organisms lived in family groups or alone, what they ate, and what kind of environment they lived in. Most fossils are from extinct organisms. ☑

Reading Check

2. **Identify** Name two things scientists can learn about organisms from fossils.

____

____

____

More Clues About Evolution

Besides fossils, there are other clues about evolution. Some kinds of evolution can be observed today. The development of penicillin-resistant bacteria is a direct observation of evolution. Another direct observation of evolution is the development of insect species that are resistant to pesticides.

What is embryology?

The study of embryos and their development is called **embryology** (em bree AH luh jee). An embryo is the earliest growth stage of an organism. The embryos of many different species are similar. The embryos of fish, birds, reptiles, and mammals have tails. As the organisms grow, the fish, birds, and reptiles keep their tails, but many mammals do not. Because the embryos of vertebrates are similar, scientists hypothesize that vertebrates come from a common ancestor.

What are homologous structures?

Body parts that are similar in origin and structure are called **homologous** (hoh MAH luh gus). Some homologous structures have the same function, but others do not. If two or more species have homologous structures, they probably have common ancestors. The figure below shows several homologous structures.

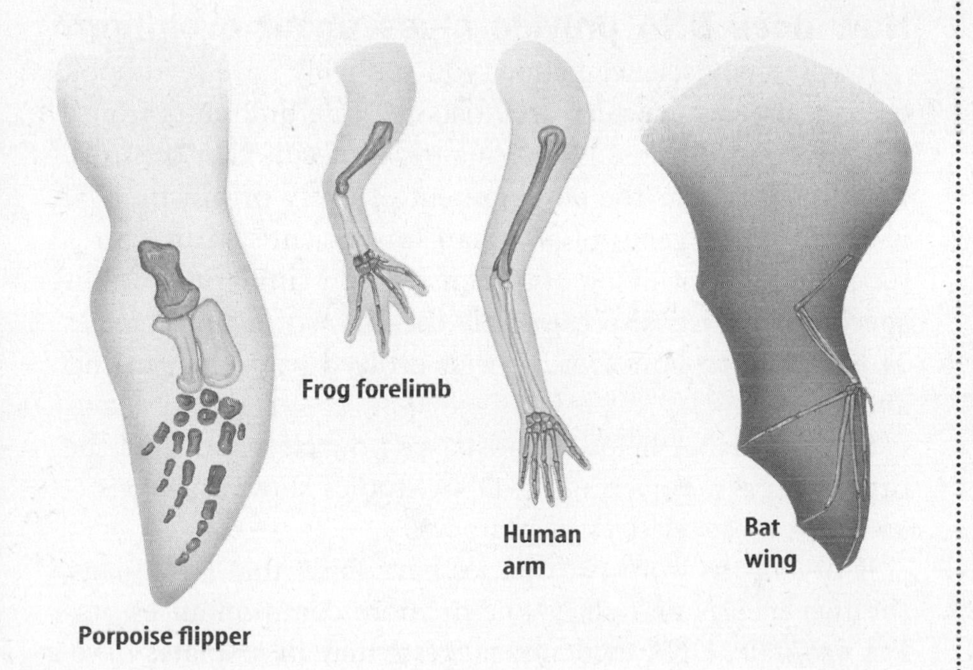

FOLDABLES™

D List Make a half-book Foldable, as shown below, to list examples of evolution and explanations of how the examples show evidence of evolution.

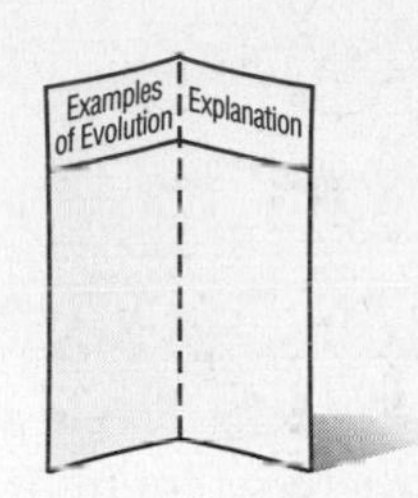

Picture This

3. **Describe** Above each structure, list one way the organism uses that structure.

What are vestigial structures?

The bodies of some organisms have structures known as vestigial (veh STIH jee ul) structures. **Vestigial structures** do not seem to have any use, or function. Vestigial structures provide evidence for evolution. Scientists hypothesize that vestigial structures are body parts that were useful in an ancestor. Humans have three small muscles around each ear that are vestigial. The figure below shows the location of these muscles. In some mammals, such as horses, these muscles are large. They allow a horse to turn its ears toward the source of a sound. ☑

Reading Check

4. **Identify** Name a vestigial structure that humans have?

____

____

____

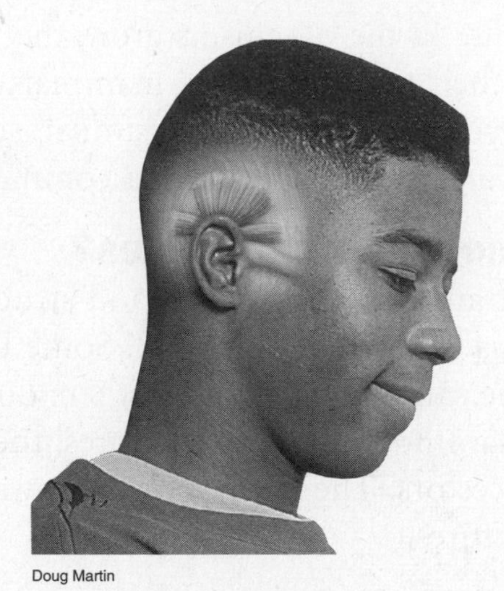

Doug Martin

How does DNA provide clues about evolution?

If you enjoy science fiction, you probably have read books or seen movies in which scientists recreate dinosaurs from DNA taken from fossils. DNA is the molecule that controls heredity. It directs the development of every organism. DNA is found in the genes of all organisms. Scientists can compare the DNA of living organisms to find similarities among species. Scientists also can study the DNA of extinct species. They can learn how some species evolved from their extinct ancestors.

Studying DNA helps scientists see how closely related the organisms are. For example, DNA studies show that dogs are the closest relatives of bears. ☑

If organisms from two species have DNA that is similar, the two species may share one or more common ancestors. For example, DNA evidence suggests that all primates have a common ancestor. Primates include chimpanzees, gorillas, orangutans, and humans.

Reading Check

5. **Explain** What clues does DNA provide about evolution?

____

____

____

After You Read

Mini Glossary

embryology (em bree AH luh jee): the study of embryos and their development

homologous (hoh MAH luh gus): body parts that are similar in origin and structure

radioactive element: an element that gives off a steady amount of radiation as it slowly changes to a nonradioactive element

sedimentary rock: rock in which most fossils are found, formed when layers of sand, silt, clay, or mud are pressed and cemented together, or when minerals are deposited from a solution

vestigial (veh STIH jee ul) structures: structures that do not seem to have a function

1. Review the terms and their definitions in the Mini Glossary. Choose one of the terms and write a sentence that explains how it provides a clue to evolution.

2. In the web diagram below, list the clues that scientists have as evidence of evolution.

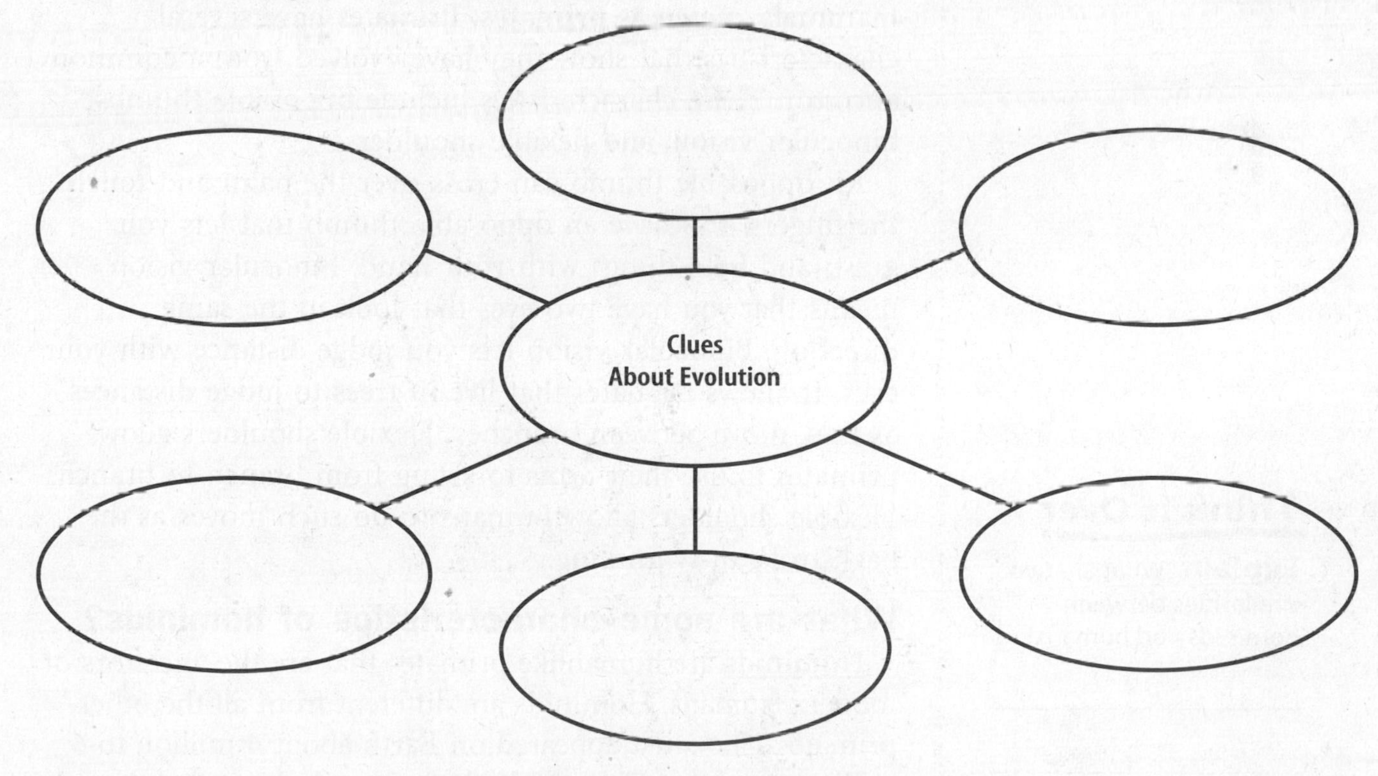

Science Online Visit **life.msscience.com** to access your textbook, interactive games, and projects to help you learn more about the clues of evolution.

End of Section

Adaptations over Time

section 3 The Evolution of Primates

What You'll Learn

- the differences among living primates
- the adaptations of primates
- the evolutionary history of modern primates

Before You Read

Describe the appearance and behavior of a primate such as monkeys and gorillas.

____

____

Mark the Text

Identify Main Points Underline the main idea of each paragraph. Then circle one supporting detail.

Read to Learn

Primates

Humans, monkeys, and apes belong to a group of mammals known as **primates.** Primates have several characteristics that show they have evolved from a common ancestor. These characteristics include opposable thumbs, binocular vision, and flexible shoulders.

An opposable thumb can cross over the palm and touch the fingers. You have an opposable thumb that lets you grasp and hold things with your hand. Binocular vision means that you have two eyes that look in the same direction. Binocular vision lets you judge distance with your eyes. It allows primates that live in trees to judge distances as they move between branches. Flexible shoulders allow primates to use their arms to swing from branch to branch. Flexible shoulders allow humans to do such moves as the backstroke in swimming.

Think it Over

1. **Explain** What are two similarities between hominids and humans?

____

____

____

What are some characteristics of hominids?

Hominids are humanlike primates that are the ancestors of modern humans. Hominids are different from all the other primates. They first appeared on Earth about 4 million to 6 million years ago. They ate both meat and plants and they walked upright on two legs.

Where have fossils of hominids been found?

In the 1920s, scientists discovered a fossil skull in South Africa. The skull had a small space for the brain, but it had a humanlike jaw and teeth. The fossil was named *Australopithecus.* It was one of the oldest hominids that had ever been discovered. In 1974, scientists found an almost-complete skeleton of *Australopithecus* in northern Africa. It had a small brain and may have walked upright. This fossil shows that modern hominids might have evolved from a common ancestor. ☑

Reading Check

2. Identify two characteristics of *Australopithecus.*

Who were the ancestors of early humans?

In the 1960s, scientists discovered a hominid fossil named *Homo habilis* that was estimated to be 1.5 million to 2 million years old. Scientists hypothesize that *Homo habilis* changed into another species, called *Homo erectus,* about 1.6 million years ago. These two hominids are thought to be ancestors of humans because they had larger brains and more humanlike features than *Australopithecus.*

Humans

Fossil records show that ***Homo sapiens*** evolved about 400,000 years ago. By 125,000 years ago, two early human groups probably lived in parts of Africa and Europe. These two groups were the Neanderthals (nee AN dur tawlz) and Cro-Magnon humans.

Who were the Neanderthals?

Neanderthals had short, heavy bodies with thick bones, small chins, and heavy browridges. They lived in caves in family groups. They used stone tools to hunt large animals. Neanderthals are probably not direct ancestors of modern humans.

Who were the Cro-Magnon humans?

The fossils of Cro-Magnon humans have been found in Europe, Asia, and Australia. They are between 10,000 and about 40,000 years old. Cro-Magnon humans looked very much like modern humans. They lived in caves, made stone carvings, and buried their dead. Cro-Magnon humans are thought to be direct ancestors of early humans. Early humans are called *Homo sapiens.* Modern humans are called *Homo sapiens sapiens.* Fossil evidence shows that modern humans evolved from *Homo sapiens.*

Think it Over

3. Compare Name two ways that Neanderthals and Cro-Magnon humans were similar.

After You Read

Mini Glossary

hominids: humanlike primates that lived about 4 million to 6 million years ago and were different from the other primates

Homo sapiens: direct ancestors of humans

primate: group of mammals to which humans, monkeys, and apes belong

1. Review the terms and their definitions in the Mini Glossary. Choose one term and write a sentence that describes how it is related to modern humans.

___

___

___

2. In the boxes below, show the sequence of the evolution of the ancestors of modern humans. Write down how long ago scientists believe each of the following human ancestors first appeared: hominids, *Homo habilis, Homo erectus, Homo sapiens,* Neanderthals and Cro-Magnon humans. The first box has been completed for you.

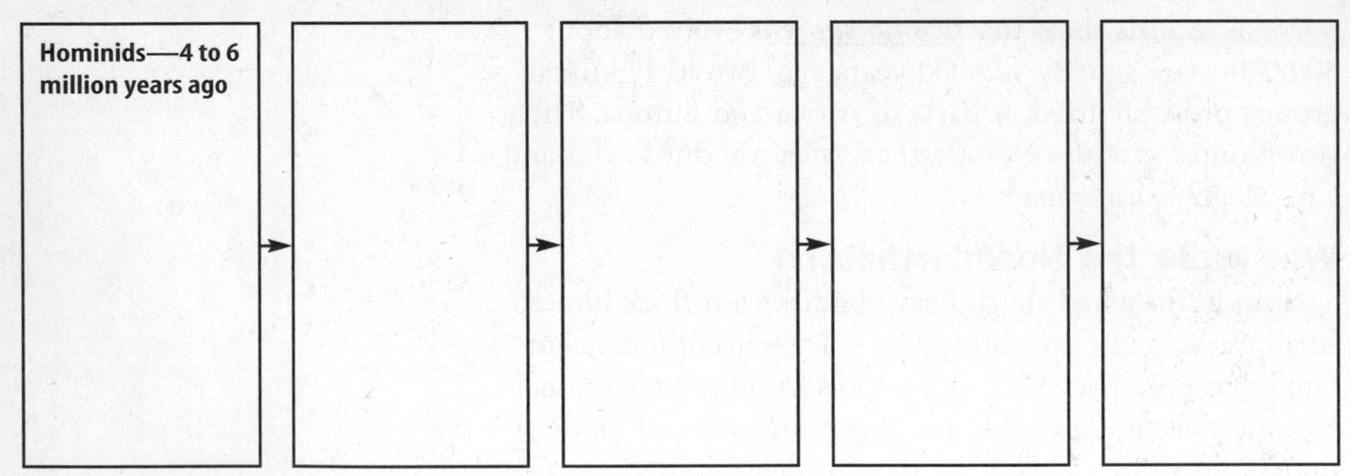

3. How did you benefit from underlining main ideas in paragraphs?

___

___

___

Science Online Visit **life.msscience.com** to access your textbook, interactive games, and projects to help you learn more about the evolution of primates.

Bacteria

section 1 What are bacteria?

Before You Read

What do you think an antibacterial soap does? Why do people use antibacterial products? Write your answer below.

____

____

____

What You'll Learn

- the characteristics of bacterial cells
- the difference between aerobic and anaerobic organisms

Read to Learn

Characteristics of Bacteria

In the 1600s, Antonie van Leeuwenhoek, a Dutch merchant, observed scrapings from his teeth under his microscope. He did not know it, but the organisms he was observing were bacteria. A hundred years later, bacteria were proven to be living cells.

Study Coach

Organize Information Take notes as you read. Organize notes into two columns. On the left, list a main idea about the material in each subhead. On the right, list the details that support the main idea.

Where do bacteria live?

Bacteria (singular, *bacterium*) live in the air, in foods, and on the surfaces of things you touch. They are found underground and deep underwater. Bacteria live on your skin and in your body. Some kinds of bacteria live in extremely hot and extremely cold environments. Very few other organisms can survive these conditions.

What are the shapes of bacterial cells?

Bacteria are found in three different shapes—spheres, rods, and spirals. Bacteria that are shaped like a sphere are called cocci (KAHK si) (singular, *coccus*). Rod-shaped bacteria are called bacilli (buh SIH li) (singular, *bacillus*). Spiral-shaped bacteria are called spirilla (spi RIH luh) (singular, *spirillum*). Bacteria are one-celled organisms that live alone or in chains or groups.

FOLDABLES™

A Describe Make a three-tab book to describe and show the three shapes of bacteria.

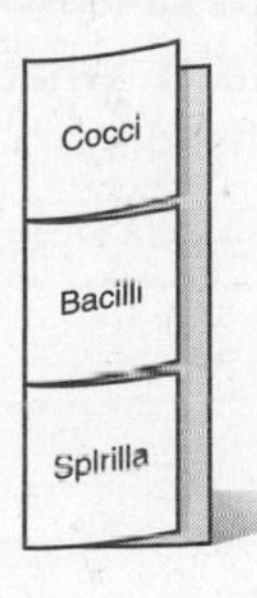

What do bacterial cells look like?

The figure below shows you what a bacterial cell looks like. A bacterial cell contains cytoplasm surrounded by a cell membrane and a cell wall. A bacterial cell is classified as prokaryotic (pro KAYR ee yah tihk) because its nucleus is not surrounded by a cell membrane. A bacterial cell's genetic material is in its one chromosome, found in the cytoplasm. A bacterial cell's cytoplasm also contains ribosomes. Ribosomes make proteins that every cell needs to survive.

Picture This

1. **Identify** Underline the names of the parts of the cell that are inside the cell's membrane.

Ribosome
Cytoplasm
Chromosome
Flagellum
Cell membrane
Gelatinlike capsule
Cell wall

What special features do bacteria have?

Some bacteria, such as the one in the figure above, have a thick, gelatinlike capsule around the cell wall. The capsule protects the bacterium from other cells that might destroy it. The capsule and the hairlike structures found on many bacteria help them stick to surfaces. Some bacteria also have an outer coating called a slime layer. The slime layer helps a bacterium stick to a surface and reduces water loss. Many bacteria that live in moist places have tails called **flagella** (singular, *flagellum*) that help them move.

2. **Explain** What results from fission in bacteria?

____

____

____

What is fission?

Bacteria usually reproduce by fission. **Fission** produces two new cells that have genetic material identical to each other and to that of the original cell. Fission is the simplest form of asexual reproduction. ☑

Do bacteria reproduce sexually?

Some bacteria reproduce through a process similar to sexual reproduction. In this process, two bacteria line up beside each other and exchange DNA through a thin tube. The cells then have different combinations of genetic material than they had before the exchange. The new combinations may improve the bacteria's chances for survival.

How do producers and consumers differ?

Some bacteria make their own food. These bacteria are called producers. Bacteria that contain chlorophyll make their food using energy from the Sun. Other bacteria use energy from chemical reactions to make food.

Most bacteria do not make their own food. They get their food from the environment. These bacteria are called consumers. Some consumers break down dead organisms to get energy. Others live as parasites of living organisms and get food from their hosts. ☑

Reading Check

3. Identify What is the difference between producer bacteria and consumer bacteria?

____

____

____

What are aerobes and anaerobes?

Most organisms need oxygen to break down food. They obtain energy through a process called respiration. An organism that uses oxygen for respiration is called an **aerobe** (AY rohb). Humans and most bacteria are aerobic organisms. An organism that lives without oxygen is called an **anaerobe** (AN uh rohb). Several kinds of anaerobic bacteria live in human intestines. Some bacteria cannot survive in places with oxygen.

Eubacteria

Bacteria are classified into two kingdoms—eubacteria (yew bak TIHR ee uh) and archaebacteria (ar kee bak TIHR ee uh). Most eubacteria are grouped by the following characteristics. ☑

- The shape and structure of the cell
- How they get food
- The kind of food they consume
- The wastes they produce
- How they move
- Whether they are aerobic or anaerobic

Reading Check

4. Determine Into what two kingdoms are bacteria classified?

____

____

What bacteria live in intestines of humans?

Many different kinds of bacteria can live in the intestines of humans and other animals. The figure below identifies bacteria based on the foods they use and the wastes they produce.

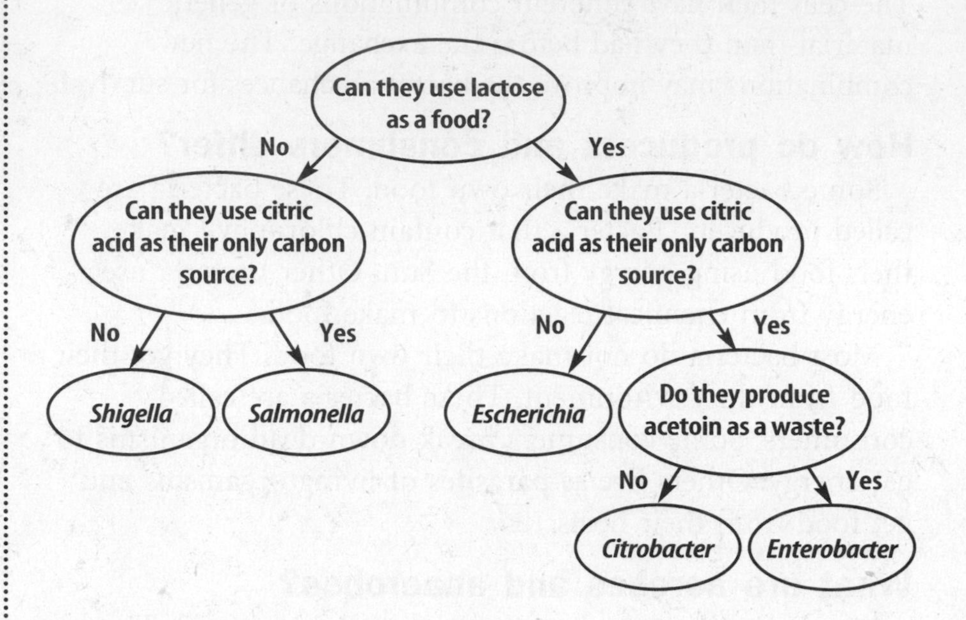

Picture This

5. Identify Which bacteria do not use lactose as a food?

____

____

What are cyanobacteria?

An important group of producer eubacteria is the cyanobacteria (si an oh bak TIHR ee uh). Cyanobacteria live in water and use carbon dioxide, water, and energy from sunlight to make their own food. They produce oxygen as waste. Cyanobacteria have chlorophyll and a blue pigment, or coloring.

Think it Over

6. Explain Why are cyanobacteria producers?

____

____

____

Why are cyanobacteria important?

Some cyanobacteria live together in long chains. They are covered with a gelatinlike substance that helps them live in groups called colonies. Cyanobacteria are food for some organisms that live in ponds, lakes, and oceans. Other water organisms use the oxygen that cyanobacteria produce.

However, cyanobacteria can harm water organisms. When a pond has large amounts of nutrients, the number of cyanobacteria increases. When the cyanobacteria population gets large enough, a mat of bubbly green slime appears on the surface of the water. This slime is called a bloom. The cyanobacteria use up the nutrients in the water and they die. Then other bacteria that are aerobic consumers feed on the dead cyanobacteria and use up the oxygen in the water. Because there is less oxygen in the water, fish and other organisms die.

How are consumer eubacteria categorized?

There are two categories of consumer eubacteria. The categories are based on the result of the Gram's stain. Gram-positive cells stain purple because they have thicker cell walls. Gram-negative cells stain pink because their cell walls are thinner.

Doctors and veterinarians use antibiotics (an ti bi AH tihks) to treat infections caused by bacteria. Some gram-positive bacteria are harder to treat with antibiotics than gram-negative bacteria.

One group of eubacteria does not have cell walls. Because they don't have cell walls, their shapes change. They can't be described as coccus, bacillus, or spirillum. The bacteria that cause pneumonia in humans do not have cell walls.

Archaebacteria

Kingdom Archaebacteria contains certain kinds of bacteria that live in extreme conditions, such as hot springs. The conditions in which some archaebacteria live today are similar to conditions found during Earth's early history. Archaebacteria is grouped according to where the bacteria live or how they get energy. ☑

Where do archaebacteria live?

One group lives in salty environments, such as the Great Salt Lake in Utah. Other groups live in environments that are acidic or hot, such as in hot springs. The temperature of the water in hot springs is more than 100°C.

How do archaebacteria get energy?

Some archaebacteria use carbon dioxide for energy and give off methane gas as waste. Methane producers are anaerobic. They live in swamps, in the intestines of cattle, and in humans. These archaebacteria are used in sewage treatment. The bacteria break down the waste material that has been filtered from sewage water.

Think it Over

7. **Infer** Why do you think some antibiotics are more effective against the gram-negative bacteria than the gram-positive bacteria?

____

____

____

✔ Reading Check

8. **Explain** how scientists classify archaebacteria.

____

____

____

After You Read

Mini Glossary

aerobe (AY rohb): an organism that uses oxygen for respiration

anaerobe (AN uh rohb): an organism that is adapted to live without oxygen

fission: a reproductive process that produces two new cells with genetic material identical to each other and that of the original cell

flagella: whiplike tails that help bacteria move

1. Review the terms and their definitions in the Mini Glossary. Write a sentence that explains the difference between aerobes and anaerobes.

____

____

____

2. Select one of the question headings and write it below. Then write an answer to the question on the lines that follow.

Write your question here.

____

____

____

3. You made an outline as you read this section. How did it help you understand bacteria?

____

____

____

Science Online Visit **life.msscience.com** to access your textbook, interactive games, and projects to help you learn more about what bacteria are.

Bacteria

section 2 Bacteria in Your Life

Before You Read

Doctors sometimes prescribe antibiotics for their patients. What is the purpose of the antibiotics? Write your answer on the lines below.

____

____

____

What You'll Learn

- how some bacteria are helpful
- why nitrogen-fixing bacteria are important
- how some bacteria cause human disease

Read to Learn

Beneficial Bacteria

Only a few bacteria cause diseases. Most bacteria are more important because of the benefits they provide.

What kinds of bacteria help you?

Bacteria are important for keeping you healthy. Large numbers of bacteria are found in the large intestine of your digestive system. They are generally harmless and help you stay healthy. For example, some bacteria in your intestine produce vitamin K, which you need for blood clotting.

Some bacteria produce antibiotics. **Antibiotics** are chemicals that slow or stop the growth of other bacteria. Many bacterial diseases in humans and animals can be treated with antibiotics.

How do bacteria help the environment?

Consumer bacteria help keep balance in nature. Without bacteria, there would be layers of dead material all over Earth. A **saprophyte** (SAP ruh fite) is any organism that uses dead organisms for food and energy. Saprophytes help recycle nutrients that other organisms can use. Most sewage-treatment plants use saprophytes. They break down wastes into carbon dioxide and water.

Mark the Text

Identify Key Words As you read this section, circle the ways bacteria are helpful. Underline the ways bacteria are harmful.

FOLDABLES™

B Describe Make a two-tab book, as shown below. Describe helpful and harmful bacteria.

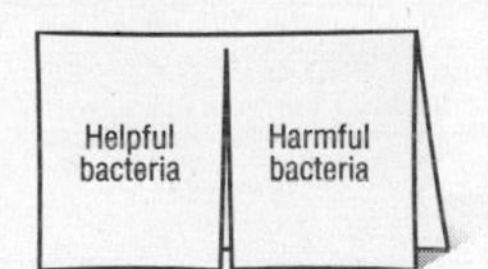

Think it Over

1. **Infer** Why are nitrogen-fixing bacteria important to humans?

______

______

______

How do plants and animals obtain nitrogen?

Plants and animals must use nitrogen to make proteins and nucleic acids. Animals can eat plants or other animals that contain nitrogen. Plants have to take in nitrogen from the soil or air. Although the air is about 78 percent nitrogen, plants and animals cannot use it directly. **Nitrogen-fixing bacteria** change the nitrogen from the air into forms that plants and animals can use. The roots of some plants such as peanuts and peas develop structures that contain nitrogen-fixing bacteria.

What is bioremediation?

Bioremediation uses organisms to clean up or remove pollutants from the environment. One kind of bioremediation uses bacteria to break down wastes into harmless compounds. Other bacteria use pollutants as food. Bioremediation has been used to clean up oil spills.

How is bacteria used to make food?

Bacteria have been used to make foods for a long time. One of the first uses of bacteria was to make yogurt. Bacteria break down substances in milk to make dairy products such as cheeses and buttermilk. Foods such as sauerkraut, pickles, and soy sauce also are made with the help of bacteria. ✔

✔ Reading Check

2. **Explain** why bacteria are used to make dairy products such as cheese.

______

______

______

How do industries use bacteria?

Industries use bacteria to make products, such as medicines, cleansers, and adhesives. Methane gas that is released as waste by certain bacteria can be used as fuel. In landfills, methane-producing bacteria break down plant and animal material. The amount of methane gas released is so large that some cities collect and burn it.

Harmful Bacteria

Some bacteria, known as pathogens, are harmful. A **pathogen** is any organism that causes disease. One bacterial pathogen, for example, causes strep throat.

✔ Reading Check

3. **Determine** How can bacterial pathogens enter the body?

______

______

______

How do pathogens make you sick?

Bacterial pathogens cause illness and disease in several ways. They can enter your body when you breathe. They also can enter through a cut in the skin. Once these pathogens are inside the body, they can multiply, damage cells, and cause illness and disease. ✔

What are toxins?

Toxins are poisonous substances produced by some pathogens. Botulism is a type of food poisoning caused by a toxin-producing bacteria. Botulism is able to grow and produce toxins inside sealed cans of food.

Some bacteria, like the one that causes botulism, can survive unfavorable conditions by making thick-walled structures called **endospores.** Endospores can exist for hundreds of years before they start growing again. If botulism endospores are in canned food, they can develop into regular bacterial cells and make toxins again. Canned foods that you buy in the store go through a process that uses steam under high pressure, which kills bacteria and most endospores.

What is pasteurization?

All food contains bacteria. You can kill the bacteria by sterilizing it with heat. Heating food to high temperatures can change the taste of the food. Pasteurization is a way of heating food to a temperature that kills most harmful bacteria but causes little change in the food's taste. The photo below shows some of the foods that are pasteurized.

Amanita Pictures

How are bacterial diseases treated?

Antibiotics are used to treat many bacterial diseases. Penicillin is an antibiotic that works by preventing bacteria from making cell walls. Some kinds of bacteria cannot survive without cell walls.

Vaccines are produced to treat many bacterial diseases. A **vaccine** can be made from damaged particles taken from bacterial cell walls or from killed bacteria. When the body is injected with a vaccine, white blood cells in the blood recognize that kind of bacteria. If the same kind of bacteria enters the body at a later time, the white blood cells attack them. ☑

FOLDABLES™

C Identify Use a half sheet of notebook paper, as shown below, to identify important facts about pasteurization.

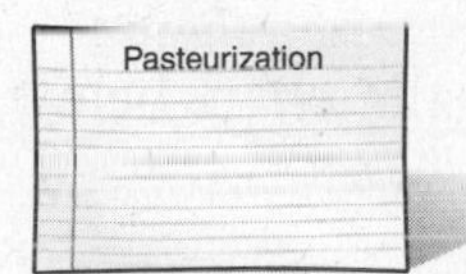

Reading Check

4. **Identify** From what is a vaccine made?

____

____

____

After You Read

Mini Glossary

antibiotics: chemicals, produced by bacteria, that limit the growth of other bacteria

endospores: thick-walled structures produced by a pathogen when conditions are not favorable for survival

nitrogen-fixing bacteria: bacteria that change nitrogen from the air into forms that plants and animals can use

pathogen: any organism that causes disease

saprophyte: any organism that uses dead organisms for food and energy

toxins: poisonous substances produced by some pathogens

vaccine: a treatment for some kinds of bacterial diseases, made from damaged particles taken from bacterial cell walls or from killed bacteria

1. Review the terms and their definitions in the Mini Glossary. Choose one term that refers to helpful bacteria and write a sentence describing how it is helpful.

____

____

____

2. In the chart below, list the ways bacteria are helpful to humans and the environment.

Helpful Roles of Bacteria

To humans	To the environment

Science Online Visit **life.msscience.com** to access your textbook, interactive games, and projects to help you learn more about bacteria in your life.

Protists and Fungi

section 1 Protists

Before You Read

Describe what you comes to mind when you hear the words *slime, mold,* and *mildew.* Where are these organisms usually found?

____

____

What You'll Learn
- the characteristics of protists
- the three groups of protists
- why protists are hard to classify

Read to Learn

What is a protist?

A **protist** is a one-celled or many-celled organism that lives in moist or wet places. Look at the organisms below. They have one thing in common—they belong to the protist kingdom. All protists have eukaryotic cells. Eukaryotic cells have a nucleus and other internal structures that are surrounded by membranes.

Study Coach

Make Flash Cards Write a quiz question on one side of a flash card. Write the answer on the other side. Use the flash cards to quiz yourself until you know the answers.

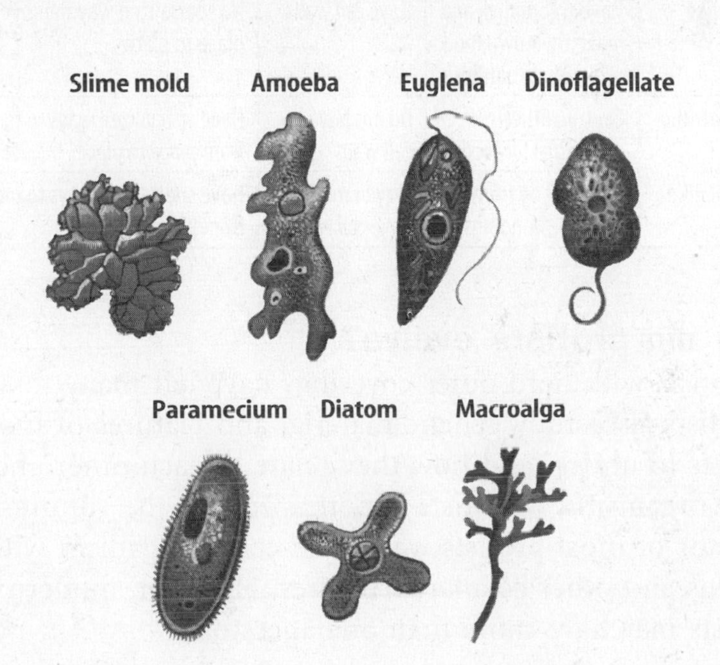

Picture This

1. **Compare and Contrast** Name two things these cells have in common.

____

____

FOLDABLES

A Describe Make a three-tab Foldable, as shown below. Describe each of the three types of protists.

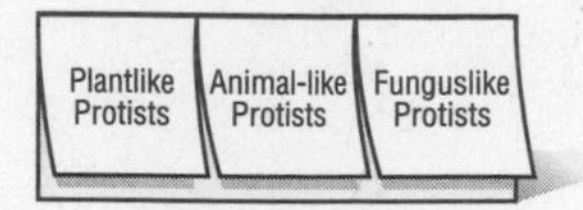

What are characteristics of protists?

Some protists have plantlike features. They contain chlorophyll and make their own food. Other protists have animal-like features. They do not have chlorophyll and they can move. Some protists have a hard shell-like covering on the outside of their bodies.

How do protists reproduce?

Protists usually reproduce asexually by cell division. The hereditary material in the nucleus is duplicated and the nucleus divides. Two genetically identical copies of the cell result. Asexual reproduction of many-celled protists occurs by regeneration. Parts of the protist break off and grow into new protists that are genetically identical.

Most protists also can reproduce sexually. A process called meiosis produces sex cells. Two sex cells, one from each of two different protists, join to form a new protist. The new protist is genetically different from the two protists that provided the sex cells. How and when sexual reproduction in protists happens depends on the specific type of protist.

What are the three groups of protists?

Protists are usually divided into three groups—plantlike, animal-like, and funguslike. The table below shows characteristics that help scientists classify protists.

Picture This

2. Use Tables Cut small pieces of paper to cover each of the nine blocks that describe characteristics of protist groups. Quiz yourself on characteristics. Uncover the cell to check your answer.

Characteristics of Protist Groups

	Food Source	Cell Structure	Movement
Plantlike	Have chlorophyll and make their own food using photosynthesis	Have cell walls	No specialized way to move from place to place
Animal-like	Capture other organisms for food	Do not have cell walls	Have specialized ways to move from place to place
Funguslike	Absorb food from their surroundings	May or may not have cell walls	Have specialized ways to move from place to place

How did protists evolve?

Protists with hard outer coverings have left many fossils. Scientists also study genetic material and features of modern protists to understand how they relate to each other and to other organisms. Scientists hypothesize that the common ancestor of most protists was a one-celled organism with a nucleus and other cellular structures. However, modern protists may have more than one ancestor.

Plantlike Protists

Like plants, protists in this group have chlorophyll in chloroplasts. Chlorophyll is a green pigment, or coloring, that traps light energy. Plantlike protists use the light energy to make their own food. These protists do not have roots, but some have structures that hold them in place. Some have cell walls like plants.

Plantlike protists are known as **algae** (AL jee) (singular, *alga*). Some algae have one cell; others have many cells. All algae have chlorophyll, but not all of them look green. Many have other pigments that cover up the chlorophyll.

What are diatoms?

Diatoms live in both freshwater and salt water. Their golden-brown pigment covers up the green chlorophyll. Diatoms form glasslike boxes around themselves. The boxes sink when the protists die. Over thousands of years, these boxes collect and form deep layers. ☑

What are dinoflagellates?

Dinoflagellates are algae. They move using two long, thin, whiplike structures, called a **flagellum** (plural, *flagella*). One flagellum circles the cell like a belt. The other flagella is attached to one end of the organism like a tail. As the two flagella move, they cause the organism to spin. Their name means "spinning flagellates." Most dinoflagellates live in salt water, and most have chlorophyll. Dinoflagellates without chlorophyll feed on other organisms.

What are euglenoids?

Euglenoids (yoo GLEE noydz) are plantlike protists that also have characteristics of animals. Many of these one-celled algae have chloroplasts and use chlorophyll to produce their own food. Euglenoids that do not have chlorophyll feed on bacteria and other protists. Although euglenoids have no cell wall, they do have a strong flexible layer inside the cell membrane that helps it move and change shape. Many euglenoids have flagella that help them move. ☑

How are red, green, and brown algae alike?

Most of the many-celled red, green, and brown algae are called seaweeds. These algae usually live in water. Green algae are the most plantlike. In fact, some scientists propose that plants evolved from green algae.

✔ Reading Check

3. **Identify** Where do diatoms live?

____

____

____

✔ Reading Check

4. **Determine** What is the name of protists that have characteristics of both plants and animals?

____

Brown Algae Brown algae vary greatly in size. One brown alga, called kelp, is an important food source for many fish and invertebrates. Giant kelp is the largest protist and can grow to be 100 m long. The chart below compares algae.

Picture This

5. **Identify** List the features that red, green, and brown algae share.

Comparisons of Green, Red, and Brown Algae					
	Common Names	Number of Cells	Contain Chlorophyll	Main Pigment	Places Found
Red	seaweed	many	yes	red	salt water, up to 200 m deep
Green	green algae; seaweed	one or many	large amounts	green	in water, damp tree trunks, wet sidewalks
Brown	seaweed; kelp	many	yes	brown	cool, saltwater environments

Importance of Algae

Many animals that live on land, such as cattle and deer, depend on grasses for food. Algae are sometimes called the grasses of the ocean. Most animals that live in the oceans eat algae or eat other animals that eat algae. One-celled diatoms and dinoflagellates are an important food source for many organisms that live in oceans. Euglenoids are an important food source for freshwater organisms.

How do algae affect the environment?

Algae produce oxygen as a result of photosynthesis. The oxygen produced by green algae is important for most organisms on Earth, including you. Under certain conditions, algae can reproduce rapidly and form an algal bloom. The color of the water where the algal bloom forms appears to change. The color change occurs because of the large number of organisms. Dinoflagellates form blooms along the east and Gulf coasts of the United States. They produce toxins that can cause other organisms to die and can cause health problems in humans.

Think it Over

6. **Describe** one way that algae help the environment and one way that they harm the environment.

How do humans use algae?

Many people around the world eat some species of red and brown algae. A substance in the cell walls of red algae is used to make cosmetics and food products. The substance also gives toothpaste, puddings, and salad dressings their smooth, creamy texture. A substance in the cell walls of brown algae helps thicken some foods and is used in making rubber tires and hand lotion.

Uses of Diatoms The glasslike remains of ancient diatoms are mined and used in insulation, filters, and road paints. The cell walls of diatoms produce the sparkle that makes some road lines visible at night and the crunch you might feel in toothpaste.

Animal-like Protists

Animal-like protists that have one cell are known as **protozoans.** Protozoans usually are classified by how they move. They live in or on other living or dead organisms. Many protozoans have specialized vacuoles for digesting food and getting rid of excess water.

What are ciliates?

Some protists have short, threadlike structures called **cilia** (SIH lee uh) that stick out of the cell membrane. Ciliates have cilia that move back and forth similar to rowboat oars. The cilia help ciliates move quickly in any direction. ☑

The *Paramecium* shown in the figure below is a common ciliate. *Paramecium* has two nuclei—a large nucleus, the macronucleus, and a small nucleus, the micronucleus. The micronucleus is involved in reproduction. The macronucleus controls feeding, the exchange of oxygen and carbon dioxide, the amount of water and salts entering and leaving *Paramecium,* and other functions of the organism.

Reading Check

7. **Describe** What do cilia do?

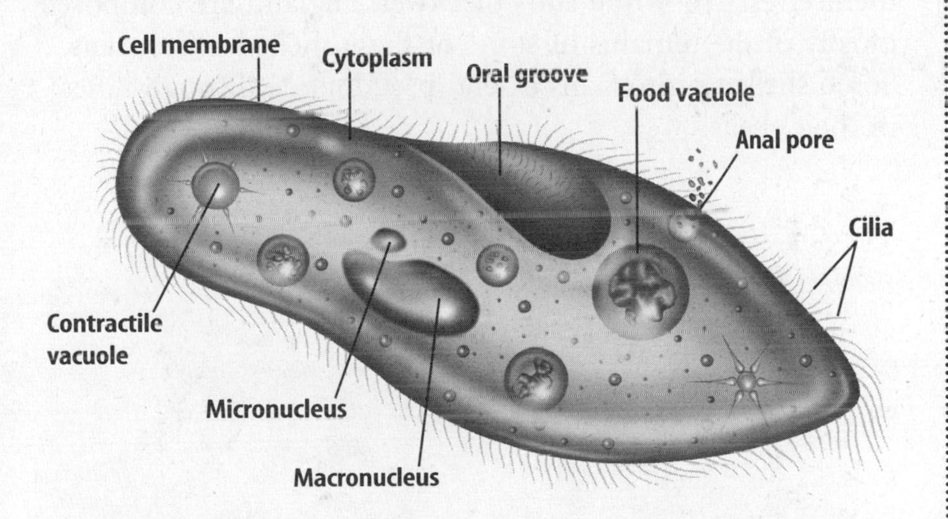

Picture This

8. **Identify** Circle the name of the structure involved in reproduction.

What do ciliates eat?

Ciliates usually feed on bacteria that are swept into the oral groove by cilia. A vacuole forms around the bacteria and the food is digested. Wastes are removed from the ciliate through the anal pore.

What are flagellates?

Protozoans called flagellates move through the water by whipping their long flagella. Many flagellates live in freshwater. Some are parasites that harm their hosts. Some flagellates may live in colonies. The colonies consist of many cells that are similar in structure to cells found in animals called sponges. The cells perform different functions. ☑

Reading Check

9. **List** two characteristics of flagellates.

____

____

____

____

What are pseudopods?

Some protozoans use temporary extensions of their cytoplasm called **pseudopods** (SEW duh pahdz) to move through their environment and to get food. The word *pseudopod* means "false foot." These organisms seem to flow along as they extend their pseudopods. They are found in freshwater and saltwater environments. Certain types are parasites in animals. The amoeba is a member of this group.

How do pseudopods work?

To get food, an amoeba extends the cytoplasm of a pseudopod on either side of the food, such as a bacterium. Then the two parts of the pseudopod flow together, trapping the food in the organism. A vacuole forms around the trapped food, and digestion takes place inside the vacuole. A food vacuole is shown in the amoeba below.

Some members of this group, such as the amoeba, have no outer covering. Others secrete hard shells around themselves. The white cliffs of Dover, England are composed mostly of the remains of some of these shelled protozoans. Some shelled protozoans extend pseudopods through holes in their shells.

Picture This

10. **Identify** Circle the name of the structure that changes temporarily to form a pseudopod.

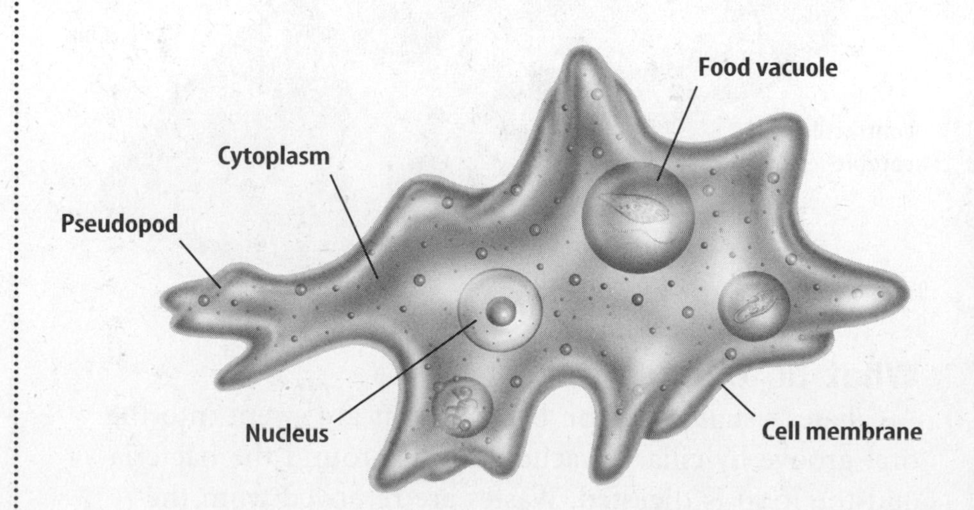

What other protozoans exist?

One group of protozoans has no way of moving on their own. All of the organisms in this group are parasites of humans and other animals. These protozoans have complex life cycles. The life cycle involves both sexual and asexual reproduction. They often live part of their lives in one animal and part of their lives in another animal. The parasite that causes malaria belongs to this group of protozoans. Asexual reproduction of the malaria parasite takes place inside a human host. Sexual reproduction takes place in the intestine of a mosquito.

Importance of Protozoans

Some protozoans are an important source of food for larger organisms. Other protozoans are indicator species for petroleum reserves. When these shelled protozoans die, their shells sink to the bottom of the water and build up over many years. Large buildups of protozoan shells mean that petroleum reserves might be found in the area.

One type of flagellated protozoan lives in the digestive tract of termites. Termites feed on wood. The protozoans make enzymes that help the termites break down the wood. Termites would not be able to use chemical energy from the wood as well if the protozoans did not live in their digestive tracts. ☑

What diseases do protozoans cause?

In tropical environments, parasitic flagellates enter humans when humans are bitten by mosquitoes or other insects. Other parasitic flagellates live in water that is contaminated with wastes from humans or animals. If humans drink this water, they might get a diarrhea-causing parasite.

Some amoebas also are parasites that cause disease. One parasitic amoeba, found in ponds and streams, can cause a brain infection that kills the host.

Funguslike Protists

Funguslike protists include several small groups of organisms such as slime molds, water molds, and downy mildews. All funguslike protists produce spores like fungi. Most have pseudopods, like the amoeba, to move from place to place. Funguslike protists cannot make their own food. They must take in food from outside sources.

Reading Check

11. Apply How does one type of protozoan help termites?

Think it Over

12. Draw Conclusions Why do many travelers avoid drinking tap water?

What are slime molds?

Slime molds have some protozoan features. During part of their life cycle, slime molds move using pseudopods and behave like amoebas.

Most slime molds live on rotting logs or dead leaves in moist, cool, shady environments. Slime molds form weblike structures on the surface of their food supply. These structures often are brightly colored. ☑

Reading Check

13. Identify In what environment are most slime molds found?

____

____

____

What are water molds and downy mildews?

Most members of this group of funguslike protists live in water or moist places. They grow as a mass of threads over a plant or animal. They absorb nutrients from the organism they live on. The spores of a water mold or downy mildew have flagella. The spore's cell walls are more like the cell walls of plants than those of fungi.

Some water molds are parasites of plants, and others feed on dead organisms. Most water molds look like fuzzy, white growths on rotting matter.

Downy mildews form when days are warm and nights are cool and moist. Downy mildews weaken plants and even can kill them.

Importance of Funguslike Protists

Some funguslike protists are helpful because they help break down dead organisms. Most funguslike protists are harmful and cause diseases in plants and animals. Water molds cause disease in some organisms that live in water.

Do funguslike protists affect the economy?

Downy mildews have caused widespread damage several times throughout history. In the 1840s, downy mildews were responsible for the Irish potato famine. Potatoes were Ireland's main crop and the main source of food. Potatoes infected with downy mildew rotted in the fields, leaving people with no food. A similar problem occurred in the 1870s in France. The mildew affected grapes and nearly destroyed the French wine industry. Downy mildews continue to infect crops, such as lettuce, corn, cabbage, avocados, and pineapples. ☑

Reading Check

14. Explain How did the downy mildew affect the Irish potato crop?

____

____

____

After You Read

Mini Glossary

algae: many-celled, plantlike protists that have chlorophyll and produce their own food; can be different colors

cilia (SIH lee uh): short, threadlike structures that stick out from the cell membrane and help the organism move quickly in any direction

flagellum: long, thin, whiplike structure of some protists that helps them move from place to place

protist: a one-celled or many-celled eukaryotic organism that lives in moist or wet places

protozoans: one-celled, animal-like protists that live in water or soil, or on living and dead organisms

pseudopods (SEW duh pahdz): temporary extensions of cytoplasm in some protists that help them move through their environment and get food

1. Review the terms and their definitions in the Mini Glossary. List the three terms that describe structures or ways that a protist moves.

____

____

2. Place each of the items in the following list in the correct category of the table.

amoebas
brown algae
ciliates
diatoms
dinoflagellates
downy mildews
euglenoids
flagellates
green algae
Paramecium
protozoans
red algae
slime molds
water molds

Plantlike Protists	Animal-like Protists	Funguslike Protists

Science Online Visit **life.msscience.com** to access your textbook, interactive games, and projects to help you learn more about protists.

End of Section

chapter 8 Protists and Fungi

section 2 Fungi

What You'll Learn

- the characteristics of fungi
- how to classify fungi
- how to distinguish imperfect fungi from all other fungi

Before You Read

Think about the places that you have seen mushrooms growing. What do those places have in common?

____

____

Mark the Text

Identify Main Ideas Read each of the question heads. As you read the paragraphs related to each head, underline the answer to the question.

Read to Learn

What are fungi?

Mushrooms are common fungi. The yeasts used to make some breads and cheeses are a type of fungus. Fungus may grow on a loaf of bread or on your shower curtain.

How did fungi evolve?

Some fossils of fungi have been found, but they do not help scientists determine how fungi are related to other organisms. Some scientists hypothesize that fungi share an ancestor with ancient, flagellated protists and slime molds. Other scientists hypothesize that their ancestor was a green or red alga.

What are hyphae?

Most species of fungi are many-celled. The body of a fungus usually is made up of many-celled threadlike tubes called **hyphae** (HI fee). The figure below shows the inside structure of hyphae.

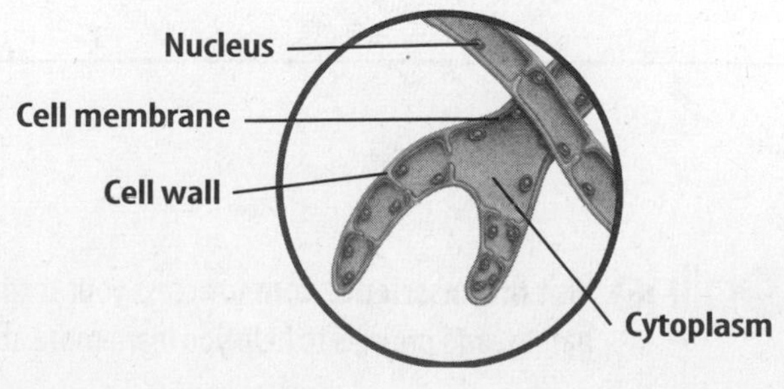

Nucleus
Cell membrane
Cell wall
Cytoplasm

Picture This

1. **Infer** How do you know these hyphae are made up of many cells?

____

____

How do most fungi get food?

Hyphae produce enzymes that help break down food that the fungus absorbs from another organism. Most fungi are **saprophytes** (SAP ruh fites), meaning they get food by absorbing dead or decaying tissues of other organisms. ☑

What characteristics do fungi share?

Some fungi grow anchored in soil and have a cell wall around each cell. Fungi do not have specialized tissues and organs such as leaves and roots. Fungi do not have chlorophyll and do not make their own food. Fungi grow best in warm, damp areas, such as tropical forests or between toes.

How do fungi reproduce?

Fungi reproduce both asexually and sexually. For both types of reproduction, fungi produce spores. A **spore** is a waterproof reproductive cell that can grow into a new organism. In asexual reproduction, the cells divide to produce spores. These spores grow into new fungi that are genetically identical to the fungus from which the spores came.

Fungi are not identified as either male or female. For sexual reproduction to take place, the hyphae of two genetically different fungi of the same species grow close together. If the hyphae join, a reproductive structure, such as the one in the figure below, forms. Meiosis, or cell division that produces sex cells, results in spores that will grow into new fungi. These fungi are genetically different from either of the two fungi whose hyphae joined together.

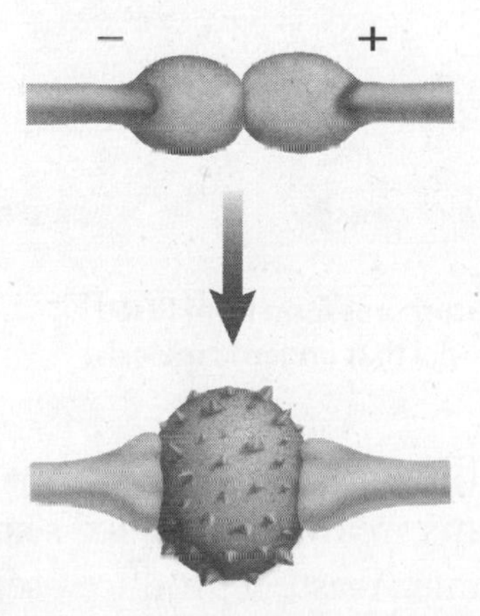
A Two hyphae fuse.
–
+
B Reproductive structure forms.

✔ Reading Check

2. Identify What is a saprophyte?

____

____

____

Picture This

3. Identify Circle the reproductive structure formed when two genetically different fungi reproduce.

![FOLDABLES]

B Describe Make a three-tab Foldable, as shown below. Describe each of the three main groups of fungi.

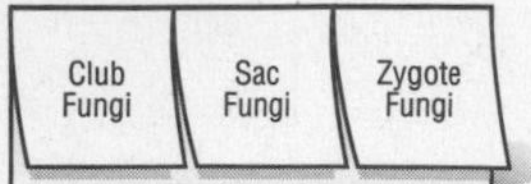

How are fungi classified?

Fungi are classified into three main groups. The groups are identified by the type of structure formed when the hyphae join together.

Steve Austin; Papilio/CORBIS

Club Fungi

Mushrooms, such as those shown above, are examples of club fungi. The mushroom is the reproductive structure of the fungus. Most of the fungus grows as hyphae in the soil or on the surface of its food source. The spores of club fungi are produced in a club-shaped structure called a **basidium** (buh SIH dee uhm) (plural, *basidia*).

Sac Fungi

This varied group of fungi includes yeasts, molds, and truffles. There are more than 30,000 different species of sac fungi. The spores of sac fungi are produced in a little, saclike structure called an **ascus** (AS kus), as shown in the figure below.

Picture This

4. Identify Highlight the ascus in the figure.

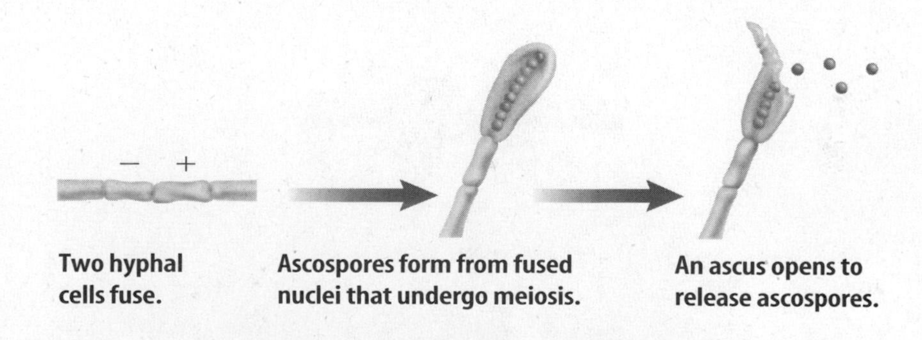

– +
Two hyphal cells fuse.
Ascospores form from fused nuclei that undergo meiosis.
An ascus opens to release ascospores.

Although most fungi are many-celled, yeasts are one-celled organisms. Yeasts reproduce sexually by forming spores like other fungi. Yeasts reproduce asexually by **budding**, in which a new organism forms on the side of the parent organism. The two organisms are genetically identical.

Zygote Fungi and Other Fungi

Black molds that you might see growing on old bread or old fruit are a type of zygospore fungus. Zygospore fungi produce spores in a round spore case called a **sporangium** (spuh RAN jee uhm) (plural, *sporangia*). Sporangia form on the tips of some hyphae. When a sporangium splits open, hundreds of spores are released into the air, as shown in the figure below. Each spore that lands on a warm, moist surface will grow and reproduce if it has a food source.

Picture This

5. **Describe** what happens to the spores when a sporangium splits open.

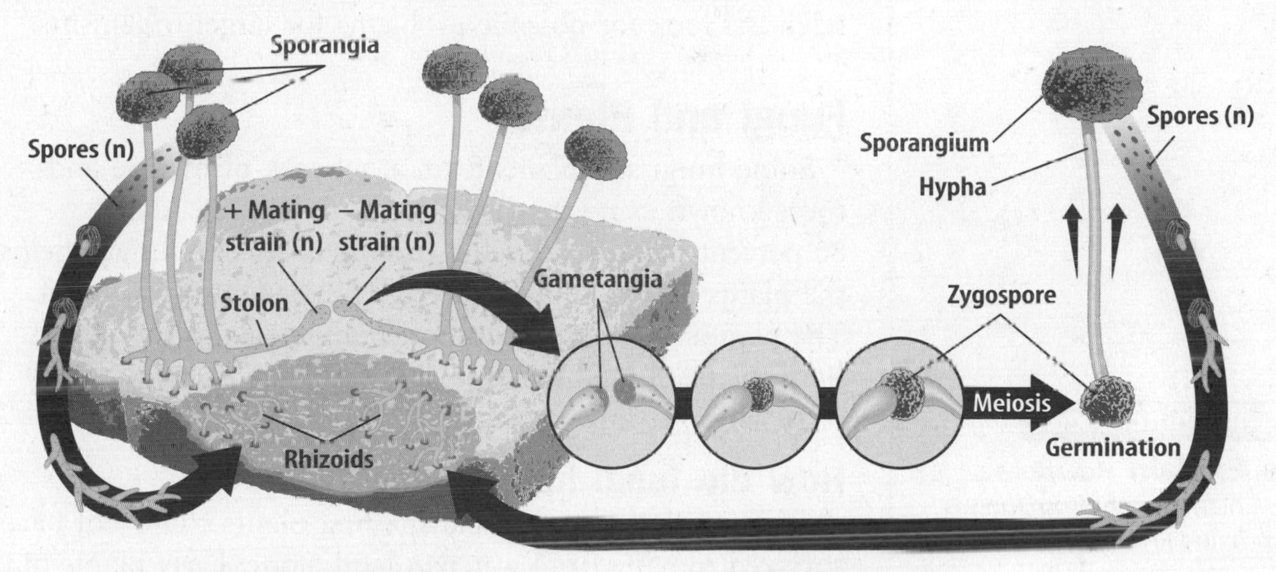

What are imperfect fungi?

Some fungi either never reproduce sexually or never have been seen reproducing sexually. They usually are called imperfect fungi because there is no evidence that their life cycle has a sexual reproduction stage. ☑

Some scientists classify *Penicillium* as an imperfect fungi. Other scientists classify it as a sac fungi because of the type of spores it produces during asexual reproduction.

Reading Check

6. **Explain** Why are some fungi called "imperfect"?

Lichens

A **lichen** (LI kun) is an organism made up of a fungus and either a green alga or a cyanobacterium. These two organisms have a relationship that benefits both of them. The alga or cyanobacterium lives among the threadlike strands of the fungus. The fungus gets food made by the green algae or cyanobacterium. The green alga or cyanobacterium gets a moist, protected place to live.

Think it Over

7. **Explain** how lichens help with the weathering process.

Why are lichens important?

Lichens are an important food source for many animals, including caribou and musk oxen. Lichens also are important because they help produce soil. Lichens grow on bare rock and produce acids as part of their metabolism. The acids help break down the rock. This process is known as weathering. Soil forms from the bits of rock and lichens that have died.

Many species of lichens are sensitive to pollution. When these organisms decline in their health or die quickly, scientists look for possible problems for larger organisms.

Fungi and Plants

Some fungi and plants form a network of hyphae and roots known as **mycorrhizae** (mi kuh RI zee). About 80 percent of plants develop mycorrhizae. The fungus helps the plants absorb more of certain nutrients from the soil. The plant supplies food and other nutrients to the fungus. Both the plants and the fungi benefit. Some plants cannot grow without mycorrhizae. ☑

Reading Check

8. **Explain** How does a plant benefit from fungus living around its roots?

How did fungi help plants evolve?

Scientists have known that the first plants could not have survived moving from water to land alone. Early plants did not have specialized roots to absorb nutrients. Also, tubelike cells used to transport water and nutrients to leaves were too simple to survive on land.

In 1999, scientists discovered a fossilized fungus in a 460 million-year-old rock. The fossil was a type of fungus that forms relationships with plant roots. Scientists have hypothesized that fungi attached themselves to the roots of early plants, passing along nutrients taken from the soil to the plant. Scientists suggest that this relationship began about 500 million years ago.

Importance of Fungi

Mushrooms, one type of fungi, are an important food crop. However, some wild mushrooms are poisonous and should never be eaten.

Fungi are used to make some cheeses and breads. Yeasts use sugar for energy and produce alcohol and carbon dioxide as waste products. The carbon dioxide causes bread dough to rise.

What problems do fungi cause in plants and animals?

Many fungi cause disease in plants and animals. Many sac fungi damage or destroy plant crops. Diseases caused by sac fungi include Dutch elm disease and apple scab. Smuts and rust are club fungi. They damage billions of dollars worth of food crops each year.

Ringworm and athlete's foot are skin infections caused by species of imperfect fungi. Some respiratory infections are caused by inhaling fungi or their spores.

How are fungi helpful to animals and humans?

Some fungi naturally produce antibiotics (an ti bi AH tihks) to help keep bacteria from growing near them. The antibiotic penicillin is produced by the imperfect fungi *Penicillium. Penicillium* is grown commercially, and the antibiotic is sold to help humans and other animals fight infections caused by bacteria. The drug cyclosporine comes from a fungus. Cyclosporine helps fight the body's rejection of transplanted organs.

There are many more examples of breakthroughs in medicine as a result of studying fungi. Scientists worldwide continue to study fungi to find more useful drugs.

Why are fungi called nature's recyclers?

Fungi's most important role is as decomposers. Fungi break down, or decompose, organic material such as food scraps and dead plants and animals. As these materials decompose, they release chemicals into the soil, where plants can reuse them. Fungi, along with bacteria, are nature's recyclers. They keep Earth from becoming buried under mountains of organic wastes.

Think it Over

9. **List** one problem that fungi cause and one benefit that they provide.

After You Read

Mini Glossary

ascus (AS kus): little saclike reproductive structures in which sac fungi produce spores

basidium (buh SIH dee uhm): club-shaped reproductive structures in which club fungi produce spores

budding: form of asexual reproduction in which a new, genetically identical organism forms on the side of the parent organism

hyphae (HI fee): many-celled, threadlike tubes that form the body of a fungus

lichen (LI kun): an organism made up of a fungus and either a green alga or a cyanobacterium

mycorrhizae (mi kuh RI zee): a network of hyphae and roots that helps plants absorb more of certain nutrients from the soil

saprophyte (SAP ruh fite): organism that gets food by absorbing dead and rotting tissues of other organisms

sporangium (spuh RAN jee uhm): round spore case of a zygospore fungus

spore: a waterproof reproductive cell of a fungus that can grow into a new organism

1. Review the terms and their definitions above. Fill in the table below with the term from the Mini Glossary that describes the reproductive structure of each type of fungus.

Club Fungi	Sac Fungi	Zygote Fungi

2. Select one of the question headings from this section and write it below. Then write an answer to the question on the lines that follow.

Write your question here.

____

____

3. How did underlining the answers to the question heads help you learn about fungi?

____

____

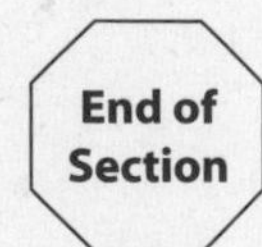

Science Online Visit **life.msscience.com** to access your textbook, interactive games, and projects to help you learn more about fungi.

Plants

section 1 An Overview of Plants

Before You Read

What are your favorite plants? Why are they your favorites?

____

____

____

What You'll Learn

- the characteristics common to all plants
- the adaptations that make it possible for plants to live on land
- how vascular and nonvascular plants are similar and different

Read to Learn

What is a plant?

Plants include trees, flowers, vegetables, and fruits. More than 260,000 plant species have been identified. Scientists expect more species will be found, mostly in tropical rain forests. Plants are important sources of food for humans. Most life on Earth would not be possible without plants.

All plants are made of cells and need water to live. Many have roots that hold them in the ground or onto an object such as a rock. Plants come in many sizes and live in almost every environment on Earth. Some grow in cold, icy regions. Others grow in hot, dry deserts.

Study Coach

Identify Answers Read each question heading aloud. When you have finished reading the section, read the question heading again. Answer the question based on what you have just read.

What are the parts of a plant cell?

Every plant cell has a cell wall, a cell membrane, a nucleus, and other cell structures. A cell wall surrounds every plant cell. The cell wall gives the plant structure and provides protection. Animal cells do not have cell walls. ☑

Many plant cells have the green pigment, or coloring, called chlorophyll (KLOR uh fihl). Most green plants use chlorophyll to make food through a process called photosynthesis. Chlorophyll is found in cell structures called chloroplasts. The green parts of a plant usually have cells that contain many chloroplasts.

Reading Check

1. **Explain** What surrounds every plant cell?

____

____

____

Central Vacuole Most of the space inside a plant cell is taken up by a large structure called the central vacuole. The central vacuole controls the water content of the cell. Many other substances also are stored in the central vacuole, including the pigments that make some flowers red, blue, or purple.

Origin and Evolution of Plants

The first land plants probably could survive only in damp areas. Their ancestors may have been green algae that lived in the sea. Green algae are one-celled or many-celled organisms that use photosynthesis to make food. Because plants and green algae have the same type of chlorophyll, they may have come from the same ancestor.

Plants do not have bones or other hard parts that can become fossils. Plants usually decay instead. But there is some fossil evidence of plants. The oldest fossil plants are about 420 million years old. Scientists hypothesize that some of these early plants evolved into the plants that live today.

Plants that have cones, such as pine trees, probably evolved from plants that lived about 350 million years ago. Plants that have flowers most likely did not exist until about 120 million years ago. Scientists do not know the exact beginning of flowering plants.

Think it Over

2. **Conclude** What do plants and green algae have in common?

____

____

____

Life on Land

Life on land has some advantages for plants. One advantage is that more sunlight and carbon dioxide are available on land than in water. Plants need sunlight and carbon dioxide for photosynthesis. During photosynthesis, plants give off oxygen. Over millions of years, as more plants grew on land, more oxygen was added to Earth's atmosphere. Because of this increase in oxygen, Earth's atmosphere became an environment in which land animals could live.

Think it Over

3. **Conclude** How would a drought affect green algae?

____

____

____

Adaptations to Land

Algae live in water or in very moist environments. Like green plants, algae make their own food through photosynthesis. To stay alive, algae need nutrients that are dissolved in the water that surrounds them. The water and dissolved nutrients enter and leave through the algae's cell membranes and cell walls. If the water dries up, the algae will die. Land plants have adaptations that allow them to conserve water.

How are land plants supported and protected?

Plants cannot live without water. Plants that live on land have adaptations that help them conserve water. The stems, leaves, and flowers of many land plants are covered with a **cuticle** (KYEW tih kul). The cuticle is a waxy, protective layer that slows the loss of water. The cuticle is a structure that helps plants survive on land. ☑

Land plants also have to be able to support themselves. The cell walls that surround all plant cells contain **cellulose** (SEL yuh lohs). Cellulose is a chemical compound that plants can make out of sugar. Long chains of cellulose molecules form fibers in plant cell walls. These fibers give the plant structure and support.

The cell walls of some plants contain other substances besides cellulose. These substances help make the plant even stronger. Trees, such as oaks and pines, could not grow without very strong cell walls. Wood from trees can be used for building because of strong cell walls.

Life on land means that each plant cell is not surrounded by water. Land plants have tubelike structures that deliver water, nutrients, and food to all plant cells. These structures also help provide support for the plant.

How do plants reproduce on land?

Land plants reproduce by forming spores or seeds. These structures can survive dryness, cold, and other harsh conditions. They grow into new plants when the environmental conditions are right.

Classification of Plants

Plants can be classified into two major groups, vascular (VAS kyuh lur) and nonvascular plants. **Vascular plants** have tubelike structures that carry water, nutrients, and other substances to all the cells of the plant. **Nonvascular plants** do not have these tubelike structures. ☑

Scientists give each plant species its own two-word name. For example, the scientific name for a pecan tree is *Carya illinoiensis* and the name for a white oak is *Quercus alba*. In the eighteenth century a Swedish scientist, Carolus Linnaeus, created this system for naming plants.

Reading Check

4. Identify the part of the plant that slows the loss of water.

____

____

____

FOLDABLES™

A Define Make a four-tab book Foldable, as shown below. List each vocabulary word on the tabs. Inside, write a complete sentence definition of the word.

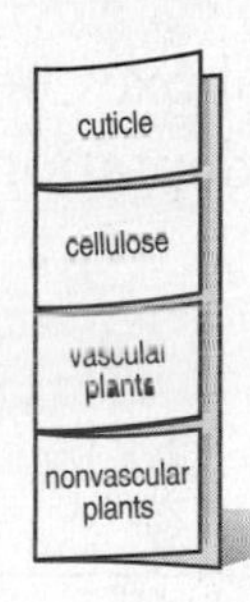

Reading Check

5. Recall the two major groups of plants.

____

____

____

After You Read

Mini Glossary

cellulose: a chemical compound that forms the walls of plants; plants make it out of sugar

cuticle: a waxy, protective layer on the surface of the plant

nonvascular plants: plants without tubelike structures; move water and other substances through the plant in other ways

vascular plants: plants that have tubelike structures to carry water, nutrients, and other substances to the cells of the plant

1. Review the terms and their definitions in the Mini Glossary. Write a sentence that explains the difference between vascular and nonvascular plants.

2. In the boxes below, describe four adaptations in plants that allow them to live on land. One adaptation is supplied for you.

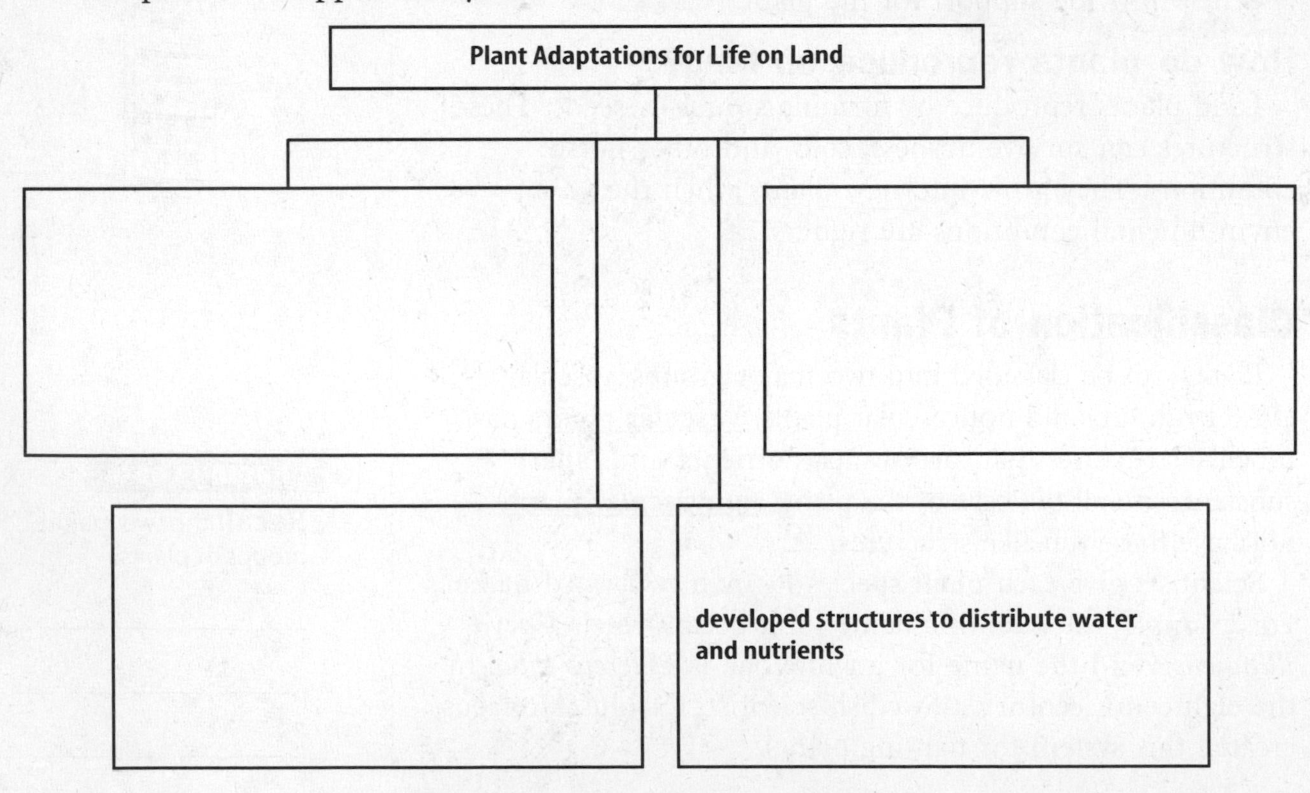

End of Section

Science Online Visit **life.msscience.com** to access your textbook, interactive games, and projects to help you learn more about plants.

Plants

section 2 Seedless Plants

Before You Read

Ferns are a type of seedless plant that people grow as house plants. What do you think you would need to do to keep a fern alive indoors?

____

____

____

____

What You'll Learn

- the differences between seedless nonvascular plants and seedless vascular plants
- the importance of some nonvascular and vascular plants

Read to Learn

Seedless Nonvascular Plants

Nonvascular plants are small and not always easy to notice. They include mosses, which you may have seen as green clumps on moist rocks or stream banks. Some other nonvascular plants are called hornworts and liverworts.

What are characteristics of seedless nonvascular plants?

Nonvascular plants do not grow from seeds. Instead, they reproduce by forming spores. They also do not have all of the parts that plants that grow from seed have. Nonvascular plants are usually only a few cells thick. They are not very tall, usually about 2 cm to 5 cm high. Nonvascular plants have structures that look like stems and leaves. Nonvascular plants do not have roots. Instead, they have **rhizoids** (RI zoydz). Rhizoids are threadlike structures that help to anchor the plants where they grow. Most nonvascular plants grow in damp places. They absorb water through their cell membranes and cell walls. ☑

Study Coach

Summarize As you read, make an outline to summarize the information in the section. Use the main headings in the section as the main headings in the outline. Complete the outline with the information under each heading in the section.

✔ Reading Check

1. **Identify** How do rhizoids help a plant?

____

____

____

FOLDABLES™

B Organize On quarter sheets of notebook paper, make notes about characteristics of vascular and nonvascular plants.

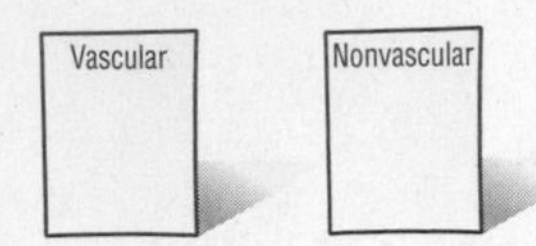

Mosses Most nonvascular plants are mosses. Mosses have green, leaflike growths arranged around a stalk. They also have rhizoids that anchor them to the ground. Moss rhizoids are made up of many cells. Mosses often grow on tree trunks, rocks, or the ground. Although most mosses live in damp places, some can live in deserts. Like all nonvascular plants, mosses reproduce by forming spores. In many moss species, a stalk grows up from the plant when it is ready to reproduce. Spores form in a cap at the top of the stalk.

Liverworts Liverworts got their name because people who lived during the ninth century used them to treat diseases of the liver. Liverworts have flattened, leaflike bodies. They usually have one-celled rhizoids.

Hornworts Hornworts have flattened, leaflike bodies like liverworts. Hornworts are usually less than 2.5 cm in diameter. Hornworts have one chloroplast in each of their cells. They get their name from the structures that produce spores, which look like tiny cattle horns.

How are nonvascular plants important?

Nonvascular plants need damp conditions to grow and reproduce. However, many species can withstand long, dry periods. Nonvascular plants can grow in thin soil and in soils where other plants cannot grow.

The spores of mosses, liverworts, and hornworts are carried by the wind. When a spore lands on the ground, it will grow into a new plant only if there is enough water and if other growing conditions are right.

Mosses, such as those pictured below, often are the first plants to grow in a new or disturbed environment, such as after a forest fire. Organisms that are the first to grow in new or disturbed areas are called **pioneer species.** As pioneer plant species die, they decay. As more and more plants grow and die, the decayed matter builds up. The decaying material and slow breakdown of rocks build soil. After enough soil is made, other organisms can move into the area.

Aaron Haupt

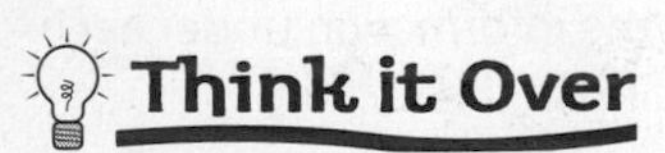

Think it Over

2. Draw Conclusions Why are mosses pioneer species?

Seedless Vascular Plants

Both ferns and mosses reproduce by spores instead of seeds. But ferns are different from mosses because ferns have vascular tissues. Their long, tubelike cells carry water, minerals, and food to cells throughout the plant. Vascular plants can grow larger and thicker than nonvascular plants because the vascular tissue carries water and nutrients to all plant cells. ☑

What are the types of seedless vascular plants?

Seedless vascular plants include ferns, ground pines, spike mosses, and horsetails. Many species of seedless vascular plants are known only from fossils because they are now extinct. These plants covered much of Earth 360 million to 286 million years ago.

What are ferns?

Ferns are the largest group of seedless vascular plants. Ferns have stems, leaves, and roots. Fern leaves are called fronds as shown in the figure to the right. Spores form in structures found on the underside of the fronds. Although thousands of species of ferns are found on Earth today, many more species existed long ago. Scientists have used clues from rock layers to learn that 360 million years ago much of Earth was covered with steamy swamps. The tallest plants were species of ferns that grew as tall as 25 m. The tallest ferns today are 3 m to 5 m tall and grow in tropical areas.

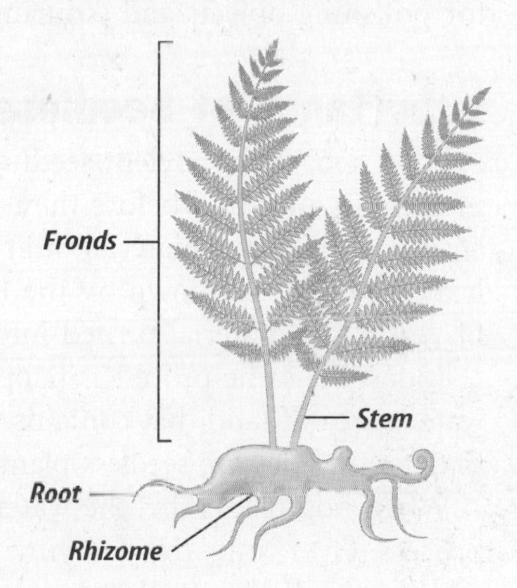

What are club mosses?

Ground pines and spike mosses are groups of plants that often are called club mosses. Club mosses are more closely related to ferns than to mosses. Club mosses have needle-like leaves. Their spores form at the end of the stems in structures that look like tiny pinecones. Ground pines grow in cold and hot areas. Ground pines are endangered in some places. They have been over-collected to make decorations such as wreaths.

☑ Reading Check

3. Explain How is having vascular tissue an advantage for plants?

____

____

____

Picture This

4. Identify Circle the name of the structure where spores are found.

Spike mosses look a lot like ground pines. One species of spike moss, the resurrection plant, lives in desert areas. When there is not enough water, the plant curls up and looks dead. When water becomes available, the resurrection plant unfolds its green leaves and begins making food again. The plant can curl up again whenever conditions make it necessary.

How are horsetails different from other vascular plants?

Horsetails have a stem structure that is different from other vascular plants. The stem has a hollow center surrounded by a ring of vascular tissue. The stem also has joints. Leaves grow out around the stem at each joint. Horsetail spores form in conelike structures at the tips of some stems. The stems of horsetails contain silica, a gritty substance found in sand. In the past, horsetails were used for polishing objects and scouring cooking utensils. ☑

Reading Check

5. **Explain** How do horsetails differ from other vascular plants?

____

____

____

Importance of Seedless Plants

Long ago, when ancient seedless plants died, they sank into water and mud before they decayed. Over time, many layers of this plant material built up. Top layers became heavy and pressed down on the layers below. Over millions of years, this material turned into coal.

Today, the same process is happening in bogs. A bog is a watery area of land that contains decaying plants. Most plants that live in bogs are seedless plants like mosses and ferns.

When bog plants die, the watery soil slows the decaying process. Over time, the decaying plants are pressed into a substance called peat. Peat is mined from bogs to use as a low-cost fuel in places such as Ireland and Russia. Scientists hypothesize that over time, if the peat remains in the bog, it will become coal.

How are seedless vascular plants used?

Peat is used to enrich garden soil. Many people keep ferns as houseplants. Ferns also are sold as landscape plants for shady outdoor areas. Ferns sometimes are woven into baskets.

The rhizomes and fronds of some ferns can be eaten. The dried stems of one kind of horsetail can be ground into flour. Some seedless plants have been used as medicines for hundreds of years. For example, ferns have been used to treat bee stings, burns, and fevers. ☑

Reading Check

6. **Identify** two ways seedless plants are used.

____

____

____

● After You Read

Mini Glossary

pioneer species: organisms that are the first to grow in new or disturbed areas

rhizoid: threadlike structures that anchor nonvascular plants

1. Review the terms and their definitions in the Mini Glossary. Write a sentence to explain the importance of pioneer species to the environment.

____

____

2. Complete the Venn diagram below to help you compare nonvascular and vascular seedless plants. Include phrases that describe how the plant cells get nutrients and how the plants reproduce.

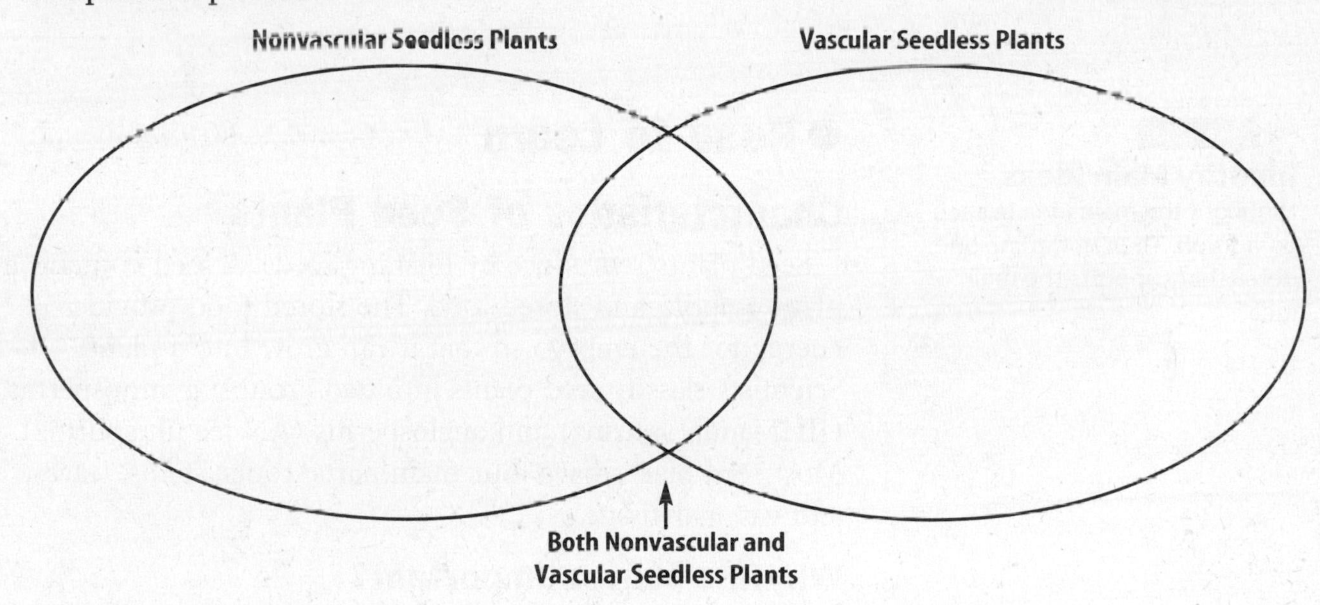

3. How did summarizing the information in this section help you learn about nonvascular and vascular seedless plants?

____

____

____

Visit **life.msscience.com** to access your textbook, interactive games, and projects to help you learn more about seedless plants.

Plants

section 3 Seed Plants

What You'll Learn
- the characteristics of seed plants
- how roots, stems, and leaves function
- the characteristics of gymnosperms and angiosperms
- how monocots and dicots are different

Before You Read

What are your favorite fruits? Where do these fruits come from?

Study Coach

Identify Main Ideas Highlight the main idea in each paragraph. Then underline one detail that supports the main idea.

FOLDABLES™

C Classify Make a three-tab Foldable to write notes about the importance of plant leaves, stems, and roots.

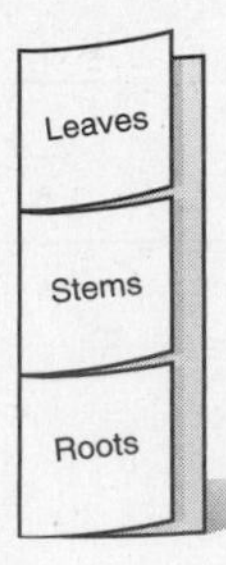

Read to Learn

Characteristics of Seed Plants

Seed plants reproduce by forming seeds. A seed contains a plant embryo and stored food. The stored food provides energy for the embryo so that it can grow into a plant. Scientists classify seed plants into two groups: gymnosperms (JIHM nuh spurmz) and angiosperms (AN jee uh spurmz). Most seed plants have four main parts: roots, stems, leaves, and vascular tissue.

Why are leaves important?

The leaves of seed plants are the organs where food is made. The food-making process is called photosynthesis. Leaves come in many shapes, sizes, and colors.

What are the cell layers of a leaf?

A leaf has several layers of cells. A thin layer of cells called the epidermis covers and protects the top and bottom of the leaf. The epidermis of some leaves is covered with a waxy cuticle. Most leaves have small openings in the epidermis called **stomata** (STOH muh tuh) (singular, *stoma*). The stomata allow carbon dioxide, water, and oxygen to enter and exit the leaf. **Guard cells** located around each stoma open and close the stoma.

The palisade layer of a leaf is located just below the upper epidermis. This layer has long, narrow cells that contain chloroplasts. Plants make most of their food in the palisade cells.

The spongy layer is found between the palisade layer and the lower epidermis. The spongy layer is made of loosely arranged cells separated by air spaces. The veins of a leaf are made of vascular tissue and are located in the spongy layer. All the parts of the leaf can be seen in the figure below.

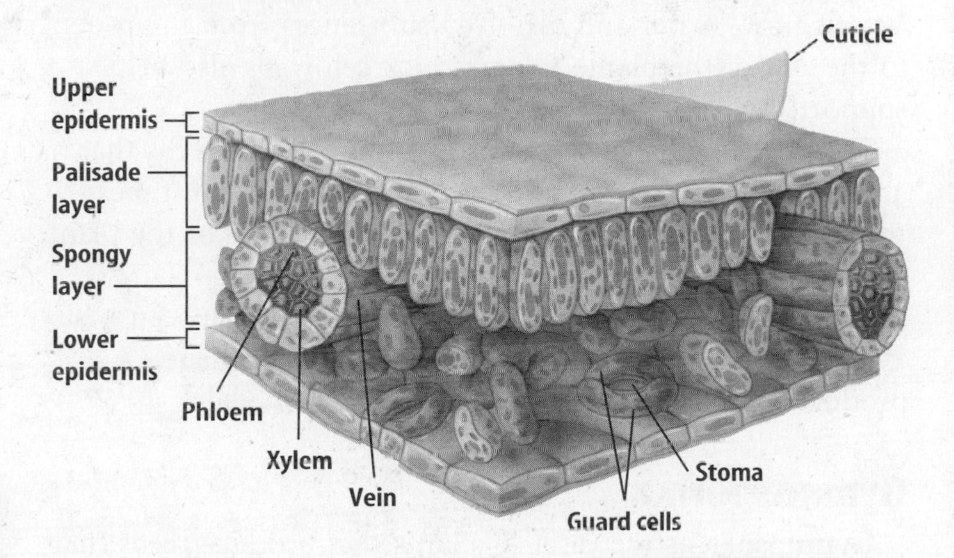

Picture This

1. Identify Color in blue the plant layer that contains the chloroplasts. Color in red the plant layer that protects the leaf. Finally, underline the name of the part of the leaf that allows carbon dioxide, water, and oxygen to enter and exit the leaf.

What is the purpose of a plant's stem?

Plant stems are usually found above the ground. They support the branches, leaves, and reproductive structures of the plant. Materials move between the leaves and roots through vascular tissues in the stem. The stems of some plants also store food and water.

Plant stems can be woody or herbaceous (hur BAY shus). Herbaceous stems are soft and green, like those of a tulip. Woody stems are hard and rigid, like those of trees and shrubs. The trunk of a tree is a stem.

What do plant roots do?

The root system of most plants is the largest part of the plant. Roots contain vascular tissue. Water and dissolved substances move from the soil into the roots, and on up through the stems to the leaves. Roots also anchor plants and prevent them from being blown or washed away. Roots support the parts of the plant that are above ground—the stem, branches, and leaves. ☑

Reading Check

2. Identify two things roots do for a plant.

____

____

____

Roots can store food and water. They can take in oxygen that the plant needs for the process of respiration. For plants that grow in water, part or all of a plant's roots may grow above ground. Water does not have as much oxygen as air. The roots take in more oxygen from the air.

What are vascular tissues made of?

The vascular system in a seed plant contains three kinds of tissue—xylem, phloem, and cambium. **Xylem** (ZI lum) tissue is made of hollow, tubelike cells that are stacked one on top of the other to form a structure called a vessel. Vessels move water and dissolved substances from the roots to the rest of the plant. Xylem's thick cell walls also help support the plant.

Phloem (FLOH em) tissue is made of tubelike cells that are stacked to form structures called tubes. Phloem tubes move food from where it is made to other parts of the plant where the food is used or stored. ☑

Some plants have a layer of cambium tissue between xylem and phloem. **Cambium** (KAM bee um) tissue produces most of the new xylem and phloem cells.

Reading Check

3. **Describe** What does phloem tissue do?

____

____

____

Gymnosperms

Gymnosperms are vascular plants that produce seeds that are not protected by a fruit. Gymnosperms do not have flowers. The leaves of gymnosperms are usually shaped like needles or scales. Many gymnosperms are called evergreens because some green leaves always stay on their branches.

The gymnosperms are divided into four divisions. These four divisions are conifers, cycads, ginkgoes, and gnetophytes (NE tuh fites). The conifers are the most familiar gymnosperm division. Pines, firs, spruces, redwoods, and junipers are conifers. Conifers produce two types of cones—male and female. Seeds develop only on the female cone.

Angiosperms

An **angiosperm** is a vascular plant that forms flowers and produces one or more seeds that are protected inside a fruit. Peaches, apples, and tulips are examples of angiosperms. Angiosperms are common in all parts of the world. More than half of all known plant species are angiosperms.

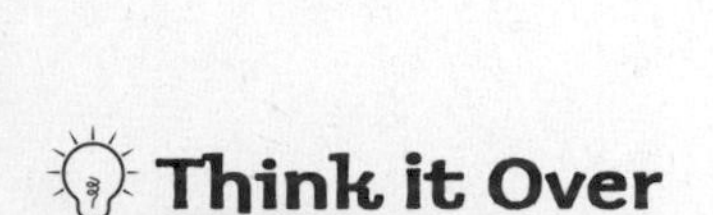

4. **Compare** What is the difference between gymnosperms and angiosperms?

____

____

____

What are the flowers of angiosperms like?

The flowers of angiosperms come in different shapes, sizes, and colors. Some parts of a flower grow into a fruit. Most fruits have seeds inside, like an apple. Some fruits have seeds on the surface, like a strawberry. Angiosperms are divided into two groups—monocots and dicots.

How do monocots and dicots differ?

A cotyledon (kah tul EE dun) is the part of a seed that stores food for the new plant. **Monocots** are angiosperms that have one cotyledon inside their seeds. **Dicots** are angiosperms that have two cotyledons inside their seeds.

Many foods come from monocots, including corn, rice, and wheat. Bananas and pineapples also are monocots. Familiar foods such as peanuts, peas, and oranges come from dicots. Most shade trees, such as oaks and maples, are dicots.

FOLDABLES™

D Compare Make notes listing the characteristics of monocots and dicots in a two-tab Foldable. Include ways in which humans use each.

Monocots | Dicots

What is the life cycle of an angiosperm?

All organisms have life cycles—a beginning and an end. The angiosperm's life cycle begins with the seed and ends when the mature plant flowers and/or produces seed. Some angiosperms grow from seeds to maturity in less than a month. Some plants take as long as 100 years to grow from seed to maturity. Plants that complete their life cycles in one year are called annuals. Annuals must be grown from new seeds each year.

Plants that complete their life cycles in two years are called biennials (bi EH nee ulz). Biennials produce flowers and seeds only during the second year of growth. Angiosperms with life cycles that take longer than two years are called perennials. Most trees and shrubs are perennials.

Importance of Seed Plants

Gymnosperms are used for many purposes. Conifers are the most commonly used gymnosperm. Most of the wood used in building comes from conifers. Resin used to make chemicals found in soap, paint, and varnish also comes from conifers. ☑

Angiosperms are widely used by humans. Many of the foods you eat come from seed plants. Angiosperms are the source of many of the fibers used in making clothes. Paper is made from wood pulp that comes from trees. Desks and chairs are made from wood.

Reading Check

5. Explain Why are conifers important to the economy?

____

____

____

After You Read

Mini Glossary

angiosperm: vascular plant that flowers and produces one or more seeds inside a fruit

cambium: plant tissue that produces most of the new xylem and phloem cells

dicot: angiosperm that has two cotyledons inside its seeds

guard cells: cells that surround a stoma and open and close it

gymnosperm: vascular plant that produces seeds that are not protected by fruit

monocot: angiosperm that has one cotyledon inside its seeds

phloem: plant tissue made up of tubelike cells that are stacked to form tubes; tubes move food from where it is made to parts of the plant where it is used

stomata: small openings in the epidermis of the leaf

xylem: plant tissue made up of hollow, tubelike cells that are stacked one on top of the other to form vessels; vessels transport water and dissolved substances from the roots to all other parts of the plant

1. Review the terms and their definitions in the Mini Glossary. Write two sentences that explain what xylem and phloem do.

____

____

____

2. Complete the chart below to list the four main parts of seed plants and describe what they do.

Parts of Seed Plants	What They Do

Visit **life.msscience.com** to access your textbook, interactive games, and projects to help you learn more about seed plants.

Plant Reproduction

section 1 Introduction to Plant Reproduction

Before You Read

List four things you need to survive. Then circle the items on your list that you think plants also need to survive.

____

____

____

What You'll Learn
- the differences between the two types of plant reproduction
- the two stages in a plant's life cycle

Read to Learn

Mark the Text

Identify Main Ideas Underline the important ideas in this section. Review these ideas as you study the section.

Types of Reproduction

What do humans and plants have in common? Both need water, oxygen, energy, and food to grow. Like humans, plants reproduce and make similar copies of themselves. Most plants can reproduce in two different ways—by sexual reproduction and by asexual reproduction.

What happens in asexual plant reproduction?

Asexual reproduction does not require the production of sex cells. Instead, one organism produces a new organism that is genetically identical to it. Under the right conditions, an entire plant can grow from one leaf or part of a stem or root. When growers use these methods to start new plants, they must make sure that the plant part has plenty of water and anything else it needs to survive. The stems of lawn grasses grow underground and produce new grass plants asexually along the length of the stem.

Think it Over

1. **Analyze** A cutting from a plant can be placed in water and roots grow. Is this an example of asexual or sexual reproduction? Explain your answer.

____

____

____

What is sexual plant reproduction?

Sexual reproduction in plants requires the production of sex cells—usually called sperm and eggs—in reproductive organs. The organism produced by sexual reproduction is genetically different from either parent organism.

Fertilization An important part of sexual reproduction is fertilization. Fertilization happens when a sperm and egg combine to produce the first cell of the new organism, the zygote. In plants, water, wind, or animals help bring the sperm and the egg together.

What reproductive organs do plants have?

A plant's female reproductive organs produce eggs. The male reproductive organs produce sperm. Some plants have both reproductive organs. A plant with both reproductive organs can usually reproduce by itself. Other plants have either female or male reproductive organs. For fertilization to happen, the male and female plants must be near each other.

Plant Life Cycles

A plant has a life cycle with two stages—the gametophyte (guh MEE tuh fite) stage and the sporophyte (SPOHR uh fite) stage. The figure below shows the two stages.

FOLDABLES™

A Explain Make a shutterfold book, as shown below. Explain the two stages of the plant life cycle.

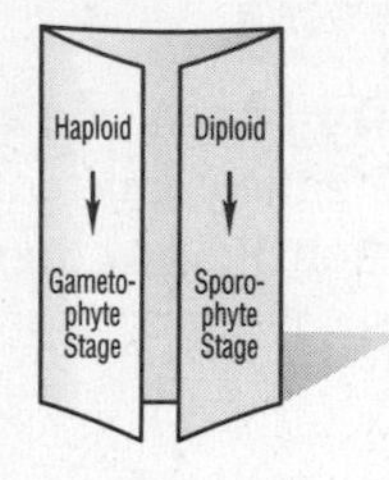

Gametophyte Stage When reproductive cells undergo meiosis and produce haploid cells called **spores**, the **gametophyte stage** begins. Spores divide by cell division to form plant structures or an entire new plant. The cells in these structures or plants are haploid and have half a set of chromosomes. Some of these cells undergo cell division and form sex cells.

Sporophyte Stage Fertilization—the joining of haploid sex cells—begins the **sporophyte stage**. Cells formed in this stage are diploid and have the full number of chromosomes. Meiosis in some of these cells forms spores, and the cycle repeats.

Picture This

2. **Identify** Write the name of the stage that begins with meiosis below the word *Meiosis*. Write the name of the stage that begins with fertilization above the word *Fertilization*.

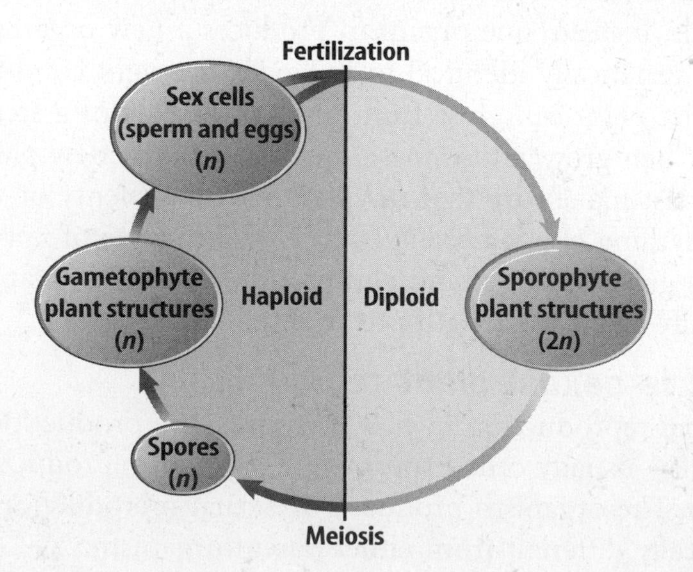

After You Read

Mini Glossary

gametophyte (guh MEE tuh fite) stage: the stage in plant reproduction when reproductive cells undergo meiosis

spores: haploid cells produced in the gametophyte stage

sporophyte (SPOHR uh fite) stage: the stage in plant reproduction when fertilization begins

1. Review the terms and their definitions in the Mini Glossary. Write a sentence that explains the difference between the gametophyte stage and the sporophyte stage.

____

____

____

2. Choose one of the question headings in the Read to Learn section. Write the question in the space below. Then write your answer to that question on the lines that follow.

Write your question here

____

____

____

3. Fill in the table below with either "yes" or "no" to compare asexual and sexual reproduction in plants.

	Asexual Reproduction	Sexual Reproduction
a. Requires production of sex cells?		
b. Produces organism that is genetically identical to parent?		
c. Requires fertilization?		

ScienceOnline Visit **life.msscience.com** to access your textbook, interactive games, and projects to help you learn more about the basics of plant reproduction.

chapter 10

Plant Reproduction

section 2 Seedless Reproduction

What You'll Learn

- the life cycles of a moss and a fern
- why spores are important to seedless plants
- the structures ferns use for reproduction

Before You Read

Describe what happens when you suddenly open a bag of candy or a bag of chips.

____

____

Study Coach

Read and Discuss the main topics of the section with a partner. Read a paragraph and then take turns saying something about what you have learned. Continue until you both understand the main ideas.

Read to Learn

The Importance of Spores

If you want to grow ferns and moss plants, you can't go to a garden store and buy a package of seeds. Ferns and moss plants don't produce seeds. They reproduce by forming spores. The sporophyte stage of these plants produces haploid spores in structures called spore cases.

If you break open a bag of candy, the candy may spill out. In the same way when a spore cases break open, the spores spill out and are spread by wind or water. The spores can grow into plants that will produce sex cells. Seedless plants include all nonvascular plants and some vascular plants.

Nonvascular Seedless Plants

A nonvascular plant does not have structures that transport water and materials throughout the plant. Instead, water and materials move from cell to cell. Mosses, liverworts, and hornworts are all nonvascular plants. They cover the ground or grow on fallen logs in damp, shaded forests.

The sporophyte stage of most nonvascular plants is very small. Moss plants have a life cycle typical of how sexual reproduction occurs in this plant group.

FOLDABLES™

B Compare Make a two-tab concept map Foldable. Write the facts related to all nonvascular plants and some vascular plants.

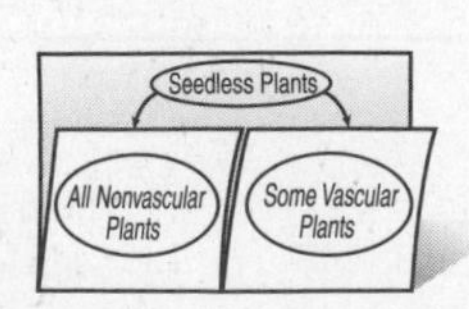

How do nonvascular plants reproduce sexually?

A moss is a green, low-growing plant when in the gametophyte stage. When brownish stalks grow up from the tip of the plant, moss is in the sporophyte stage. The sporophyte stage does not carry on photosynthesis. It depends on the gametophyte for nutrients and water. On the tip of the stalk is a tiny spore case where millions of spores have been produced. Under the right environmental conditions, the spore case opens and the spores are released.

New moss gametophytes can grow from each spore and the cycle repeats. This process is shown in the figure below.

Picture This

1. **Explain** Highlight each of the following words in the captions of the figure: *meiosis, gametophyte, fertilization, sporophyte.* Use each of these words as you explain to a partner the life cycle of a moss.

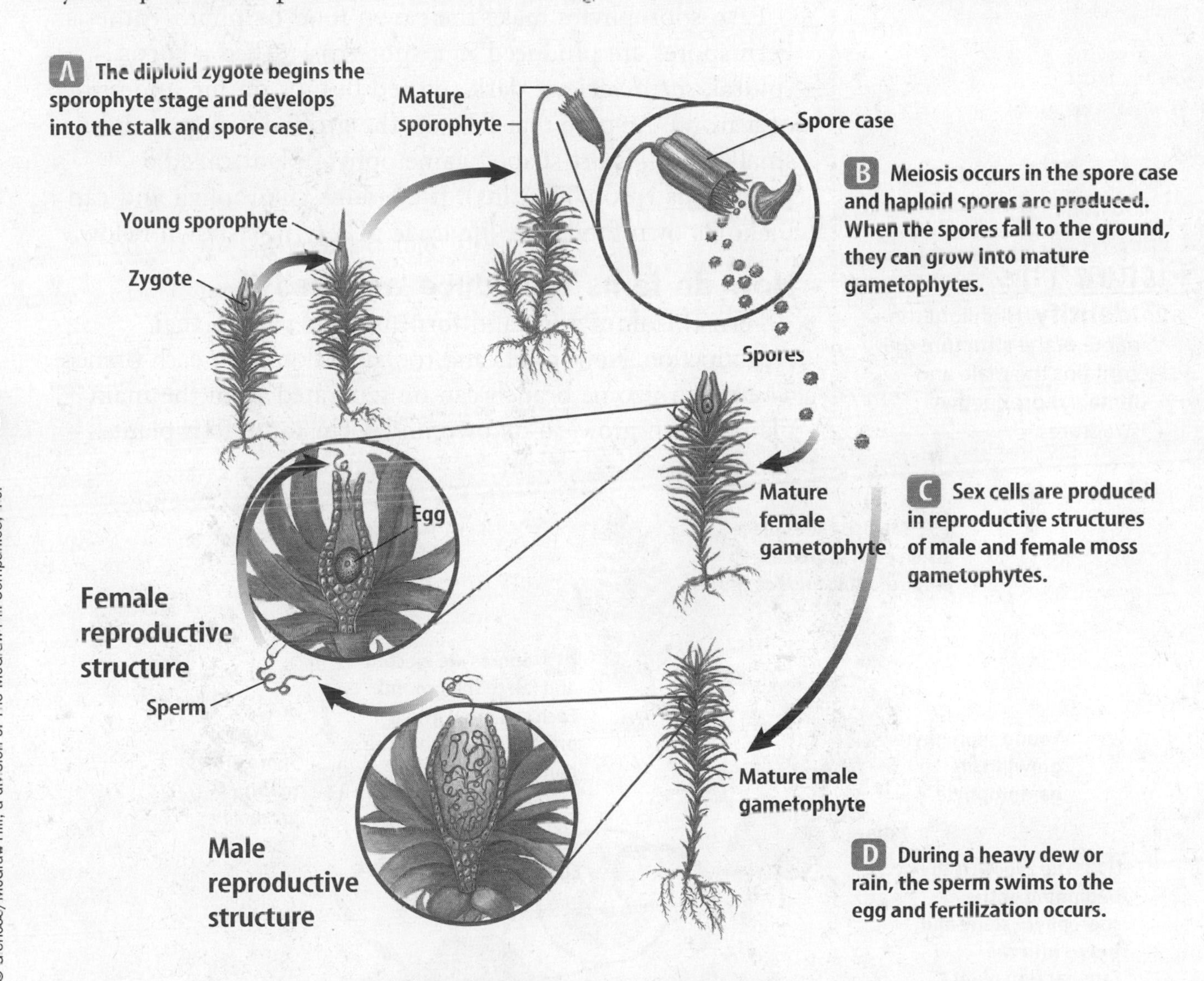

A The diploid zygote begins the sporophyte stage and develops into the stalk and spore case.
Mature sporophyte
Spore case
B Meiosis occurs in the spore case and haploid spores are produced. When the spores fall to the ground, they can grow into mature gametophytes.
Young sporophyte
Zygote
Spores
Egg
Mature female gametophyte
C Sex cells are produced in reproductive structures of male and female moss gametophytes.
Female reproductive structure
Sperm
Mature male gametophyte
Male reproductive structure
D During a heavy dew or rain, the sperm swims to the egg and fertilization occurs.

How do nonvascular plants reproduce asexually?

If a piece of moss gametophyte plant breaks off, it can grow into a new plant. Liverworts can form small balls of cells on the surface of the gametophyte plant. These can be carried away by water and grow into new gametophyte plants.

Vascular Seedless Plants

Vascular plants have tubelike cells that transport water and materials throughout the plant. Most vascular seedless plants are ferns. Horsetails and club mosses are other vascular seedless plants. Unlike nonvascular plants, the gametophyte of a vascular seedless plant is the part that is small.

How do ferns reproduce sexually?

A fern leaf is called a **frond** and grows from an underground stem called a **rhizome**. Roots grow from the rhizome. Roots anchor the plant and absorb water and nutrients.

Fern sporophytes make their own food by photosynthesis. Fern spores are produced in a spore case called a **sorus** (plural, *sori*). Sori are dark colored bumps on the underside of a frond. A spore that falls on the ground can grow into a small, green, heart-shaped gametophyte plant called a **prothallus** (proh THA lus). It contains chlorophyll and can make its own food. The life cycle of a fern is shown below.

How do ferns reproduce asexually?

Fern rhizomes grow and form branches in asexual reproduction. New fronds and roots develop from each branch. The new rhizome branch can be separated from the main plant. It can grow on its own and form more fern plants.

Picture This

2. Identify Highlight the name of the structure that contains the male and female reproductive structures.

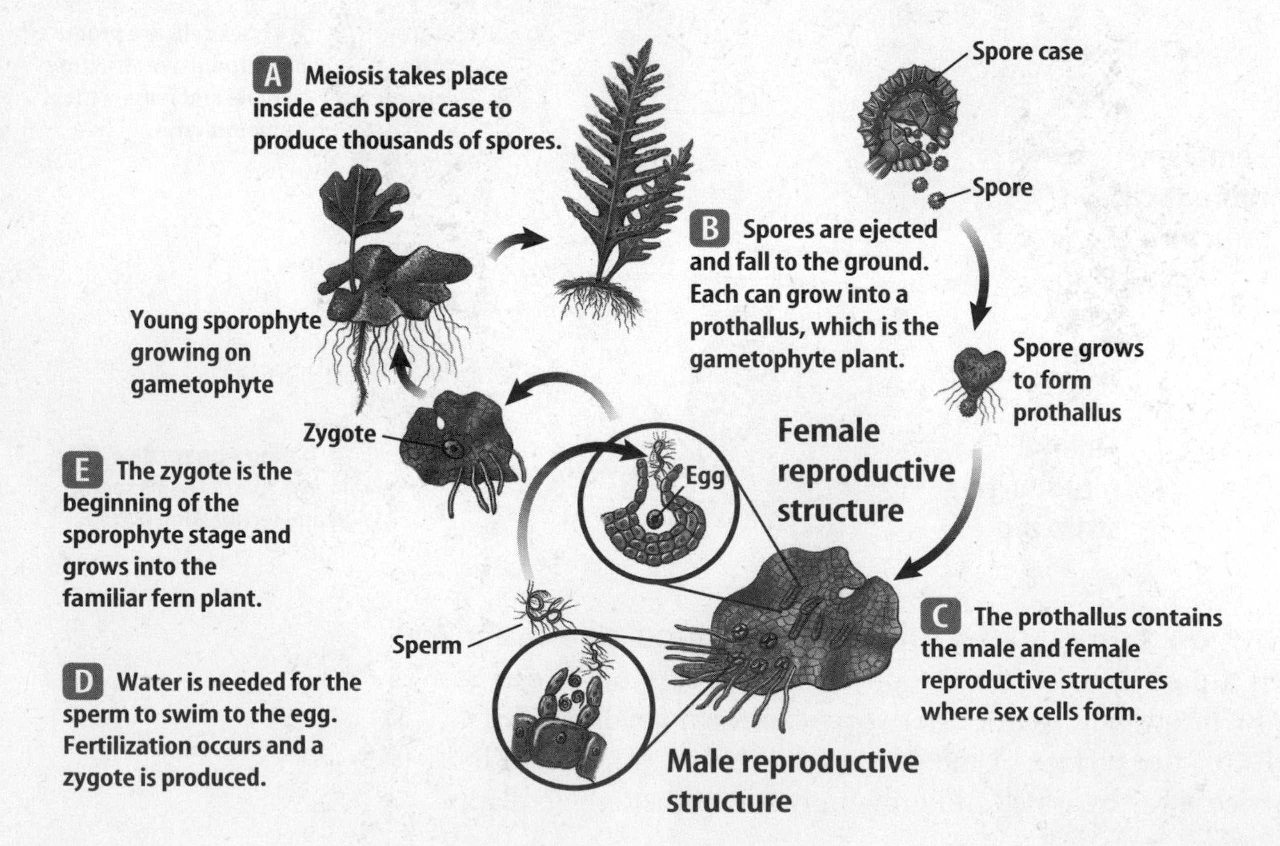

A Meiosis takes place inside each spore case to produce thousands of spores.
Spore case
Spore
B Spores are ejected and fall to the ground. Each can grow into a prothallus, which is the gametophyte plant.
Spore grows to form prothallus
Young sporophyte growing on gametophyte
Zygote
Egg
Female reproductive structure
E The zygote is the beginning of the sporophyte stage and grows into the familiar fern plant.
Sperm
C The prothallus contains the male and female reproductive structures where sex cells form.
D Water is needed for the sperm to swim to the egg. Fertilization occurs and a zygote is produced.
Male reproductive structure

After You Read

Mini Glossary

frond: a fern leaf

prothallus (proh THA lus): a small, green, heart-shaped gametophyte fern plant

rhizome: underground stem of a fern

sorus: a spore case where fern spores are produced

1. Review the terms and their definitions in the Mini Glossary. Write a sentence that explains the relationship between a frond and a sorus.

____

____

2. Complete the flow chart below to show the life cycle of a moss.

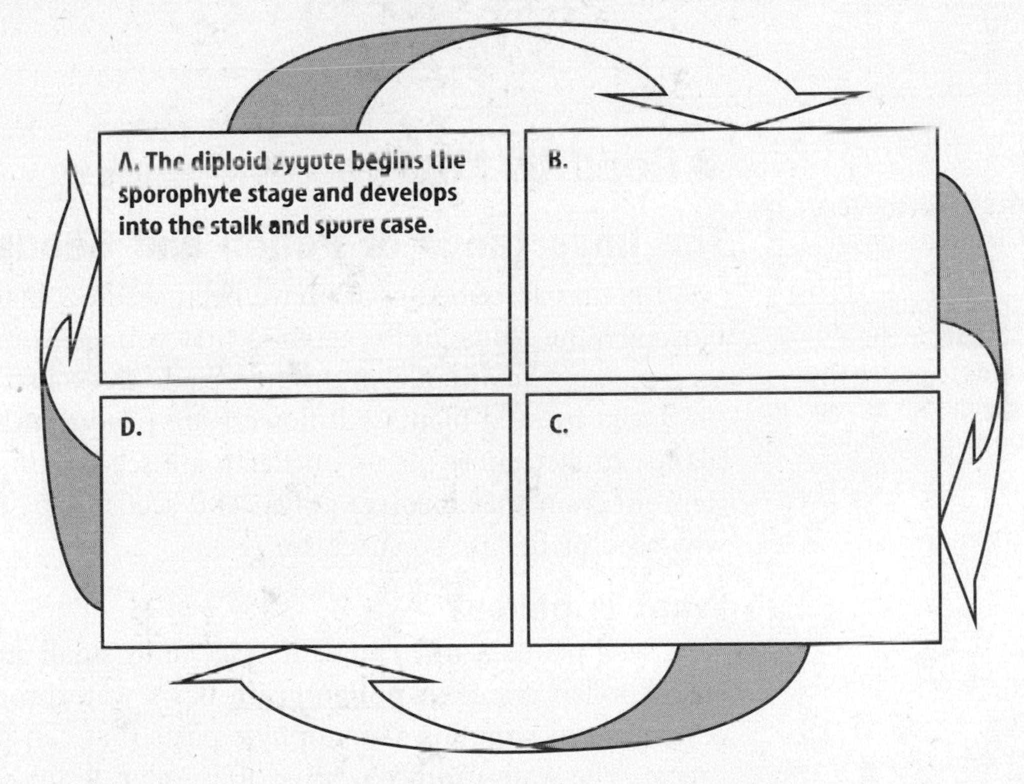

A. The diploid zygote begins the sporophyte stage and develops into the stalk and spore case.
B.
C.
D.

3. How does discussing what you have read help you remember the important ideas?

____

____

Visit **life.msscience.com** to access your textbook, interactive games, and projects to help you learn more about seedless reproduction in plants.

chapter 10

Plant Reproduction

section 3 Seed Reproduction

What You'll Learn

- the life cycles of most gymnosperms and angiosperms
- the structure and function of the flower
- the ways seeds are scattered

Before You Read

On the lines below, write the names of three fruits or vegetables. Next to each name, describe its seed.

____

____

____

Study Coach

Define Terms Skim the text and write each key term on an index card. As you read the section, write the definition of each term on another index card. Use the cards to match the terms to their definitions as you review the important words in this section.

Read to Learn

The Importance of Pollen and Seeds

All plants described so far have been seedless plants. However, the fruits and vegetables that you eat come from seed plants. Oak, maple, and other shade trees also are produced by seed plants. All flowers are produced by seed plants. In fact, most plants on Earth are seed plants. Reproduction that involves pollen and seeds helps explain why seed plants are so successful.

What is pollen?

In seed plants, some spores develop into small structures called pollen grains. A **pollen grain** has a waterproof covering and contains gametophyte parts that can produce sperm. The waterproof covering of a pollen grain can be used to identify the plant that the pollen grain came from.

The sperm of seed plants are carried as part of the pollen grain by gravity, wind, water, or animals. The transfer of pollen grains to the female part of the plant is called **pollination**. ☑

After the pollen grain reaches the female part of the plant, a pollen tube is produced. The sperm moves through the pollen tube, then fertilization can happen.

☑ Reading Check

1. **Identify** one way a pollen grain reaches the female part of the plant.

____

What are the three main parts of a seed?

After fertilization, the female part can develop into a seed. As shown in the figure below, a seed has three main parts, an embryo, stored food, and a protective seed coat. The embryo will grow to become the plant's stem, leaves, and roots. The stored food gives the embryo energy when it begins to grow into a plant. Because a seed contains an embryo and stored food, a new plant develops faster from a seed than from a spore.

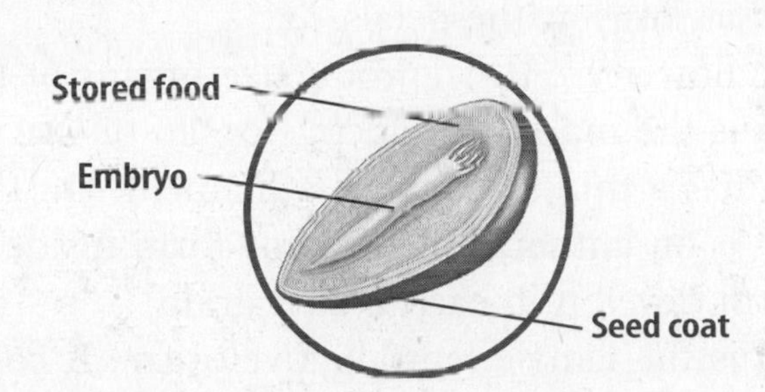

Picture This

2. Infer Circle the part of the seed that will become a plant.

Gymnosperms and Angiosperms Gymnosperms (JIHM nuh spurmz) and angiosperms are the two groups of seed plants. In gymnosperms, seeds usually develop in cones. In angioperms, seeds develop in flowers and fruit.

Gymnosperm Reproduction

Cones are the reproductive structures of gymnosperms. Gymnosperm plants include pines, firs, cedars, cycads, and ginkgoes. Each kind of gymnosperm has a different cone.

A pine tree is a gymnosperm. The way pines produce seeds is typical of most gymnosperms. The pine is a sporophyte plant that produces both male cones and female cones. Male and female gametophyte structures are produced in the cones, but they are very small. A mature female cone is made up of woody scales on a short stem. At the base of each scale are two ovules. The egg is produced in the **ovule**. Pollen grains are produced in the smaller male cones. In the spring, clouds of pollen are released from the male cones.

How are gymnosperm seeds produced?

Pollen is carried from the male cones to the female cones by the wind. The pollen must land between the scales of a female cone to be useful. There it can be trapped in the sticky fluid given off by the ovule. If the pollen grain and the female cone are the same species, fertilization can take place. It can take from two to three years for the seed to develop.

FOLDABLES™

C Categorize Make a folded chart, as shown below. Write facts in each block to describe the reproduction of gymnosperms and angiosperms.

Seed Reproduction	Reproductive Organs	Seeds
Gymnosperm Reproduction		
Angiosperm Reproduction		

Reading Check

3. **Identify** the four main parts of a flower.

____

____

____

____

Angiosperm Reproduction

Most seed plants are angiosperms. All angiosperms have flowers, which are the reproductive organs. Flowers have gametophyte structures that produce sperm or eggs for sexual reproduction.

Most flowers have four main parts—petals, sepals, stamen, and pistil, as shown in the figure below. The petals usually are the most colorful parts of the flower. Sepals often are small, green, leaflike parts. In some flowers, the sepals are as colorful and as large as the petals. ☑

Inside the flower are the reproductive organs of the plant. The **stamen** is the male reproductive organ of the plant. The stamen has a thin stalk called a filament. On the end of the filament is an anther. Pollen grains form inside the anther. Sperm develop in each pollen grain.

The **pistil** is the female reproductive organ. It consists of a stigma, a long stalklike style, and an ovary. Pollen grains land on the stigma and move down the style to the ovary. The **ovary** is the swollen base of the pistil where the ovules are found. Eggs are produced in the ovules. Not all flowers have both male and female reproductive parts.

Picture This

4. **Locate** Highlight the male reproductive structure. Circle the female reproductive structure.

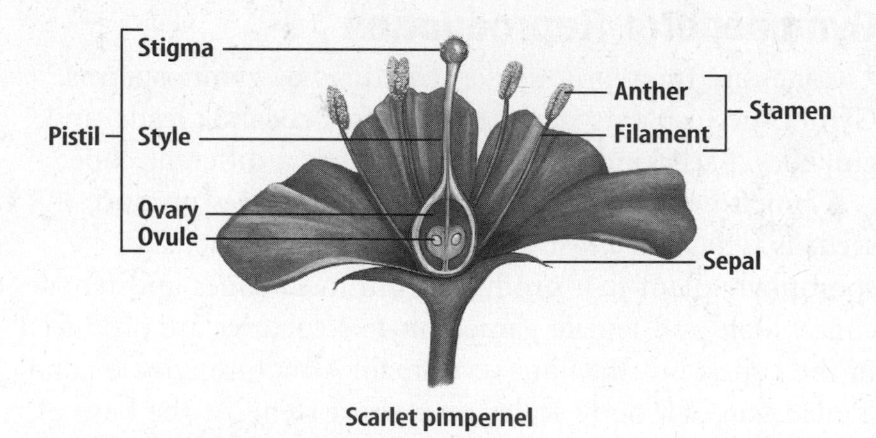

Scarlet pimpernel

How is pollen spread?

Insects and other animals eat the flower, its nectar, or pollen. As insects and other animals move about the flower, they get pollen on their body parts. These animals spread the flower's pollen to other plants they visit. Some flowers depend on the wind, rain, or gravity to spread their pollen. Following pollination and fertilization, the ovules of flowers can develop into seeds.

How do angiosperm seeds develop?

A flower is pollinated when pollen grains land on a pistil. A pollen tube grows from the pollen grain. The pollen tube enters the ovary and reaches an ovule. The sperm then travels down the pollen tube and fertilizes the egg in the ovule. A zygote forms and grows into a plant embryo. ☑

Parts of the ovule develop into the stored food and the seed coat that surround the embryo, and a seed is formed. The seeds of some plants, like beans and peanuts, store food in structures called cotyledons. The seeds of other plants, like corn and wheat, store food in a tissue called endosperm.

Reading Check

5. Identify Where is the egg fertilized?

____

____

Seed Dispersal

Some plant seeds are spread by gravity. They fall off the parent plant. Other seeds have attached structures, like wings or sails, which help the wind carry them.

Some seeds are eaten by animals and spread after the seeds are digested. Other seeds are stored or buried by animals. Raindrops can knock seeds out of dry fruit. Some fruits and seeds float on flowing water or ocean currents.

What is germination?

A series of events that results in the growth of a plant from a seed is called **germination**. Seeds will not germinate until the environmental conditions are right. Conditions that affect germination include temperature, light, water, and oxygen. Germination begins when seed tissues absorb water. This causes the seed to get larger and the seed coat to break open.

As you can see in the figure below, a root eventually grows from the seed. Then a stem and leaves grow. Once the plant grows above the soil, photosynthesis begins. Photosynthesis provides food as the plant continues to grow.

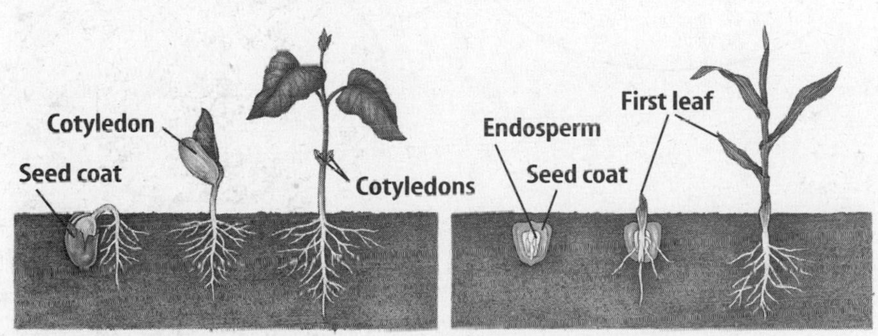

In beans, the cotyledons rise above the soil. As the stored food is used, the cotyledons shrivel and fall off.

In corn, the stored food in the endosperm remains in the soil and is gradually used as the plant grows.

Picture This

6. Locate Highlight the structure in each plant that stores food.

After You Read

Mini Glossary

germination: a series of events that result in the growth of a plant from a seed

ovary: the swollen base of the pistil where ovules are found

ovule: the place where eggs are produced

pistil: the female reproductive organ in the flower of an angiosperm

pollen grain: a small structure in seed plants that has a waterproof covering and that contains gametophyte parts that can produce sperm

pollination: the transfer of pollen grains to the female part of the plant

stamen: the male reproductive organ in the flower of an angiosperm

1. Review the terms and their definitions in the Mini Glossary. Write a sentence that describes either the male or the female reproductive organs of a flower.

___

___

___

2. Complete the concept web below to identify the ways seeds are spread.

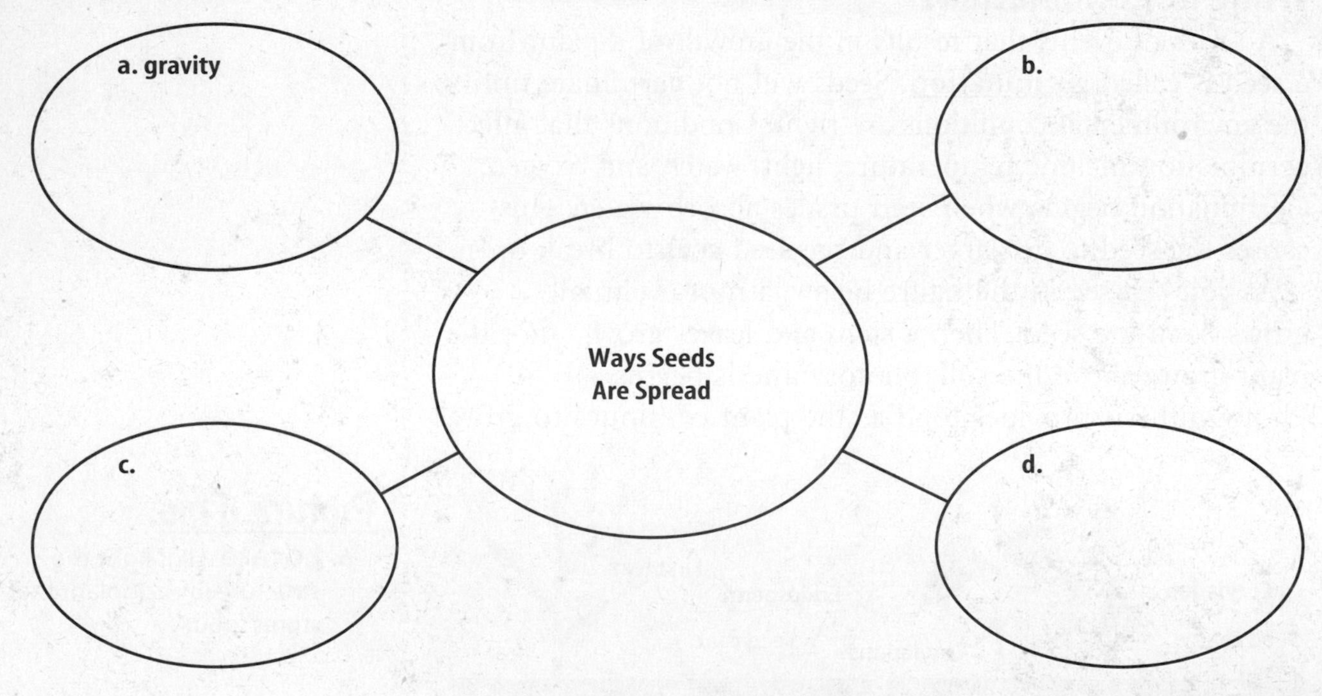

End of Section

Science Online Visit **life.msscience.com** to access your textbook, interactive games, and projects to help you learn more about seed reproduction in plants.

Plant Processes

section 1 Photosynthesis and Respiration

Before You Read

Name the parts of a plant that you have seen recently. For one of the parts, describe its function.

___

___

___

What You'll Learn

- how plants take in and give off gases
- the differences and similarities between photosynthesis and respiration
- why photosynthesis and respiration are important

Read to Learn

Study Coach

Summarize Main Ideas Read the section. Recall and write down the main ideas. Go back and check the main ideas to make sure they are accurate. Then use your notes to summarize the main ideas of this section.

Taking in Raw Materials

Plants make their own food using the raw materials water, carbon dioxide, and inorganic chemicals in the soil. Plants also produce wastes.

Which plant structures move water into the plant?

The figure below shows the plant structures that take in raw materials. Most of the water used by plants is taken in through the roots and moves through the plant to where it is used.

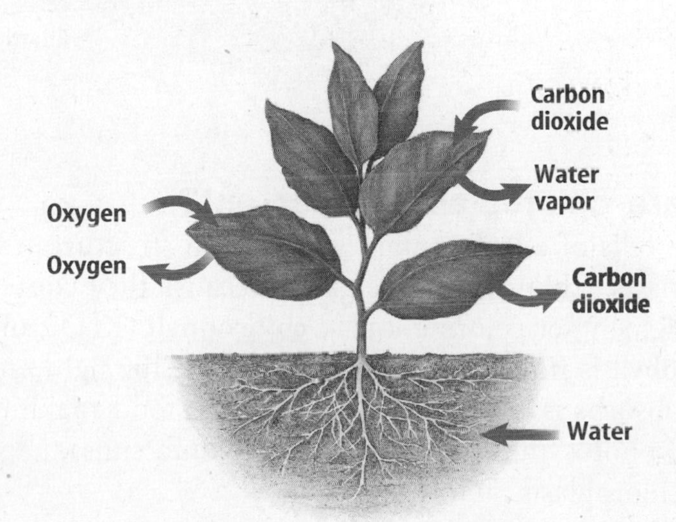

Picture This

1. Identify Circle the raw materials that a plant takes in.

What is the function of a leaf?

Gas is exchanged in the leaves. Most of the water taken in by the roots of a plant exits the plant through its leaves. Carbon dioxide, oxygen, and water vapor enter and exit the plant through openings in the leaves.

What is the structure of a leaf?

A leaf is made up of many different layers. The outer layer of the leaf is called the epidermis. The epidermis is nearly transparent and allows sunlight, which is used to make food, to reach the cells inside the leaf.

The epidermis has many small openings called **stomata** (stoh MAH tuh) (singular, *stoma).* Raw materials such as carbon dioxide, water vapor, and waste gases enter and exit the leaf through the stomata. Many plants have stomata on their stems. Guard cells surround each stoma to control how much water enters and exits the plant. Stomata close when a plant is losing too much water.

As you can see in the figure below, the inside of a leaf is made up of a spongy layer and a palisade layer. Carbon dioxide and water vapor fill the spaces of the spongy layer. Most of the plant's food is made in the palisade layer.

Picture This

2. Identify Circle the part of the leaf where most of the plant's food is made.

Why are chloroplasts important?

Some cells of a leaf contain small green structures called chloroplasts. Chloroplasts are green because they contain a green pigment, or coloring, called **chlorophyll** (KLOR uh fihl). Chlorophyll is important to plants because the light energy that it absorbs is used to make food. This food-making process, photosynthesis (foh toh SIHN thuh suhs), happens in the chloroplasts. ☑

3. Describe What happens in the chloroplasts?

____

The Food-Making Process

Photosynthesis is the process during which a plant's chlorophyll traps light energy and sugars are produced. In plants, photosynthesis occurs only in cells with chloroplasts. For example, photosynthesis occurs only in a carrot plant's green leaves. The carrot's root cells do not have chlorophyll, so they cannot perform photosynthesis. But excess sugar produced in the leaves is stored in the root. The familiar orange carrot you eat is the root of the carrot plant. When you eat a carrot, you benefit from the energy stored as sugar in the plant's root.

Plants need light, carbon dioxide, and water for photosynthesis. The chemical equation for photosynthesis is shown below.

$$6CO_2 + 6H_2O + \text{light energy} \xrightarrow{\text{chlorophyll}} C_6H_{12}O_6 + 6O_2$$

carbon dioxide — water — glucose — oxygen

What are light-dependent reactions?

Chemical reactions that occur during photosynthesis that need light are called the light-dependent reactions. During light-dependent reactions, chlorophyll and other pigments trap light energy that will be stored in sugar molecules. Light energy causes water molecules, which were taken up by the roots, to split into oxygen and hydrogen. The oxygen exits the plant through the stomata. This is the oxygen that you breathe. The hydrogen produced when water is split is used in light-independent reactions.

What are light-independent reactions?

Chemical reactions that occur during photosynthesis that do not need light are called light-independent reactions. The light energy trapped during the light-dependent reactions is used to combine carbon dioxide and hydrogen to make sugars, such as glucose. The chemical bonds that hold glucose and other sugars together are stored energy.

What happens to the oxygen and glucose that are made during photosynthesis?

Most of the oxygen produced during photosynthesis is a waste product and is released through the stomata. Glucose is the main form of food for plant cells. A plant usually produces more glucose than it can use. The extra glucose is stored in plants as other sugars and starches. When you eat carrots or potatoes, you are eating the stored product of photosynthesis.

FOLDABLES™

A Describe Use two quarter-sheets of notebook paper, as shown below, to take notes about photosynthesis.

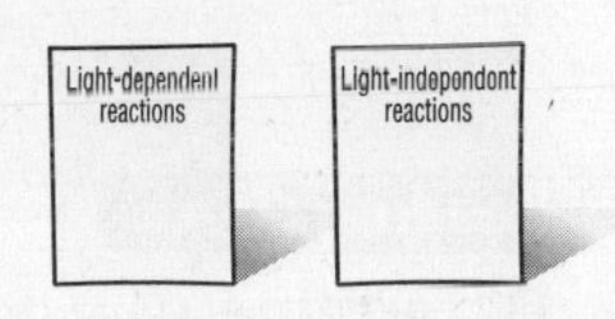

Think it Over

4. **List** two other foods that are stored products of photosynthesis.

____

____

How does a plant use glucose?

Glucose also is the basis of a plant's structure. Plants grow larger by taking in carbon dioxide gas and changing it to glucose. Cellulose, an important part of plant cell walls, is made from glucose. Leaves, stems, and roots are made of cellulose and other materials produced using glucose.

Why is photosynthesis important?

Photosynthesis produces food. Photosynthesis uses carbon dioxide and releases oxygen. This removes carbon dioxide from the atmosphere and adds oxygen to it. Most organisms need oxygen to live. About 90 percent of the oxygen in the atmosphere today is a result of photosynthesis. ☑

Reading Check

5. **Determine** What does photosynthesis add to Earth's atmosphere?

____

The Breakdown of Food

Respiration is a series of chemical reactions that breaks down food molecules and releases energy. Respiration that uses oxygen to break down food chemically is called aerobic respiration. The overall chemical equation for aerobic respiration is shown below.

$$\underset{\text{glucose}}{C_6H_{12}O_6} + \underset{\text{oxygen}}{6O_2} \rightarrow \underset{\text{carbon dioxide}}{6CO_2} + \underset{\text{water}}{6H_2O} + \text{energy}$$

Where does aerobic respiration occur?

Before aerobic respiration begins, glucose molecules in the cytoplasm are broken down into two smaller molecules. These molecules enter a mitochondrion, where aerobic respiration takes place. Oxygen is used to break down the molecules into water and carbon dioxide and to release energy. The figure below shows aerobic respiration in a plant cell.

Picture This

6. **Identify** Circle the part of the cell where aerobic respiration occurs. Highlight the waste products of respiration.

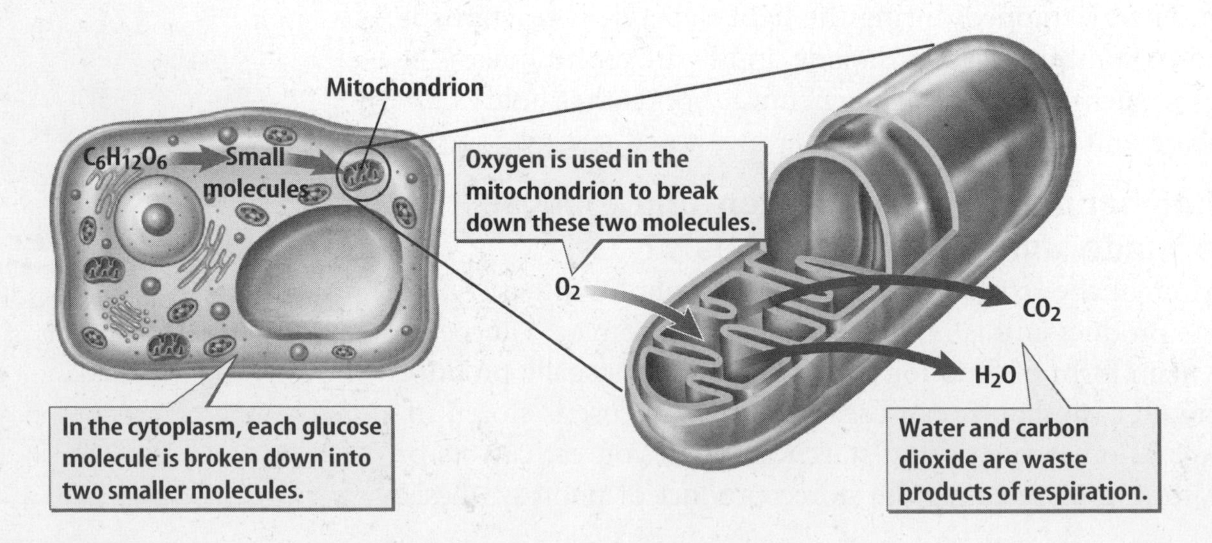
Mitochondrion
$C_6H_{12}O_6$
Small molecules
Oxygen is used in the mitochondrion to break down these two molecules.
O_2
CO_2
H_2O
In the cytoplasm, each glucose molecule is broken down into two smaller molecules.
Water and carbon dioxide are waste products of respiration.

Why is respiration important?

Food contains energy. But it is not in a form that can be used by cells. Respiration changes food energy into a form that cells can use. This energy drives the life processes of almost all organisms on Earth.

Plants use energy produced by respiration to transport sugars, to open and close stomata, and to produce chlorophyll. When seeds sprout, they use energy from the respiration of stored food in the seed. The figure below shows some uses of energy in plants.

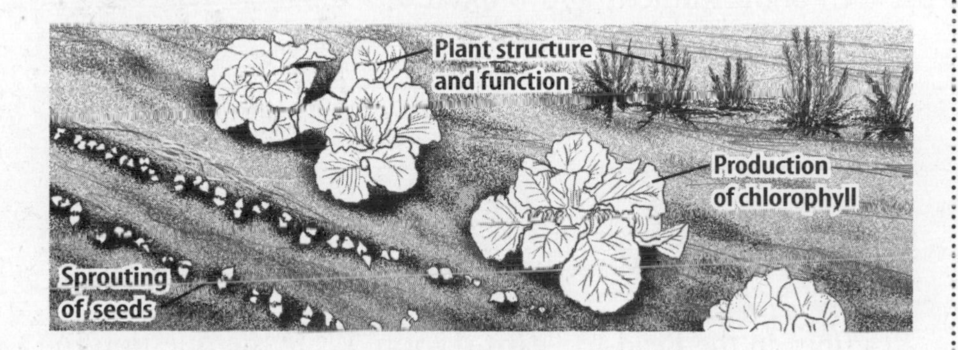

The waste product carbon dioxide also is important. Aerobic respiration returns carbon dioxide to the atmosphere, where plants and some other organisms use it for photosynthesis.

Comparison of Photosynthesis and Respiration

Aerobic respiration is almost the reverse of photosynthesis. Photosynthesis combines carbon dioxide and water by using light energy. The end products are glucose (food) and oxygen. Aerobic respiration combines oxygen and food to release the energy in the chemical bonds of the food. The end products of aerobic respiration are energy, carbon dioxide, and water. Look at the table below to compare the differences between photosynthesis and aerobic respiration.

Comparing Photosynthesis and Aerobic Respiration

	Photosynthesis	Aerobic Respiration
Energy	stored	released
Raw materials	water and carbon dioxide	glucose and oxygen
End products	glucose and oxygen	water and carbon dioxide
Where	cells with chlorophyll	cells with mitochondria

Picture This

7. List two uses of energy produced by respiration in plants.

___

___

___

Picture This

8. Compare and Contrast Highlight water and carbon dioxide for each process in one color and glucose and oxygen in another color.

After You Read

Mini Glossary

chlorophyll (KLOR uh fihl): a green pigment found in chloroplasts

photosynthesis (foh toh SIHN thuh suhs): the process during which a plant's chlorophyll traps light energy and sugars are produced

respiration: a series of chemical reactions that breaks down food molecules and releases energy

stomata (stoh MAH tah): small openings in the leaf epidermis, which act as doorways for raw materials to enter and exit the leaf

1. Review the terms and their definitions in the Mini Glossary. Write one or two sentences that explain the difference between photosynthesis and respiration.

____

____

____

2. Choose one of the question headings in the Read to Learn section. Write the question in the space below. Then write your answer to that question on the lines that follow.

Write your question here.

____

____

____

3. How did your notes help you summarize what you read in this section?

____

____

____

Science Online Visit **life.msscience.com** to access your textbook, interactive games, and projects to help you learn more about photosynthesis and respiration.

Plant Processes

section 2 Plant Responses

Before You Read

Have you ever been suddenly surprised? On the lines below, describe what surprised you and how your body responded to the surprise.

____

____

____

What You'll Learn

- the relationship between a stimulus and a tropism in plants
- about long-day and short-day plants
- how plant hormones and responses are related

Read to Learn

Study Coach

Create a Quiz Write a quiz question for each paragraph. Answer the question with information from the paragraph.

What are plant responses?

A stimulus is anything in the environment that causes a response in an organism. A stimulus may come from outside (external) or inside (internal) the organism. An outside stimulus could be something that startles or surprises you. An inside stimulus is usually a chemical produced by the organism. Many of these chemicals are hormones. Hormones are substances made in one part of an organism for use somewhere else in the organism. The response to the stimulus often involves movement toward or away from the stimulus. All living organisms, including plants, respond to stimuli.

Tropisms

Some plant responses to external stimuli are called tropisms (TROH pih zumz). A **tropism** can be seen as movement caused by a change in growth. It can be positive or negative. A positive tropism would be growth toward a stimulus. A negative tropism would be growth away from a stimulus. You may have seen plants responding to touch, light, and gravity. Plants also can respond to electricity, temperature, and darkness.

FOLDABLES™

B Compare Make a two-tab Foldable, as shown below, to compare inside and outside stimuli.

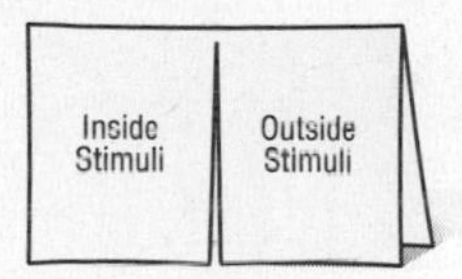

FOLDABLES™

C Explain Make a two-tab concept map Foldable, as shown below, to describe and list positive and negative tropisms.

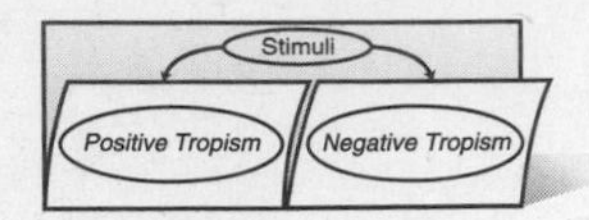

Touch If a pea plant touches a solid object, it responds by growing faster on one side of its stem than on the other side. As a result, the stem bends and twists around any object it touches.

Light When a plant responds to light, the cells on the side of the plant opposite the light get longer than the cells facing the light. Because of this, the plant bends toward the light. The leaves turn and can absorb more light. This positive response to light is called positive phototropism.

Gravity Plants respond to gravity. The downward growth of plant roots is a positive response to gravity. A stem growing upward is a negative response to gravity.

Plant Hormones

Plants have hormones that control the changes in growth that result from tropisms and affect other plant growth. These hormones include ethylene, auxin, gibberellin, cytokinin, and abscisic acid.

How does ethylene affect plants?

Many plants produce the hormone ethylene (EH thuh leen) gas and release it into the air around them. This hormone helps fruits ripen. Ethylene also causes a layer of cells to form between a leaf and the stem. The cell layer causes the leaf to fall from the stem.

What is auxin?

The plant hormone **auxin** (AWK sun) causes a positive response to light in stems and leaves. The figure below shows the effect of auxin.

When light shines on a plant from one side, the auxin moves to the shaded side of the stem where it causes a change in growth. Auxin causes plants to grow toward light.

Picture This

1. **Describe** Use the figure to describe to a partner how auxin affects a plant's response to light.

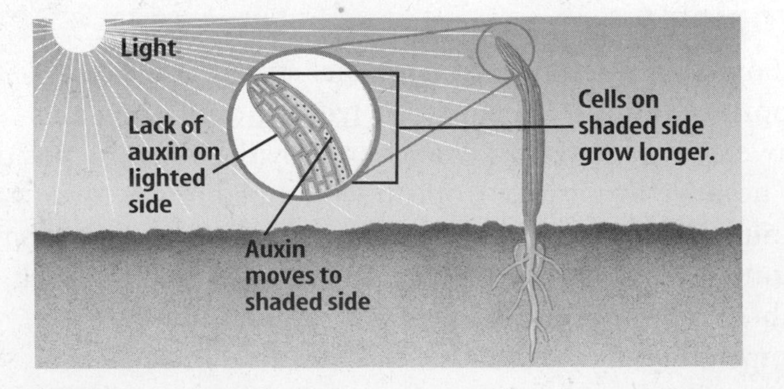

How do gibberellins and cytokinins affect plant growth?

Two other groups of plant hormones also affect a plant's growth. Gibberellins (jih buh REH lunz) can be mixed with water and sprayed on plants and seeds to stimulate plant stems to grow and seeds to germinate. Cytokinins (si tuh KI nunz) promote growth by causing faster cell division. Cytokinins can keep stored vegetables fresh longer.

How does abscisic acid affect plant growth?

Abscisic (ab SIH zihk) acid is a plant hormone that keeps seeds from sprouting and buds from developing during the winter. This hormone also causes stomata to close in response to water loss on hot summer days. ☑

Photoperiods

A plant's response to the number of hours of daylight and darkness it receives daily is called **photoperiodism** (foh toh PIHR ee uh dih zum). Because Earth is tilted about 23.5° from a line perpendicular to its orbit, the hours of daylight and darkness change with the seasons. These changes in the length of daylight and darkness affect plant growth.

How does darkness affect flowers?

Many plants must have a certain length of darkness to flower. Plants that need less than 10 h to 12 h of darkness to flower are called **long-day plants.** These plants include spinach, lettuce, and beets. Plants that need 12 or more hours of darkness to flower are called **short-day plants.** These plants include poinsettias, strawberries, and ragweed. If a short-day plant receives less darkness than it needs to flower, it will produce larger leaves instead of flowers.

What are day-neutral plants?

Plants that do not need a set amount of darkness to flower are called **day-neutral plants.** They can flower within a range of hours of darkness. These plants include dandelions and roses. Knowing the photoperiods of plants helps farmers and gardeners know which plants will grow best in the area where they live.

Reading Check

2. Explain How does abscisic acid affect plants?

______

______

______

Think it Over

3. Infer Why could you produce long-day flowering plants in a greenhouse during winter months?

______

______

______

After You Read

Mini Glossary

auxin (AWK sun): a plant hormone that causes plant stems and leaves to exhibit positive response to light

day-neutral plant: a plant that does not have a specific photoperiod to flower

long-day plant: a plant that needs less than 10 h to 12 h of darkness to flower

photoperiodism (foh toh PIHR ee uh dih zum): a plant's response to the number of hours of daylight and darkness it receives daily

short-day plant: a plant that needs 12 h or more of darkness to flower

tropism (TROH pih zum): a response of a plant to an external stimulus

1. Review the terms and their definitions in the Mini Glossary. Write one or two sentences to explain the differences among the long-day, short-day, and day-neutral plants.

____

____

____

2. Complete the cause-and-effect chart below to show how plant hormones affect plant growth.

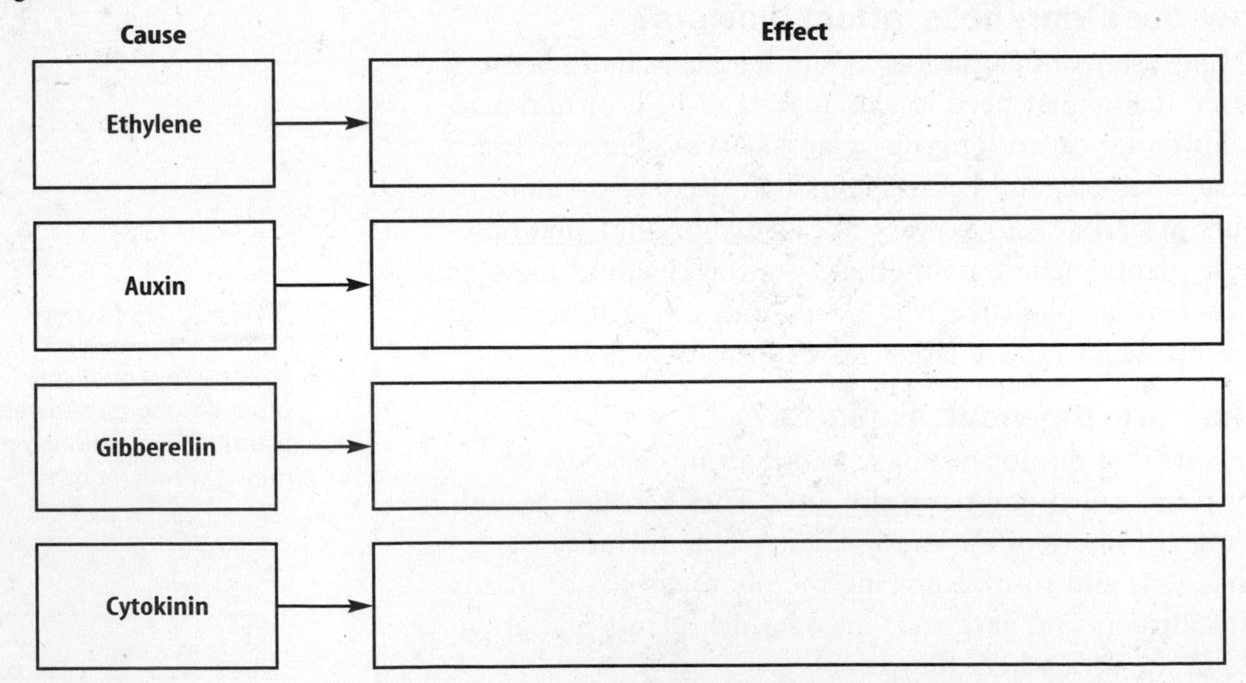

Science Online Visit **life.msscience.com** to access your textbook, interactive games, and projects to help you learn more about plant responses.

Introduction to Animals

section 1 Is it an animal?

Before You Read

List five different animals on the lines below. Read over your list. Name two things all the animals have in common.

___

___

___

What You'll Learn

- what most animals have in common
- how animals get what they need
- the difference between invertebrates and vertebrates

Read to Learn

Animal Characteristics

There are many different kinds of animals but all animals share certain characteristics. Animals can be identified by the following characteristics.

1. Animals are made of many cells.
2. Animal cells have a nucleus and other parts inside the cells called organelles. The nucleus directs all cell activities.
3. Animals get their food from other living things in their environment. Some animals eat plants. Some animals eat other animals. Some animals eat both plants and animals.
4. All animals digest their food. During digestion, the carbohydrates, proteins, and fats in the food are broken into small particles that can move into the animal's cells. Once inside the cell, some of the particles give the cell energy.
5. Many animals move from place to place. Animals move to escape from their enemies. Animals move to find food, mates, and shelter. Some animals move slowly or not at all. These animals have ways, called adaptations, to survive while living in one place.
6. All animals can reproduce sexually. Some animals also can reproduce asexually.

Mark the Text

Identify the Main Point Underline the important idea in each paragraph. Review the main ideas at the end of your reading.

Think it Over

1. **Explain** one way that an animal differs from a plant.

___

___

___

How Animals Meet Their Needs

Animals adapt, or change, over time. Adaptations are passed down from generation to generation. Adaptations help animals survive in a changing environment. An adaptation might be a body structure, a process, or a behavior.

How do adaptations help animals get energy?

Food is a basic need. Without food, living things die. Animals have adaptations that allow them to find, eat, and digest different foods.

Herbivores Animals that eat only plants are called **herbivores.** Deer, some fish, and many insects are herbivores. Plants do not give animals as much energy as other kinds of foods. To get the energy they need, herbivores usually eat more food than other animals do.

Carnivores Animals that eat other animals are called **carnivores.** Most carnivores, like lions and red-tailed hawks, catch and kill other animals for food.

Some carnivores eat only the remains of other animals. These carnivores are called scavengers. A buzzard is a scavenger that can be seen eating animals that may have been hit by cars.

The meat from animals supplies more energy than plants supply. For this reason, carnivores do not need to eat as much or as often as herbivores do.

Omnivores Animals that eats both plants and animals are called **omnivores.** Bears, raccoons, robins, and humans are examples of omnivores.

Millipedes and many beetles eat tiny bits of decaying matter called detritus (dih TRI tus). These animals are called detritivores (dih TRI tih vorz).

FOLDABLES™

A Describe Make a four-tab Foldable, as shown below. As you read, list facts that describe how different animals eat different foods.

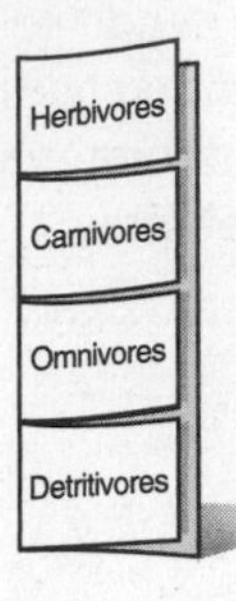

Picture This

2. **List** one more example of an omnivore.

____

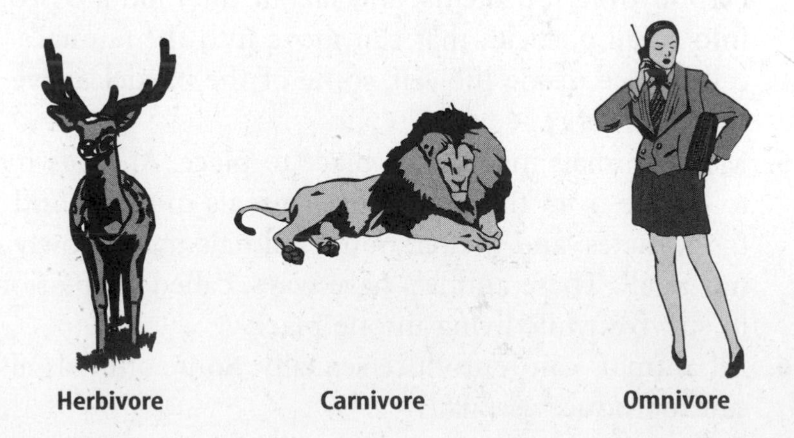

What physical adaptations help animals survive?

Animals that capture and eat other animals are called predators. Some animals have physical features that help them avoid predators. A turtle has a hard shell that protects it. A porcupine has sharp quills that keep it safe from predators.

Size Many animals avoid danger because of their size. Large animals are usually safer than small animals. Moose and bison are large animals. Because they are so large, most predators will not attack them. ☑

Mimicry Some animals avoid predators because they act or look like other animals. The scarlet king snake looks like the coral snake. The coral snake is poisonous and will kill predators that attack it. The scarlet king snake is not poisonous. Because the two snakes look alike, predators usually leave both kinds of snakes alone. The scarlet king snake is adapted to help it survive. This type of adaptation is called mimicry (MIHM ih kree).

Camouflage Some animals hide from predators by blending in with their environment. For example, a trout has a speckled back. From above, its back looks like the gravel bottom of the stream it lives in. Some animals can change their coloring to match their surroundings. A lizard can have light green skin while resting on a light green leaf. The same lizard can have dark brown skin when resting on a dark brown tree. Any marking or coloring that helps an animal hide from other animals is called camouflage.

What physical adaptations do predators have?

Camouflage is an adaptation that helps some predators sneak up on their prey. A tiger has stripes that help it hide in tall grasses. A killer whale is black on top and white underneath. From above, the black looks like the dark ocean. From underneath, the white looks like the light of the sky above. The killer whale's coloring helps it hunt its prey. ☑

How do behavioral adaptations help animals survive?

Many animals have behaviors that help them avoid predators or catch their prey. Some animals use chemicals produced by their bodies. A skunk sprays a bad smelling liquid at predators. An octopus squirts ink when it is in danger. The ink darkens the water and confuses predators, giving the octopus time to escape.

✔ Reading Check

3. **Explain** How can size help some animals avoid predators?

____

____

____

✔ Reading Check

4. **Name** a physical adaptation that some predators have.

____

____

____

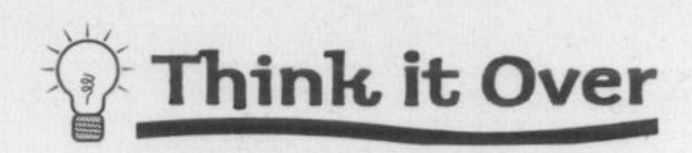

5. **List** three behavioral adaptations that help animals survive.

____

____

____

Run from Predators Some animals run away from predators. The Thomson's gazelle can run faster than a lion and usually escapes.

Live in Groups Some animals live in groups. This is a behavior used by both predators and prey. Herring swim in groups called schools. Predators do not attack the school because the group appears to be one large fish.

Some predators also travel in groups. Wolves live and travel in groups called packs. A pack of wolves works together to hunt, exhaust, and kill prey. The pack hunts larger prey than one wolf could capture on its own.

Animal Classification

Scientists have identified and named more than 1.8 million species of animals. Scientists estimate that there are millions more animals to be identified and named. To help organize so many animals, scientists have ways to sort, name, and group them.

How do scientists classify animals?

Picture This

6. **Identify** Write one characteristic of an invertebrate on the line provided. Then write one characteristic of a vertebrate on the line provided.

Scientists classify animals into two main groups, vertebrates and invertebrates. A **vertebrate** is an animal with a backbone. Fish, frogs, snakes, birds, and humans are vertebrates. An **invertebrate** is an animal without a backbone. Sponges, jellyfish, worms, insects, and clams are all invertebrates. The figure below shows how animals are classified and lists several classes of invertebrates.

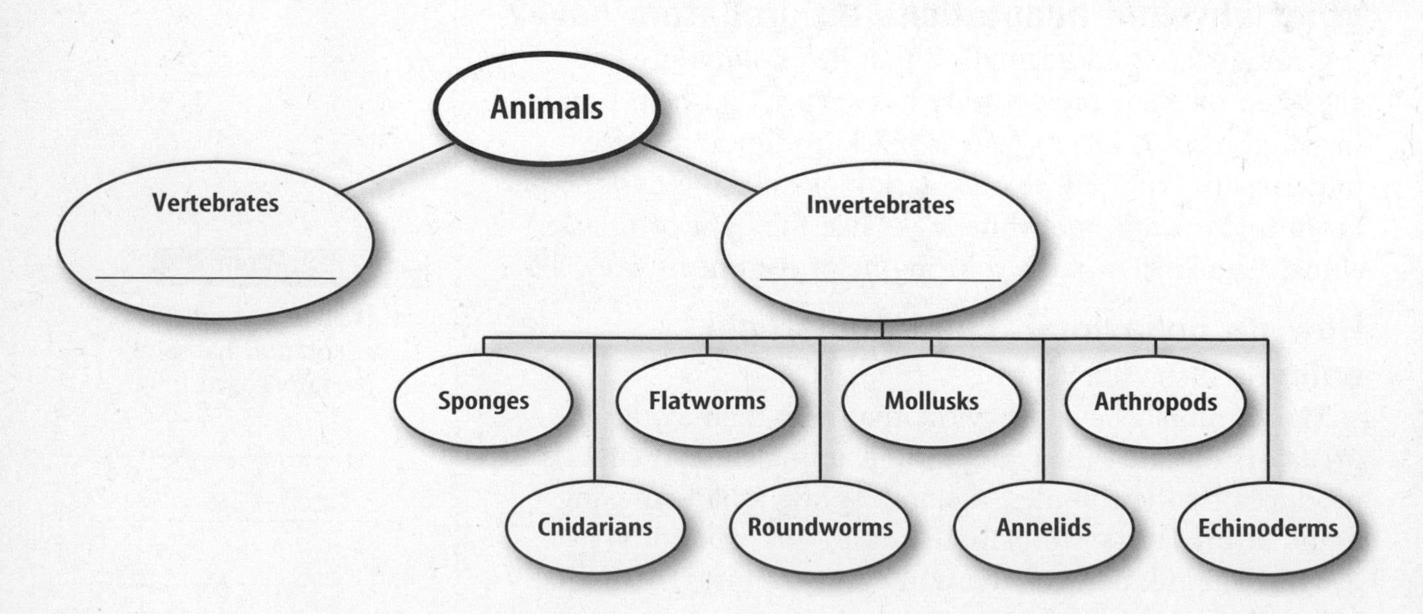

Invertebrates About 97 percent of all animal species are invertebrates. Many invertebrates have outer coverings to protect them. Some invertebrates have shells. Some have skeletons on the outsides of their bodies. Others have spiny outer coverings.

What does a backbone do?

A backbone is a stack of structures called vertebrae. The backbone supports the animal. It also protects and covers the spinal cord. The spinal cord contains nerves that carry messages from the brain to other parts of the body. It also carries messages back to the brain. ☑

Reading Check

7. State the purpose of the backbone in vertebrates.

____

____

____

How is symmetry used to classify animals?

Animals are first classified as invertebrates or vertebrates. Next, animals are grouped by symmetry (SIH muh tree). Symmetry is how the body parts of an animal are arranged. Some animals have no definite shape. An animal with no definite shape is called asymmetrical. Most sponges are asymmetrical.

Radial symmetry is the arrangement of body parts around a center point, like spokes on a bicycle wheel. Jellyfish, anemones, and sea urchins have radial symmetry.

Most animals have bilateral symmetry. An animal with **bilateral symmetry** has a body that can be divided into right and left halves. Each half is nearly a mirror image of the other. Crayfish, humans, butterflies, and birds have bilateral symmetry. The figure below shows the types of symmetry.

Picture This

8. Identify Write the type of symmetry shown on the line below each animal.

Butterfly

____

Sea anemone

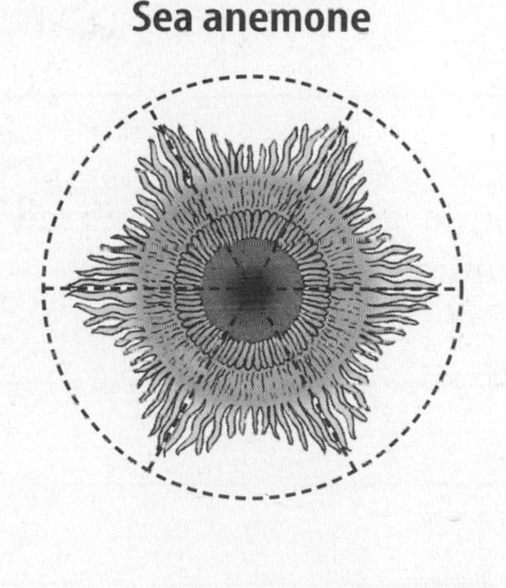

____

After You Read

Mini Glossary

bilateral symmetry: arrangement of body parts divided into right and left halves; each half is nearly a mirror image of the other

carnivore: an animal that eats only animals

herbivore: an animal that eats only plants

invertebrate: an animal without a backbone

omnivore: an animal that eats both plants and animals

radial symmetry: arrangement of body parts around a center point

vertebrate: an animal with a backbone

1. Review the terms and their definitions in the Mini Glossary. Write a sentence that compares two of the terms.

____

____

____

2. Fill in the tables below with appropriate animal examples to review physical and behavioral adaptations in animals.

Physical Adaptations	
Adaptation	**Animal**
size	
mimicry	
camouflage	

Behavioral Adaptations	
Adaptation	**Animal**
chemicals produced by their own bodies	
run from predators	
travel in groups	

3. How did reviewing the main ideas after you have read the section help you remember the important ideas about animals?

____

____

____

Visit **life.msscience.com** to access your textbook, interactive games, and projects to help you learn more about animal characteristics.

Introduction to Animals

section 2 Sponges and Cnidarians

Before You Read

The artificial sponge you use for dishwashing is similar to a dried natural sponge. On the lines below, describe how a sponge looks and feels.

____

____

____

____

What You'll Learn

- the characteristics of sponges and cnidarians
- how sponges and cnidarians get food and oxygen
- why living coral reefs are important

Read to Learn

Sponges

Sponges are invertebrates that live in water. They play many roles. Worms, shrimp, snails, and sea stars live on, in, and under sponges. Sponges are an important source of food for some snails, sea stars, and fish. Some sponges contain organisms that provide oxygen to the sponge and remove its wastes.

For many years, humans have dried sponges and used them for bathing and cleaning. Today, most sponges you see are artificial. Scientists are finding new uses for natural sponges. Chemicals found in sponges may be used to make drugs that help fight diseases. ☑

Study Coach

Create a Chart Make a two-column chart. Label the left column *Sponges* and the right column *Cnidarians.* As you read about each, list the facts you learn in the correct column.

Reading Check

1. **Identify** two uses for natural sponges.

____

____

When did sponges first appear on Earth?

Fossil records show that sponges appeared on Earth about 600 million years ago. Because sponges have little in common with other animals, many scientists have concluded that sponges developed separately from all other animals. Today's sponges are similar to sponges that lived 600 million years ago.

Characteristics of Sponges

There are more than 5,000 species of sponges. Most live in salt water near coastlines. Others live deep in the ocean. A few species are found in freshwater. ☑

Sponges are different shapes, sizes, and colors. Saltwater sponges are bright red, orange, yellow, or blue. Freshwater sponges are usually a dull brown or green. Some sponges have radial symmetry, but most are asymmetrical. Sponges can be smaller than a marble or larger than a compact car.

Adult sponges are sessile (SE sile). A **sessile** organism remains attached to one place during its lifetime. Sponges often live in groups called colonies.

Because sponges do not move around, early scientists classified sponges as plants. Later scientists found that sponges could not make their own food. Because animals cannot make their own food, scientists reclassified sponges as animals.

Reading Check

2. **Describe** Where do sponges live?

How is a sponge's body organized?

The figure below shows the body of a sponge. Sponges have a simple body structure. They do not have tissues, organs, or organ systems. A sponge's body is a hollow tube. It is closed at the bottom and open at the top. A sponge has many tiny openings in its body called pores. Water moves through the pores.

Many sponge bodies have sharp, pointed structures called spicules (SPIH kyewlz). You can see these sharp structures in the figure above. Some sponges have skeletons made of a fiber-like material called spongin. Spicules and spongin provide support for a sponge and protect it from predators.

Picture This

3. **Identify** Circle the name of the structure that provides support and protection for the sponge. Underline the name of the structure that allows water to enter the sponge.

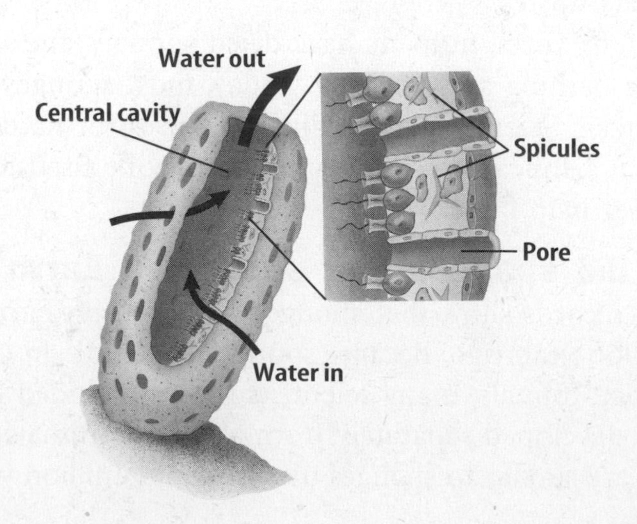

How do sponges get food and oxygen?

Sponges pull water into their bodies through their pores. The water contains tiny food particles. Cells filter out bits of food from the water. Oxygen also is removed from the water. The filtered water carries away wastes through the opening in the top of the sponge.

How do sponges reproduce?

Sponges can reproduce sexually as shown in the figure below. Most sponges are hermaphrodites (hur MA fruh dites). A **hermaphrodite** is an animal that produces both sperm and eggs in the same body.

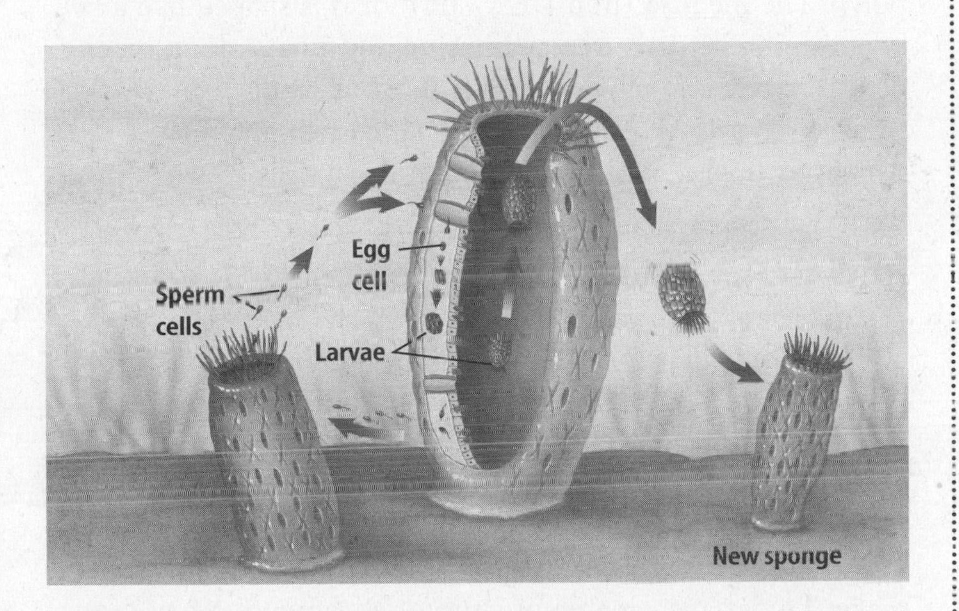

Sponges release sperm into the water. The sperm float until they are drawn into another sponge. A sperm fertilizes an egg. From this, a larva develops inside the sponge. A larva looks different from an adult sponge. A sponge larva has short, threadlike structures, called cilia, that allow it to swim. The larva swims from the sponge, and eventually settles on a surface. It will slowly grow into an adult.

Sponges also can reproduce asexually. A bud forms on a parent sponge, then drops off and grows on its own. Sponges also can grow by regeneration from small pieces of a sponge. In regeneration, an organism grows new body parts to replace lost or damaged ones. Sponge growers cut sponges into small pieces, then throw the pieces into the ocean to regenerate.

FOLDABLES™

B Explain Make a folded table Foldable, as shown below. Use the table to explain how sponges and cnidarians obtain oxygen and food.

	Sponges	Cnidarians
obtain oxygen		
obtain food		

Picture This

4. Circle the names in the figure that indicate that the sponge is a hermaphrodite.

Cnidarians

Cnidarians (ni DAR ee uhnz) are another group of invertebrates that live in water. Cnidarians include corals, jellyfish, sea anemones, hydras, and the Portuguese man-of-war.

Where are cnidarians found?

Most cnidarians live in salt water, although many types of hydras live in freshwater. Sea anemones and jellyfish live as single organisms. Hydras and corals often live in colonies.

What body forms do cnidarians have?

Cnidarians have two different body forms—medusa and polyp. The **medusa** (mih DEW suh) form is shaped like a bell or an umbrella. It is free-swimming and floats along on the ocean currents. A jellyfish spends most of its life as a medusa.

The **polyp** (PAH lup) form is shaped like a vase and usually is sessile. Sea anemones, corals, and hydras live most of their lives as polyps.

FOLDABLES™

C Compare Make a three-tab Foldable using notebook paper, as shown below. Use it to compare the polyp and medusa forms of cnidarians.

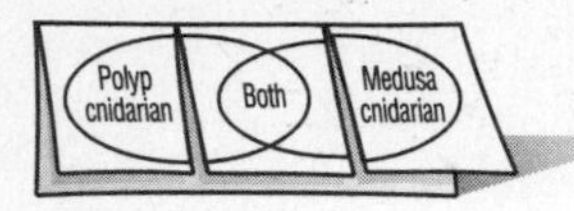

How is the cnidarian body organized?

Cnidarians have radial symmetry. A cnidarian has two layers of cells that form tissues and a digestive area. In this two-cell-layer body plan, all of the cnidarian's cells are close to the water. In the cells, oxygen from the water is exchanged for carbon dioxide and other cell wastes.

A cnidarian has a system of nerve cells called a nerve net. The nerve net carries messages to all parts of the body. This makes cnidarians capable of simple movements. They can somersault away in response to danger. ☑

Reading Check

5. **Name** the structure that allows a cnidarian to make simple movements.

____

____

Most cnidarians have tentacles (TEN tih kulz) around their mouths. **Tentacles** are armlike structures used for getting food.

How do cnidarians get food?

A cnidarian has one body opening, a mouth, through which food enters and undigested food is removed. Cnidarians are predators. They have stinging cells on their tentacles. A **stinging cell** has a capsule with a threadlike structure with toxins that helps the cnidarian capture food. When prey lightly touch or swim near the stinging cells, the thread goes into the prey, and the toxin stuns it. The tentacles then move the prey into the cnidarian's mouth.

Some fish live among the tentacles of large sea anemones. The fish are not harmed by the sea anemone's sting because they are protected by a special covering.

How do cnidarians reproduce?

Cnidarians reproduce sexually and asexually. Polyp forms reproduce asexually by producing buds that fall off the cnidarian and develop into new polyps.

Polyp forms also reproduce sexually. Eggs are produced and released into the water. Sperm are released into the water and fertilize the eggs, which develop into new polyps.

Medusa (plural, *medusae*) forms have two stages of reproduction—a sexual stage and an asexual stage. The stages are shown in the figure below. During the sexual stage, free-swimming medusae produce eggs or sperm. The eggs and sperm are released into the water. The sperm from one medusa fertilize the eggs from another medusa. The fertilized eggs grow into larvae.

The larvae grow into polyps. During the asexual stage, the polyp forms buds that become tiny medusae. The medusae buds off the polyp, and the cycle continues.

Picture This

6. Identify Circle the two body forms of the cnidarian.

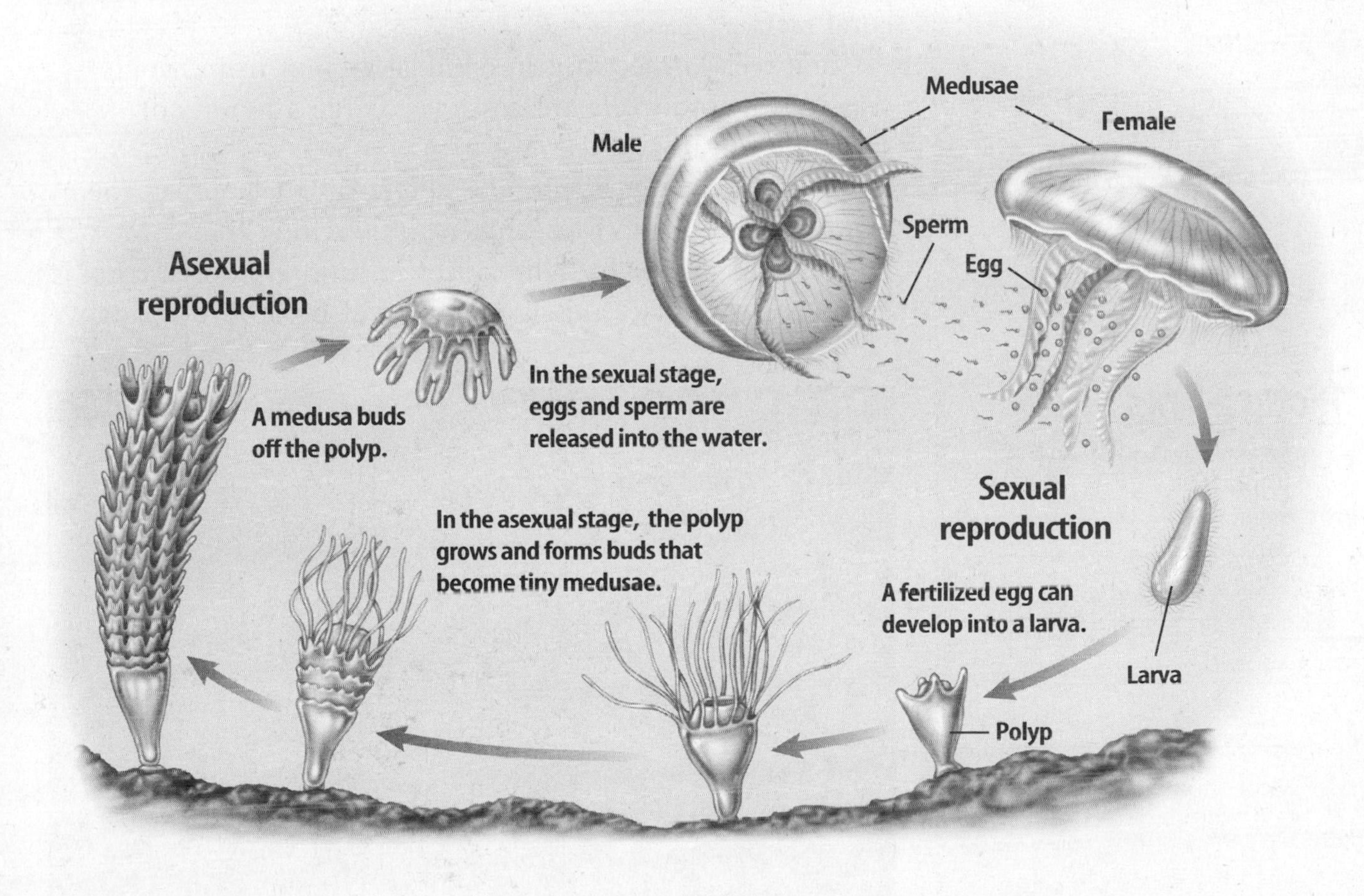
Medusae
Male
Female
Sperm
Egg
Asexual reproduction
A medusa buds off the polyp.
In the sexual stage, eggs and sperm are released into the water.
Sexual reproduction
In the asexual stage, the polyp grows and forms buds that become tiny medusae.
A fertilized egg can develop into a larva.
Larva
Polyp

FOLDABLES™

D Explain Make a two-tab concept map Foldable using notebook paper, as shown below. Write notes to explain the causes and effects of coral reef destruction.

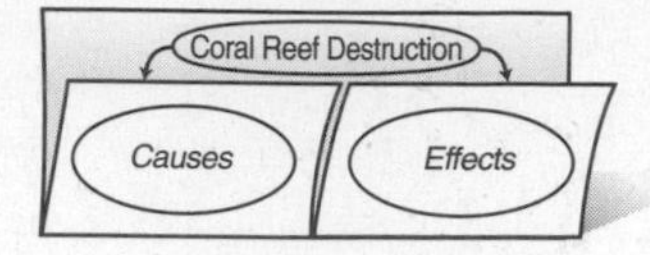

Origin of Cnidarians

Scientists hypothesize that cnidarians have been on Earth for more than 600 million years. The medusa body was probably the first form. Polyps may have formed from medusae larvae that became permanently attached to a surface. Most of the cnidarian fossils are corals.

Corals

Corals live in colonies called coral reefs. Coral polyps form hard, protective shells or skeletons. A reef forms as each new generation of coral polyps builds on top of existing coral skeletons. It takes millions of years for large reefs to form.

Why are corals important?

Coral reefs are important in the ecology of tropical waters. The diversity of life in coral reefs is similar to that in tropical rain forests. Some of the most beautiful and interesting animals in the world live in the formations of coral reefs.

Coral reefs protect beaches and shorelines from rough seas. When coral reefs are destroyed, large amounts of shoreline can be washed away.

Corals, like sponges, produce chemicals that protect them from disease. Medical researchers are learning that some of these chemicals might fight cancer in humans. Some coral is even used to replace missing sections of bone in humans.

Picture This

7. **List** two reasons coral is important.

____

____

____

____

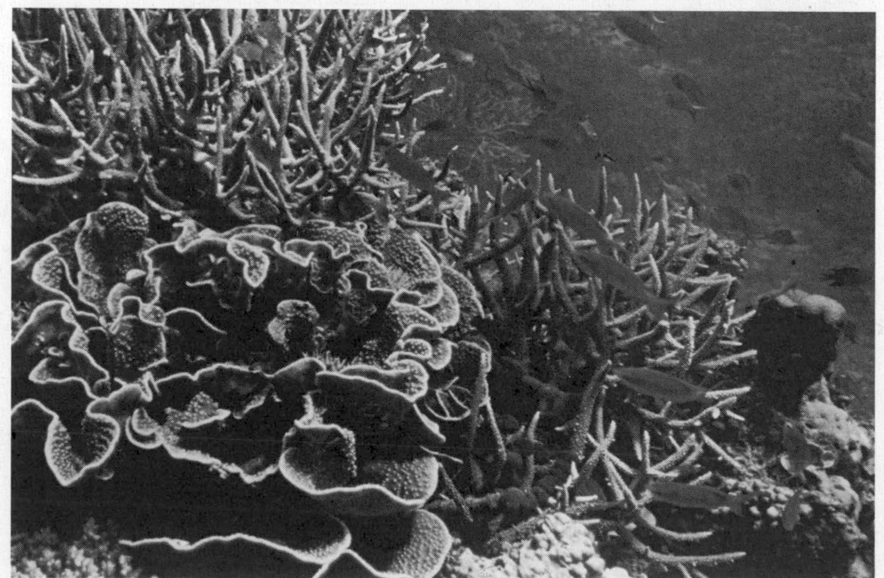

N. Sefton/Photo Researchers

After You Read

Mini Glossary

hermaphrodite (hur MA fruh dite): an animal that produces both sperm and eggs in the same body

medusa (mih DEW suh): a free-floating bell-shaped cnidarian body form

polyp (PAH lup): a sessile vase-shaped cnidarian body form

sessile (SE sile): staying attached to one place for most of an animal's life

stinging cells: cells with coiled threads that help cnidarians get food

tentacle (TEN tih kul): armlike structure used for obtaining food

1. Review the terms and their definitions in the Mini Glossary. Use a term related to reproduction in sponges and cnidarians in a sentence.

____

____

2. Use the Venn diagram below to compare and contrast the body forms and structures of sponges and cnidarians.

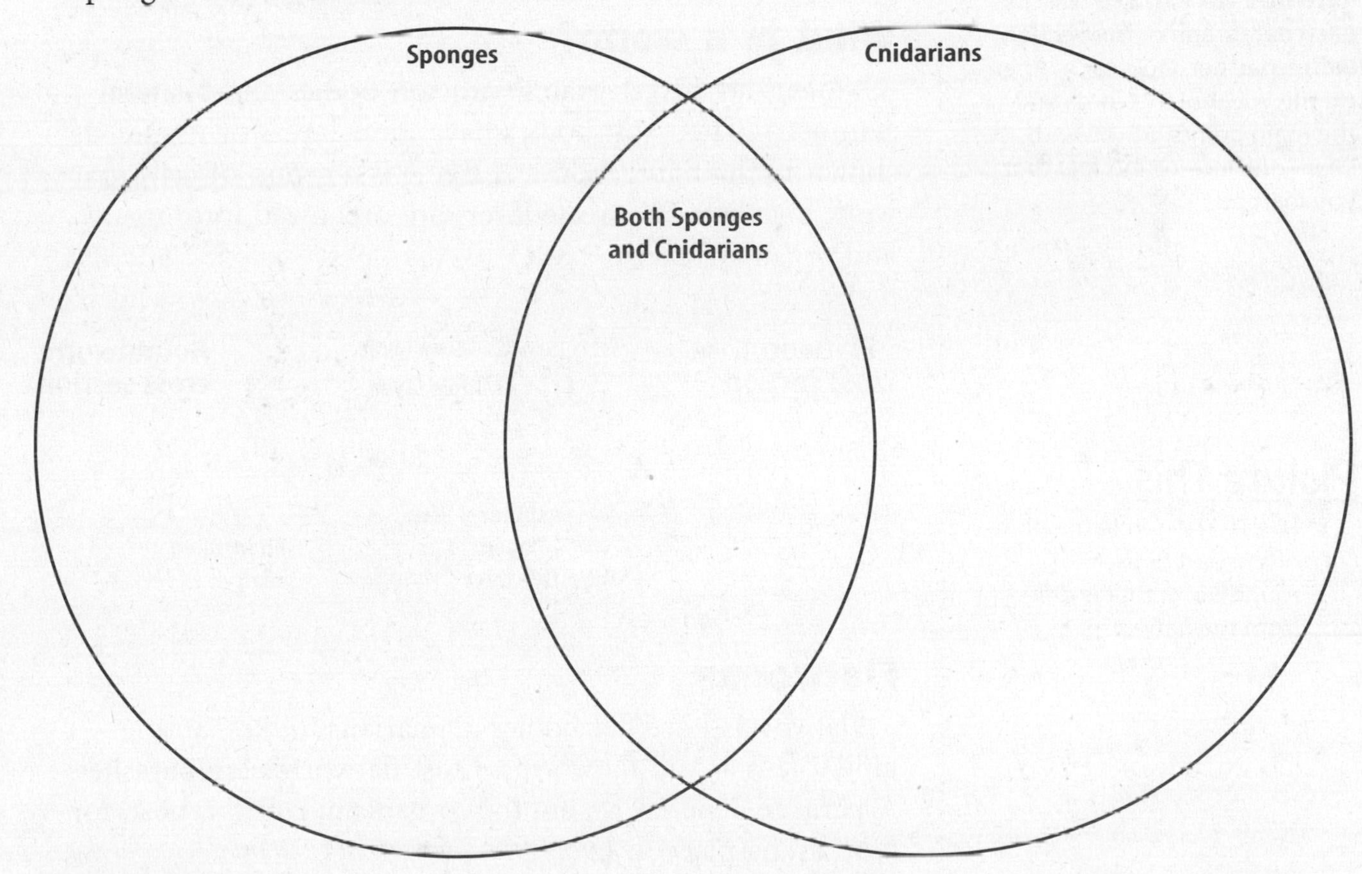

Science Online Visit **life.msscience.com** to access your textbook, interactive games, and projects to help you learn more about sponges and cnidarians.

End of Section

chapter 12

Introduction to Animals

section 3 Flatworms and Roundworms

What You'll Learn

- about flatworms and roundworms
- how free-living and parasitic organisms are different
- the diseases caused by flatworms and roundworms

Before You Read

List four things you know about worms on the lines below.

____

____

____

Study Coach

Think-Pair-Share Discuss each paragraph of this section with a partner. Quiz each other on the vocabulary words and the main points. At the end, share one interesting fact that you learned.

Read to Learn

What is a worm?

Worms are invertebrates with soft bodies and bilateral symmetry. Their soft bodies have three layers of tissues, as shown in the figure below. A tissue is a group of cells that work together. The tissue layers are organized into organs and organ systems.

Picture This

1. **Identify** Circle the name of the part of the roundworm that is different from the flatworm.

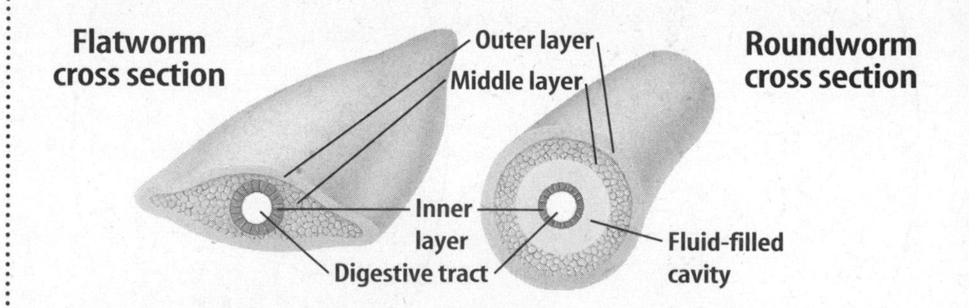

Flatworms

Flatworms have flat bodies. Planarians, flukes, and tapeworms are all flatworms. Most flatworms are parasites. A parasite depends on another organism, called a host, for food and a place to live.

Some flatworms are free-living organisms. **Free-living organisms** do not depend on another organism for food or a place to live.

What is a planarian?

A planarian is a free-living flatworm. Most planarians live under rocks, on plant materials, or in freshwater. Planarians range in size from 3 mm to 30 cm. They have triangle-shaped heads with two eyespots, as shown in the figure below.

Planarian

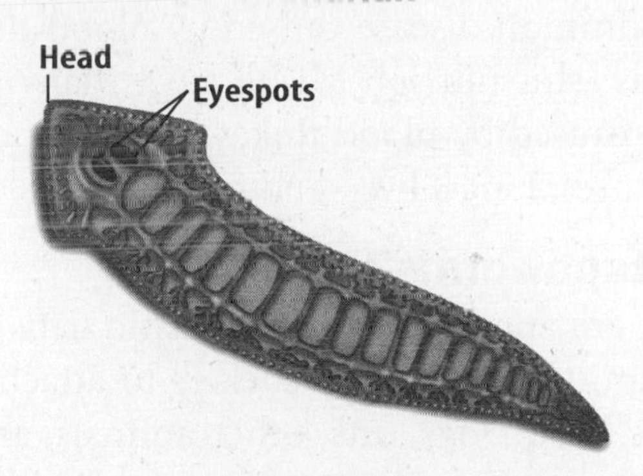

How does a planarian eat?

A planarian has one opening—a mouth. The opening is on the underside of the body. A planarian feeds on small organisms and the dead bodies of larger organisms. A muscular tube called the pharynx connects the mouth to the digestive tract. The digestive tract breaks down food into nutrients for the worm.

How does a planarian move?

A planarian's body is covered with cilia, which are fine, hairlike structures. As the cilia move, the worm slides along a slimy mucous track. The mucus is secreted from the underside of the planarian.

How does a planarian reproduce?

Planarians reproduce asexually and sexually. Planarians reproduce asexually by dividing in two. They also can regenerate, like sponges. If a planarian is cut in two, each piece will grow into a new worm.

Planarians reproduce sexually by producing eggs and sperm. Most planarians are hermaphrodites. They exchange sperm with each other and then lay fertilized eggs that hatch in a few weeks.

What are flukes?

Flukes are parasitic flatworms that feed on the blood, cells, and other fluids of their hosts. A fluke's life cycle requires more than one host.

Picture This

2. **Trace** the shape of the head with a colored marker or pen. What shape does it form?

____

FOLDABLES™

E Identify Make a three-tab Foldable using notebook paper. Write notes that help you identify planarians, flukes, and

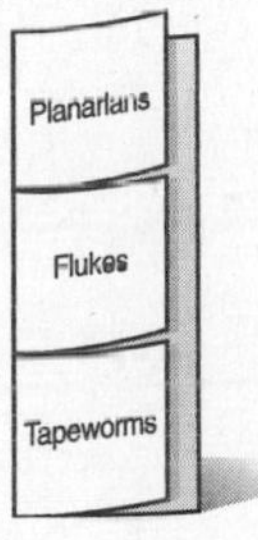

How do flukes reproduce?

Most flukes reproduce sexually. The male worm deposits sperm in the female worm. The female lays fertilized eggs inside the host. The eggs leave the host in its waste products. Fertilized eggs that end up in water usually infect snails. The eggs grow into young worms. The young worms leave the snail and enter the body of a new host through the skin.

The most common disease caused by blood flukes is schistosomiasis (shis tuh soh MI uh sus). Many people die from diseases caused by blood flukes. Other types of flukes can infect the eyes, lungs, liver, and other organs of their host.

FOLDABLES™

F Explain Make a two-tab concept map Foldable using notebook paper, as shown below. List the diseases caused by flatworms and roundworms.

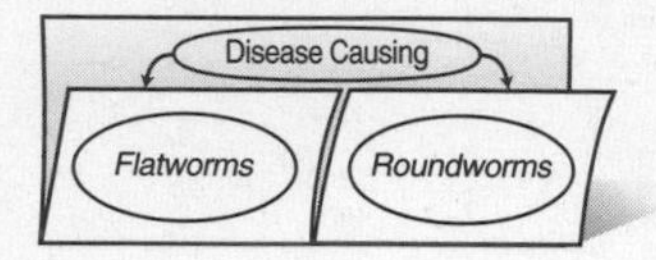

What are tapeworms?

Tapeworms are another type of parasitic flatworm. An adult tapeworm uses hooks and suckers to attach itself to the intestine of its host. Dogs, cats, other animals, and humans are the hosts for tapeworms. A tapeworm absorbs food that is being digested by the host from the host's intestine.

How does a tapeworm reproduce?

The figure below shows the reproductive cycle of a tapeworm. A tapeworm's body segments have both male and female reproductive organs. Sperm in a segment often fertilizes the eggs in the same segment. When a mature segment at the end of the worm is full of fertilized eggs, the segment breaks away. This segment passes out of the host's body with the host's other wastes. If another host eats a fertilized egg, the egg hatches and grows into an immature tapeworm called a bladder worm.

Picture This

3. Explain Use the diagram to explain to a classmate how tapeworms reproduce.

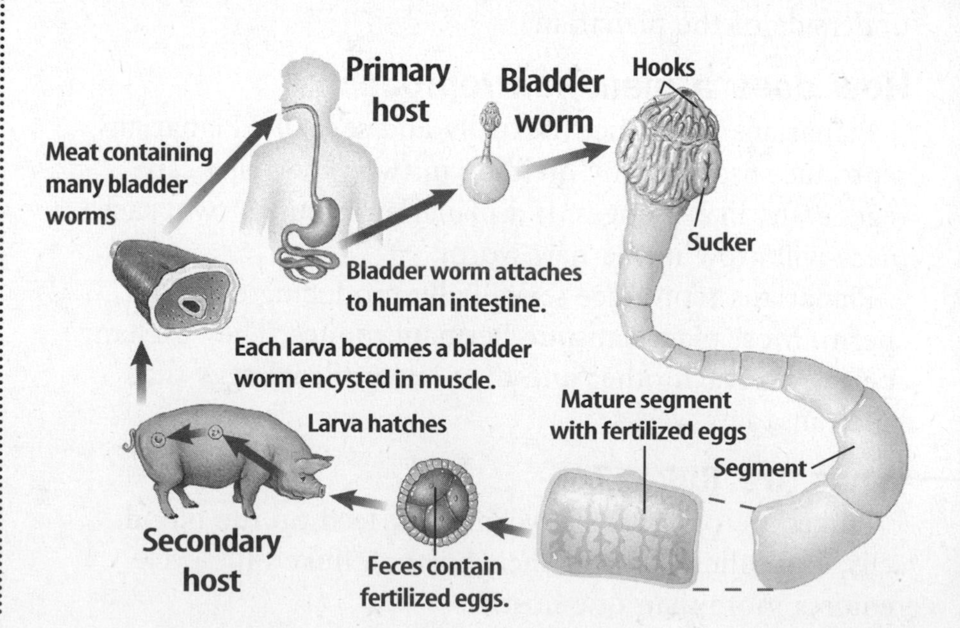

Origin of Flatworms

Scientists do not know when flatworms first appeared on Earth. Evidence does suggest that flatworms are the first animal group to have bilateral symmetry with nerves and senses around the head area. Flatworms also are probably the first animals to have a third tissue layer that develops into organs and organ systems. Some scientists hypothesize that flatworms and cnidarians may have had a common ancestor.

Roundworms

Roundworms also are called nematodes. More roundworms live on Earth than any other type of many-celled organism. More than half a million species of roundworm are found in soil, in animals, in plants, in freshwater, and in salt water. Some are parasitic, but most are free-living. Most nematode species have male and female worms and reproduce sexually. ☑

A roundworm is long and thin like a piece of thread. Its body tapers like a carrot tip on both ends. The roundworm's body is a tube with another tube inside. Fluid separates the two tubes. Roundworms have two body openings, a mouth and an anus. An **anus** is an opening at the end of the digestive tract through which wastes leave the body.

What is the origin of roundworms?

Roundworms first appeared on Earth about 550 million years ago. They were the first animals to have a digestive system with a mouth and an anus. They may be closely related to arthropods.

Why are roundworms important?

Roundworms can cause diseases in humans and animals. Some roundworms cause damage to plants, fabrics, crops, and food.

Not all roundworms are a problem for humans. Some roundworm species feed on fleas, ticks, ants, beetles, and other insects that cause damage to crops and humans. Researchers are studying roundworms that kill deer ticks that cause Lyme disease.

Roundworms are important to the health of soil. They break down organic material and put nutrients in the soil. They also help in cycling elements such as nitrogen.

Reading Check

4. **Identify** What is another name for roundworms?

____

FOLDABLES™

G Describe Use quarter sheets of notebook paper, as shown below, to write notes on how roundworms are helpful and harmful.

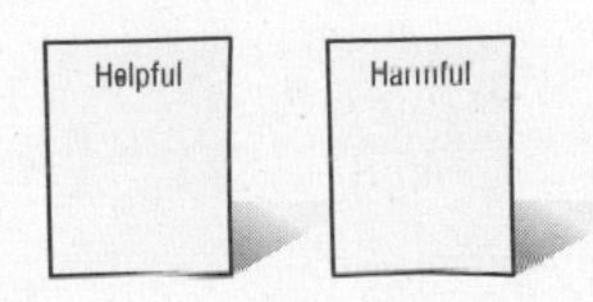

After You Read

Mini Glossary

anus: an opening at the end of the digestive tract through which waste leaves the body

free-living organism: an organism that does not depend on another organism for food or a place to live

1. Review the terms and their definitions in the Mini Glossary. Write a sentence using one of the terms.

____

____

____

2. Fill in facts for each row to describe flatworms and roundworms.

	Flatworms	Roundworms
Number of body openings		
Type of symmetry		
Food sources	Parasites: Free-living:	
Reproduction		

End of Section

ScienceOnline Visit life.msscience.com to access your textbook, interactive games, and projects to help you learn more about flatworms and roundworms.

Mollusks, Worms, Arthropods, Echinoderms

section 1 Mollusks

Before You Read

Look at the picture on this page. Have you ever seen this kind of animal? If so, where?

____

____

What You'll Learn

- the characteristics of mollusks
- about gastropods, bivalves, and cephalopods
- why mollusks are important

Read to Learn

Mark the Text

Build Vocabulary Read all the headings for this section and circle any word you cannot define. At the end of the section, review the circled words and underline the part of the text that helps you define them.

Characteristics of Mollusks

Mollusks (MAH lusks) are soft-bodied invertebrates. They have bilateral symmetry and usually one or two shells. The organs are in a fluid-filled cavity. Most mollusks live in water, but some live on land. Snails, clams, and squid are mollusks.

What do the bodies of mollusks look like?

Mollusks have a thin layer of tissue called a mantle. As you can see in the figure below, the **mantle** covers the body organs. It secretes the shell or protects the body if the mollusk doesn't have a shell. Between the soft body and the mantle is a space called the mantle cavity. It contains **gills**—the organs in which carbon dioxide from the mollusk is exchanged for oxygen in the water.

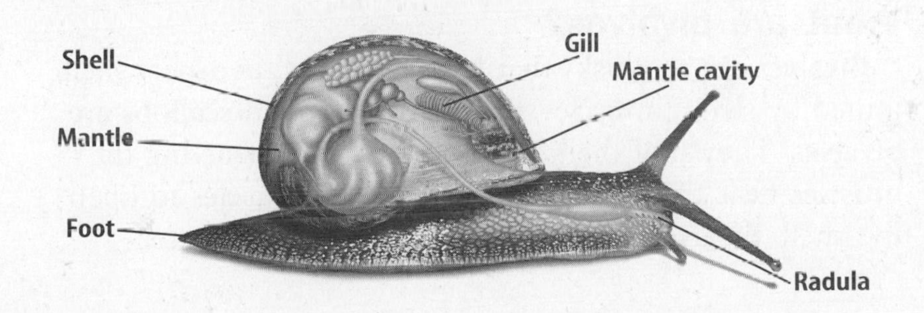

Picture This

1. **Highlight** the part of the mollusk that covers the body organs.

What is an open circulatory system?

The circulatory system of most mollusks is an open system. In an **open circulatory system,** the heart moves blood out into the open spaces around the body organs. The blood completely surrounds and nourishes the body organs. ☑

Reading Check

2. Explain What type of circulatory system do most mollusks have?

____

____

____

What other features do mollusks have?

Most mollusks have a head with a mouth and some sensory organs. Some mollusks, such as squid, have tentacles. The muscular foot located on the underside of a mollusk is used for movement.

Classification of Mollusks

Mollusks are classified according to whether or not they have a shell. Mollusks with a shell are then classified by the kind of shell and kind of foot that they have. Gastropods, bivalves, and cephalopods are the three most common groups of mollusks.

What are gastropods?

Gastropods make up the largest group of mollusks. The group includes snails, conchs, and garden slugs. All gastropods, except slugs, have a single shell. Many have a pair of tentacles with eyes at the tips.

Gastropods use a **radula** (RA juh luh), which is a tonguelike organ with rows of teeth, to get food. The radula works like a file to scrape and tear food materials. That's why snails are helpful in an aquarium. They use their radula to scrape the algae off the walls of the tank.

Slugs and many snails can live on land. The muscular foot helps them move. Glands in the foot secrete a layer of mucus on which they slide. Slugs do not have shells. They are protected by a layer of mucus instead, so they must live in moist places. Slugs and land snails damage plants when they eat leaves and stems.

FOLDABLES™

A Describe Make a three-tab book using notebook paper, as shown below. Use the Foldable to describe gastropods, bivalves, and cephalopods.

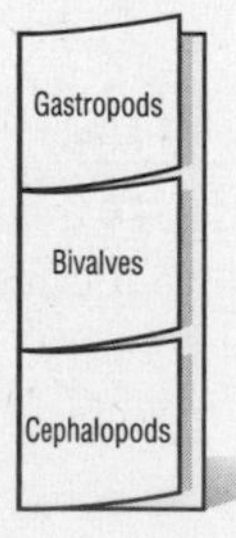

What are bivalves?

Bivalves are mollusks that have a hinged, two-part shell joined by strong muscles. Clams, oysters, and scallops are bivalves. They pull their shells closed by contracting the muscles near the hinge. They relax these muscles to open the shell. Bivalves are well adapted to living in water.

What are cephalopods?

Cephalopods (SE fuh luh pawdz) are the most specialized and complex mollusks. Squid and octopuses belong to this group. A cephalopod has a large, well-developed head. Its foot is divided into many tentacles with strong suction cups or hooks for capturing prey. Cephalopods are predators. They feed on fish, worms, and other mollusks. ☑

Squid and octopuses have a well-developed nervous system and large eyes. Unlike other mollusks, cephalopods have a closed circulatory system. In a **closed circulatory system**, blood containing food and oxygen moves through the body in a series of closed vessels, just as blood moves through blood vessels in a human body.

Reading Check

3. **List** Name two cephalopods.

____

____

How do cephalopods propel themselves?

All cephalopods live in oceans and are adapted for swimming. They have a water-filled cavity between an outer muscular covering and the internal organs. When the cephalopod tightens its muscular covering, water is forced out through an opening near the head. The jet of water propels, or moves, the cephalopod backwards, and it moves away quickly. The figure below compares a squid's movement to releasing air from a balloon.

A squid can propel itself at speeds of more than 6 m/s. However, it can only maintain this speed for a few seconds. Octopuses also can swim by jet propulsion. However, they usually use their tentacles to move slowly over the ocean floor.

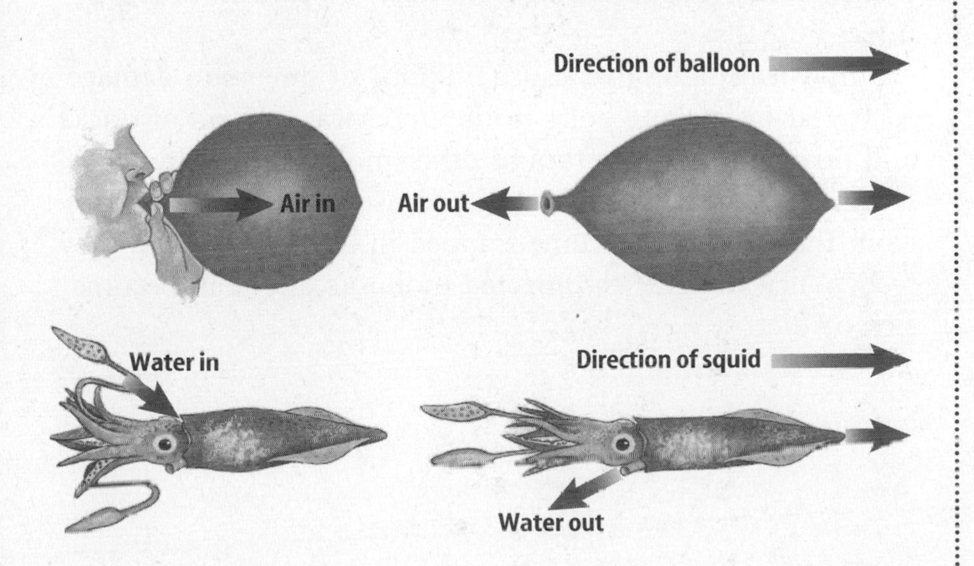

Picture This

4. **Explain** Use the figure to explain to a partner how a squid moves.

When did mollusks first appear on Earth?

Some species of mollusks have changed little from their ancestors. Mollusk fossils date back more than 500 million years. Many species of mollusks became extinct about 66 millions years ago. Today's mollusks are descendants of ancient mollusks.

Value of Mollusks

Mollusks have many uses. They are food for fish, birds, and humans. Many people make their living raising or collecting mollusks to sell for food. Other invertebrates, such as hermit crabs, use empty mollusk shells as their homes. Mollusk shells are used for jewelry and decoration. Several species of mollusks make pearls, but most pearls come from pearl oysters. ☑

Reading Check

5. Identify two uses of mollusks.

Mollusk shells provide information about the conditions in an ecosystem. Scientists use the shells to find the source and distribution of water pollutants. The internal shell of a cuttlefish is called the cuttlebone. Cuttlebones are used to provide calcium to caged birds. Squid and octopuses are able to learn new tasks, so scientists are studying their nervous systems to understand how learning takes place and how memory works.

What problems do mollusks cause?

Although they are helpful in many ways, mollusks can cause problems for humans. Land slugs and snails damage plants. Some species of snails are hosts of parasites that infect humans.

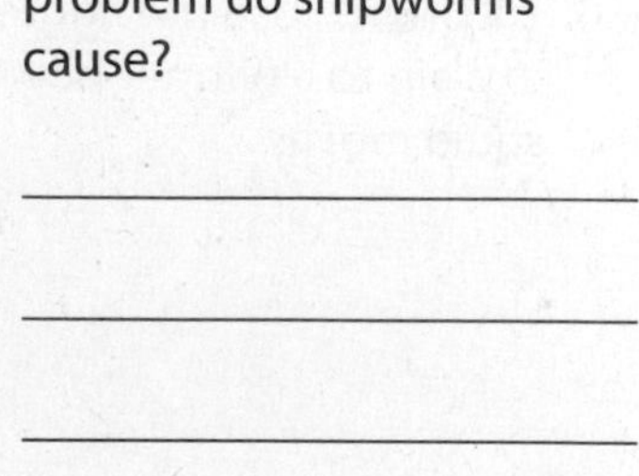

6. Determine What problem do shipworms cause?

Shipworms, a bivalve, cause millions of dollars in damage each year by making holes in the underwater wood of docks and boats. Clams, oysters, and other mollusks are filter feeders. Because of this, bacteria, viruses, and toxic protists from the water can become trapped in these animals as they feed. When humans eat infected mollusks, they can become sick or die.

After You Read

Mini Glossary

closed circulatory system: system in which blood containing food and oxygen moves through the body in a series of closed vessels

gill: organ in which carbon dioxide from the mollusk is exchanged for oxygen in the water

mantle: a thin layer of tissue that covers the body organs of a mollusk

open circulatory system: system in which the heart moves blood out into the open spaces around the body organs

radula (RA juh luh): a tonguelike organ with rows of teeth used to obtain food

1. Review the terms and their definitions in the Mini Glossary. Write a sentence that explains the difference between a closed circulatory system and an open circulatory system.

____

____

____

2. Complete the chart below by listing the characteristics of the three main groups of mollusks.

Type of Mollusk	Characteristics
Gastropods	
Bivalves	
Cephalopods	

Science Online Visit **life.msscience.com** to access your textbook, interactive games, and projects to help you learn more about mollusks.

Mollusks, Worms, Arthropods, Echinoderms

section 2 Segmented Worms

What You'll Learn
- the characteristics of segmented worms
- the structures of an earthworm
- why segmented worms are important

Before You Read

Where can you find worms in your community?

____

____

Create a Quiz Write a quiz question about the information you read on each page. Be sure to write the answers.

Read to Learn

Segmented Worm Characteristics

The worms that you see crawling on sidewalks after it rains are called annelids (A nuh ludz). They have tube-shaped bodies that are divided into many segments. On the outside of each segment are bristlelike structures called **setae** (SEE tee). Segmented worms use these structures to hold on to the soil and to move.

Segmented worms have bilateral symmetry and a body cavity that holds the organs. They also have two body openings—a mouth and an anus. Annelids are found in freshwater, salt water, and moist soil.

FOLDABLES™

B Identify Make a three-tab Foldable using notebook paper, as shown below. Write notes on the characteristics of earthworms, marine worms, and leeches.

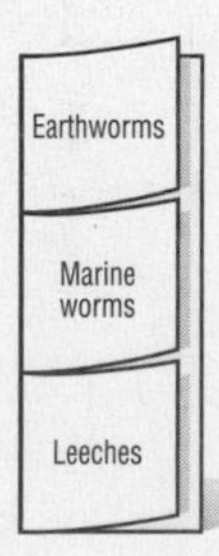

Earthworm Body Systems

Earthworms are the most well-known annelids. They have a front end, a back end, and more than 100 body segments. Except for the first and last segments, each body segment has four pairs of setae. Earthworms move by using their setae and two sets of muscles in the body wall. When an earthworm contracts one set of muscles, some of the segments bunch up and the setae stick out. This holds the worm to the soil. When the earthworm contracts the other set of muscles, the setae are pulled in and the worm moves forward.

How do digestion and excretion happen in earthworms?

The figure below shows the parts of an earthworm. As the earthworm burrows through the soil, it takes soil into its mouth. It gets energy from the leaves and other organic matter found in the soil. ☑

The soil that the worm takes in moves to the **crop**, which is a sac used for storage. Behind the crop is a muscular structure called the **gizzard**, which grinds the soil and the bits of organic matter. The ground material passes to the intestine. There the organic matter is broken down and nutrients are absorbed by the blood. Wastes leave the body through the anus. The wastes pile up at the openings to their burrows. The piles are called castings. Castings help fertilize the soil.

Reading Check

1. **Analyze** From what does an earthworm get energy?

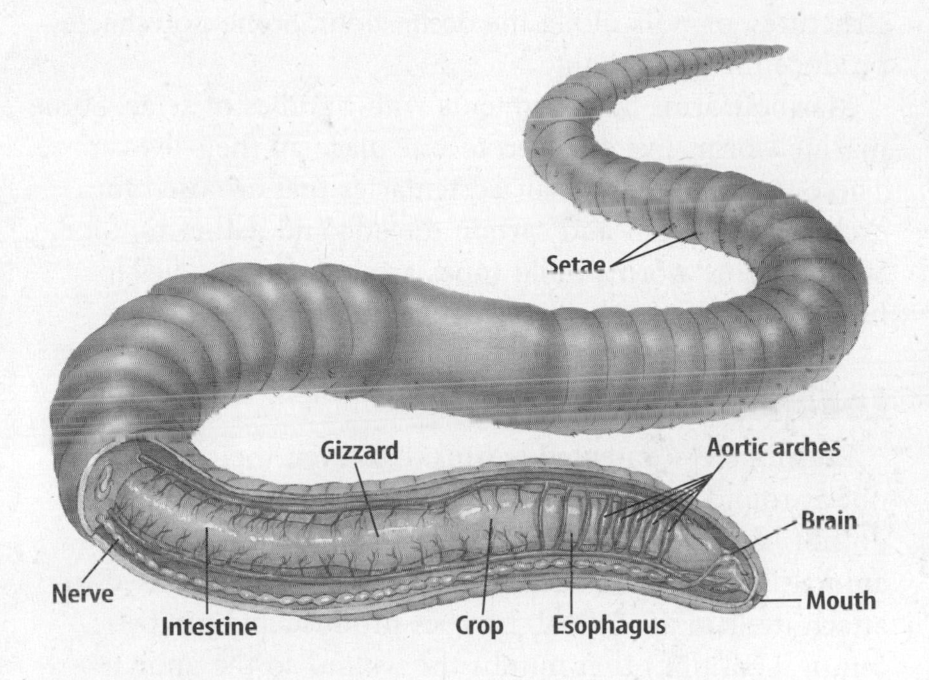

Picture This

2. **Sequence** Highlight the names of the four digestive structures of the worm. Number the structures 1 to 4 to show the order in which food moves from the mouth to the intestine.

What kind of circulatory system do earthworms have?

Earthworms have a closed circulatory system. There are two blood vessels located along the top of the body and one along the bottom. They meet in the front end of the earthworm, where they connect to heartlike structures called aortic arches. These structures pump blood through the body. Smaller vessels go into each body segment.

Earthworms have no lungs or gills. They exchange oxygen and carbon dioxide through their skin. The skin is covered with a thin film of watery mucus. If the mucus layer is removed, the earthworm could suffocate.

How do earthworms respond and reproduce?

Earthworms have a small brain located in the front segment. Each segment has nerves that join to form a nerve cord that connects to the brain. Earthworms respond to light, temperature, and moisture.

Earthworms are hermaphrodites (hur MA fruh dites). That means they produce eggs and sperm in the same body. A worm cannot fertilize its own eggs. It needs to receive sperm from another earthworm in order to reproduce. ☑

Reading Check

3. **Define** What is a hermaphrodite?

____

____

____

Marine Worms

There are more than 8,000 species of marine worms, or polychaetes (PAH lee keets). This is more than any other kind of annelid. Marine worms float, burrow, build structures, or walk along the ocean floor. Some polychaetes produce their own light.

Marine worms have segments with bundles of setae. Some marine worms live attached to one place all their lives. These worms have specialized tentacles that are used for exchanging oxygen and carbon dioxide and gathering food. Some marine worms build tubes around their bodies to hide in when something startles them.

Leeches

Leeches are segmented worms. However, their bodies are not as round or as long as those of earthworms. They also do not have setae. They feed on the blood of other animals. A sucker on each end of a leech's body is used to attach itself to an animal. Leeches produce an anesthetic (an us THE tihk) that numbs the wound so the animal won't feel the bite. After it attaches itself, the leech cuts into the animal and sucks out blood. They also can survive by eating aquatic insects and other organisms.

Reading Check

4. **Explain** why leeches are sometimes used on people after surgery.

____

____

____

Leeches and Medicine

Sometimes leeches are used after surgery to keep blood flowing to the surgical site. As the leeches feed on the blood, chemicals in their saliva prevent the blood from clotting. Other chemicals dilate the blood vessels, improving blood flow and helping the wound heal more quickly. Scientists are studying ways to use the chemicals that leeches produce to treat people with heart disease and arthritis. ☑

Value of Segmented Worms

Different kinds of segmented worms are helpful to other animals in a variety of ways. Earthworms help aerate, or add air, to soil by burrowing through it. By grinding and partially digesting soil and plant material, earthworms speed up the return of nitrogen to the soil for plants to use. ☑

Scientists are developing drugs based on the chemicals in the leeches' saliva. They hope the drugs will prevent blood clots. Marine worms are food for many fish, invertebrates, and mammals.

Reading Check

5. **Describe** how earthworms help aerate the soil.

____

____

Origin of Segmented Worms

Some scientists hypothesize that segmented worms evolved in the sea. The fossil record for segmented worms is limited because of their soft bodies. The tubes of marine worms are the most common fossils of the segmented worms. Some of these fossils date back about 620 million years.

There are similarities between mollusks and segmented worms. Scientists use these similarities to suggest that mollusks and segmented worms could have a common ancestor. These groups were the first animals to have a body cavity with space for body organs. Mollusks and segmented worms have a one-way digestive system with a separate mouth and anus. Their larvae, shown in the figure below, are similar. This provides the best evidence that they have a common ancestor.

Picture This

6. **Identify** Look at the figure. What do the similarities between the two larvae suggest?

____

____

____

Stomach
Mouth
Ring of cilia
Anus

Mollusk larva

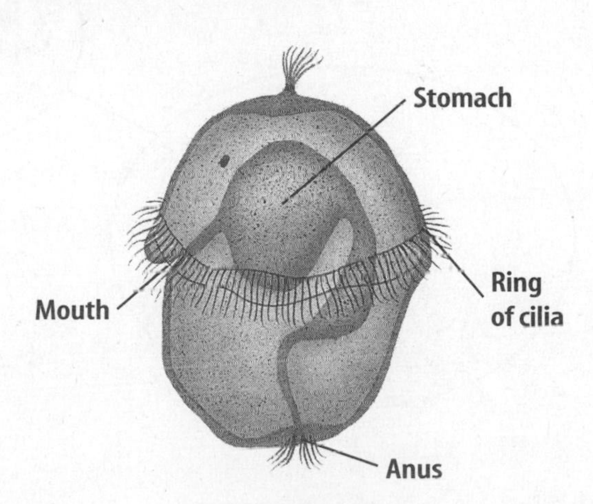

Annelid larva

After You Read

Mini Glossary

crop: storage sac to which ingested soil moves

gizzard: muscular structure behind the crop, which grinds the ingested soil and organic matter

setae (SEE tee): bristlelike structures on the outside of each body segment of earthworms and marine worms; used to hold on to the soil and to move

1. Review the terms and their definitions in the Mini Glossary. Choose one of the terms and write a sentence explaining its role in the earthworm's body.

____

____

____

2. Complete the web diagram below by describing the importance of segmented worms to people and other organisms.

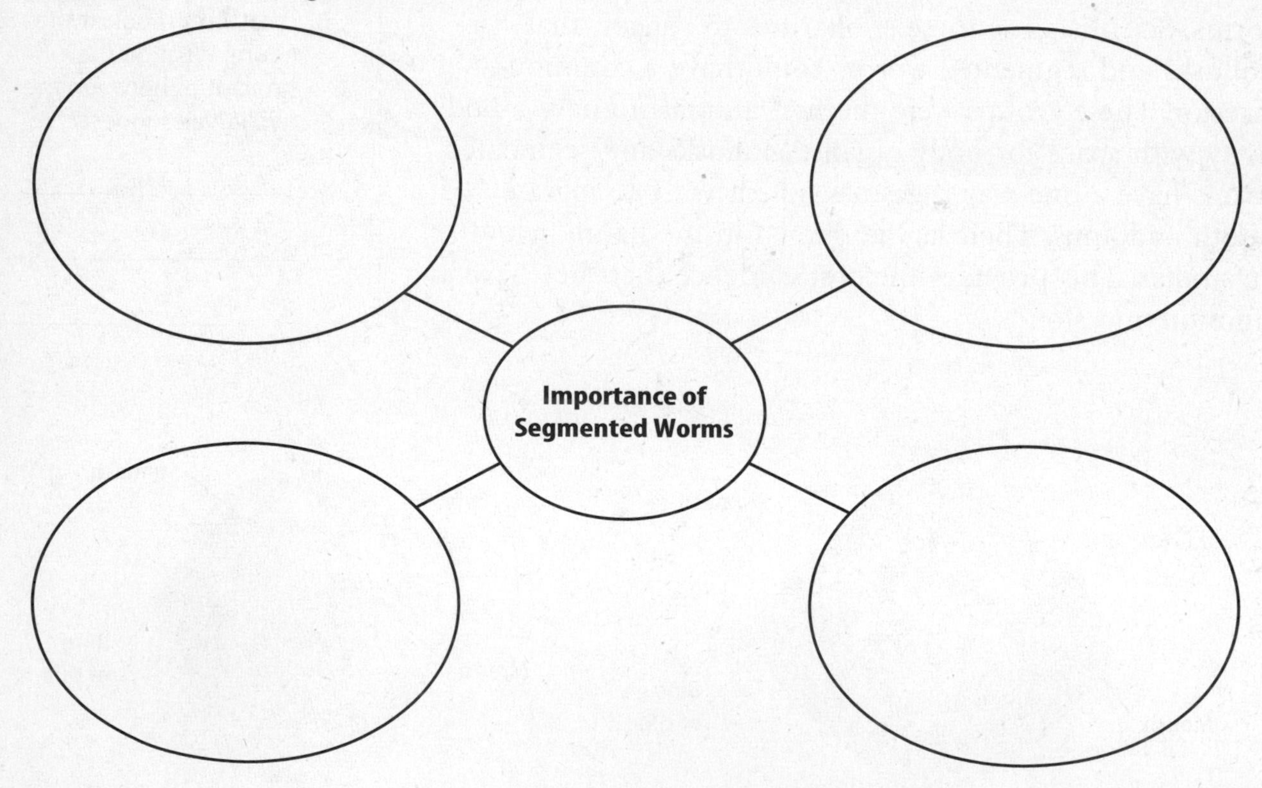

Science Online Visit **life.msscience.com** to access your textbook, interactive games, and projects to help you learn more about segmented worms.

Mollusks, Worms, Arthropods, Echinoderms

section 3 Arthropods

Before You Read

On the lines below, write the name of an insect and describe it.

____

____

____

What You'll Learn

- the characteristics used to classify arthropods
- the structure and function of the exoskeleton
- the difference between the two types of metamorphosis

Read to Learn

Characteristics of Arthropods

There are more than a million different species of arthropods (AR thruh pahdz). They are the largest group of animals. The jointed **appendages** of arthropods include legs, antennae, claws, and pincers. The appendages are adapted for moving around, capturing prey, feeding, mating, and sensing the environment.

Arthropods have bilateral symmetry. They have segmented bodies, an exoskeleton, a body cavity, a digestive system with two openings, and a nervous system. Most arthropods have separate sexes and reproduce sexually. ☑

Arthropods are adapted to live in almost every environment. They vary in size from microscopic dust mites to the large Japanese spider crab that can have a leg span of more than 3 m.

How are the bodies of arthropods divided?

Arthropods have segments like those of segmented worms. The segments of some arthropods are fused together to form body regions, such as those of insects and spiders.

Study Coach

Create an Outline Make an outline to summarize the information in this section. Use the main headings in the section as the main headings in the outline. Complete the outline with the information under each heading in the section.

Reading Check

1. **Identify** What type of symmetry do arthropods have?

____

What is the purpose of exoskeletons?

All arthropods have a hard, outer covering called an **exoskeleton.** It covers, supports, and protects the internal body. It also provides places for muscles to attach. In many land-dwelling arthropods, such as insects, the exoskeleton has a waxy layer to prevent water loss.

An exoskeleton does not grow as the animal grows. From time to time, the exoskeleton is shed and replaced by a new one in a process called **molting.** While they are molting, the arthropod's exoskeleton is soft and offers little protection from predators. Before the new exoskeleton hardens, the animal swallows air or water to increase its size. This way the new exoskeleton allows room for growth. ☑

Reading Check

2. **Analyze** Why do arthropods shed and replace their exoskeletons?

____

____

____

Insects

There are more species of insects than all other animal groups put together. Insects have three body regions—a head, a thorax, and an abdomen.

What does an insect's head look like?

An insect's head has a pair of antennae, eyes, and a mouth. Insects use the antennae for touch and smell. The eyes are simple or compound. Simple eyes can recognize light and dark. Compound eyes have many lenses and can detect colors and movement. The kind of mouthpart an insect has depends on what it eats.

What body parts are attached to the thorax?

The thorax has three pairs of legs and one or two pairs of wings. Some insects do not have wings. Other insects have wings only for part of their lives. Insects are the only invertebrate animals that can fly. Flying helps insects to find mates, food, and places to live. It also helps them escape from their predators.

FOLDABLES™

C Describe Make a two-tab Foldable using notebook paper, as shown below. Write notes that describe the characteristics of insects and other arthropods.

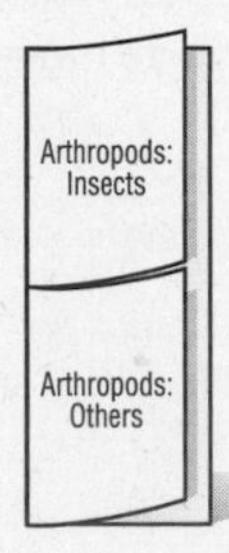

What is the purpose of an insect's abdomen?

An insect's reproductive structures are found in the abdomen. Females lay thousands of eggs. But only a small number of the eggs develop into adults. Insects have an open circulatory system. It carries digested food to cells and removes wastes. However, insect blood does not carry oxygen. Instead, insects have openings called **spiracles** (SPIHR ih kulz) on the abdomen and thorax. Air enters and waste gases leave the insect's body through these openings.

How do insects grow?

The body forms of many insects change as they grow. This series of changes is called **metamorphosis** (me tuh MOR fuh sihs). Insects such as grasshoppers and crickets go through incomplete metamorphosis. The first figure below shows the stages of incomplete metamorphosis. They are egg, nymph, and adult. In incomplete metamorphosis, nymphs are smaller versions of their parents. The nymph form molts several times before becoming an adult.

Many insects, such as butterflies, beetles, ants, and bees, go through complete metamorphosis. The second figure shows the stages of complete metamorphosis. They are egg, larva, pupa, and adult. Caterpillar is the common name for the larva of a moth or butterfly. Other insect larvae are called grubs, maggots, or mealworms. Only larval forms molt.

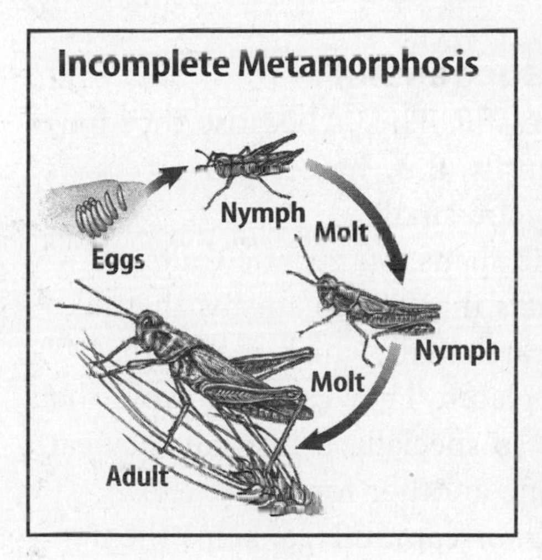

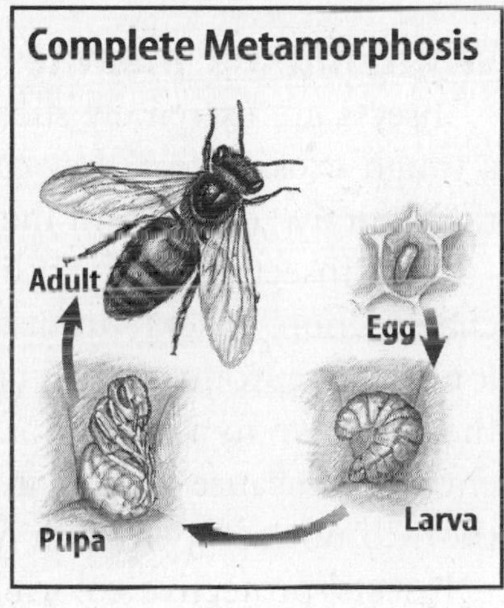

Picture This

3. Identify Highlight the names of the stages that are the same for complete and incomplete metamorphosis.

How do insects get food?

Insects feed on plants, the blood of animals, flower nectar, decaying materials, wood, and clothes. Insects have varied mouthparts, depending on what they feed upon. Grasshoppers and ants have large mandibles (MAN duh bulz) for chewing plants. Butterflies and honeybees have siphons for drinking in nectar in flowers. Praying mantises eat other animals. Some moth larvae eat wool clothing. Mosquitos, fleas, and lice drink the blood and body fluids of other animals. The mouthparts of grasshoppers, butterflies, and mosquitoes are shown at the top of the next page. ☑

4. Explain Why do the mouthparts of insects vary widely?

____

____

____

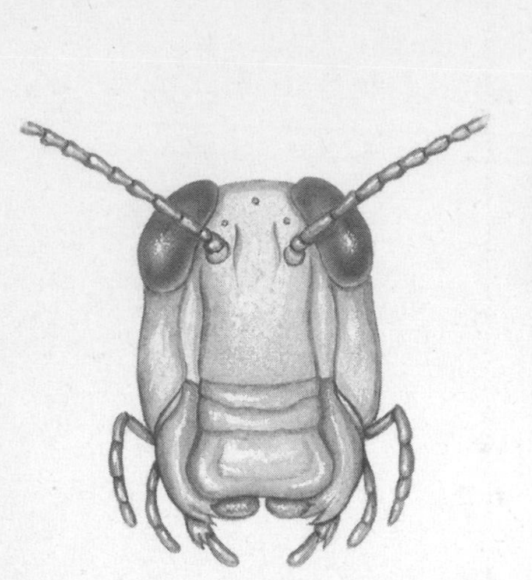

Grasshopper

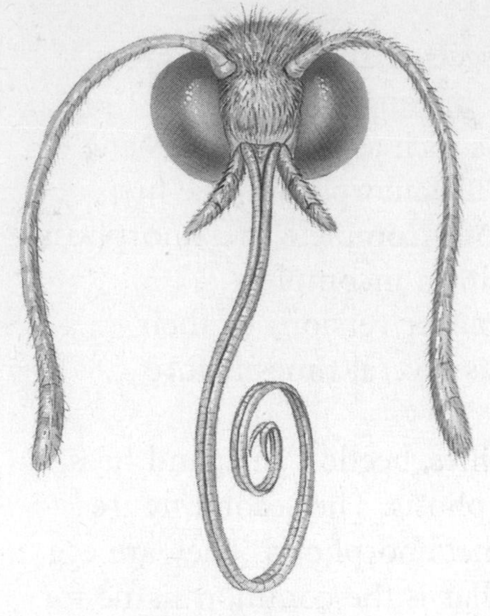

Butterfly

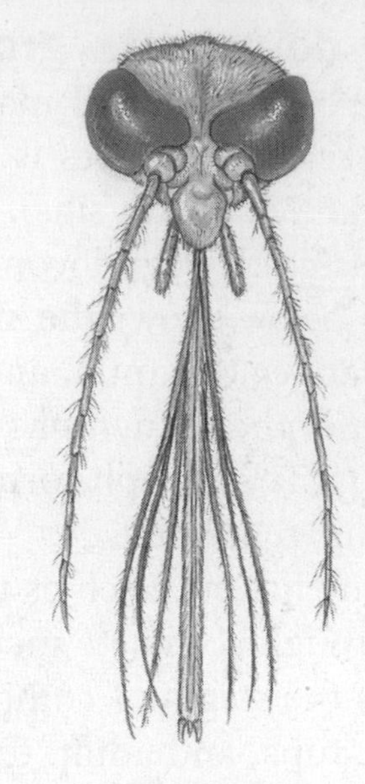

Mosquito

Picture This

5. Identify Match the following foods with the insect that feeds on them: blood, nectar, plants. Write your choices on the lines below the names of the insects.

What makes insects successful?

Insects are extremely successful. This is because they have a tough exoskeleton, they can fly, they have rapid reproductive cycles, and they are small.

Most insects have short life spans. So genetic traits can change more quickly in insects than in organisms that take longer to reproduce. Insects are small, which means that they can live in a variety of places. They can also avoid their enemies. Because insects are so specialized in what they eat, they do not compete with one another for food.

Insects' protective coloring, or camouflage, helps them blend in with their surroundings. Some moths resting on trees look like part of the bark. Some caterpillars look like twigs. When a leaf butterfly folds its wings, it looks like a dead leaf.

Arachnids

Spiders, scorpions, mites, and ticks are examples of arachnids (uh RAK nudz). They have two body regions—a head-chest region that is called the cephalothorax (se fuh luh THOR aks) and an abdomen. Arachnids have four pairs of legs but no antennae. Some arachnids kill prey with venom, or poison, glands. Some use stingers or fangs to kill prey. Other arachnids are parasites. ☑

Reading Check

6. List the two body regions of arachnids.

What are scorpions like?

A scorpion has a sharp, venom-filled stinger at the end of its abdomen. The venom from the stinger paralyzes the prey. Scorpions have a pair of appendages, called pincers, with which they catch their prey. A scorpion sting can be fatal to humans. ☑

Reading Check

7. Describe How do scorpions use their pincers?

What are spiders like?

Spiders cannot chew their food. Instead they release enzymes, or chemicals, into their prey that help digest it. The spider then sucks the liquid into its mouth.

Oxygen and carbon dioxide are exchanged in a spider's book lungs, shown in the figure below. Air circulates between the moist folds of the book lungs bringing oxygen to the blood. Openings on the abdomen allow the gases to move into and out of the lungs.

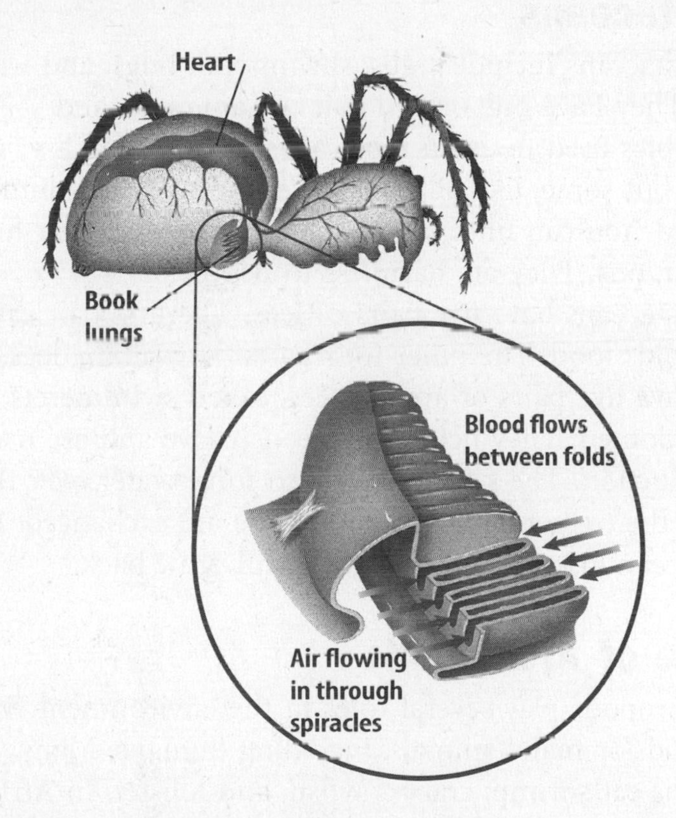

Picture This

8. Identify In what body region are the book lungs located?

What are mites and ticks like?

Most mites are so small that they look like tiny specks to the human eye. Most mites are animal or plant parasites. All ticks are animal parasites. They attach to the skin of their hosts and take blood from their hosts through specialized mouthparts. Ticks often carry bacteria and viruses that cause disease in humans and other animals. Lyme disease and Rocky Mountain spotted fever are carried by ticks.

Centipedes and Millipedes

Centipedes and millipedes are two groups of arthropods that have long bodies with many segments and many legs. They have antennae and simple eyes. They live in damp places, such as in woodpiles and in basements. Centipedes and millipedes reproduce sexually. They lay eggs in nests and stay with them until they hatch. Centipedes have one pair of legs per segment.

Centipedes are predators—they capture and eat their prey, which includes snails, slugs, and worms. They have claws that they use to inject venom into their prey. Their pinches are painful to humans, but are not fatal.

Millipedes have two pairs of legs per segment. They feed on plants and decaying material. They are often found under damp plant material.

Think it Over

9. Compare Name one difference between centipedes and millipedes.

____

____

____

Crustaceans

Crustaceans include crabs, shrimp, pill bugs, and water fleas. They have one or two pairs of antennae and mandibles used to crush food. Most crustaceans live in water, but some, like pill bugs, live in moist environments on land. You can find pill bugs in gardens and near house foundations. They are harmless to people.

Crustaceans have five pairs of legs. The first pair catches and holds food. The other four pairs are walking legs. They also have five pairs of appendages, called swimmerets, on the abdomen. They help crustaceans move and are used in reproduction. The swimmerets also force water over the gills where the oxygen and carbon dioxide are exchanged. If a crustacean loses an appendage, it will grow back.

Value of Arthropods

Arthropods play several roles in the environment. They are food for many animals, including humans. Many humans eat shrimp, crab, crayfish, and lobster. In Africa and Asia, many people eat insect larvae and insects such as grasshoppers, termites, and ants. These insects are excellent sources of protein.

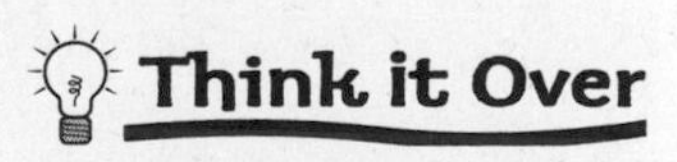

10. Determine What are two ways arthropods benefit humans?

____

____

Agriculture would not be possible without bees and other insects that pollinate crops. Bees make honey, and silkworms make silk. Many insects and spiders are predators of harmful species, such as stableflies. Some arthropods provide useful chemicals. Bee venom is used to treat rheumatoid arthritis.

What problems do arthropods cause?

Some arthropods are not useful to people. Almost every crop has some insect pest that feeds on it. Many arthropods—mosquitoes, tsetse flies, fleas, and ticks—carry diseases that harm humans and other animals. Other arthropods, such as carpenter ants, moths, and termites, destroy food, clothing, and property. However, insects are important to ecosystems. Removing all insects from an ecosystem would cause more harm than good. ☑

How are insects controlled?

Insecticides are used to control problem insects. But many insecticides also kill helpful insects. In addition, many poisonous substances that kill insects stay in the environment. They can build up in the bodies of animals that eat them. As other animals eat the contaminated animals, the insecticides find their way into human food. The toxins can harm people.

Different types of bacteria, fungi, and viruses are being used to control some insect pests. In some cases, natural predators of insect pests have been successful in controlling them. Other ways of controlling insect pests include using chemicals that interfere with the reproduction or behavior of insect pests.

What is the origin of arthropods?

Because of their hard body parts, arthropod fossils are among the oldest and best-preserved fossils of many-celled animals. One of the most recognized types of fossils is an arthropod—the trilobite. The figure below shows one of the more than 15,000 species of trilobites that have been classified. Some arthropod fossils are more than 500 million years old. Scientists hypothesize that arthropods came from an ancestor of segmented worms. This is because both earthworms and leeches have individual body segments. Over time, groups of body segments fused and became adapted for locomotion, feeding, and sensing the environment. The hard exoskeleton and walking legs allowed arthropods to be among the first successful land animals.

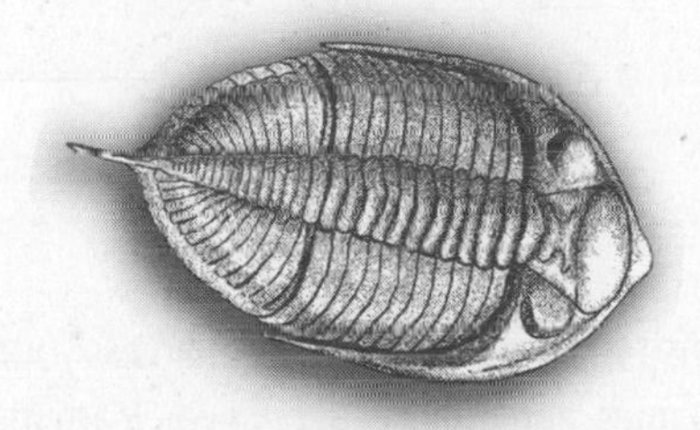

Reading Check

11. Explain In what ways are arthropods harmful to people?

____

____

____

Picture This

12. Apply What characteristics of arthropods made them among the first successful land animals?

____

____

____

After You Read

Mini Glossary

appendages: in arthropods, includes legs, antennae, claws, and pincers
exoskeleton: hard, outer covering
metamorphosis (me tuh MOR fuh sihs): a series of changes in body form
molting: process in which exoskeleton is shed and replaced by a new one
spiracle (SPIHR ih kul): opening on the abdomen and thorax of an arthropod through which air enters and waste gases leave the insect's body

1. Review the terms and their definitions in the Mini Glossary. Choose one of the terms above and write a sentence that explains the role of the term in an arthropod's body.

____

____

____

2. Complete the web diagram to identify five types of arthropods discussed in this section.

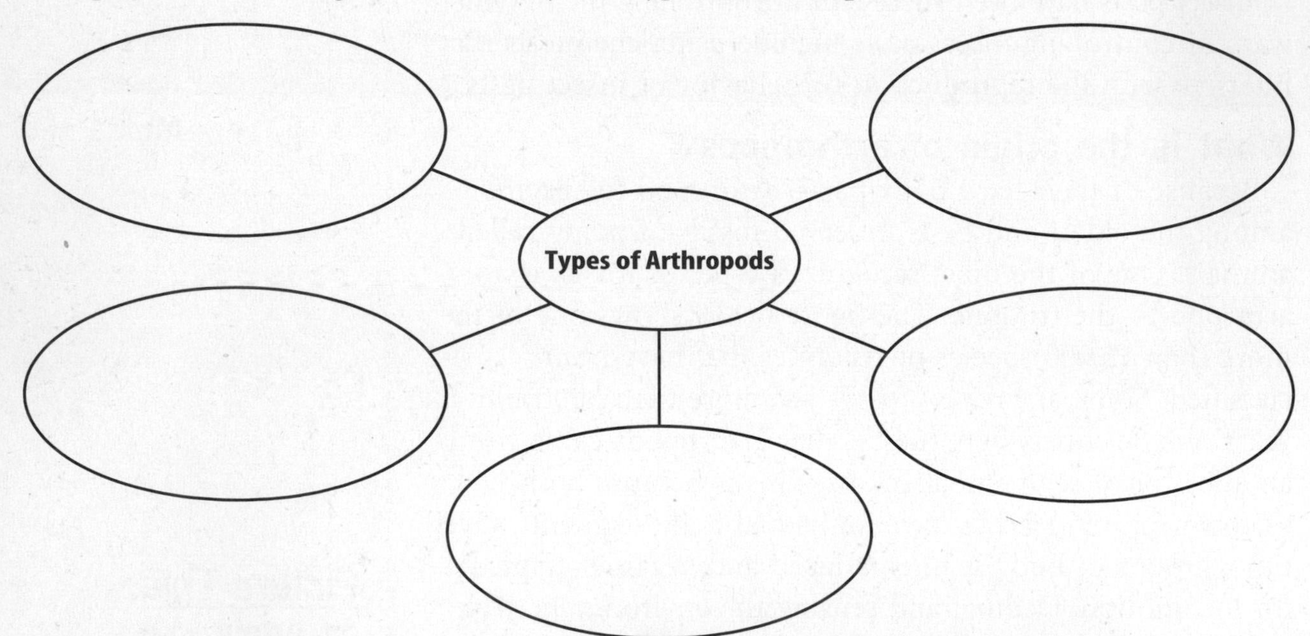

3. How did the outline help you organize what you learned about arthropods?

____

____

Science Online Visit **life.msscience.com** to access your textbook, interactive games, and projects to help you learn more about arthropods.

Mollusks, Worms, Arthropods, Echinoderms

section 4 Echinoderms

Before You Read

If you were walking on a beach by an ocean, what animals would you expect to see? After you read this section, see if any of the animals you named are echinoderms.

____

____

What You'll Learn
- the characteristics of echinoderms
- how sea stars obtain and digest food
- why echinoderms are important

Read to Learn

Echinoderm Characteristics

Echinoderms (ih KI nuh durmz) are found in oceans. They have a hard endoskeleton covered by a thin, bumpy, or spiny skin. Because they have radial symmetry, they can sense things in their environment from all directions.

All echinoderms have a mouth, stomach, and intestines, as shown in the figure below. They feed on plants and animals. Echinoderms have no head or brain. They do have a nerve ring that surrounds the mouth. They also have cells that respond to light and touch.

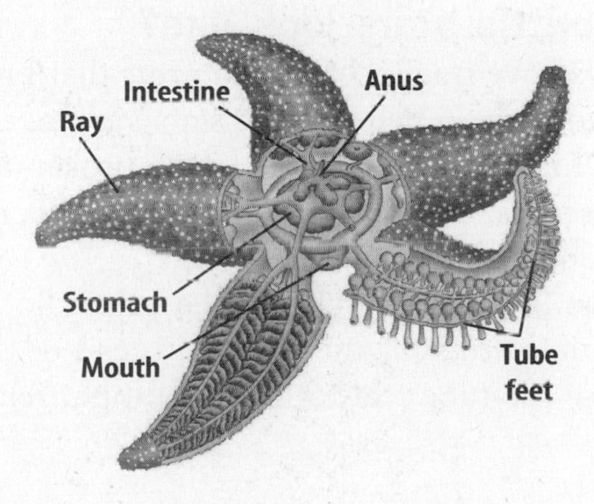

Study Coach

Identify the Main Point Read each subhead. Then work with a partner to write questions about the information found under each subhead. Take turns asking and answering the questions. Use the questions as a study guide about echinoderms.

Picture This

1. **Identify** Shade the mouth and intestines of the sea star.

Reading Check

2. Explain What is the main purpose of tube feet?

____

____

____

What is the water-vascular system?

Echinoderms have a **water-vascular system**, which is a network of water-filled canals with thousands of tube feet connected to it. The water-vascular system allows echinoderms to move, exchange carbon dioxide and oxygen, capture food, and release wastes.

Tube feet are hollow, thin-walled tubes that end in suction cups. As the pressure in the tube feet changes, the animal is able to move along by pushing out and pulling in its tube feet. ☑

Types of Echinoderms

There are about 6,000 species of echinoderms living today. More than one third are sea stars. Other groups of echinoderms include brittle stars, sea urchins, sand dollars, and sea cucumbers.

FOLDABLES™

D Describe Make a five-tab book using notebook paper, as shown below. Describe the characteristics of five types of echinoderms.

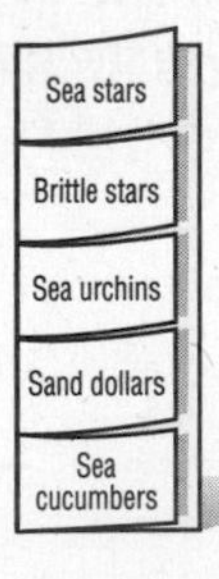

What do sea stars look like?

Sea stars have at least five arms arranged around a central point. The arms have thousands of tube feet. Sea stars use the tube feet to open the shells of their prey. When the shell opens a little, the sea star pushes its stomach through its mouth and into its prey. The sea star's stomach surrounds the soft body of the prey and gives off enzymes that help digest it. When the meal is over, the sea star pulls its stomach back into its own body.

Sea stars reproduce sexually. Females release eggs and males release sperm into the water. Females can produce millions of eggs in one season.

Sea stars can grow new body parts through regeneration. If a sea star loses an arm, a new one will grow. If enough of the center disk is left attached to a severed arm, a whole new sea star can grow from the piece of arm.

What do brittle stars look like?

Brittle stars have fragile, branched arms that break off easily. This adaptation helps brittle stars survive attacks by predators. While a predator is eating the broken arm, the brittle star escapes. The broken part grows back, or regenerates quickly.

Brittle stars live under rocks on the ocean floor. They use their flexible arms for movement instead of their tube feet. They use the tube feet to move food particles into their mouth.

What do sea urchins and sand dollars look like?

Another group of echinoderms includes sea urchins, sea biscuits, and sand dollars. They are disk-shaped animals covered with spines. They do not have arms, but sand dollars have a five-pointed pattern on their surface.

Sand dollars have stiff, hairlike spines and sea urchins have long, pointed spines that protect them from predators. Some sea urchins have sacs near the end of the spines that hold toxic fluid that is injected into predators. The spines also help the animals move and burrow. Sea urchins have five toothlike structures around their mouth.

What do sea cucumbers look like?

Sea cucumbers are soft-bodied echinoderms with a leathery covering. They have tentacles around their mouth and rows of tube feet on their upper and lower surfaces. When sea cucumbers are threatened, they force out their internal organs. These organs grow back in a few weeks. Some sea cucumbers feed on dead and decaying matter called detritus (de TRI tus) found on the ocean floor.

Value of Echinoderms

Echinoderms are important to ocean environments because they feed on dead organisms and help recycle materials. Sea urchin eggs and sea cucumbers are used for food in some places. Many echinoderms are used in research and some might be possible sources of medicines. Sea stars are predators that control the populations of other animals. However, because sea stars eat oysters and clams, they also destroy millions of dollars' worth of mollusks each year. ☑

What is the origin of echinoderms?

A good fossil record of echinoderms exists. Echinoderms date back more than 400 million years. The earliest echinoderms might have had bilateral symmetry as adults and may have been attached to the ocean floor by stalks. Many larval forms of modern echinoderms have bilateral symmetry.

Scientists hypothesize that echinoderms more closely resemble animals with backbones than any other group of invertebrates. This is because echinoderms have complex body systems and an embryo that develops the same way that embryos of animals with backbones develop.

Think it Over

3. **Compare** What is one difference between sea stars and sand dollars?

____

____

____

Reading Check

4. **Explain** In what way are echinoderms important to the ocean environment?

____

____

____

After You Read

Mini Glossary

tube feet: hollow, thin-walled tubes that each end in a suction cup

water-vascular system: a network of water-filled canals with thousands of tube feet connected to it

1. Review the terms and their definitions in the Mini Glossary. Write a sentence explaining how echinoderms use the water-vascular system.

2. Choose one of the question headings in the Read to Learn section. Write the question in the space below. Then write your answer to that question on the lines that follow.

Write your question here.

3. Complete the diagram below by identifying five types of echinoderms.

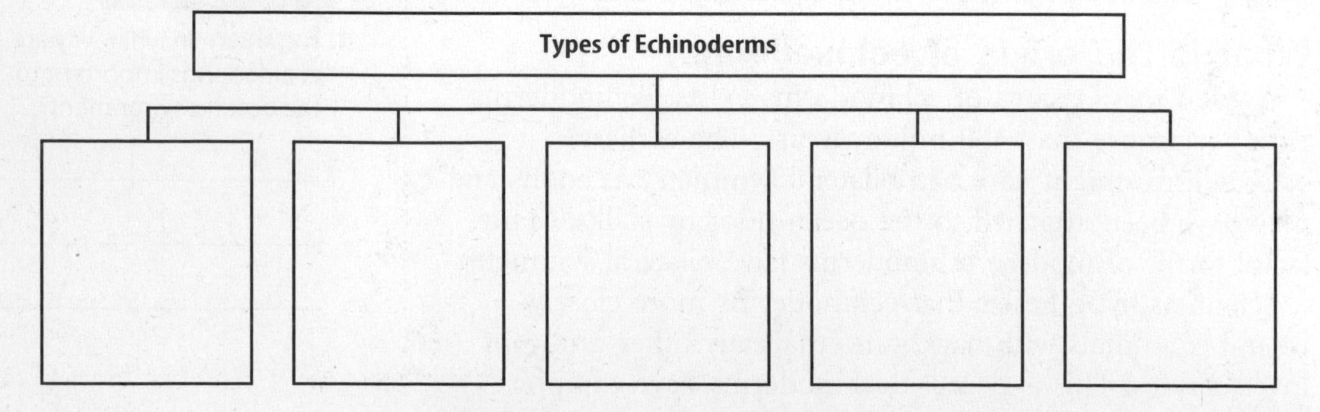

End of Section

Science Online Visit **life.msscience.com** to access your textbook, interactive games, and projects to help you learn more about echinoderms.

Fish, Amphibians, and Reptiles

section 1 Chordates and Vertebrates

Before You Read

Describe what happens to your body temperature when you go outside on a cold day or on a hot day.

____

____

____

What You'll Learn

- the characteristics of chordates
- the characteristics of vertebrates
- the difference between ectotherms and endotherms

Read to Learn

Chordate Characteristics

Many types of animals are classified as chordates. A **chordate** (KOR dayt) is an animal that has four characteristics present at some stage of its development—a notochord, postanal tail, nerve cord, and pharyngeal pouches. These are shown in the figure below.

Study Coach

Write a Summary of the section using all the vocabulary words. Be sure your summary includes the main idea from the section.

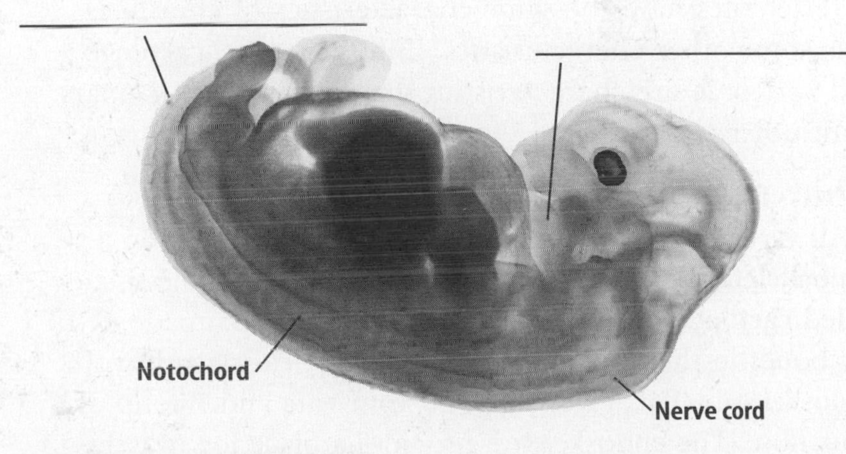

Picture This

1. **Identify** Select one of the four characteristics of chordates identified in the figure. Skim the section to find the function of this feature. Write the function on the line below the label.

FOLDABLES™

A Describe Make a four-tab Foldable, as shown below, to describe the four chordate characteristics.

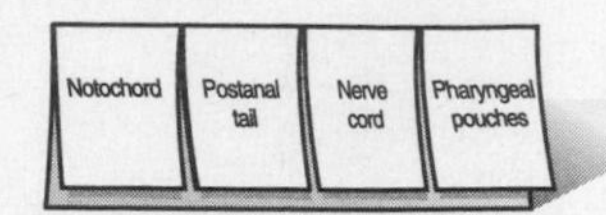

Reading Check

2. Explain What happens to the nerve cord as the chordate develops?

____

____

____

What is a notochord?

The **notochord** is located inside the chordate and supports the animal and extends along the upper part of the animal's body. The notochord is flexible but firm. Some chordates, such as fish, amphibians, reptiles, birds, and mammals develop backbones that partly or completely replace the notochord. These animals are called vertebrates. Some animals, such as the sea squirt, keep the notochord into adulthood.

What is the postanal tail?

The notochord extends into the postanal tail. The **postanal tail** is a muscular structure at the end of the developing chordate.

Why is the nerve cord important?

The **nerve cord** is a tubelike structure along the length of the developing chordate's body. As most chordates develop, the front end of the nerve cord enlarges to form the brain and the rest becomes the spinal cord. The brain and the spinal cord become the central nervous system. ☑

Where are the pharyngeal pouches?

The **pharyngeal pouches** are found between the mouth and the digestive tube. They are pairs of openings to the outside of a developing chordate. Ancient chordates used them for filter feeding. Some chordates today, such as lancelets, still use pharyngeal pouches for filtering food. In humans, pharyngeal pouches are present only as the embryo develops. One pair of these pouches becomes the tubes that go from the ears to the throat.

Vertebrate Characteristics

Vertebrates have the same characteristics of chordates plus some other characteristics. Endoskeletons, cartilage, and vertebrae are characteristics that set vertebrates apart from other chordates.

What is the structure of vertebrates?

All vertebrates have an internal framework called an **endoskeleton**. It is made up of bone and/or flexible tissue called **cartilage**. In humans, the endoskeleton is made of all the bones in the body. There also is some cartilage in your endoskeleton. This gives shape to your ears and the tip of your nose. The endoskeleton provides a place for muscle attachment and supports and protects the body's organs.

What is the backbone?

Part of the endoskeleton is the flexible column called the backbone. It is a stack of **vertebrae** alternating with cartilage. The backbone surrounds and protects the spinal nerve cord. Vertebrates also have a head with a skull that encloses and protects the brain. Most of a vertebrate's internal organs are in the central part of the body. A vertebrate has skin covering its body. Sometimes hair, feathers, scales, or horns grow from the skin.

What are the main vertebrate groups?

The table below shows the seven main groups of vertebrates found on Earth today. Vertebrates are either ectotherms or endotherms.

Group	Estimated Number of Species	Examples
Jawless fish	60	lamprey, hagfish
Jawed cartilaginous fish	500 to 900	shark, ray, skate
Bony fish	20,000	salmon, bass, guppy, sea horse, lungfish
Amphibians	4,000	frog, toad, salamander
Reptiles	7,970	turtle, lizard, snake, crocodile, alligator
Birds	8,700	stork, eagle, sparrow, turkey, duck, ostrich
Mammals	4,600	human, whale, bat, mouse, lion, cow, otter

Picture This

3. Classify Circle the name of the vertebrate group that has the most species.

An ectotherm is a cold-blooded animal. An **ectotherm** has an internal body temperature that changes with the temperature of its surroundings. Fish, amphibians, and reptiles are ectotherms.

An endotherm is a warm-blooded animal. An **endotherm** has an internal temperature that changes little. Birds and mammals are endotherms. When you go outside on a hot day or cold day, your body temperature does not change much. You are an endotherm.

FOLDABLES™

B Identify Make a two-tab Foldable, as shown below, to list facts about ectotherms and endotherms.

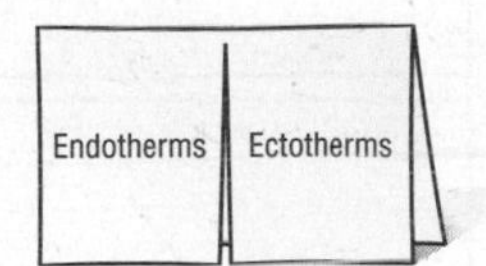

What fossil records of vertebrates exist?

There are fossils of vertebrates that lived about 420 million years ago (mya). The oldest known amphibian fossils date to about 370 mya. Reptile fossils have been found in deposits that are about 350 million years old. Mammals first appeared about 190 mya.

After You Read

Mini Glossary

cartilage: tough, flexible tissue that joins vertebrae and makes up all or part of the vertebrate endoskeleton

chordate (KOR dayt): an animal that has four characteristics present at some stage of its development—a notochord, postanal tail, nerve cord, and pharyngeal pouches

ectotherm: a cold-blooded animal that has an internal body temperature that changes with the temperature of its surroundings

endoskeleton: an internal framework in all vertebrates

endotherm: a warm-blooded animal that has an internal body temperature that changes little

nerve cord: a tubelike structure above the notochord and along the length of the developing chordate's body, which eventually develops into the central nervous system in most chordates

notochord: an internal structure that supports an animal and extends along the upper part of its body

pharyngeal pouches: pairs of openings to the outside of a developing chordate

postanal tail: a muscular structure at the end of the developing chordate

vertebrae: the stack of bones, alternating with cartilage, that form the column of the backbone

1. Review the terms and their definitions in the Mini Glossary. Write a sentence to explain the difference between an ectotherm and an endotherm.

____

____

2. Complete the concept map below to show the characteristics of vertebrates.

Chordates

- **1.**
- **2.**
- **3.**
- **4.**

Vertebrates

- **1.**
- **2.**
- **3.**
- **4.**
- **5.**

Science Online Visit **life.msscience.com** to access your textbook, interactive games, and projects to help you learn more about chordates and vertebrates.

Fish, Amphibians, and Reptiles

section 2 Fish

Before You Read

On each line below, write the name of a type of fish. Next to each name, describe some characteristics of that species.

____

____

____

What You'll Learn
- the characteristics of the three classes of fish
- how fish obtain food and oxygen and reproduce
- the importance and origin of fish

Read to Learn

Fish Characteristics

There are more species of fish than species of any other vertebrate group. All fish are ectotherms and live in freshwater or salt water. Some fish, such as salmon, live in both freshwater and salt water. Fish are found at different depths, from shallow pools to deep oceans.

Mark the Text

Identify Main Ideas Underline the main idea in each paragraph. Circle the details that support the main ideas. Use this information to study the section.

Why can fish move quickly through the water?

A streamlined shape, a muscular tail, and fins allow most fish to move quickly through the water. **Fins** are fanlike structures attached to the endoskeleton. Fish use fins to steer, balance, and move. Paired fins on the sides allow fish to move right, left, forward, and backward. Fins on the top and bottom of the body give the fish stability. Most fish secrete a slimy mucus that also helps them move through water.

What are scales?

Most fish have scales. **Scales** are hard, thin plates that cover the skin and protect the body, like shingles on the roof of a house. Most fish scales are made of bone. Scales can be tooth shaped, diamond shaped, cone shaped, or round. The shape of the scales can be used to classify fish. ☑

Reading Check

1. **Explain** What is the purpose of scales?

____

____

____

How do fish sense their surroundings?

All fish have highly developed sensory systems. Most fish have a lateral line system that is made up of a shallow, canal-like structure that runs the length of the fish's body and is filled with sensory organs. This system allows fish to sense their environment and to detect movement. Some fish, such as sharks, have a strong sense of smell. A fish has a two-chambered heart in which oxygen-filled blood mixes with carbon dioxide-filled blood.

How do fish get oxygen?

Most fish have organs called gills that exchange carbon dioxide and oxygen. Gills are located on both sides of the fish's head and contain many tiny blood vessels. When a fish takes water into its mouth, the water passes over the gills, where oxygen from the water is exchanged with carbon dioxide in the blood. The water then passes out through openings on each side of the fish.

Think it Over

2. Infer What organ in humans has a similar function as gills do in fish?

____

How do fish feed?

Fish get food in different ways. Some of the largest sharks swim with their mouths open. They take in small fish as they swim. Parrot fish use their hard beaks to bite off pieces of coral. An electric eel produces a powerful electric shock to stun its prey before it eats it. The archerfish shoots down insects by spitting drops of water at them. Some fish have strong teeth, but most do not chew their food. Instead, they use their sharp teeth to catch their food or to tear off chunks of food.

How do fish reproduce?

Fish reproduce sexually. Most female fish release large numbers of eggs into the water. Males then swim over the eggs and release sperm. This way of reproducing is called spawning. The joining of the egg and sperm cells outside the female's body is called external fertilization. The joining of the egg and sperm cells inside the female's body is called internal fertilization. Some sharks and rays have internal fertilization and lay fertilized eggs. Some other fish, such as guppies and other sharks, have internal fertilization but the eggs develop and hatch inside the female's body. After they hatch, they leave her body. ☑

Reading Check

3. Explain What is spawning?

____

____

____

____

Fish that do not care for their young release hundreds or even millions of eggs. Fish that care for their young lay fewer eggs.

Types of Fish

Fish are different in size, shape, color, living environments, and in many other ways. Even though fish have many differences, there are only three groups of fish—jawless fish, jawed cartilaginous (kar tuh LA juh nuss) fish, and bony fish.

Jawless Fish

Jawless fish have round, toothed mouths. Their long, tubelike bodies are covered with slimy skin but no scales. The endoskeletons of jawless fish are flexible and made of cartilage.

Lampreys and hagfish are jawless fish. Most lampreys attach to other fish with their suckerlike mouths. They are parasites that feed on the blood and body fluids of the host fish. Hagfish feed on dead or dying fish. They also eat other water animals. Some lamprey species live in salt water, while other species live in freshwater. Hagfish live in salt water.

Jawed Cartilaginous Fish

Jawed cartilaginous fish have endoskeletons made of cartilage. They have movable jaws that usually have well-developed teeth. Their bodies are covered with tiny scales. Sharks, skates, and rays are jawed cartilaginous fish.

Bony Fish

The fish you named at the beginning of the section are probably bony fish. About 95 percent of all species of fish are bony fish. Bony fish have skeletons made of bone. A bony flap covers and protects the gills. It closes as water moves into the mouth and over the gills. When the bony flap opens, water leaves the gills. The figure below shows the basic body structure of bony fish.

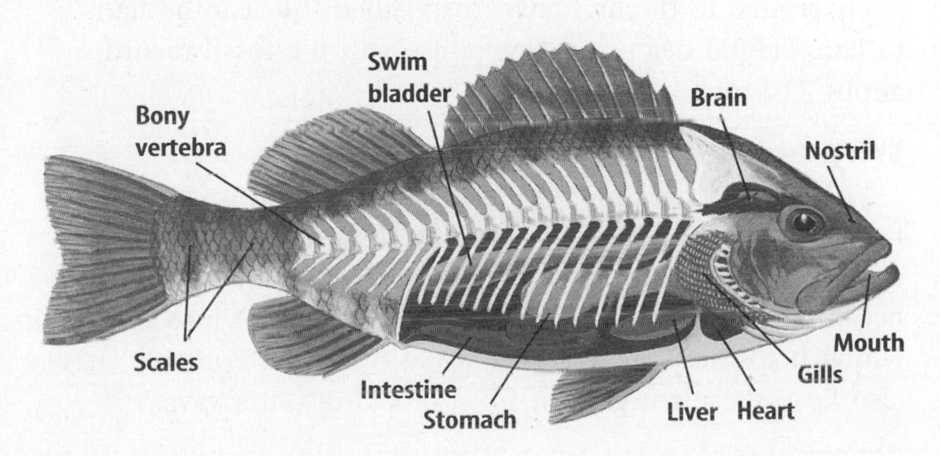

FOLDABLES™

C Describe Make a three-tab Foldable, as shown below, to record descriptions of each of the three groups of fish.

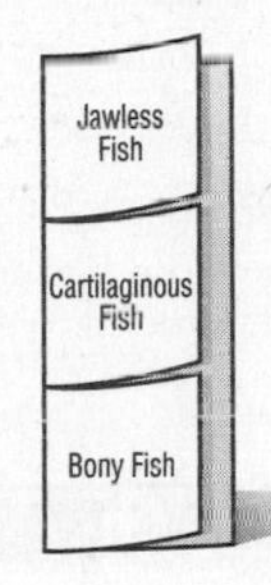

Picture This

4. Compare Highlight the body parts that fish and humans have in common.

What is a swim bladder?

Most bony fish have a swim bladder. A swim bladder is an air sac that allows the fish to adjust its density in response to the density of the surrounding water. If a fish is denser than the surrounding water, it will sink. If a fish is less dense than the surrounding water, it will float on top of the water. If a fish is the same density as the surrounding water it will not sink and it will not float to the top of the water.

How does the swim bladder work?

The swim bladder is like a balloon. It inflates and deflates depending on how much gas is in it. The exchange of gases between the swim bladder and the blood causes the swim bladder to inflate and deflate. As the swim bladder fills with gases, the fish's density decreases and it rises in the water. When the swim bladder deflates, the fish's density increases and it sinks. Glands in the fish control the amount of gas in the swim bladder, so the fish is able to stay at a certain depth in the water. ☑

5. **Explain** the main purpose of the swim bladder.

____

____

____

What are the three types of bony fish?

The three types of bony fish are the lobe-finned fish, the lungfish, and the ray-finned fish. Lobe-finned fish have fins that are lobelike and fleshy. Scientists hypothesize that fish similar to the lobe-finned fish are the ancestors of amphibians.

Lungfish have one lung and gills. This adaptation allows them to live in shallow waters that have little oxygen.

Most bony fish are ray-finned fish. They have fins made of long, thin bones covered with skin. Salmon, tuna, and swordfish are examples of ray-finned fish.

What do fossils tell us about fish?

The earliest fossils of fish are those of jawless fish that lived about 450 million years ago. Today's bony fish are most likely related to the first jawed fish called the acanthodians (a kan THOH dee unz). They appear in the fossil record about 410 mya. ☑

Reading Check

6. **Identify** When do acanthodians appear in the fossil record?

____

____

Why are fish important?

Fish are food for many animals, including humans. Fish farming and commercial fishing are important to the U.S. economy. Many people enjoy fishing as recreation. Fish also help the environment. They eat large amounts of insect larvae, which helps keep the insect population under control. They also keep the plant growth from blocking waterways.

After You Read

Mini Glossary

fin: a fanlike structure attached to the endoskeleton of fish

scales: hard, thin plates that cover the skin and protect the body of fish

1. Review the terms and their definitions in the Mini Glossary. Select one term and use it in a sentence to describe a structure of fish.

___

___

___

2. Complete the concept web below to show the characteristics of most fish.

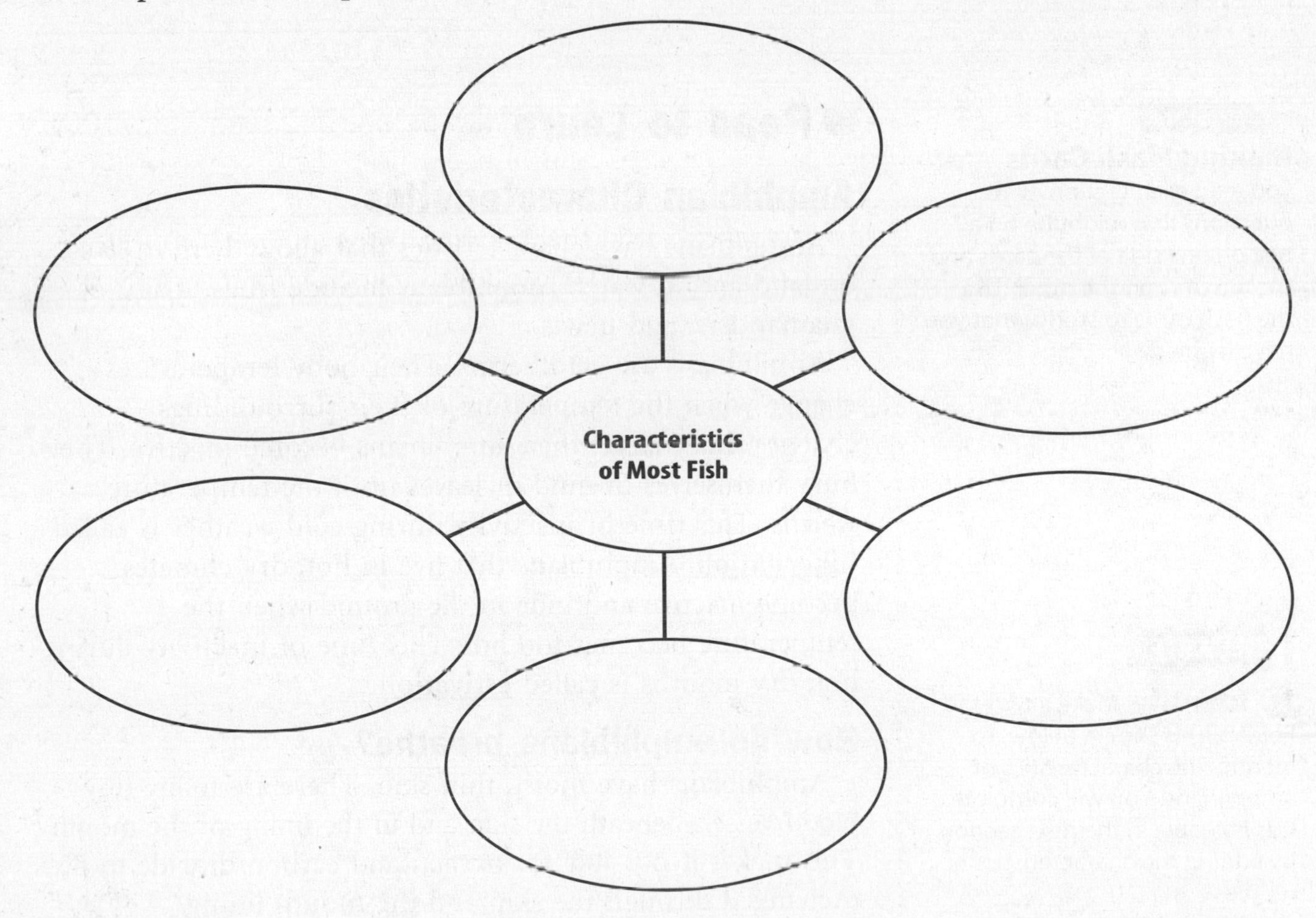

Science Online Visit **life.msscience.com** to access your textbook, interactive games, and projects to help you learn more about fish.

End of Section

Fish, Amphibians, and Reptiles

section 3 Amphibians

What You'll Learn

- the adaptations amphibians have for living in water and living on land
- the characteristics of various kinds of amphibians
- how amphibians reproduce and develop

Before You Read

Describe three characteristics of frogs and toads on the lines below.

____

____

____

Study Coach

Making Flash Cards After you read each section, write questions that might be on a test on one side of the cards and the answers on the other. Use the flash cards to study what you have learned.

Read to Learn

Amphibian Characteristics

Amphibians have characteristics that allow them to live on land and in water. Amphibians include frogs, toads, salamanders, and newts.

Amphibians are ectotherms. Their body temperatures change when the temperature of their surroundings changes. In cold weather, amphibians become inactive. They bury themselves in mud or leaves until the temperature warms. This time of inactivity during cold weather is called **hibernation**. Amphibians that live in hot, dry climates become inactive and hide in the ground when the temperature becomes too hot. This time of inactivity during hot, dry months is called **estivation**.

FOLDABLES™

D Identify Make a two-tab Foldable, as shown below, to identify the characteristics of amphibians. You will complete this Foldable in the next section by adding the characteristics of reptiles.

Characteristics of Amphibians	Characteristics of Reptiles

How do amphibians breathe?

Amphibians have moist, thin skin. There are many tiny blood vessels beneath the skin and in the lining of the mouth. This makes it possible for oxygen and carbon dioxide to be exchanged through the skin and the mouth lining. Amphibians also have small, simple, saclike lungs for the exchange of oxygen and carbon dioxide. Some salamanders have no lungs and breathe only through their skin.

What kind of circulatory system do amphibians have?

Amphibians have three-chambered hearts. One chamber receives oxygen-filled blood from the lungs and skin. Another chamber receives carbon dioxide-filled blood from the body tissues. Blood moves from both of these chambers to the third chamber, which pumps oxygen-filled blood to body tissues and carbon-dioxide filled blood back to the lungs. Limited mixing of the two bloods occurs.

How do amphibians reproduce?

Amphibians depend on water for reproduction. Amphibian eggs are fertilized externally. As the eggs come out of the female's body, the male releases sperm over them. In most amphibian species, the female lays eggs in a pond or other body of water. Some amphibians have adaptations that allow them to reproduce away from water. For example, red-eyed tree frogs lay eggs in thick jellylike material on the underside of leaves that hang over water. After the tadpoles hatch, they fall into the water, where they continue to develop.

How do amphibians develop?

Most amphibians go through a four-stage developmental process called metamorphosis (me tuh MOR fuh sus). The first stage for a frog is laying and fertilizing eggs. Then the fertilized eggs hatch into tadpoles that live in water. Tadpoles have fins, gills, and a two-chambered heart. As tadpoles grow into adults, they develop legs, lungs, and a three-chambered heart. The adult frog can live and move about on land. The figure below shows the stages of this process for frogs.

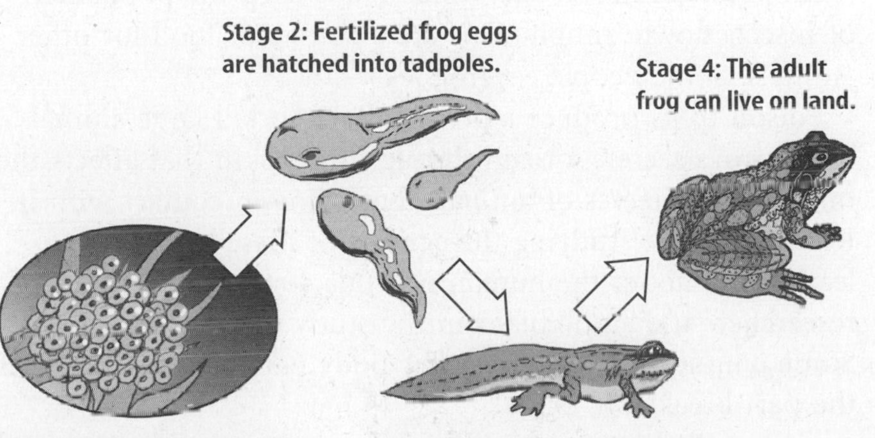

Stage 2: Fertilized frog eggs are hatched into tadpoles.
Stage 4: The adult frog can live on land.
Stage 1: Amphibian eggs are laid and fertilized in water.
Stage 3: Tadpoles begin to grow into adults.

Think it Over

1. **Analyze** From what two structures do amphibians get oxygen?

______

______

______

Picture This

2. **Infer** Circle the stage at which frogs and toads look the most like fish.

Frogs and Toads

Adult frogs and toads have short, wide bodies with four legs. They do not have a neck or tail. They use their strong hind legs for swimming and jumping. Their large eyes and nostrils on top of their heads let frogs and toads see and breathe while the rest of their body is under water. On each side of the head, just behind the eyes, are round membranes that frogs and toads use to hear. These membranes vibrate somewhat like an eardrum in response to sounds.

How do frogs and toads catch their food?

Most frogs and toads have tongues attached at the front of their mouths. When they see prey, their tongues flip out and contact the prey. The prey gets caught in the sticky saliva on the tongue. The tongue flips back into the mouth. Frogs and toads eat mostly insects, worms, and spiders.

Salamanders

Think it Over

3. Contrast Name one way salamanders and newts differ from frogs and toads.

____

____

____

Most species of salamanders and newts live in North America. They have long, slender bodies and short legs. Species of salamanders and newts that live on land are usually found near water. They hide under leaves and rocks during the day to stay out of the drying heat of the Sun. They use their well-developed senses of smell and sight to find and eat such things as worms and insects.

Many species of salamanders reproduce on land using internal fertilization. Water species of salamanders and newts release and fertilize their eggs in the water.

Importance of Amphibians

Amphibians eat insects, which helps keep the population of insects down. Amphibians are a source of food for other animals. Some people eat frog legs.

Reading Check

4. Explain What two things are scientists studying about amphibians?

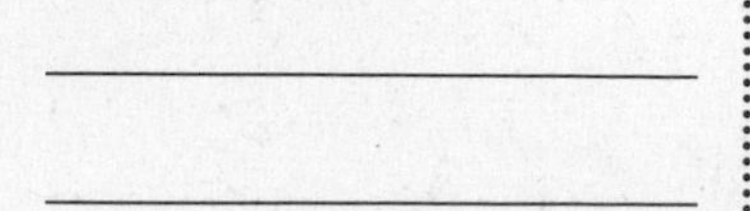

Poison frogs produce a poison that can kill large animals. These frogs secrete a toxin through their skin that affects the muscles and nerves of animals that come in contact with it. Researchers are studying the actions of these toxins to learn more about the human nervous system. Other researchers use amphibians in the study of regeneration. Some amphibians can grow new body parts, such as a tail, if the part breaks off. ☑

How do amphibians tell us about the environment?

Because amphibians live on land and reproduce in the water, they are affected by changes in the environment such as pollution. As a result, amphibians are considered biological indicators. Biological indicators are species whose overall health reflects the health of the ecosystem in which they live. Beginning in 1995, deformed frogs, such as the one below, were found. Scientists hypothesize that an increase in the number of deformed frogs could indicate environmental problems for other organisms. ☑

Reading Check

5. **Identify** What is a biological indicator?

Rob and Ann Simpson/Visuals Unlimited

When did amphibians first live on Earth?

Amphibians probably evolved from lobe-finned fish about 350 million years ago. Few animals competed with amphibians for insects, spiders, and other food. With few predators, amphibians reproduced in large numbers and many species evolved. For about 100 million years, amphibians were the dominant land animals. ☑

Reading Check

6. **Explain** Why were amphibians able to become dominant land animals?

After You Read

Mini Glossary

estivation: a time of inactivity during hot, dry months

hibernation: a time of inactivity during cold weather

1. Review the terms and their definitions in the Mini Glossary. Write a sentence that compares hibernation and estivation.

2. Fill in the events-chain concept map below to show the stages of frog metamorphosis. Draw a picture in the box of each stage and write a short description of each picture to explain what is happening in that stage.

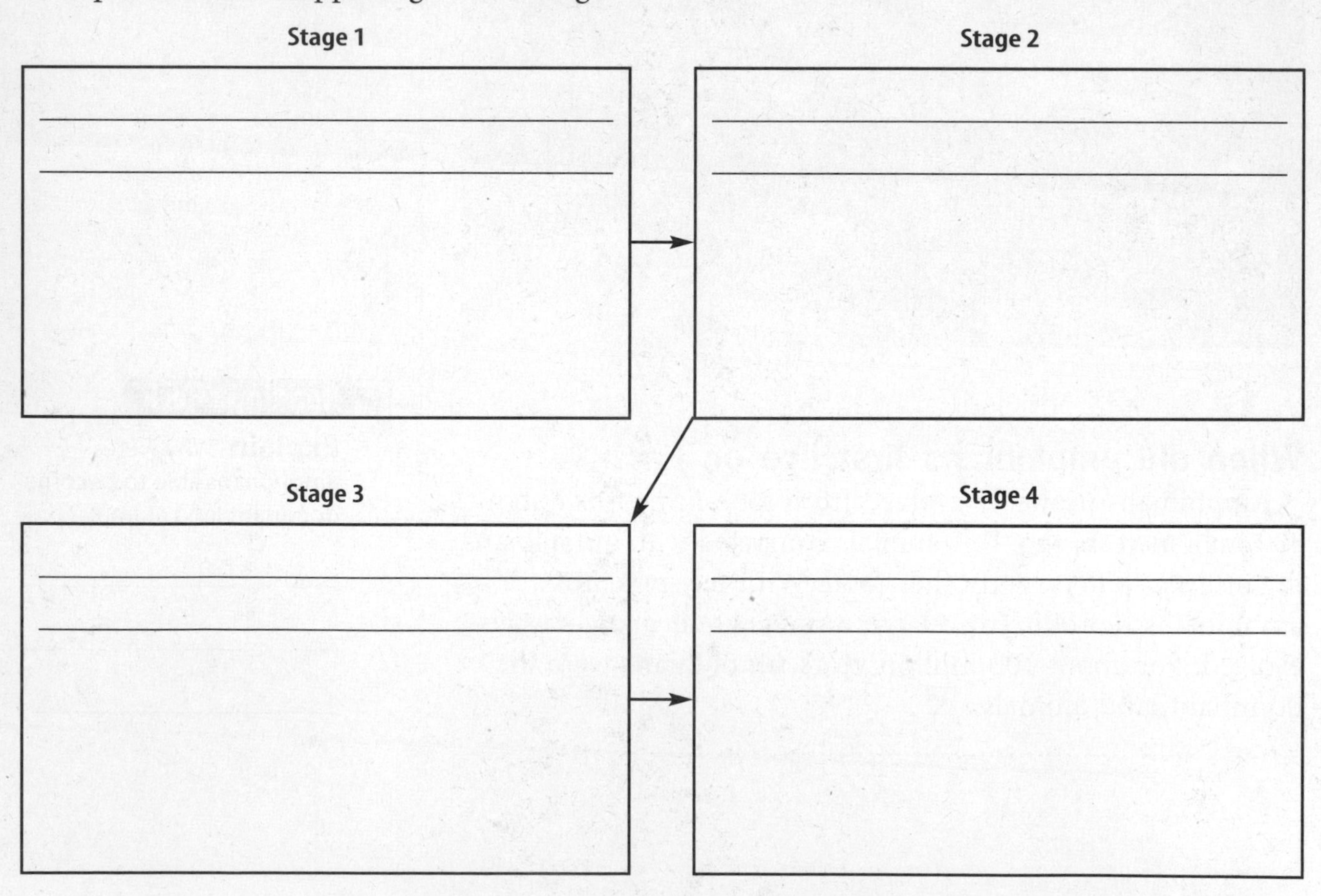

End of Section

Science Online Visit **life.msscience.com** to access your textbook, interactive games, and projects to help you learn more about amphibians.

Fish, Amphibians, and Reptiles

section 4 Reptiles

Before You Read

On the lines below, describe how a snake is different from a fish or an amphibian.

____

____

What You'll Learn

- the characteristics of reptiles
- how reptile adaptations enable them to live on land
- the importance of the amniotic egg

Read to Learn

Reptile Characteristics

Reptiles have thick, dry, waterproof skin. Their skin is covered with scales. The scales help reduce water loss and protect them from injury. Even though reptiles are ectotherms, they can change their internal body temperature by changes in their behavior. For example, when the weather is cold, they lie in the Sun, which warms them. When the weather is warm, they move into the shade to stay cool.

Study Coach

Discuss What You Read Work with a partner. Read a paragraph to yourselves. Then discuss what you learned in the paragraph. Continue until you and your partner understand the main ideas of this section.

How do reptiles move?

Some reptiles, like turtles, crocodiles, and lizards, move on four legs. They use claws to dig, climb, and run. Other reptiles, such as snakes and some lizards, move without legs.

How do reptiles breathe?

Reptiles breathe with lungs. Reptiles that live in water, like turtles and sea snakes, must come to the surface to breathe.

What is the circulatory system of reptiles like?

Most reptiles have a three-chambered heart with a partial wall inside the main chamber. This type of circulatory system provides more oxygen to all parts of the body. Crocodilians have a four-chambered heart that completely separates the oxygen-filled blood and the carbon dioxide-filled blood. ☑

Reading Check

1. **Compare** How is a reptile heart different from the heart of an amphibian?

____

____

____

Picture This

2. Determine What is the purpose of the egg membrane and the yolk sac?

Why are reptiles able to lay their eggs on land?

Eggs of reptiles are fertilized internally. Many female reptiles lay eggs that are covered by tough shells. The shells keep the eggs from drying out. This adaptation allows reptiles to lay their eggs on land.

The **amniotic egg** provides a complete environment for the embryo's development. The figure at the right shows the parts of the amniotic egg. The egg membrane protects and cushions the embryo and helps it get rid of wastes. The yolk gives the embryo a food supply. Tiny holes in the egg's shell, called pores, allow for the exchange of oxygen and carbon dioxide. When it hatches, the reptile looks like a small adult.

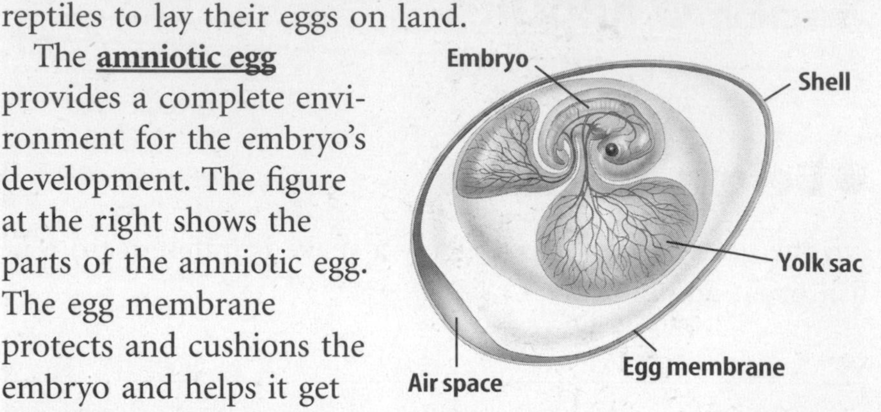

Types of Modern Reptiles

Reptiles live on every continent except Antarctica. They live in all the oceans except those in polar regions. Reptiles are different sizes, shapes, and colors. The three living groups of reptiles are lizards and snakes, turtles, and crocodilians.

What is the largest group of reptiles?

The largest group of reptiles is lizards and snakes. These reptiles have an unusual type of jaw. The jaw has a joint that unhinges and increases the size of their mouths. This lets them swallow their prey whole.

Reading Check

3. Identify three characteristics of lizards.

Lizards Lizards have movable eyelids. Their ears are on the outside and have legs with clawed toes on each foot. Lizards eat plants, other reptiles, insects, spiders, worms, and mammals. ☑

Snakes Snakes move without legs. They have poor hearing and most have poor eyesight. Snakes sense vibrations in the ground through the lower jawbone. These vibrations are interpreted as sound by the snake's brain. Snakes eat meat. Some snakes wrap around and constrict their prey. Other snakes inject their prey with poison called venom.

Most snakes lay eggs after they are fertilized internally. In some snakes, eggs develop and hatch inside the female's body and then leave her body soon after.

What are the characteristics of turtles?

Turtles have a two-part shell made of hard, bony plates. The vertebrae and ribs are fused to the inside of the top part of the shell. The muscles attach to the lower and upper part of the inside of the shell. Most turtles can bring the head and legs into the shell for protection. ☑

Turtles have powerful jaws with a beaklike structure to crush food. They eat insects, worms, fish, and plants. Turtles that live on land are called tortoises. Like most reptiles, turtles do not care for their young. The female digs out a nest, lays her eggs, covers the nest, and leaves. Turtles hatch fully formed and live on their own immediately.

What are the characteristics of crocodilians?

Crocodilians are among the largest living reptiles on Earth. They have a lizardlike shape. Their backs have large, deep scales. Crocodiles have a narrow head with a triangular-shaped snout. Alligators have a wide head with a rounded snout. Another kind of crocodilian, called a gavial, has a very slender snout with a rounded growth on the end.

Crocodiles are aggressive. They can attack prey as large as cattle. Alligators are less aggressive than crocodiles. They eat fish, turtles, and waterbirds. Gavials mainly eat fish. Crocodilians care for their young. The female guards the nest and both the male and female protect their young.

The Importance of Reptiles

Reptiles can be important predators. In farming areas, snakes eat rats and mice that destroy grains. Small lizards eat insects. Large lizards eat small animals that are pests. In many parts of the world, humans eat reptiles and their eggs. ☑

The number of reptile species is getting smaller in areas where swamps and other lands are being developed. Coastal nesting sites of sea turtles are being destroyed by development or harmed by pollution. Today, laws protect most species of turtles and their habitats.

What does the fossil record tell us about reptiles?

Fossil records show that the earliest reptiles lived about 345 mya. Dinosaurs, descendants of the early reptiles, lived about 200 mya. Then they died out about 65 mya. Some of today's reptiles, such as crocodiles and alligators, have not changed much from their ancestors.

Reading Check

4. Explain Name one purpose of a turtle's shell.

____

____

____

Think it Over

5. Contrast Name one difference between turtles and crocodilians.

____

____

____

Reading Check

6. Explain What are two ways that reptiles are important?

____

____

____

After You Read

Mini Glossary

amniotic egg: the complete environment for the development of an embryo

1. Review the term and its definition in the Mini Glossary. Write a sentence that explains the importance of the amniotic egg.

2. Complete the concept map below to show the main characteristics of reptiles.

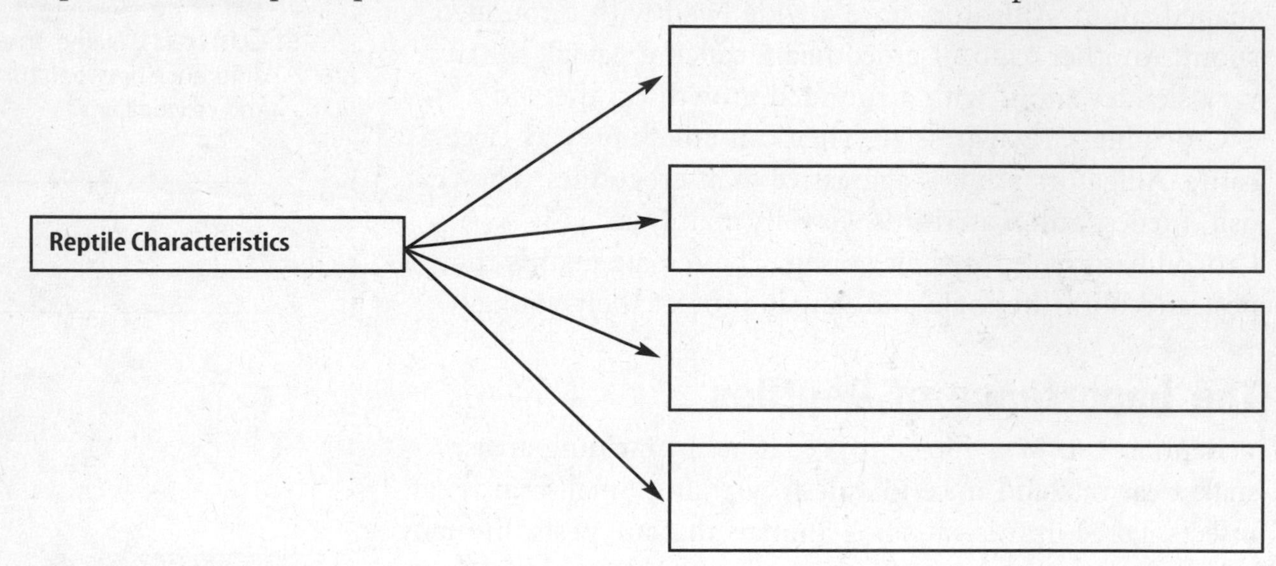

3. Which ideas that you and your partner discussed were hardest to understand? Write a question and the answer on the lines below to show you understand the idea.

Visit **life.msscience.com** to access your textbook, interactive games, and projects to help you learn more about reptiles.

Birds and Mammals

section 1 Birds

Before You Read

Did you ever wish you could fly like a bird? What would you need to be able to fly like a bird?

____

____

____

What You'll Learn
- the features of birds
- the adaptations birds have for flying
- how birds reproduce and develop

Read to Learn

Mark the Text

Identify the Main Point Underline the important ideas in this section. This will help you remember what you read.

Bird Characteristics

Birds are versatile animals. They live in some of the warmest and the coldest places on Earth. Some birds fly high in the air. Some swim deep underwater. Some birds are very heavy, while others are very light. Birds eat meat, fish, insects, fruit, seeds, and nectar.

What are birds' eggs made of?

Birds lay amniotic (am nee AH tik) eggs with hard shells. An amniotic egg provides the developing embryo with a moist environment. The hard shell is made up of calcium carbonate, the same chemical that makes up seashells and marble.

The egg is fertilized before the shell forms around it. The female bird lays one or more eggs in a nest. A group of eggs is called a clutch. One or both parents incubate the eggs, or keep them warm, until they hatch. One or both parents care for the young.

FOLDABLES™

A Classify Make a layered-look book, as shown below. As you read, list facts on the Foldable that describe the characteristics, body systems, origin, and importance of birds.

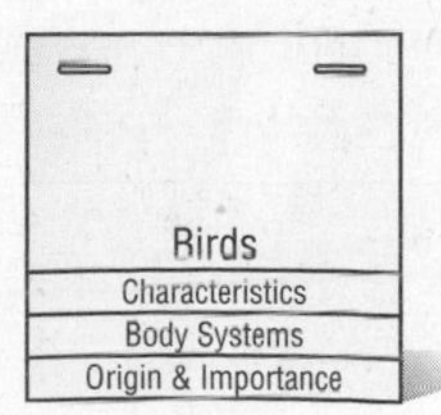

Why can birds fly?

A bird can fly because of its skeleton, wings, and feathers. A bird also has strong muscles and a strong respiratory system. Birds have sharp eyesight and large amounts of energy, which also are needed for flying.

What kind of bones do birds have?

A bird has a skeleton that is different from most other animals. The figure below shows what a bird's skeleton looks like. A bird's bones are almost hollow. The hollow spaces are filled with air. Some birds have bones that are joined together, making them strong for flying. A large breastbone supports the chest muscles that are needed for flight. The last bones of the spine support the tail feathers. These feathers help birds steer and balance while flying and landing.

Picture This

1. **Describe** Label the following features of a bird on the figure to the right: skeleton, wings, and feathers.

What kinds of feathers do birds have?

Birds are the only animals with feathers. They have two main types. Strong, lightweight **contour feathers** give birds their coloring and shape. They help the birds steer while flying. Soft, fluffy **down feathers** give adult birds a layer of protection next to their skin. Down feathers cover the bodies of young birds. Birds are endotherms. **Endotherms** have a constant body temperature. Feathers help birds keep their body temperature constant no matter what the air or water temperature.

2. **Explain** What helps keep a bird's body temperature constant?

____

____

Feathers grow like your hair does. Each feather grows from a small opening in the skin called a follicle (FAH lih kul). When a feather falls out, a new one grows in its place. A bird has an oil gland found just above the base of its tail. A bird uses its beak or bill to rub the oil from the gland over its feathers in a process called **preening**. The oil helps the feathers last longer.

How do wings help birds fly?

Most birds have wings that allow them to fly. The wings are attached to strong chest muscles. When a bird flaps its wings, it gets the power to go forward and the lift to stay in the air. A bird's wings move up and down, as well as back and forth.

The shape of a bird's wings helps it fly. As you can see in the figure below, the wings are curved on top and flat or slightly curved on the bottom. When a bird flies, air moves more slowly across the bottom of its wings than across the top. The slow-moving air has greater pressure than the fast-moving air. This causes an upward push called lift.

Not all birds fly, but wings are important for nonflying birds, too. Penguins use their wings to swim underwater. Ostriches use their wings to keep their balance when they walk or run.

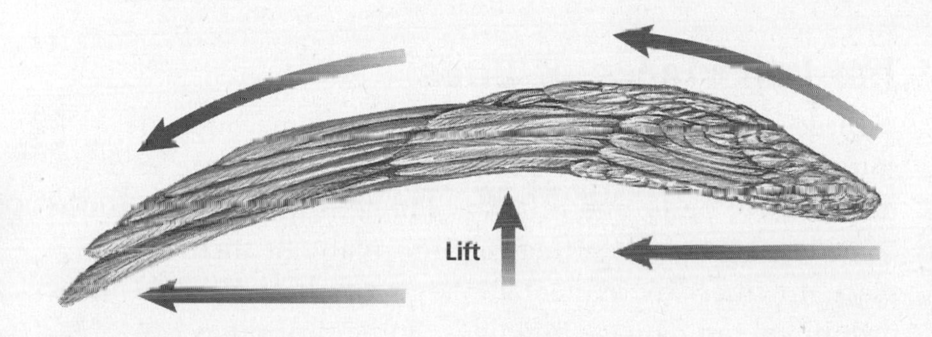

Picture This

3. **Describe** to a classmate how the shape of a bird's wing helps it fly.

Body Systems

Birds are extremely active. They have body systems that help them to be active.

What does the digestive system do?

Because birds use a lot of energy when they fly, they need large amounts of high energy foods. These foods are nuts, seeds, insects, and meat. A bird's digestive system breaks down food quickly to supply this energy. A bird takes food into its mouth. From there, unchewed food passes into an organ called the crop. There the food takes in moisture to help it move on to the stomach. The food is partly digested in the stomach before it moves to the muscular gizzard. In the gizzard, the food is crushed by small stones that the bird has swallowed. The food then moves to the intestine, where its nutrients move into the bloodstream. ☑

Reading Check

4. **Explain** why a bird's diet includes large amounts of high-energy foods.

What does the respiratory system do?

Oxygen combines with the energy in food to make body heat. A bird's respiratory system gets oxygen from the air. Oxygen helps to change food into energy needed to power the flight muscles. A bird has two lungs. Each lung is connected to air sacs. The air a bird inhales passes into the air sacs. When a bird exhales, air with oxygen passes from the air sacs into the lungs. A bird gets air with oxygen both when it inhales and exhales. This gives the flight muscles a constant supply of oxygen. ☑

Reading Check

5. Identify What are the two main parts of a bird's respiratory system?

____

____

What is the circulatory system made up of?

A bird's circulatory system is made up of a heart, arteries, capillaries, and veins. A bird's heart is large compared to the rest of its body. Blood filled with oxygen is kept separate from blood filled with carbon dioxide as both move through a bird's heart and blood vessels. A bird's heart beats rapidly so enough oxygen-filled blood is carried to the bird's muscles.

The Importance of Birds

Birds have important roles in nature. Some birds are sources of food. Other birds are kept as pets. Some birds help control pests, and others pollinate flowers. Birds can be considered pests when there are too many of them in one place. In cities where there are large numbers of birds, their droppings can damage buildings. Some droppings can contain microorganisms that cause disease in humans. ☑

Reading Check

6. Analyze Why is large numbers of birds living in cities a problem?

____

____

____

What are some uses of birds?

For many years, people hunted birds for food and for their colorful feathers. Some wild birds such as chickens and turkeys were tamed to provide humans with eggs and meat. Mattresses, pillows, and clothing are made from birds' down feathers. Droppings of some birds are used as fertilizer. Birds, such as parakeets, parrots, and canaries, are kept as pets.

How did the first birds develop?

Scientists learn how most living things began by studying their fossils. However, scientists have found very few fossils of birds. Scientists hypothesize that birds developed from reptiles millions of years ago. Some characteristics of birds, such as the scales on their feet and legs, are similar to those of reptiles.

After You Read

Mini Glossary

contour feathers: strong, lightweight feathers that give birds their color and shape and help birds steer

down feathers: soft, fluffy feathers that give a layer of protection next to the skin of adult birds and cover the bodies of young birds

endotherms: animals that have a constant body temperature

preening: process in which a bird rubs oil from a gland over its feathers which helps them last longer

1. Review the terms and their definitions in the Mini Glossary. Choose one type of bird feather and describe it in a sentence.

2. Use the web chart below to identify five characteristics that help birds fly.

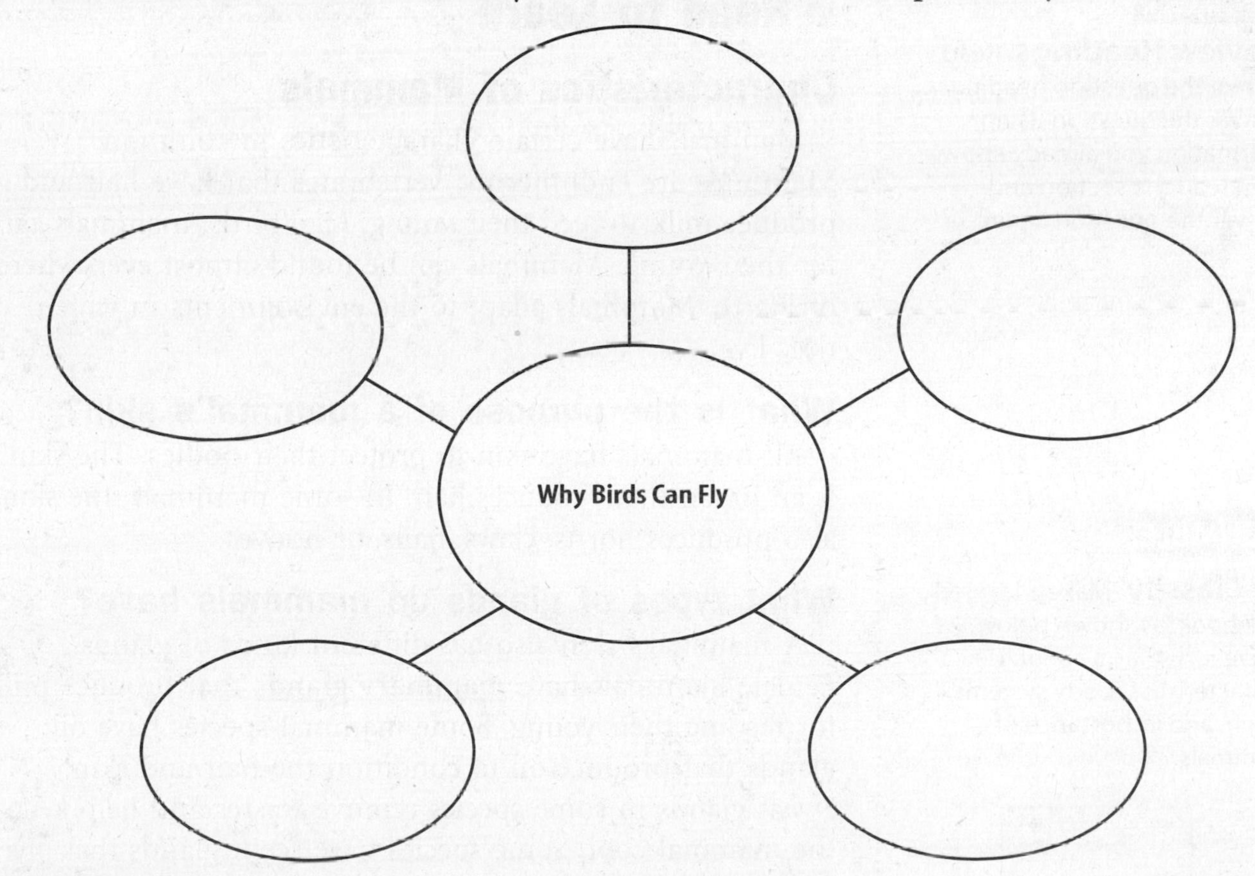

ScienceOnline Visit **life.msscience.com** to access your textbook, interactive games, and projects to help you learn more about birds.

End of Section

Birds and Mammals

section 2 Mammals

What You'll Learn

- how the features of mammals have helped them adapt
- the differences among monotremes, marsupials, and placentals
- why some mammals are in trouble today

Before You Read

How many teeth do you have? Are they all the same? Why do you think people have different kinds of teeth?

____

____

____

Study Coach

Preview Headings Read each of the question headings. Answer the question using information you already know. Then read the section and answer the question again.

Read to Learn

Characteristics of Mammals

Mammals have certain characteristics in common. **Mammals** are endothermic vertebrates that have hair and produce milk to feed their young. Like birds, mammals care for their young. Mammals can be found almost everywhere on Earth. Mammals adapt to the environments in which they live.

What is the purpose of a mammal's skin?

All mammals have skin to protect their bodies. The skin is an organ that produces hair. In some mammals, the skin also produces horns, claws, nails, or hooves.

What types of glands do mammals have?

A mammal's skin also has different kinds of glands. Female mammals have **mammary glands** that produce milk for feeding their young. Some mammal species have oil glands that produce oil to condition the hair and skin. Sweat glands in some species remove wastes and help keep the mammal cool. Some species have scent glands that give off a scent to mark their territory, attract mates, or help the mammals defend themselves.

FOLDABLES™

B Classify Make a layered-look book, as shown below. As you read, list facts about the characteristics, body systems, origin, and importance of mammals.

Mammals
Characteristics
Body Systems
Origin & Importance

Why do mammals have different kinds of teeth?

Mammals have different kinds of teeth. Scientists know what a mammal eats by the kind of teeth it has. Incisors are front teeth used for biting and cutting. Canine teeth, next to the incisors, are sometimes used to grip and tear. Premolars and molars, located at the back of the mouth, grind and crush. Some mammals, including humans, have all four kinds of teeth.

Omnivores are animals that eat both plants and animals. Humans are omnivores. Mammals, such as tigers, that are **carnivores** have large canine teeth because they eat only the flesh of other animals. **Herbivores** eat only plants. Herbivores such as horses have large premolars and molars to grind the tough fibers of plants. The different kinds of teeth that omnivores, carnivores, and herbivores have are shown in the figure below.

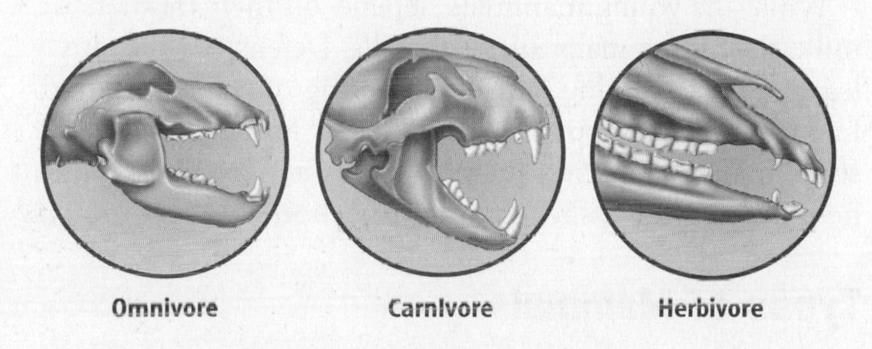

Picture This

1. **Identify** Which mammal has large canine teeth—an omnivore, a carnivore, or a herbivore?

____

How does hair protect mammals?

All adult mammals have hair on their bodies. The hair can be thick fur or just a few hairs around the mouth. Fur traps air and helps keep the mammal warm. Whiskers near the mouth help some mammals sense their environment. Whales have almost no hair. A thick layer of fat under their skin, called blubber, helps to keep them warm. Porcupine quills are a kind of hair that protects porcupines from other animals.

Body Systems

A mammal's body systems are important for its activities and its survival. A mammal has a heart with four chambers that pumps oxygen-filled blood throughout the body in blood vessels. A mammal's lungs are made up of millions of air sacs. These air sacs allow greater exchange of oxygen and carbon dioxide. ☑

Reading Check

2. **Describe** What does a four-chambered heart do?

____

____

____

Mammal Nervous System A mammal's nervous system is made up of a brain, spinal cord, and nerves. The brain is involved in learning. It also controls the muscles.

The digestive systems of mammals vary depending on the kinds of food they eat. Herbivores have longer digestive systems than carnivores because plants take longer to digest than meat.

Reading Check

3. **Apply** Why do mammals have different kinds of digestive systems?

____

____

____

How do mammals reproduce?

All mammals reproduce sexually. Most mammals give birth to live young after they develop in the uterus, the female reproductive organ. Most mammals can not take care of themselves for at least the first several days after they are born. Some can not take care of themselves for several years. Some mammals, such as deer and elephants, are able to stand when they are a few minutes old. They can travel with their constantly moving parents.

While the young mammals depend on their mothers' milk, they learn many survival skills. Defensive skills are learned while playing with other young of their own kind. In many mammal species only females raise the young. In some mammal species, such as wolves and humans, males help provide food, shelter, and protection for their young.

Types of Mammals

Mammals are classified into three groups based on how their young develop. The three groups are monotremes (MAH nuh treemz), marsupials (mar SEW pee ulz), and placentals (pluh SEN tulz).

What are monotremes?

A mammal that lays eggs with leathery shells is a **monotreme**. A platypus is a monotreme. The female incubates or keeps the eggs warm for about 10 days. After the young hatch, they nurse by licking the female's skin and hair where milk comes from the mammary glands. Monotreme mammary glands do not have nipples.

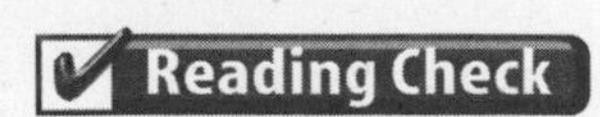

4. **Explain** How do young monotremes nurse?

____

____

____

What are marsupials?

A mammal that gives birth to immature young that usually crawl into a pouch on the female's abdomen is a **marsupial**. However, not all marsupials have pouches. An immature marsupial crawls to a nipple. It stays attached to the nipple until it is developed. In pouched marsupials, the developed young come back to the pouch for feeding and protection.

Where Marsupials Live Most marsupials, including kangaroos, wallabies, and koalas, live in Australia, New Guinea, or South America. The opossum is the only marsupial found in North America.

What are placentals?

A mammal in which an embryo completely develops inside the female's uterus is a **placental**. The time it takes for the embryo to develop is the **gestation period**. The time can range from 16 days for hamsters to 650 days for elephants. Placentals are named for the placenta. The **placenta** is an organ that develops from tissues of the embryo and tissues that line the inside of the uterus. The placenta takes in oxygen and food from the mother's blood. An umbilical cord connects the embryo to the placenta. The umbilical cord is made up of blood vessels. The blood in the **umbilical cord** moves food and oxygen from the placenta to the embryo and removes waste products from the embryo. The blood of the mother and the embryo do not mix.

Think it Over

5. **Draw Conclusions** To which group of mammals do humans belong? Why?

____

____

____

Importance of Mammals

Like other living things, mammals help keep a balance in the environment. Carnivores, such as tigers, help control the populations of other animals. Bats help pollinate flowers and control insects. However, some mammals and other animals are in danger today. Many of their habitats are being destroyed for housing, roads, and shopping centers. Many mammals are left without food, shelter, and space to survive.

What were the first kinds of mammals?

Mammals began branching out into many different species after dinosaurs became extinct about 65 million years ago. Today, more than 4,000 species of mammals have evolved from animals that existed about 200 million years ago. One example of an ancient mammal is shown in the figure below.

Picture This

6. **Identify** What animal today do you think looks similar to this ancient animal?

____

____

After You Read

Mini Glossary

carnivore: animal that eats only the flesh of other animals
gestation period: the time during which the embryo develops in the uterus
herbivore: animal that eats only plants
mammals: animals that are vertebrates, endothermic, have hair, and make milk to feed their young
mammary glands: in female mammals, glands that make milk for feeding their young
marsupials: mammals that give birth to immature young that usually crawl into a pouch on the female's abdomen
monotremes: mammals that lay eggs with leathery shells
omnivore: animal that eats both plants and animals
placenta: organ that develops from tissues of the embryo and tissues that line the inside of the uterus; takes in oxygen and food from the mother's blood
placentals: mammals in which embryos completely develop inside the female's uterus
umbilical cord: made up of blood vessels that transport food and oxygen from the placenta to the embryo and removes waste products from the embryo

1. Review the terms and their definitions in the Mini Glossary. Write a sentence explaining the difference between marsupials and placentals.

2. In the chart below, list four types of glands found in mammals and tell what each one does.

Gland	What It Does

3. How did previewing the headings before you read help prepare you to read the section?

Visit **life.msscience.com** to access your textbook, interactive games, and projects to help you learn more about mammals.

Animal Behavior

section 1 Types of Behavior

Before You Read

On the lines below, explain how you learned a new skill, such as in-line skating or jumping rope.

____

____

____

What You'll Learn

- the differences between innate and learned behavior
- how organisms use reflexes and instincts to survive
- examples of different learned behaviors

Read to Learn

Behavior

Animals are different from one another in their behavior. **Behavior** is the way an organism interacts with other organisms and its environment. Animals are born with certain behaviors, and they learn other behaviors.

Anything in the environment that causes a reaction is called a stimulus. A stimulus can be external, such as a male dog entering the territory of another male dog. A stimulus can be internal, such as hunger or thirst. The way an animal reacts to a stimulus is called a response. Getting a drink of water is a response to the internal stimulus of thirst.

Innate Behavior

A behavior that an organism is born with is called an **innate behavior**. These types of behaviors are inherited. They do not have to be learned.

Innate behavior patterns occur the first time an animal reacts to an internal or external stimulus. For birds, building a nest is an innate behavior. Although the first nest a bird builds may be messy, it is built correctly.

Mark the Text

Identify Main Ideas Highlight each question head in this section. Then use a different color to highlight the answers to the questions.

FOLDABLES™

A Describe Make a two-tab Foldable, as shown below. Describe the innate and learned behaviors of an animal that you have observed.

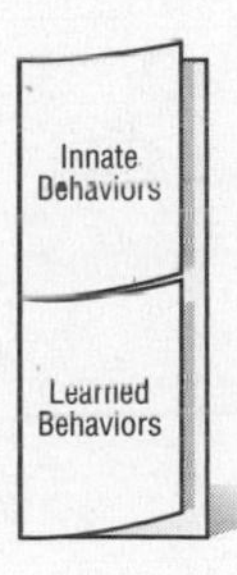

Reading Check

1. **Explain** Why do insects have mostly innate behavior?

Think it Over

2. **Identify** Which animal likely has more learned behaviors—a spider or a monkey? Why?

Why are innate behaviors important?

The behavior of animals with short life spans, such as insects, is mostly innate behavior. An insect cannot learn from its parents. By the time the insect hatches, its parents have died or moved on. Innate behavior allows animals to respond quickly. A quick response often means the difference between life and death.

Reflex actions are the simplest innate behaviors. A **reflex** is an automatic response that does not involve a message from the brain. When something is thrown at you, you blink. Blinking is a reflex action. Your body reacts on its own. You do not think about the behavior.

What is instinctive behavior?

An **instinct** is a complex pattern of innate behavior. Instinctive behavior begins when an animal recognizes a stimulus. It continues until the animal has performed all parts of the behavior. Spinning a web is an instinctive spider behavior. A spider knows how to spin a web as soon as it hatches. Instinctive behaviors take much more time to complete than reflexes. A spider may spend days building a web.

Learned Behavior

Animals also have learned behaviors. Learned behavior develops over an animal's lifetime as a result of experience or practice. Animals with more complex brains have more learned behaviors. Fish, reptiles, amphibians, birds, and mammals all learn.

Learned behavior helps animals respond to changing situations. In changing environments, an animal that can learn a new behavior is more likely to survive than an animal that cannot learn a new behavior. Learned behavior is important for animals with long life spans. The longer an animal lives, the more likely it is that its environment will change.

Can instincts change?

Learned behavior can change instincts. Some young birds instinctively crouch and freeze if they see something moving above them. They will crouch and keep still even if the object is only a moving leaf. Older birds have learned that some things that move above them, such as leaves, are not harmful. Learned behavior includes imprinting, trial and error, conditioning, and insight.

How does imprinting occur?

Imprinting occurs when an animal forms a social attachment to another organism within a short time after birth or hatching. A gosling follows the first moving object it sees after hatching. The moving object is usually an adult female goose. This behavior is important because adult geese have experience in finding food, protecting themselves, and getting along in the world. Animals that become imprinted toward animals of another species have difficulty recognizing members of their own species.

What is trial-and-error learning?

Behavior that changes with experience is called trial-and-error learning. You learned many skills through trial and error, such as feeding yourself, tying your shoes, and riding a bicycle. Once you learn a skill, you can do it without having to think about it.

How does conditioning change behavior?

Animals often learn new behaviors by conditioning. In **conditioning**, behavior is changed so that a response to one stimulus becomes linked with a different stimulus.

There are two types of conditioning. One type adds a new stimulus before the usual stimulus. Russian scientist Ivan Pavlov performed an experiment to explain how this conditioning works. He knew that hungry dogs salivate when they see and smell food. Pavlov added another stimulus, as seen in the figure below. He rang a bell before he fed the dogs. The dogs connected the sound of the bell with food. The dogs were conditioned to salivate at the sound of a bell even if they were not fed.

FOLDABLES™

B Compare Make a three-tab Foldable, as shown below. Use a Venn diagram to compare the two types of conditioning.

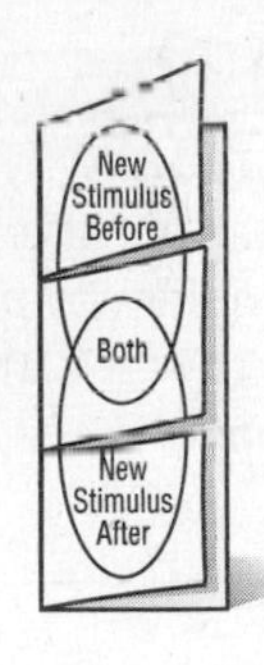

Picture This

3. **Explain** Working with a partner, take turns describing how Pavlov conditioned dogs to respond to a bell.

Think it Over

4. **Analyze** Give an example of this second kind of conditioning from your own experiences.

____

____

____

Another Type of Conditioning In the second kind of conditioning, a new stimulus is given after a behavior has happened. Getting an allowance for doing chores is an example of this type of conditioning. You do the chores because you want to get your allowance. You have learned, or been conditioned, to perform activities that you may not have done if you had not been offered a reward.

How do past experiences help solve problems?

In the problem-solving experiment shown below, bananas were placed out of a chimpanzee's reach. Instead of giving up, the chimpanzee piled up boxes found in the room, climbed them, and reached the bananas. At some time in the past, the chimpanzee must have solved a similar problem. The chimpanzee used past experiences, or insight, to solve the problem.

Insight is a form of reasoning that allows animals to use past experiences to solve new problems. When you were a baby, you learned to solve problems using trial and error. As you grow older, you use insight more often to solve problems. Much of adult human learning is based on insight.

Picture This

5. **Describe** Why was the chimpanzee able to solve the problem by piling the boxes to reach the bananas?

____

____

____

After You Read

Mini Glossary

behavior: the way an organism interacts with other organisms and its environment

conditioning: a way of changing a learned behavior so that a response to one stimulus becomes associated with a different stimulus

imprinting: a learned behavior that happens when an animal forms a social attachment to another organism within a short time after birth or hatching

innate behavior: a behavior that an organism is born with

insight: a form of reasoning that uses past experiences to solve new problems

instinct: a complex pattern of innate behavior, such as spinning a web

reflex: an automatic response that does not involve a message from the brain

1. Review the terms and their definitions in the Mini Glossary. Write a sentence that explains the difference between instinct and insight.

____

____

2. Fill in the graphic organizer below with the different types of animal behavior.

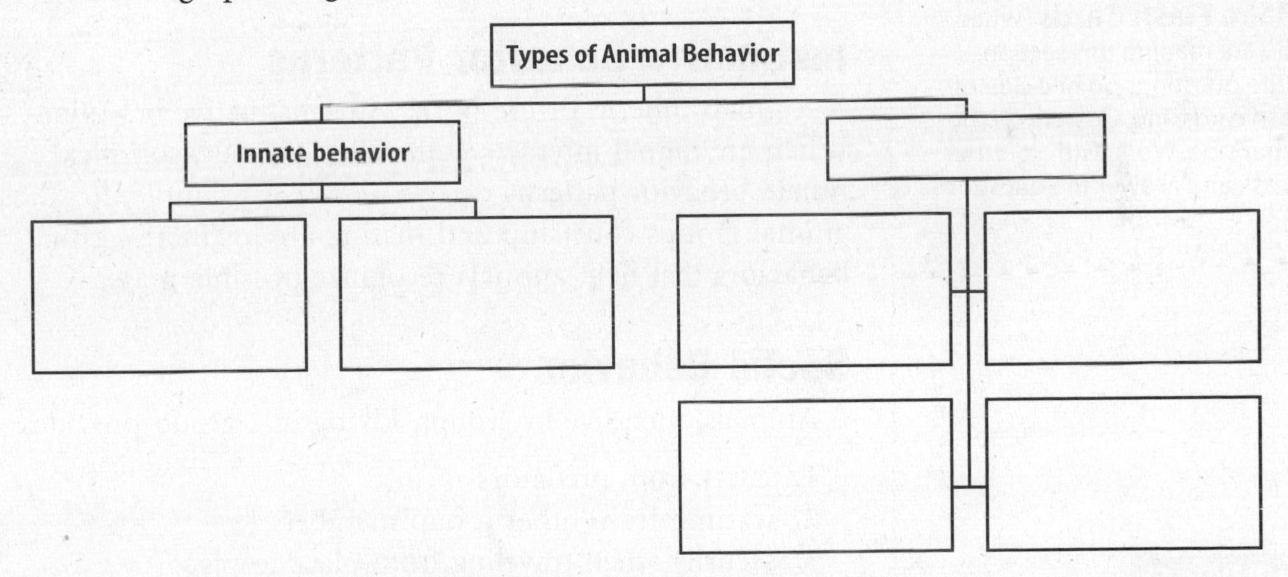

3. How did finding answers to the question heads help you learn about types of behavior?

____

____

Visit **life.msscience.com** to access your textbook, interactive games, and projects to help you learn more about the types of behavior.

chapter 16

Animal Behavior

section 2 Behavioral Interactions

What You'll Learn

- the importance of behavioral adaptations
- how courtship behavior improves reproductive success
- the importance of social behavior and cyclic behavior

Before You Read

One way that you communicate with your family and friends is by talking. On the lines below, list three other ways that you communicate.

____

____

____

Study Coach

Make Flash Cards While you are reading this section, write questions on one side of flash cards and answers on the other side. Work with a partner to ask and answer the questions.

Read to Learn

Instinctive Behavior Patterns

Animals inherit, or are born with, instinctive behavior. When an animal interacts with other animals, complex innate behavior patterns can be seen. For example, for most animal groups courtship and mating are instinctive ritual behaviors that help animals recognize possible mates.

Social Behavior

Animals often live in groups. Living in a group provides:

1. safety from predators
2. warmth from other group members
3. security when traveling from place to place

Interactions among organisms of the same species are examples of **social behavior**. Social behaviors are inherited. Social behaviors include courtship and mating, caring for the young, claiming territories, protecting each other, and getting food. Social behaviors help the species survive. For example, because zebras live in herds, lions are less likely to attack them.

FOLDABLES™

C Describe Make a three-tab concept map Foldable, as shown below. Describe three behavioral adaptations of animals.

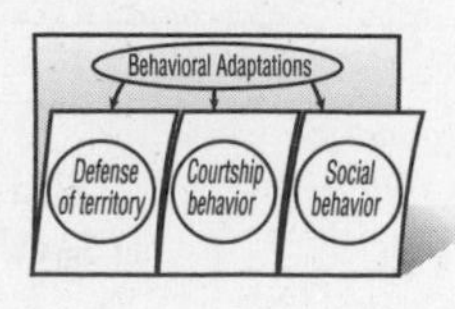

What is a society?

A **society** is a group of animals of the same species living and working together in an organized way. Insects, such as ants, live in societies. Each member of the society has a certain role. One female ant lays eggs and a male ant fertilizes the eggs. The workers do all the other jobs in the society. ☑

Some animal societies are organized by dominance. The top animal controls the other members of the society. Wolves live in packs. One female in the pack is dominant. She controls the mating of other females in the pack. This behavior controls the size of the pack and helps the pack survive.

✔ Reading Check

1. Define What is a society?

Territorial Behavior

A territory is an area that an animal defends from other members of the same species. Animals may show ownership of territories by making sounds, leaving scent marks, or attacking members of the same species who enter the territory.

Why do animals defend their territories?

Territories contain food, shelter, and possible mates. If an animal has a territory, it will be able to mate and produce offspring. Defending territories is an instinctive behavior. It improves the survival rate of an animal's offspring.

How do animals defend their territories?

In order to defend their territory, protect their young, or get food, many animals show aggression. **Aggression** is a forceful behavior used to dominate or control another animal. Animals of the same species rarely fight to the death. An animal that avoids being attacked by another animal is showing submission. In the figure below, one wolf has rolled over and made itself as small as possible to communicate submission to the dominant wolf.

Picture This

2. Draw a circle around the wolf that is showing submissive behavior.

Communication

Communication is important in all social behavior. Communication is an action by a sender that affects the behavior of a receiver. Animals in a group communicate with sounds, scents, and actions. Alarm calls, chemicals, speech, courtship behavior, aggression, and submission are types of communication.

Reading Check

3. **Describe** the purpose of courtship behaviors.

____

____

____

How do animals attract mates?

Courtship behavior allows the male and female members of a species to recognize each other. Courtship behaviors excite males and females so they are ready to mate at the same time. The courtship behavior of a male bird of paradise includes spreading its tail feathers and strutting. This behavior attracts female birds of paradise. Courtship behavior helps increase reproductive success. ☑

In most species, males are more colorful than females. The males perform the courtship activities to attract a mate. Some courtship behaviors allow males and females to find each other across distances.

FOLDABLES™

D Identify Make a half-book Foldable, as shown below. Write notes to identify ways animals communicate with chemicals, sound, and light.

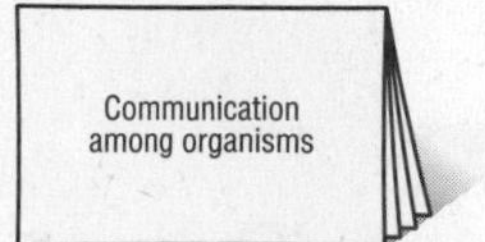

How do animals use chemicals to communicate?

Ants leave trails that other ants can follow. Male dogs urinate on objects and plants to let other dogs know they have been there. These animals are using chemicals called pheromones (FER uh mohnz) to communicate. A **pheromone** is a chemical produced by one animal to influence the behavior of another animal of the same species. Pheromones remain in the environment so that the sender and receiver can communicate without being in the same place at the same time.

Males and females use pheromones to set up territories, warn of danger, and attract mates. Some animals release alarm pheromones when hurt or in danger.

How do animals use sounds to communicate?

Male crickets rub one forewing against the other to make a chirping sound. The sound attracts female crickets. Each species of crickets makes a different sound. Rabbits thump the ground, gorillas pound their chests, and beavers slap the water with their flat tails. These sounds are forms of communication to others animals of the same species.

How is light used to communicate?

The ability of certain living things to give off light is called bioluminescence (bi oh lew muh NEH sunts). The light is produced through chemical reactions in the organism's body. A firefly gives off a flash of light to locate a possible mate. Each species has its own pattern of flashing. Certain kinds of flies, marine organisms, and beetles use bioluminescence to communicate. ☑

Reading Check

4. **Define** What is bioluminescence?

____

____

____

What are other uses of bioluminescence?

Many bioluminescent animals are found deep in oceans where sunlight does not reach. Bioluminescence helps some animals attract prey. Deep-sea shrimp give off clouds of a luminescent substance that helps them escape their predators.

Cyclic Behavior

Innate behavior that happens in a repeating pattern is called **cyclic behavior**. This type of behavior is often repeated in response to changes in the environment. Most animals have a 24-hour cycle of sleeping and wakefulness called a circadian rhythm. Animals that are active during the day are diurnal (dy UR nul). Animals that are active at night are nocturnal (nahk TUR nul). Owls are nocturnal.

FOLDABLES™

E Explain Use quarter sheets of notebook paper, as shown below, to explain the two basic cyclic behaviors found in most animals.

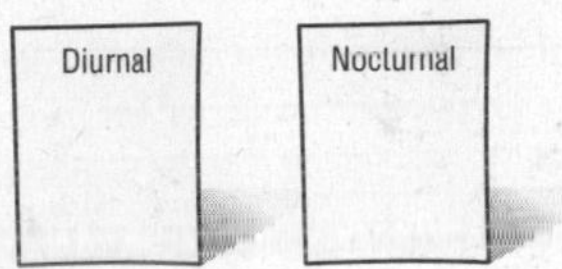

What is hibernation?

Hibernation is a cyclic behavior in which an animal responds to cold temperatures and a limited food supply. During hibernation, an animal's body temperature drops to near that of its surroundings. The animal's breathing rate slows. Animals in hibernation survive on their stored body fat and stay inactive until the weather becomes warm in the spring. Some mammals and many amphibians and reptiles hibernate.

In desertlike environments, some animals go into a period of reduced activity called estivation. Desert animals do this as a response to extreme heat, lack of food, or drought.

What is migration?

Many animals move to new locations when the seasons change. This instinctive seasonal movement is called **migration**. Most animals migrate to find food or to reproduce in environments that give their offspring a better chance for survival.

● After You Read

Mini Glossary

aggression: a forceful behavior used to dominate or control another animal

courtship behavior: behaviors that allow male and female members of a species to recognize each other and to excite each other so they are ready to mate at the same time

cyclic behavior: innate behavior that occurs in a repeating pattern

hibernation: a cyclic response to cold temperatures and limited food supplies

migration: instinctive seasonal movement of animals

pheromone (FER uh mohn): a chemical produced by one animal to influence the behavior of another animal of the same species

social behavior: interactions among organisms of the same species

society: a group of animals of the same species living and working together in an organized way

1. Review the terms and their definitions in the Mini Glossary. Write a sentence using one of the terms to explain how innate animal behavior helps species survive.

___

___

___

2. Write one fact you learned about each form of animal communication in the graphic organizer below.

Forms of Animal Communication

Courtship behavior	Chemical communication	Sound communication	Light communication
___	___	___	___
___	___	___	___
___	___	___	___
___	___	___	___
___	___	___	___

Science Online Visit **life.msscience.com** to access your textbook, interactive games, and projects to help you learn more about behavioral interactions.

Structure and Movement

section 1 The Skeletal System

Before You Read

What is your favorite sport? On the lines below, name all the body parts you use to play this sport.

____

____

____

What You'll Learn

- five functions of the skeletal system
- the differences and similarities between the movable and the immovable joints

Read to Learn

Mark the Text

Identify the Important Points Write a phrase beside each of the main headings to summarize the main point of that section.

Living Bones

The bones in your body are very much alive. Each is a living organ made of several different tissues. Like all living tissues, bone tissue is made up of cells that take in nutrients and use energy. Bone cells have the same needs as other body cells.

What are the major functions of the skeletal system?

All 206 bones in your body make up your **skeletal system**. It is your body's framework, just like the framework of a building. The skeletal system has five major functions.

1. The skeleton gives shape and support to your body.
2. Bones protect your internal organs. For example, the skull surrounds the brain.
3. Major muscles are attached to bones. These muscles help bones move.
4. Blood cells are formed in the center of many bones in soft tissue called red marrow.
5. Major amounts of calcium and phosphorus compounds are stored in the skeleton for later use. Calcium and phosphorus make bones hard. ☑

Reading Check

1. **Identify** Where are blood cells formed?

____

____

____

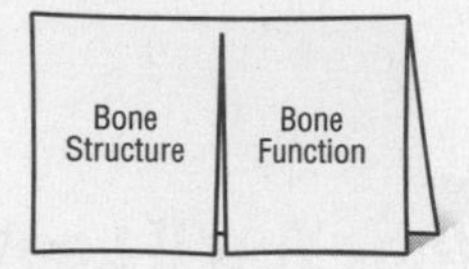
A Describe Make a two-tab book Foldable, as shown below, to organize facts about bone structure and function.

Bone Structure | Bone Function

Picture This

2. Identify Circle the name of the membrane that covers the surface of bone.

3. Explain How do bone cells get nutrients?

____

____

____

Bone Structure

Bones are different sizes and shapes. The shapes of your bones are inherited. But, a bone's shape can change when the attached muscles are used.

As you can see in the figure below, bones are not smooth. They have bumps, edges, round ends, rough spots, and many pits and holes. Muscles and ligaments attach to some of the bumps and pits. The figure below shows that blood vessels and nerves enter and leave the bone through the holes.

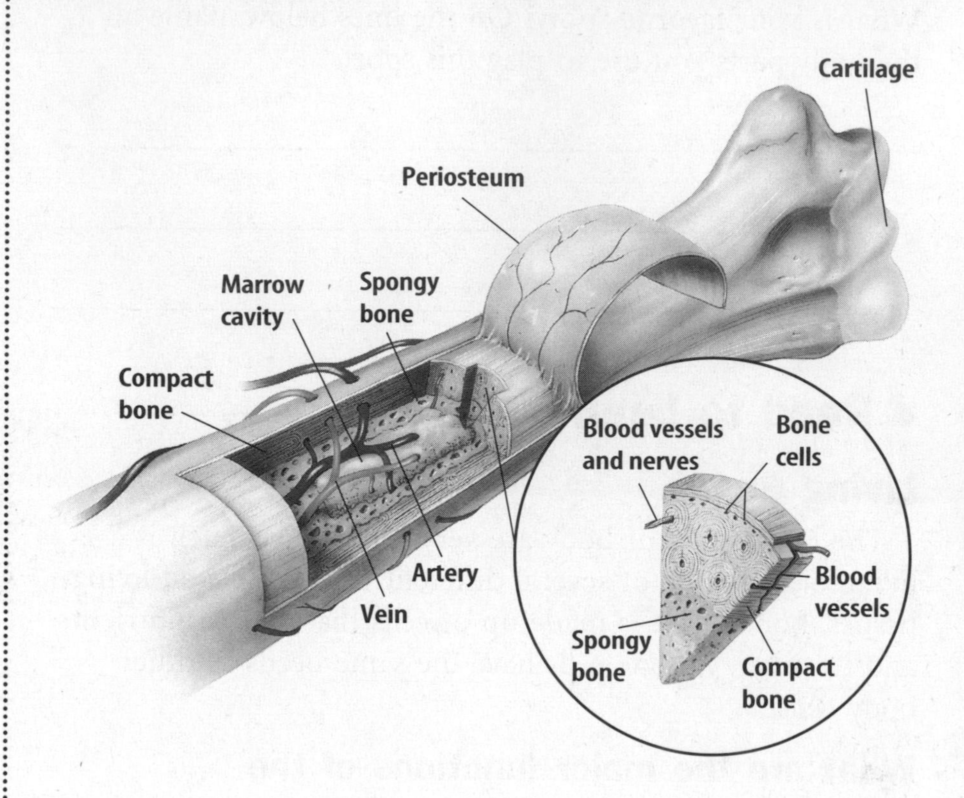

The figure above shows that the surface of a living bone is covered with a tough, tight-fitting membrane called the **periosteum** (per ee AH stee um). Small blood vessels in the periosteum carry nutrients into the bone. Its nerves signal pain. Cells involved in the growth and repair of bone also are found in the periosteum. ☑

What keeps bones from breaking?

Under the periosteum is a hard, strong layer called compact bone. This bone tissue gives bone strength. It has a framework containing deposits of calcium phosphate that makes the bone hard and keeps them from being easily broken. Bone cells and blood vessels are found in compact bone.

Spongy Bone Spongy bone is tissue found in the ends of long bones such as those in your thigh and upper arm. Spongy bone has many small, open spaces that make bones lightweight. In the centers of long bones are large openings called cavities. These cavities and the spaces in spongy bone are filled with yellow and red marrow. Yellow marrow is made up of fat cells. Red marrow makes 2 million to 3 million red blood cells per second.

Why is cartilage important in joints?

The ends of bones are covered with a smooth, slippery, thick layer of tissue called **cartilage**. Cartilage protects the joints and absorbs shock. It also makes movement easier because it lessens friction between bones. People with damaged cartilage have pain when they move, because the bones rub together. ☑

Reading Check

4. Identify two functions of cartilage.

____

____

____

Bone Formation

Before you were born, your skeleton was made of cartilage. Over time, the cartilage was replaced by bone. Bone-forming cells called osteoblasts (AHS tee oh blasts) deposit calcium and phosphorus in bones, making bones hard. Throughout your life, healthy bone tissue is being formed. Osteoblasts build up bone.

Another type of bone cell, called an osteoclast, breaks down bone tissue. Osteoclasts release calcium and phosphorus into your bloodstream. This keeps the amount of calcium and phosphorus in your blood at healthy levels.

Joints

You are able to move because your skeleton has joints. Anyplace where two or more of your bones come together is a **joint**. The bones that make up a joint are kept apart by cartilage and are held in place by a tough band of tissue called a **ligament**. Muscles move bones by moving joints. The figure below shows the different types of joints in your body.

Skull

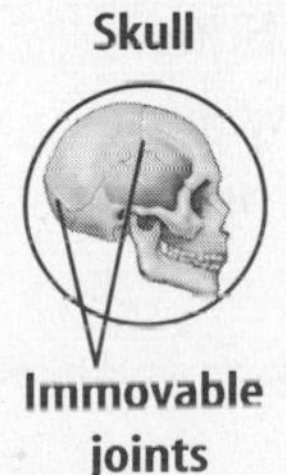

Immovable joints

Arm

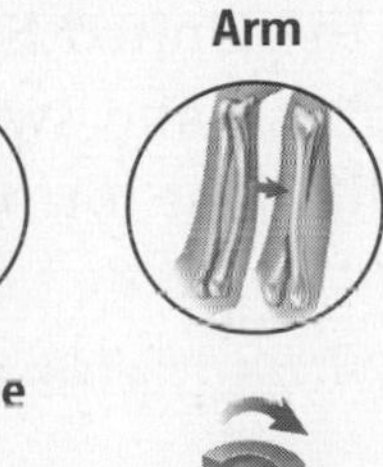

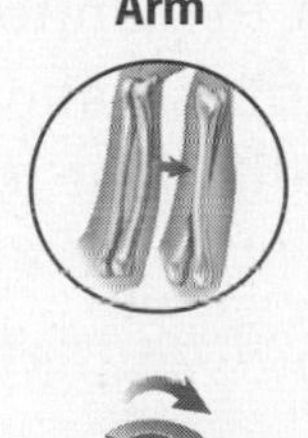

Pivot joint

Shoulder

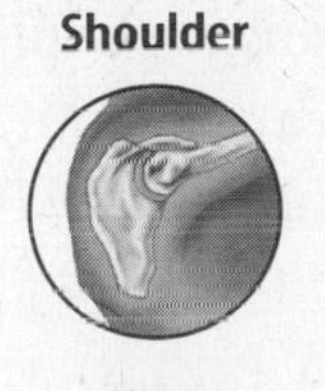

Ball-and-socket joint

Knee

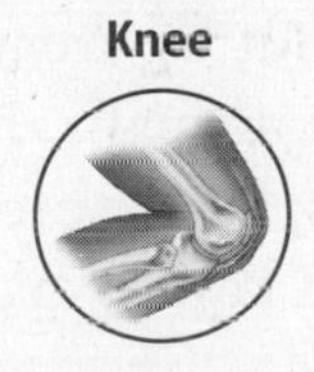

Hinge joint

Vertebrae

Gliding joint

Picture This

5. Classify Use the figure to determine which kind of joint you use in each of the following activities.

a. raise your arm

____

b. kneel

____

FOLDABLES™

B Describe Make quarter-sheet Foldables, as shown below, to organize key terms and concepts about immovable and movable joints.

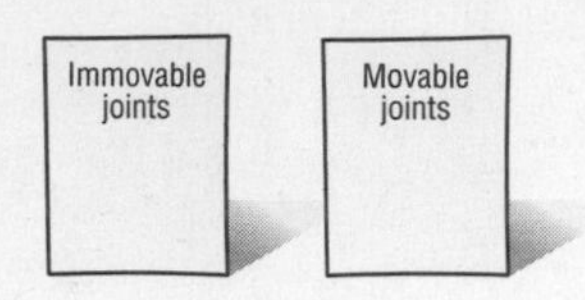

What is an immovable joint?

An immovable joint allows little or no movement. The skull has immovable joints.

What are the types of movable joints?

All movements require movable joints. As you can see in the figure on the previous page, there are four types of movable joints—pivot, ball-and-socket, hinge, and gliding.

Pivot Joints In a pivot joint, one bone rotates in a ring of another bone that does not move. Turning your head is an example of a pivot movement.

Ball-and-socket Joints Your legs and arms can swing in almost any direction because they have ball-and-socket joints. This kind of joint consists of a bone with a rounded end that fits into a cuplike cavity on another bone. This joint gives you a wide range of motion.

Hinge Joints A hinge joint has a back-and-forth movement like hinges on a door. Elbows, knees, and fingers have hinge joints. Hinge joints have a smaller range of motion than ball-and-socket joints.

Gliding Joints Your wrists, ankles, and vertebrae have gliding joints. In a gliding joint, one part of a bone slides over another bone. This joint also moves back and forth.

Why do your bones move smoothly?

If your bones did not have cartilage at the ends, they would wear away at the joints. Cartilage allows bones to slide more easily over each other by reducing friction. Pads of cartilage, called disks, are located between the vertebrae in your back. These disks act as cushions and prevent injury to your spinal cord. A fluid that comes from nearby blood vessels keeps the joint lubricated. ☑

Reading Check

6. **Describe** What are the functions of disks?

____

____

____

What is arthritis?

Arthritis is the most common joint problem in humans. About one out of every seven people in the United States suffers from arthritis. It causes pain, stiffness, and swelling of the joints. There are more than 100 different forms of arthritis that can damage the joints.

After You Read

Mini Glossary

cartilage: a smooth, slippery, thick layer of tissue that covers the ends of bones

joint: the place where two or more bones come together

ligament: a tough band of tissue that holds bones together at the joint

periosteum (per ee AH stee um): a tough, tight-fitting membrane on the surface of a living bone

skeletal system: the framework of bones in the body

1. Review the terms and their definitions in the Mini Glossary. Use two terms in the glossary to write a sentence describing some part of the skeletal system.

____

____

____

2. Complete the concept web below by naming the five functions of the skeletal system.

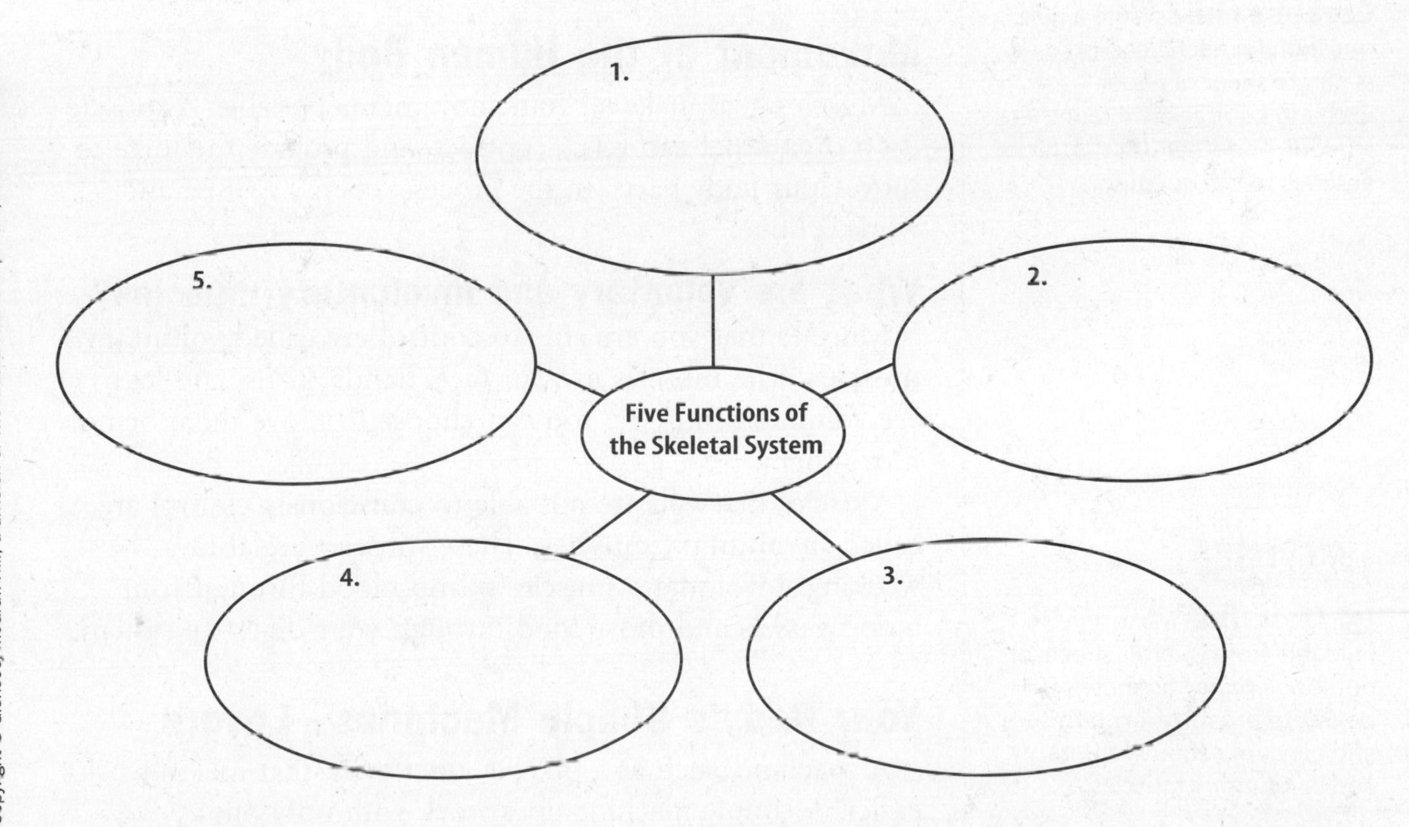

Science Online Visit **life.msscience.com** to access your textbook, interactive games, and projects to help you learn more about the skeletal system.

End of Section

Structure and Movement

chapter 17

section 2 The Muscular System

What You'll Learn
- the major function of the muscular system
- how the three types of muscles differ
- how muscles move body parts

Before You Read

On the lines below, describe a movement you make that uses muscles.

____

____

____

Study Coach

Create a Quiz Write a quiz question for each heading on a separate sheet of paper. Exchange quizzes with another student. Together discuss the answers to the quizzes.

Read to Learn

Movement of the Human Body

Muscles help make all your movements possible. A **muscle** is an organ that can relax, contract, and provide the force to move your body parts. In the process, energy is used and work is done.

What are voluntary and involuntary muscles?

Muscles that you are able to control are called **voluntary muscles.** The muscles in your face, hands, arms, and legs are voluntary muscles. You can choose to move them or not move them.

Muscles that you are not able to consciously control are called **involuntary muscles.** These muscles are always working. Involuntary muscles pump blood through your blood vessels and move food through your digestive system.

FOLDABLES™

C Describe Make a Foldable from quarter sheets of notebook paper, as shown below, to organize key terms and concepts about voluntary and involuntary muscles.

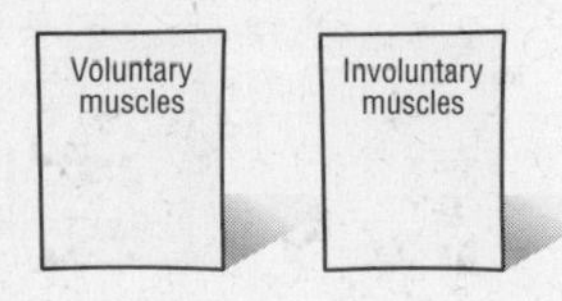

Your Body's Simple Machines—Levers

A machine, such as a bike, is any device that makes work easier. A simple machine does work with only one movement. The hammer is a type of simple machine called a **lever.** A lever is a rod or plank that pivots or turns about a point. The point is called a fulcrum.

Types of Levers The action of bones, joints, and muscles working together is like a lever. In your body, your bones are the rods and your joints are the fulcrums. The relaxation and contraction of muscles provide the force to move body parts. There are three types of levers—first-class, second-class, and third-class. The figure below shows how all three levers are used in the body when serving a tennis ball.

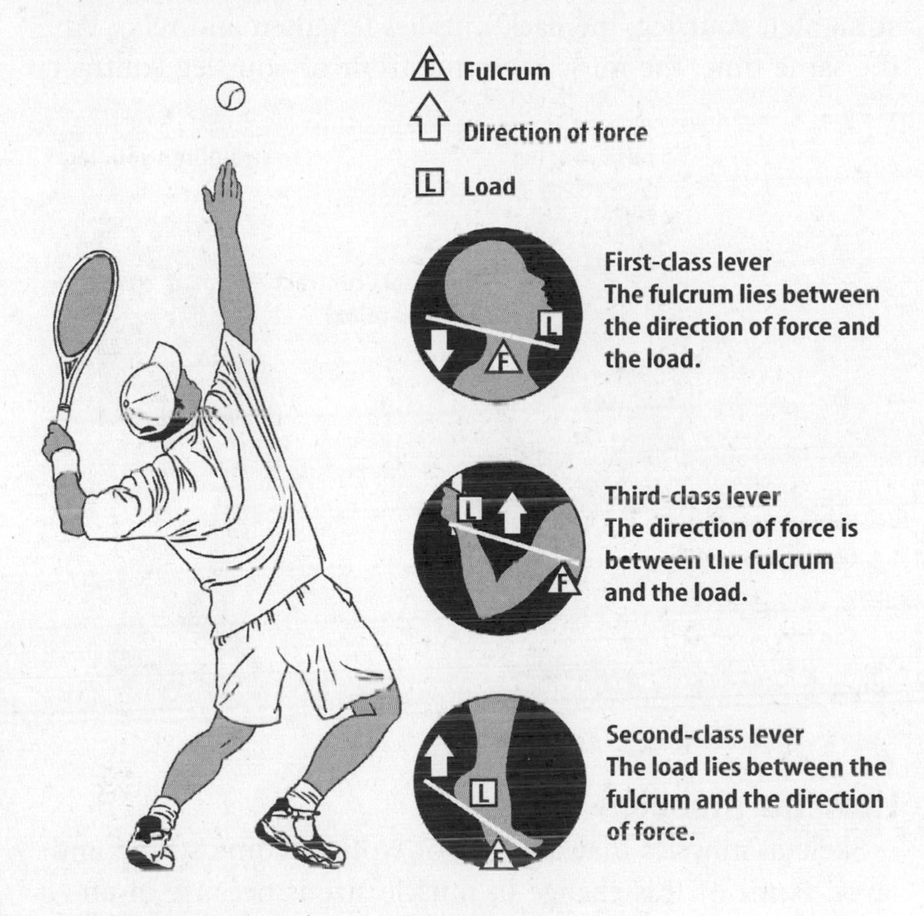

F Fulcrum
Direction of force
L Load
First-class lever
The fulcrum lies between the direction of force and the load.
Third-class lever
The direction of force is between the fulcrum and the load.
Second-class lever
The load lies between the fulcrum and the direction of force.

Picture This

1. **Identify** In the drawing of the tennis player, circle and label the three types of levers in the body that are used to serve a tennis ball.

Classification of Muscle Tissue

There are three types of muscles in your body—skeletal muscle, cardiac muscle, and smooth muscle. The muscles that move bones are **skeletal muscles.** They are the most common muscle type in your body. They are attached to bones by thick bands of tissue called **tendons**. Skeletal muscles are voluntary muscles. You choose to use them or not use them, such as when you walk or when you rest.

Cardiac muscle is found only in the heart. This involuntary muscle contracts about 70 times per minute. **Smooth muscles** are found in your intestines, bladder, blood vessels, and other internal organs. They are involuntary muscles that slowly contract and relax.

FOLDABLES™

D Organize Make a three-tab Foldable, as shown below, to organize facts about skeletal muscles, cardiac muscle, and smooth muscles.

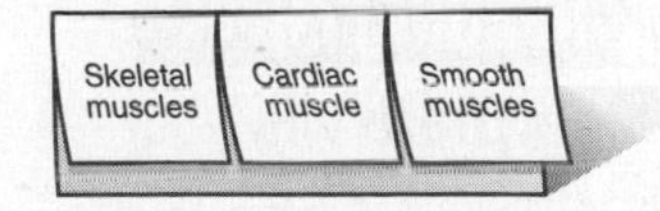

Skeletal muscles
Cardiac muscle
Smooth muscles

Working Muscles

You are able to move because skeletal muscles work in pairs, as shown in the figure below. When one muscle of a pair contracts, the other muscle relaxes, or returns to its original length. Muscles always pull. They never push. When the muscles on the back of your upper leg contract, they shorten. This pulls your lower leg back and up. When you straighten your leg, the back muscles lengthen and relax. At the same time, the muscles on the front of your leg contract.

Picture This

2. Describe Use these drawings to explain to a classmate how muscles work by pulling, rather than by pushing.

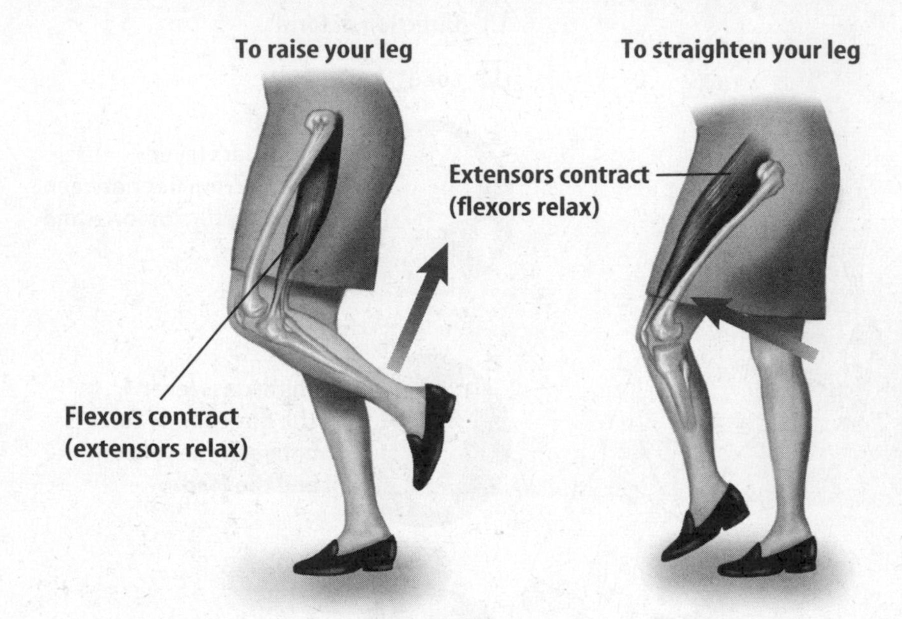

How do muscles change?

Skeletal muscles that do a lot of work become strong and large. Some of this change in muscle size is because of an increase in the number of muscle cells. Most of the change is because individual muscle cells become larger. Muscles that are not exercised become soft, flabby, and weak.

How are muscles fueled?

Your muscles need energy to contract and relax. Your blood carries energy-rich molecules to your muscle cells. As the muscle contracts, part of the chemical energy changes to mechanical energy (movement). Some of the chemical energy changes to thermal energy (heat) as muscles are used. The heat produced by muscle contractions helps keep your body temperature constant. When the supply of energy-rich molecules is used up, the muscle becomes tired and needs to rest. While your muscle rests, your blood supplies more energy-rich molecules to your muscle cells. ☑

3. Explain why your muscles need energy?

____

____

____

After You Read

Mini Glossary

cardiac muscle: muscle found only in the heart

involuntary muscle: a muscle, such as the heart muscle, that cannot be consciously controlled

lever: a rod or plank that pivots or turns about a point; a simple machine

muscle: an organ that can relax, contract, and provide the force to move your body parts

skeletal muscle: a muscle that moves the body

smooth muscle: the muscle found in the intestines, bladder, blood vessels, and other internal organs

tendons: thick bands of tissue that attach muscles to bones

voluntary muscle: a muscle, such as a leg or arm muscle, that can be consciously controlled

1. Review the terms and their definitions in the Mini Glossary. Write a sentence explaining the difference between voluntary muscles and involuntary muscles.

____

____

2. Fill in the table below to identify the three types of muscle tissues and explain the functions of each.

Types of Muscle Tissue	Function of the Muscle Tissue

2. How did the quiz help you review what you have learned about the muscular system?

____

____

Visit **life.msscience.com** to access your textbook, interactive games, and projects to help you learn more about the muscular system.

chapter 17

Structure and Movement

section 3 The Skin

What You'll Learn

- the difference between the epidermis and the dermis of the skin
- the functions of the skin
- how skin protects the body from disease
- how skin heals itself

Before You Read

After you scrape your knee or cut your finger, what happens to your skin? On the lines below, describe what happens as your skin heals itself.

____

____

____

Study Coach

Organize Information Create an outline of the section, using the headings as your main outline items. Add main ideas below the headings.

Read to Learn

Your Largest Organ

Did you know that your skin is your body's largest organ? Much of the information you receive about your environment comes through your skin. You can think of your skin as your largest sense organ. Look at the figure below. Notice that your skin is made up of three layers of tissue—the epidermis, the dermis, and a fatty layer. Refer to this figure as you read about the three layers of skin.

Picture This

1. **Identify** Highlight the three layers of skin. Use a different color for each layer.

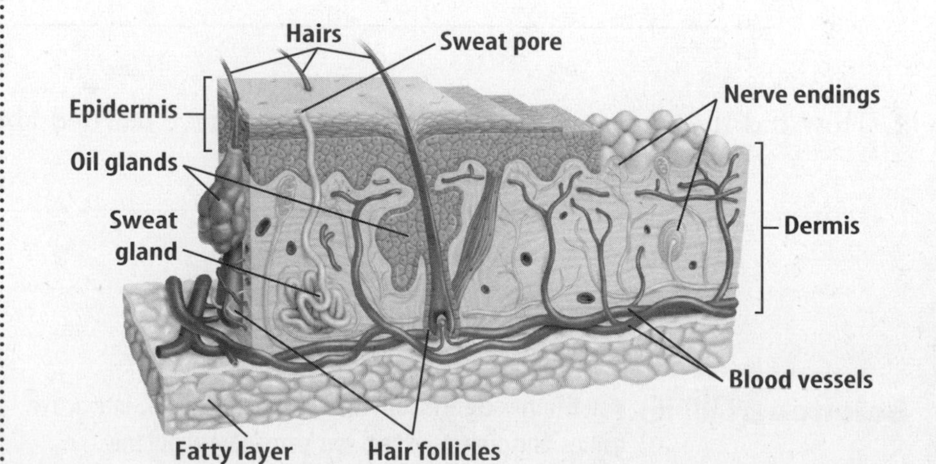

Skin Structures

The **epidermis** is the outer, thinnest layer of skin. The outermost cells of the epidermis are dead. Thousands of these cells rub off your body when you shower, shake hands, or blow your nose. New cells are constantly being made to replace the dead cells.

What causes different skin colors?

Cells in the epidermis make the chemical melanin (MEL uh nun). **Melanin** is a pigment that protects your skin and gives it color, as shown in the figure below. The different amounts of melanin made by cells cause differences in skin color. When your skin is exposed to ultraviolet (UV) rays, more melanin is made and your skin becomes darker. The lighter a person's normal skin color, the less protection that person has from the Sun. ☑

Aaron Haupt

Reading Check

2. **Explain** what causes differences in skin color.

What are the dermis and the fatty layer?

The **dermis** is the layer of skin cells right below the epidermis. The dermis contains many blood vessels, nerves, muscles, oil, and sweat glands. Below the dermis is a fatty layer, which helps keep the body warm.

Skin Functions

The skin is important to the body. Some of its most important functions include protection, sensory response, formation of vitamin D, control of body temperature, and ridding the body of wastes. ☑

Reading Check

3. **Identify** two functions of skin.

Which skin function is the most important?

The most important skin function is protection. The skin is a protective covering over the body. It stops some bacteria and other disease-causing organisms from passing through unbroken skin. The skin slows down water loss from body tissues.

Why do you know a pan is hot?

Special nerve cells in the skin are able to sense things. The cells send this information to the brain. This is why you can sense the softness of a cat or the heat of a hot pan.

How does the skin produce vitamin D?

Another important function of skin is the formation of vitamin D. Ultraviolet light produces small amounts of this vitamin from a fatlike molecule in your epidermis. Vitamin D is needed for good health because it helps your body absorb calcium from food in your digestive tract. ☑

Reading Check

4. **Explain** why your body needs Vitamin D.

____

____

How does skin control body temperature?

Humans can withstand a limited range of body temperatures. The thermometer below shows environmental changes that affect the body. Skin plays an important role in controlling body temperature. Blood vessels in the skin can help release or hold heat. When blood vessels constrict, or get smaller, blood flow slows, and less heat is released. When blood vessels expand, blood flow increases, and more heat is released.

Picture This

5. **Describe** What happens to a person's body temperature when they are excited?

____

____

____

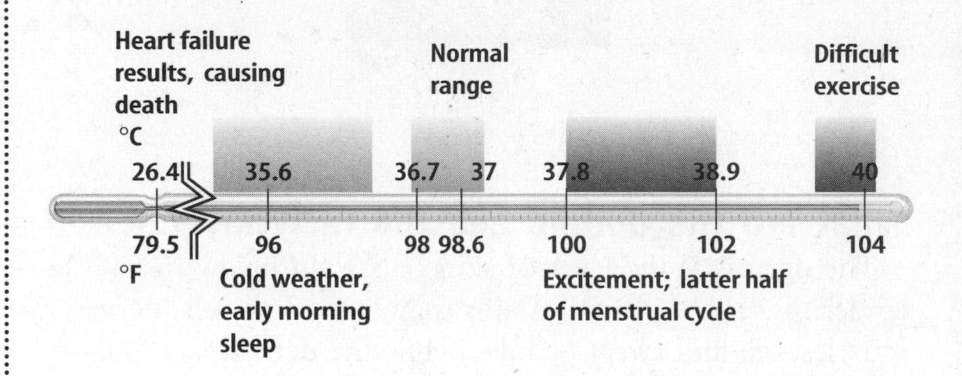

Sweat glands in the skin also help control the body's temperature. As blood vessels expand, pores open in the skin that lead to the sweat glands. Sweat moves out onto the skin. Heat is transferred from the body to the sweat on the skin. As the sweat evaporates, heat is removed, and the skin is cooled.

Sweat glands also help release wastes. As your cells use nutrients for energy, they produce wastes. Sweat glands release water, salt, and other wastes.

Skin Injuries and Repair

When skin is injured, it responds by producing new cells in the epidermis and repairing tears in the dermis. Injured skin allows disease-causing organisms to enter the body rapidly and an infection may occur.

What are bruises?

When you have a bruise, the tiny blood vessels underneath the skin have burst, releasing red blood cells. These blood cells break down and release hemoglobin. The chemical hemoglobin breaks down into its components, called pigments. The color of the pigments causes the bruised area to turn blue, red, and purple. Swelling also may occur. As the injury heals, the bruise turns yellow as the pigment in the red blood cells is broken down even more and reenters the bloodstream. After all the pigment is absorbed into the bloodstream, the skin looks normal again.

How do cuts heal?

Any tear in the skin is called a cut. Blood flows out of a cut until a clot forms over it. Then a scab forms, stopping bacteria from entering the body. Cells in the surrounding blood vessels fight infection. In time, the scab falls off, and new skin is left behind. If the cut is large, a scar may develop because of the large amounts of thick tissue fibers that form.

What are skin grafts?

When a person has a serious injury to large areas of skin, there may not be enough skin cells left that can divide to replace the skin that has been lost. This can lead to infection and possible death. Skin grafts can be used to replace the lost skin. Skin grafts are pieces of skin that are cut from one part of a person's body and then moved to the injured or burned area where there is no skin. ☑

Skin grafts are usually taken from the person's own body. When a person does not have enough healthy skin, doctors may use skin from dead humans, or cadavers, for skin grafts. The cadaver skin is used for a short time to prevent infections. Doctors then grow large sheets of epidermis from small pieces of the burn victim's own healthy skin. The cadaver skin patch is removed and the new skin is put in place.

FOLDABLES™

E Describe Make a Foldable from quarter sheets of notebook paper, as shown below, to organize facts and concepts about bruises, cuts, and grafts.

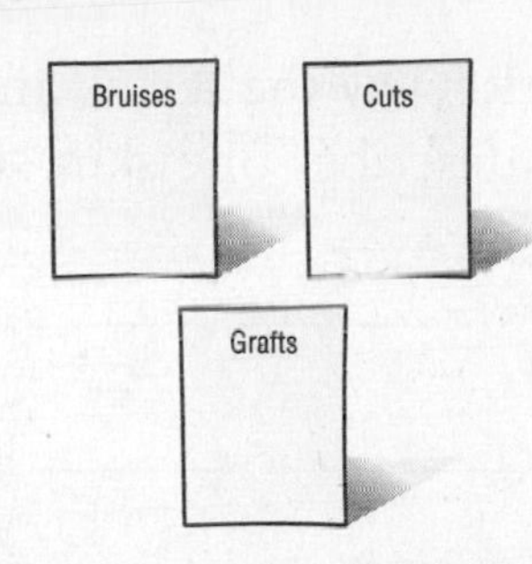

Reading Check

6. Explain Why are skin grafts used?

____

____

____

After You Read

Mini Glossary

dermis: the layer of cells directly below the epidermis, which contains blood vessels, nerves, muscles, oil and sweat glands

epidermis: the outer, thinnest layer of skin

melanin: a pigment that protects your skin and gives it color

1. Review the terms and their definitions in the Mini Glossary. Write a sentence that describes one skin structure.

____

____

____

2. Fill in the table below to identify five functions of skin.

Functions of Skin
1.
2.
3.
4.
5.

Visit **life.msscience.com** to access your textbook, interactive games, and projects to help you learn more about about skin.

Nutrients and Digestion

section 1 Nutrition

Before You Read

List on the lines below five foods that you think are good for you, or are nutritious. Explain what makes them nutritious.

____

____

What You'll Learn

- the six kinds of nutrients
- why each nutrient is important
- how your diet affects your health

Read to Learn

Why do you eat?

Your body needs energy for every activity that it performs. You need energy to run, blink your eyes, and lift your backpack. This energy comes from the foods you eat. The foods you eat also give your body the nutrients it needs. **Nutrients** (NEW tree unts) are substances in food that provide the energy and materials cells need to develop, grow, and repair themselves.

How is the energy in food measured?

The amount of energy you need depends on your body mass, age, and how active you are. The amount of energy in food is measured in Calories. A Calorie (Cal) is the amount of heat needed to raise the temperature of 1 kg of water 1°C. The number of calories in a food depends on the kinds of nutrients the food contains.

Classes of Nutrients

Six kinds of nutrients are found in food. The six nutrients are proteins, carbohydrates, fats, vitamins, minerals, and water. Proteins, carbohydrates, vitamins, and fats are organic nutrients because they contain carbon. Minerals and water are inorganic nutrients because they do not contain carbon. ☑

Study Coach

Use an Outline As you read, make an outline to summarize the information in the section. Use the main headings in the section as the main headings in the outline. Add information under each heading in the section.

☑ Reading Check

1. **Identify** three nutrients found in foods.

____

____

____

Absorption of Nutrients Foods with carbohydrates, fats, and proteins have to be digested or broken down before the body can use them. Water, vitamins, and minerals are absorbed directly into the bloodstream.

How does the body use proteins?

Proteins replace and repair body cells and help the body grow. **Proteins** are large molecules that contain carbon, hydrogen, oxygen, nitrogen, and sometimes sulfur. A protein molecule is made up of many smaller units called **amino acids**. Different foods have different amounts of protein, as shown in the table below.

Calories and Protein in Selected Food Items

Food	Calories	Protein
Pepperoni pizza (1 slice)	280	16 g
Large taco	186	15 g
Banana split	540	10 g

Picture This

2. **Determine** Which of the food choices in the table has the most Calories?

3. **Explain** What unit is used to measure protein?

4. **Identify** Which of the food choices in the table provides the least protein?

What are essential amino acids?

Your body needs 20 amino acids to make the thousands of proteins that your cells use. Most of the amino acids can be made in the body's cells. Eight of the amino acids, however, cannot be made by the body. These eight are called essential amino acids. You have to get them from the food you eat. Foods that provide all eight essential amino acids are called complete proteins. Complete proteins are found in eggs, milk, cheese, and meat. Incomplete proteins are missing one or more of the essential amino acids. Vegetarians need to eat a wide variety of protein-rich vegetables, fruits, and grains to get all eight essential amino acids.

FOLDABLES™

A Classify Make a folded table, as shown below, to explain how your body uses proteins, carbohydrates, and fats.

Nutrients	How body uses	Nutrient becomes
Proteins		
Carbo-hydrates		
Fats		

Why are carbohydrates important?

Carbohydrates (kar boh HI drayts) are the main sources of energy for your body. A carbohydrate molecule is made up of carbon, hydrogen, and oxygen atoms. Energy holds these atoms together. When carbohydrate molecules break apart in the cells, energy is released for your body to use.

What are the three types of carbohydrates?

The three types of carbohydrates are sugar, starch, and fiber. Sugars are simple carbohydrates. Table sugar is one of these sugars. Fruits, honey, and milk also contain forms of sugar. Your cells break down glucose, which is a simple sugar.

What is the difference between starch and fiber?

Starch and fiber are complex carbohydrates. Starch is found in potatoes and in foods made from grains such as pasta. Starches are made up of simple sugars strung together in long chains. Fiber is found in the cell walls of plant cells. Foods such as whole-grain breads, cereals, beans, and vegetables and fruits are good sources of fiber. You cannot digest fiber, but it is needed to keep your digestive system running smoothly.

How does the body use fats?

Fats, also called lipids, provide the body with energy and help it absorb vitamins. Fat tissue cushions the body's internal organs. A major part of every cell membrane is made up of fat. Fats release more energy than carbohydrates do. When food is being digested, fat is broken down into smaller molecules called fatty acids and glycerol (GLIH suh rawl). Fat is a good storage unit for energy. Your body takes excess energy from the foods you eat and changes it to fat that is stored for later use.

Think it Over

5. **Explain** why fats are an important part of a healthful diet.

____

____

____

What are saturated and unsaturated fats?

There are two kinds of fats, unsaturated fats and saturated fats. Unsaturated fats are usually liquid at room temperature. Vegetable oil is an example of an unsaturated fat. Saturated fats are usually solid at room temperature. Saturated fats are found in meats, animal products, and some plants. ☑

Eating too many saturated fats has been linked to high levels of cholesterol in the body. Cholesterol is part of the cell membrane in all of your cells. However, a diet that is high in cholesterol can cause deposits to form on the inside walls of blood vessels. The deposits can keep the blood supply from getting to organs. The deposits also can increase blood pressure and lead to heart disease and strokes.

Reading Check

6. **Determine** which fat is usually liquid at room temperature and which fat is solid.

____

____

____

What are vitamins?

Vitamins are nutrients that the body needs in small amounts. Vitamins help the body grow, help keep the body functioning properly, and help prevent some diseases. Most foods contain some vitamins. However, no single food has all the vitamins you need.

Think it Over

7. Describe a major difference between the two groups of vitamins.

What is the difference between water-soluble and fat-soluble vitamins?

There are two groups of vitamins, water-soluble and fat-soluble. Water-soluble vitamins dissolve easily in water. Your body does not store these vitamins, so you need to get them every day. Fat-soluble vitamins dissolve only in fat. These vitamins are stored in the body.

You get most of your vitamins from food. However, your body makes some vitamins. For example, your body makes vitamin D when your skin is exposed to sunlight.

How do minerals affect the body?

Minerals are inorganic nutrients that take part in many chemical reactions in your body. Minerals build cells, send nerve impulses throughout your body, and carry oxygen to body cells.

Your body uses about 14 minerals. Of the 14 minerals, your body uses calcium and phosphorus in the largest amounts. Calcium and phosphorus help form and maintain bones. Some minerals, such as copper, are trace minerals. The body only needs very small amounts of trace minerals. Review the table below to learn more about some of the minerals your body uses.

Picture This

8. Identify Which minerals does your body use to help conduct nerve impulses?

Health Benefits of Minerals from Your Food

Mineral	Health Effect	Food Sources
Calcium	builds strong bones and teeth, helps blood clotting and muscle and nerve activity	milk, cheese, eggs, green leafy vegetables, soy
Phosphorus	builds strong bones and teeth, helps muscles contract, stores energy	cheese, meat, cereal
Potassium	balances water in cells, conducts nerve impulses, helps muscles contract	bananas, potatoes, nuts, meat, oranges
Sodium	balances fluid in tissues, conducts nerve impulses	meat, milk, cheese, salt, beets, carrots, nearly all foods
Iron	moves oxygen in hemoglobin by red blood cells	red meat, raisins, beans, spinach, eggs
Iodine (trace)	helps thyroid activity, stimulates metabolism	seafood, iodized salt

Why is water an important nutrient?

Next to oxygen, water is the most important thing your body needs for survival. You could live a few weeks without food but only a few days without water. Cells need water to carry out their work. Many other nutrients that the body needs have to be dissolved in water before they can be used.

Water Loss Water makes up about 60 percent of the weight of your body. Most of the body's water is located in body cells. Water is also found around cells and in blood. Your body loses water when you perspire and when you exhale. Your body also loses water when it gets rid of wastes. To replace the amount of water your body loses each day, you need to drink about 2 L of liquids. Drinking liquids is not the only way to get water. Many foods, such as apples and meats, are made up of a large amount of water.

Why do you get thirsty?

When your body needs to replace water that it lost, messages are sent to your brain that make you feel thirsty. Drinking water satisfies your thirst. Drinking water also helps to restore the body's homeostasis (hoh mee oh STAY sus). When your body is in homeostasis, or balance, it has the right amount of water and the right temperature. When homeostasis returns, the messages to the brain stop, and you no longer feel thirsty. ☑

Reading Check

9. Define What is homeostasis?

____

____

____

Food Groups

No natural food has all the nutrients your body needs. You need to eat a variety of foods. Nutritionists have set up a system to help people choose foods that supply all the nutrients the body needs for energy and growth. The system is called the food pyramid. It is shown in the figure below.

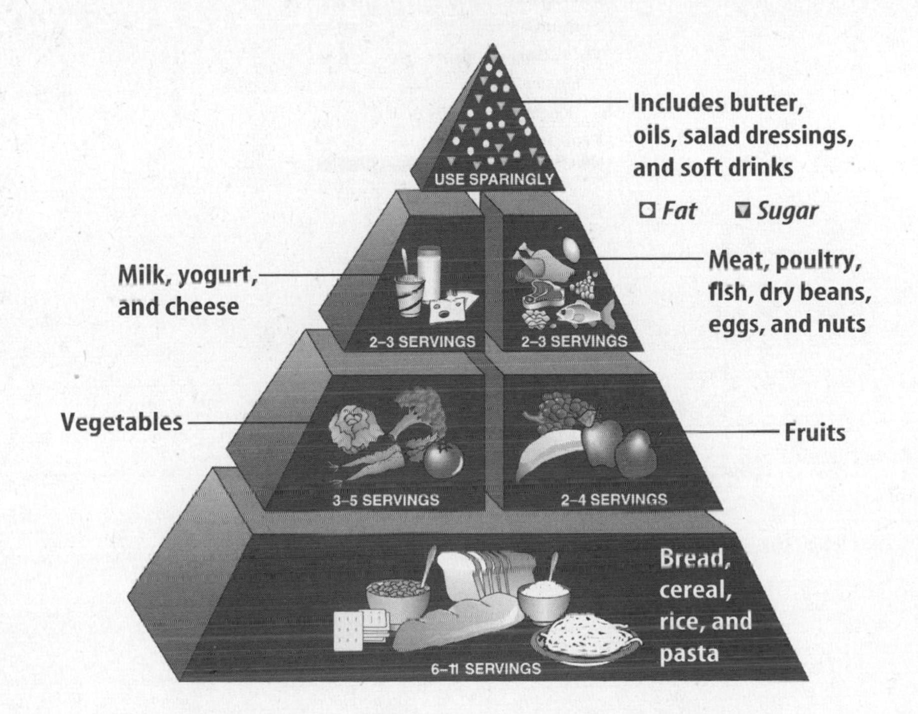

Picture This

10. Apply To the side of each block of the pyramid, write the approximate number of servings you ate in each food group yesterday.

What are the five food groups?

Foods that have the same type of nutrient belong to a **food group**. There are five food groups: bread and cereal, vegetables, fruits, milk, and meat. ☑

Reading Check

11. Define What is a food group?

What should you eat from each food group?

The food pyramid on the previous page shows the recommended number of servings from each food group that people should eat every day. Eating the recommended daily amount for each group will give your body the nutrients it needs for good health. The size of a serving is different for different food groups.

Why should you read food labels?

The food labels, such as the one below, on all packaged foods contain nutritional facts about the foods. These facts can help you make healthful food choices. The labels can help you plan meals that include the recommended amounts of nutrients.

Picture This

12. Identify Circle the total number of Calories per serving on the label.

Nutrition Facts
Serving Size 1 Meal

Amount Per Serving	
Calories 330	Calories from Fat 60
	% Daily Value*
Total Fat 7g	**10%**
Saturated Fat 3.5g	**17%**
Polyunsaturated Fat 1g	
Monounsaturated Fat 2.5g	
Cholesterol 35mg	**12%**
Sodium 460mg	**19%**
Total Carbohydrate 52g	**18%**
Dietary Fiber 6g	**24%**
Sugars 17g	
Protein 15g	
Vitamin A 15% • Vitamin C 70%	
Calcium 4% • Iron 10%	

* Percent Daily Values are based on a 2,000 calorie diet. Your daily values may be higher or lower depending on your calorie needs.

	Calories	2,000	2,500
Total Fat	Less than	65g	80g
Sat Fat	Less than	20g	25g
Cholesterol	Less than	300mg	300mg
Sodium	Less than	2,400mg	2,400mg
Total Carbohydrate		300g	375g
Dietary Fiber		25g	30g

KS Studios

After You Read

Mini Glossary

amino acid: one of the small units that make up a protein molecule

carbohydrate (kar boh HI drayt): a molecule made up of carbon, hydrogen, and oxygen atoms; nutrient that is the main source of energy for the body

fat: necessary nutrient that provides the body with energy and helps it absorb vitamins, also known as a lipid

food group: foods that have the same type of nutrient

mineral: inorganic nutrient that takes part in many chemical reactions in the body

nutrient (NEW tree unt): substances in food that provide energy and materials for cells to develop, grow, and repair themselves

protein: large molecules that contain carbon, hydrogen, oxygen, nitrogen, and sometimes sulfur; one of the six kinds of nutrients

vitamin: nutrient that is needed in small amounts to help the body grow, to regulate body functions, and to prevent some diseases

1. Review the terms and their definitions in the Mini Glossary. Write a sentence that explains the relationship between amino acids and proteins.

____

____

2. Complete the diagram below by classifying the six kinds of nutrients.

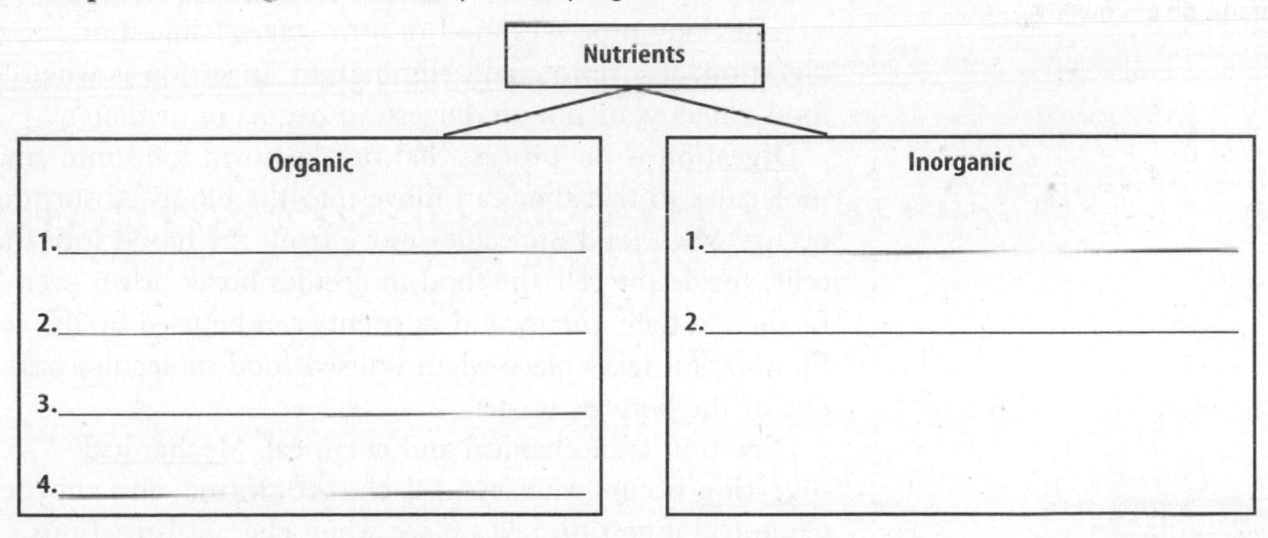

3. How does your outline help you understand ideas about nutrition?

____

____

ScienceOnline Visit **life.msscience.com** to access your textbook, interactive games, and projects to help you learn more about nutrition.

Nutrients and Digestion

section 2 The Digestive System

What You'll Learn
- the differences between mechanical digestion and chemical digestion
- the organs of the digestive system and what they do
- how homeostasis is maintained in digestion

Before You Read

Read the title of this chapter above. On the lines below, tell how you think nutrients and digestion are related.

____

____

____

Mark the Text

Identify the Main Point Underline the main idea in each paragraph as you read.

Read to Learn

Functions of the Digestive System

Your body processes food in four stages—ingestion, digestion, absorption, and elimination. Ingestion is when food enters your mouth. Digestion begins immediately.

Digestion is the process that breaks down food into small molecules so that they can move into the blood. Absorption occurs when food molecules move from the blood into the cells. Inside the cell, the food molecules break down even further so their energy and nutrients can be used by the cell. Elimination takes place when unused food molecules pass out of the body as wastes.

Digestion is mechanical and chemical. **Mechanical digestion** occurs when food is chewed, mixed, and churned. **Chemical digestion** takes place when chemical reactions occur that break down large molecules of food into smaller molecules.

FOLDABLES™

B Describe Use quarter sheets of notebook paper, as shown below, to write descriptions of mechanical and chemical digestion.

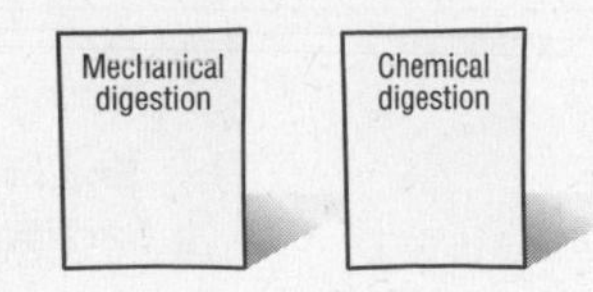

Enzymes

Enzymes (EN zimez) make chemical digestion possible. An **enzyme** is a type of protein that speeds up the rate of a chemical reaction in your body. Enzymes reduce the energy needed for a chemical reaction to start.

How do enzymes help digestion?

Enzymes help you digest different nutrients. Amylase (AM uh lays) is an enzyme made by glands near the mouth. Amylase helps speed up the breakdown of complex carbohydrates, such as starch, into simpler carbohydrates—sugars. The enzyme pepsin works in the stomach to help the chemical reactions that break down proteins. Enzymes in the small intestine help to break down proteins into amino acids. ☑

The pancreas, an organ on the back side of the stomach, releases enzymes into the small intestine. Some of these enzymes continue the starch breakdown that started in the mouth. The sugars from this breakdown are turned into glucose and are used by the body's cells. Some enzymes from the pancreas help break down fats into fatty acids. Other enzymes from the pancreas aid the breakdown of proteins.

Reading Check

1. Explain What does amylase do?

____

____

____

What else do enzymes do?

Enzymes help speed up chemical reactions that help your body grow. Muscle and nerve cells use enzymes to produce energy. Blood needs enzymes to clot.

Organs of the Digestive System

Your digestive system has two parts—the digestive tract and the accessory organs. The parts of the digestive system are shown on the next page.

What are the parts of the digestive tract?

The digestive tract includes the mouth, esophagus (ih SAH fuh guhs), stomach, small intestine, large intestine, rectum, and anus. Food passes through all of these organs. The accessory organs include the tongue, teeth, salivary glands, liver, gallbladder, and pancreas. Food does not pass through these organs. However, the accessory organs help with mechanical and chemical digestion.

What part does the mouth play in digestion?

Both mechanical and chemical digestion take place in the mouth. Mechanical digestion happens when you chew your food. Chemical digestion happens when the tongue moves food around and mixes it with saliva (suh LI vuh).

Saliva is made by three glands near the mouth. Saliva contains an enzyme that helps break down starch into sugar. After the food is swallowed, it passes into the esophagus.

FOLDABLES™

C Sequence Make a six-tab book, as shown below, to identify how food moves through the digestive system.

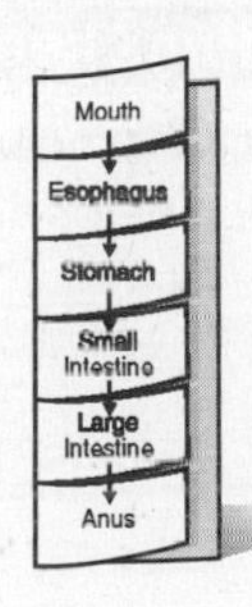

What does the esophagus do?

Food passes over the epiglottis (ep uh GLAH tus) as it moves into the esophagus. The epiglottis automatically covers the opening to the windpipe to stop food from entering it. Food moves down the esophagus to the stomach, as shown in the figure below. Mucous glands in the esophagus keep the food moist. Smooth muscles in the esophagus move the food downward with a squeezing action. These muscle contractions, called **peristalsis** (per uh STAHL sus), move food through the entire digestive tract. No digestion happens in the esophagus.

Picture This

2. Identify Circle the names of the organs of the digestive tract. Highlight the names of the accessory organs.

How does the stomach help digestion?

The stomach is a bag of muscle. Both mechanical and chemical digestion take place in the stomach. Mechanical digestion happens when food is mixed in the stomach by peristalsis. Chemical digestion happens when food in the stomach mixes with enzymes and strong digestive acids, such as hydrochloric acid solution.

The acidic solution works with the enzyme pepsin to digest protein. The acidic solution also destroys bacteria present in the food. ☑

The stomach also produces mucus, which makes food more slippery and protects the stomach from the strong digestive solutions. Food moves through the stomach and is changed into a watery liquid called **chyme** (KIME). Chyme moves out of the stomach into the small intestine.

3. Describe how an acidic solution helps digestion.

____

____

____

What fluids are added in the small intestine?

Chyme enters the first part of the small intestine called the duodenum (doo AH duh num). Most digestion takes place there. Bile, a greenish fluid from the liver, is added in the duodenum. Bile breaks the large fat particles in chyme into smaller particles. Chemical digestion of carbohydrates, proteins, and fats occurs when a digestive solution from the pancreas is mixed in. The solution neutralizes the stomach acid in the chyme.

How is food absorbed in the small intestine?

Food is absorbed from the small intestine into the bloodstream. The wall of the small intestine has many ridges and folds that are covered with fingerlike projections called villi (VIH li). Villi, shown in the figure below, increase the surface area of the small intestine, giving nutrients in the chyme more places to be absorbed. Nutrients move into the blood vessels in the villi. Peristalsis moves undigested and unabsorbed materials into the large intestine.

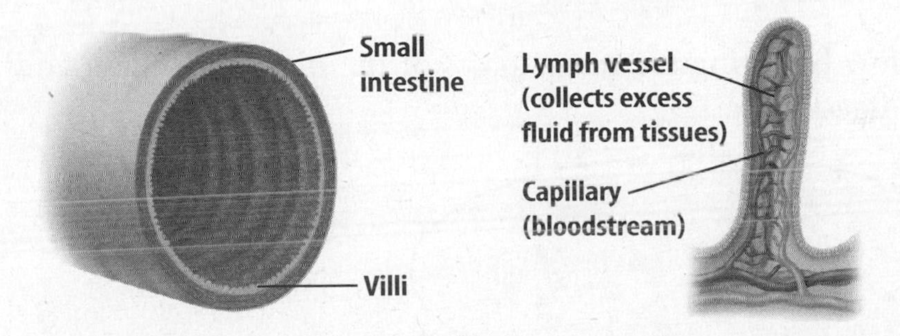

Picture This

4. **Describe** why villi are important to the digestive process.

What happens in the large intestine?

The main job of the large intestine is to absorb water from the undigested materials. This keeps large amounts of water in the body and helps maintain homeostasis. After the water is absorbed, the remaining undigested materials become more solid. Muscles in the rectum and the anus control the release of the wastes from the body in the form of feces (FEE seez). ☑

Reading Check

5. **List** the main job of the large intestine.

Bacteria Are Important

Bacteria live in many parts of the digestive tract. Bacteria in the large intestine feed on undigested material like cellulose. Bacteria make Vitamin K and two B vitamins, niacin and thiamine. Vitamin K is needed for blood clotting. The B vitamins help the nervous system and other body functions.

After You Read

Mini Glossary

chemical digestion: digestion that takes place when chemical reactions break down large molecules of food into smaller ones

chyme (KIME): food from the stomach that has been changed into a thin, watery liquid that moves into the small intestine

digestion: the process that breaks down food into small molecules so that they can be moved into the blood and absorbed by the cells

enzyme (EN zime): a type of protein that speeds up the rate of a chemical reaction in the body

mechanical digestion: digestion that takes place when food is chewed, mixed, and churned

peristalsis (per uh STAHL sus): waves of muscle contractions

villi (VIH li): fingerlike projections in the small intestine

1. Review the terms and their definitions in the Mini Glossary. Write a sentence explaining the difference between mechanical digestion and chemical digestion.

____

____

2. Complete the web diagram below by listing the major parts of the digestive system and describing the role of each in digestion.

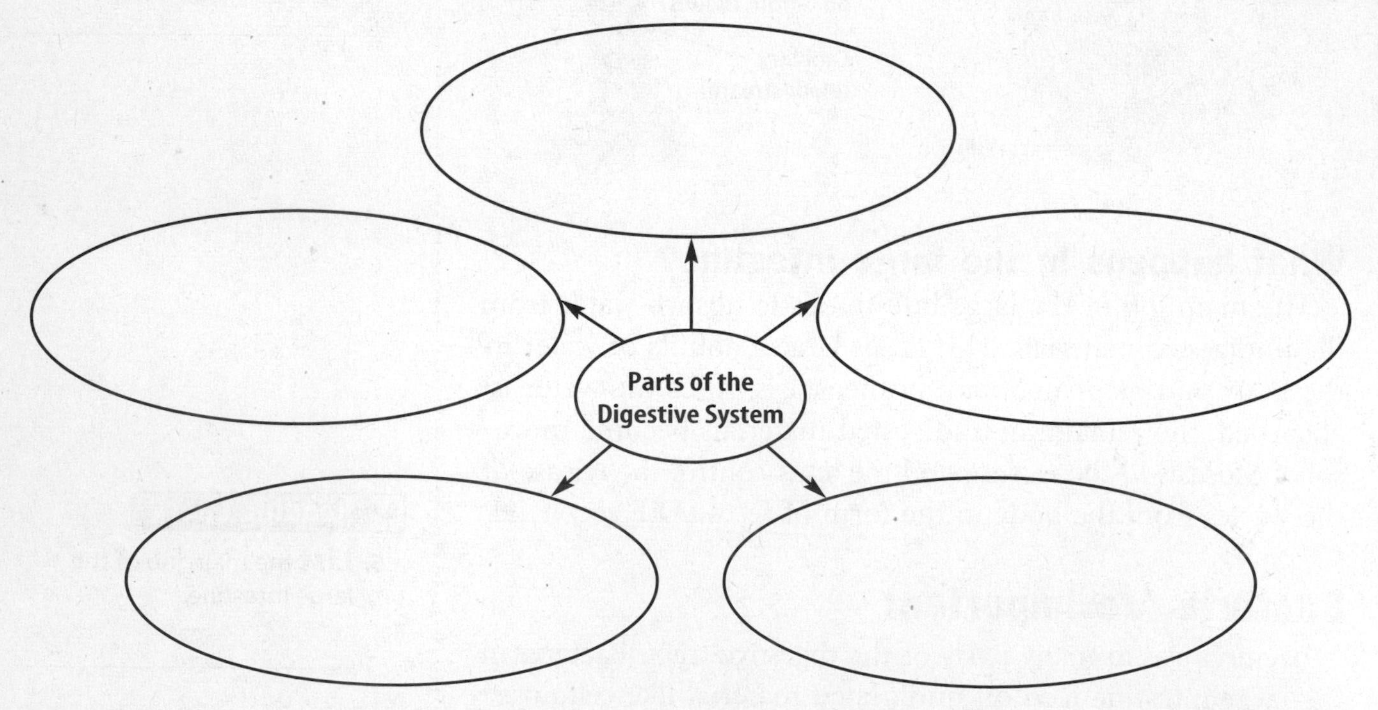

Science Online Visit **life.msscience.com** to access your textbook, interactive games, and projects to help you learn more about the digestive system.

Circulation

section 1 The Circulatory System

Before You Read

Explain the function of the plumbing system in your home. Describe how it works.

____

____

____

What You'll Learn

- the differences among arteries, veins, and capillaries
- how blood moves through the heart
- the functions of the pulmonary and systemic circulation systems

Read to Learn

How Materials Move Through the Body

The cardiovascular (kar dee oh VAS kyuh lur) system supplies materials to and removes wastes from your body cells. This system includes your heart, blood vessels, and blood.

Movement of materials into and out of your cells occurs by diffusion (dih FYEW zuhn) and active transport. Diffusion occurs when a material moves from an area where there is more of it to an area where there is less of it. Nutrients and oxygen diffuse from your blood into your body's cells. Active transport is the opposite of diffusion. Active transport needs energy from the cell to occur.

Mark the Text

Identify Main Ideas After you have read the material under each question head, highlight the answer to the question.

The Heart

Your heart is an organ made of cardiac muscle tissue. Your heart has four compartments called chambers. The two upper chambers are called the right and left **atriums** (AY tree umz). The two lower chambers are called the right and left **ventricles** (VEN trih kulz). During one heartbeat, both atriums contract at the same time. Then, both ventricles contract at the same time. A one-way valve separates each atrium from the ventricle below it.

FOLDABLES™

A Describe Make a shutterfold book, as shown below. Identify the four chambers of the heart. Include a sketch of the heart and the four chambers.

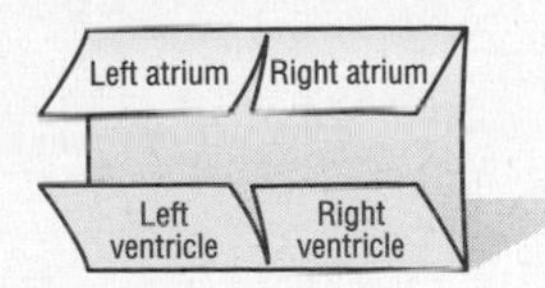

Three Sections of the Circulatory System Scientists divide the circulatory system into three sections: coronary (KOR uh ner ee) circulation, pulmonary (PUL muh ner ee) circulation, and systemic circulation. The beating of your heart controls blood flow through each section. ☑

Reading Check

1. **Identify** What controls the blood flow through the three sections of the circulatory system?

____

What is coronary circulation?

Blood vessels supply the heart with nutrients and oxygen and remove wastes. **Coronary circulation** is the flow of blood to and from the tissues of the heart.

What is pulmonary circulation?

The flow of blood through the heart to the lungs and back to the heart is **pulmonary circulation.** Use the figure below to trace the path blood takes through this part of the circulatory system.

The blood returning from the body through the right side of the heart and to the lungs contains wastes from the body's cells. Carbon dioxide is one of these wastes.

In the lungs, carbon dioxide and other gaseous wastes diffuse out of the blood, and oxygen diffuses into the blood. Then the blood returns to the left side of the heart.

In the final step of pulmonary circulation, the oxygen-rich blood is pumped from the left ventricle into the aorta (ay OR tuh). The aorta is the largest artery in your body. Next, the oxygen-rich blood flows to all parts of your body.

Picture This

2. **Explain** to a partner the flow of blood in pulmonary circulation.

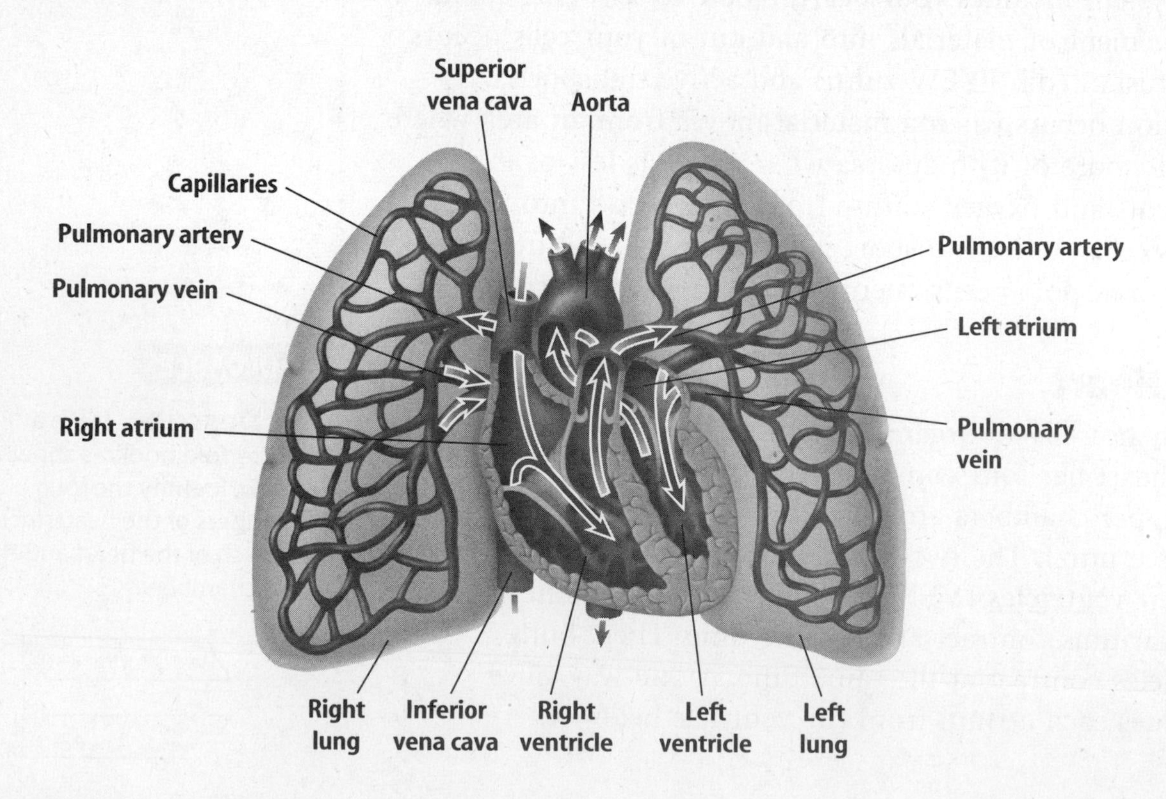

What is systemic circulation?

Oxygen-rich blood moves to all of your organs and body tissues, except the heart and lungs, by **systemic circulation**. Oxygen-poor blood returns to the heart by systemic circulation. The figure below shows the major arteries and veins (VAYNZ) of the systemic circulation system. Oxygen-rich blood flows from your heart in the arteries. Then nutrients and oxygen are delivered by blood to your body cells and exchanged for carbon dioxide and wastes, as shown below. The blood then returns to your heart in the veins.

Think it Over

3. Infer Why is systemic circulation important to your muscles?

____

____

____

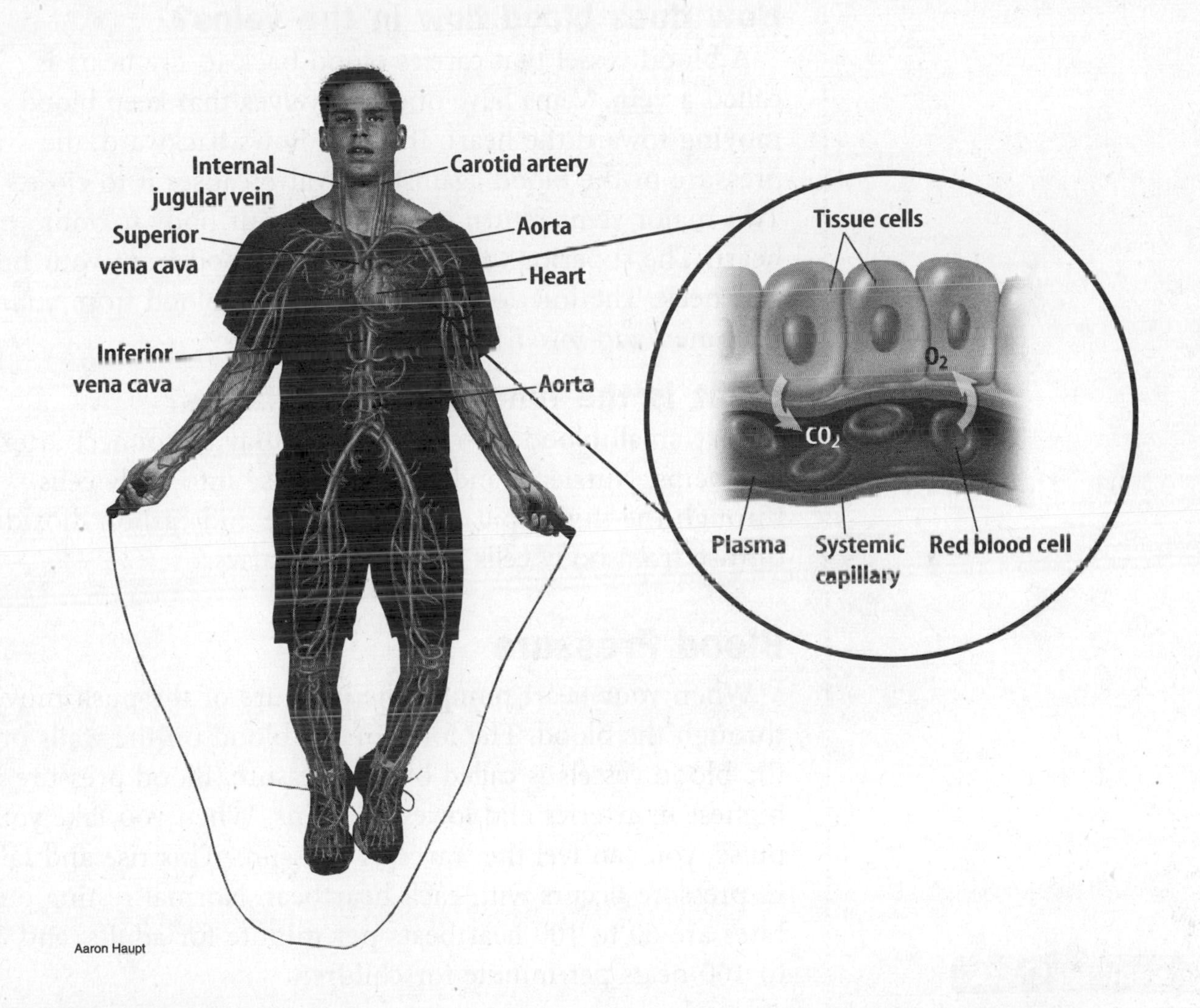

Blood Vessels

In the middle 1600s, scientists proved that blood moves in one direction in a blood vessel, like traffic on a one-way street. They discovered that blood moves by the pumping of the heart and flows from arteries to veins. They couldn't explain how blood got from arteries to veins. With the invention of the microscope, scientists discovered that capillaries (KAP uh ler eez) connect the arteries and veins.

Picture This

4. Identify Circle the name of the blood vessel in which oxygen and carbon dioxide are exchanged.

FOLDABLES™

B Explain Make a three-tab Foldable, as shown below. Label the outside of the Foldable *Circulation*. Inside take notes on arteries, capillaries, and veins.

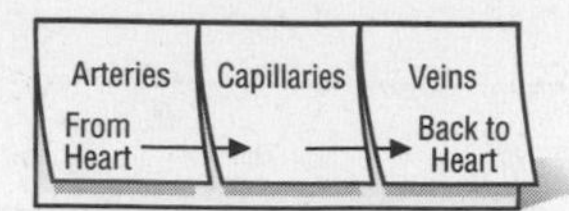

What is the function of arteries?

The blood vessels that carry blood away from the heart are called **arteries.** Arteries have thick, elastic walls made of connective tissue and smooth muscle tissue.

Each ventricle of the heart is connected to an artery. The right ventricle of the heart is connected to the pulmonary artery. The left ventricle of the heart is connected to the aorta. Every time your heart contracts, blood moves from your heart into your arteries.

How does blood flow in the veins?

A blood vessel that carries blood back to the heart is called a **vein.** Veins have one-way valves that keep blood moving toward the heart. If blood flows backward, the pressure of the blood against the valve causes it to close. Two major veins return blood from your body to your heart. The superior vena cava returns blood from your head and neck. The inferior vena cava returns blood from your abdomen and lower body.

What is the function of capillaries?

Very small blood vessels called **capillaries** connect arteries and veins. Nutrients and oxygen diffuse into body cells through the thin capillary walls. Waste and carbon dioxide diffuse from body cells into the capillaries.

Blood Pressure

When your heart pumps, the pressure of the push moves through the blood. The force of the blood on the walls of the blood vessels is called blood pressure. Blood pressure is highest in arteries and lowest in veins. When you take your pulse, you can feel the waves of pressure. This rise and fall in pressure occurs with each heartbeat. Normal resting pulse rates are 60 to 100 heartbeats per minute for adults, and 80 to 100 beats per minute for children.

✔ Reading Check

5. **Describe** What does the first number in your blood pressure measure?

____

____

____

How is blood pressure measured?

Blood pressure is measured in large arteries. Two numbers describe blood pressure, such as 120 over 80. The first number is a measure of the pressure caused when the ventricles contract and blood is pushed out of the heart. The second number is a measure of the pressure that occurs as the ventricles fill with blood just before they contract again. ☑

Blood Pressure and Heart Rate When blood pressure is higher or lower than normal, messages are sent to the brain by nerve cells in the arteries. One way the brain lowers or raises blood pressure is by speeding up or slowing down the heart rate. When blood pressure stays constant, enough blood reaches all organs and tissues in the body.

Cardiovascular Disease

There are many diseases that affect the cardiovascular system—the heart, blood vessels, and blood. Heart disease is the leading cause of death in the United States.

What is atherosclerosis?

Atherosclerosis (ah thuh roh skluh ROH sus) is a leading cause of heart disease. In this condition, deposits of fat build up on the walls of the arteries. These fat deposits can block an artery. If a coronary artery is blocked, a heart attack can occur.

What happens with hypertension?

Hypertension (HI pur TEN chun) is high blood pressure. When blood pressure is higher than normal most of the time, the heart must work harder to keep blood flowing. Atherosclerosis is one cause of hypertension. ☑

How does heart failure occur?

Heart failure occurs when the heart cannot pump blood efficiently. When the heart does not pump properly, fluid collects in the arms, legs, and lungs. A person with heart failure is usually short of breath and tired.

Can cardiovascular disease be prevented?

Cardiovascular disease can be prevented by following a diet that is low in salt, sugar, cholesterol, and saturated fats. Large amounts of body fat force the heart to pump faster. Relaxing and exercising help prevent tension and relieve stress. Exercising strengthens the heart and lungs and helps maintain proper weight. Not smoking also helps prevent heart disease. ☑

Reading Check

6. **Apply** What is another name for high blood pressure?

____

Reading Check

7. **Explain** What is one thing you can do to prevent cardiovascular disease?

____

____

____

After You Read

Mini Glossary

artery: a blood vessel that carries blood away from the heart

atriums (AY tree umz): the two upper chambers of the heart

capillaries (KAP uh ler eez): very small blood vessels that connect arteries and veins

coronary (KOR uh ner ee) circulation: the flow of blood to and from the tissues of the heart

pulmonary circulation: the flow of blood through the heart to the lungs and back to the heart

systemic circulation: the system in which oxygen-rich blood moves to all of the organs and body tissues, except the heart and lungs, and oxygen-poor blood returns to the heart

vein: a blood vessel that carries blood back to the heart

ventricles (VEN trih kulz): the two lower chambers of the heart

1. Review the terms and their definitions in the Mini Glossary. Write a sentence that explains the difference between pulmonary circulation and systemic circulation.

___

___

___

2. Complete the concept map below to show the kinds of blood vessels and their functions.

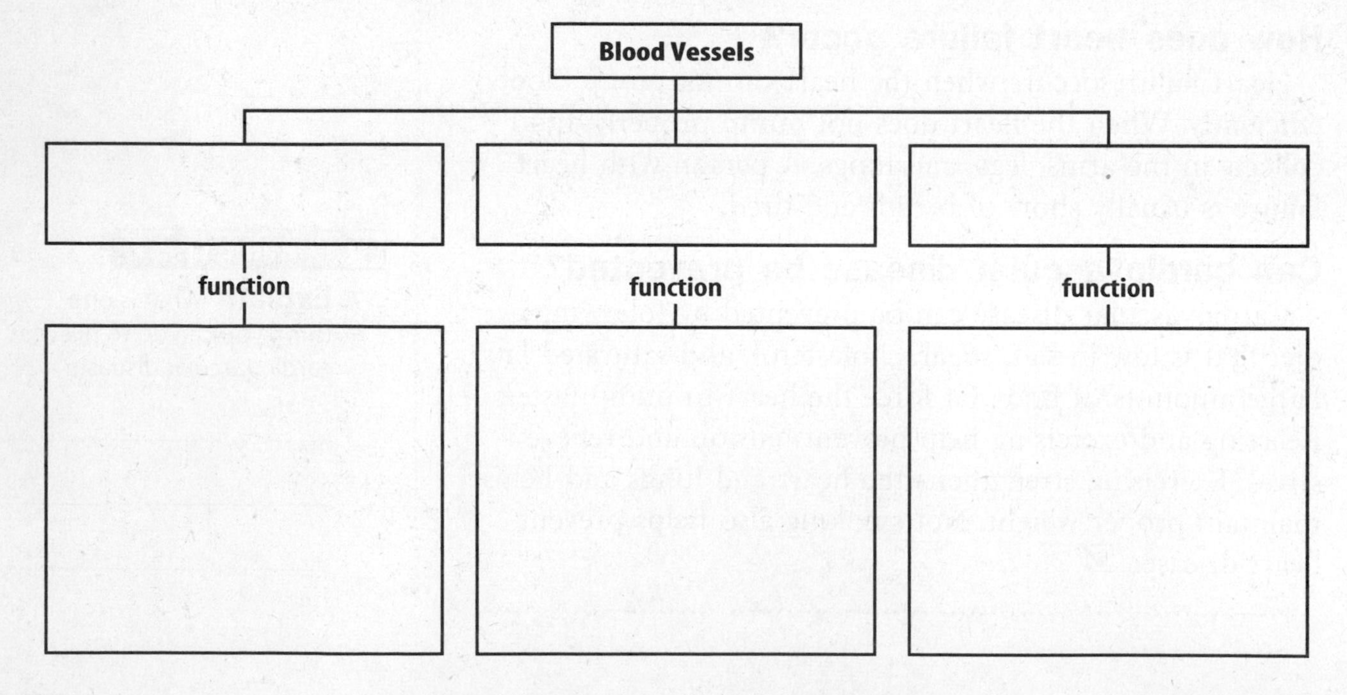

End of Section

Science Online Visit **life.msscience.com** to access your textbook, interactive games, and projects to help you learn more about the circulatory system.

Circulation

section 2 Blood

Before You Read

Have you ever fallen and scraped your knee? What happens to the wounded area? What happens while the wound is healing?

____

____

What You'll Learn

- the parts and function of blood
- why blood types are checked before a transfusion
- kinds of blood diseases

Read to Learn

Mark the Text

Identify Main Points Highlight the main idea of each paragraph. Circle the details that support the main idea.

Functions of Blood

Blood has four important functions.

1. Blood carries oxygen from your lungs to your body cells. Carbon dioxide diffuses from your body cells to your blood. Your blood carries carbon dioxide to your lungs to be exhaled.
2. Blood carries waste from your cells to your kidneys to be removed.
3. Blood carries nutrients and other materials to your body cells.
4. Cells and molecules in blood fight infections and help heal wounds.

FOLDABLES™

C Explain Make a four-tab Foldable, as shown below, to describe the four parts of blood—plasma, red blood cells, white blood cells, and platelets.

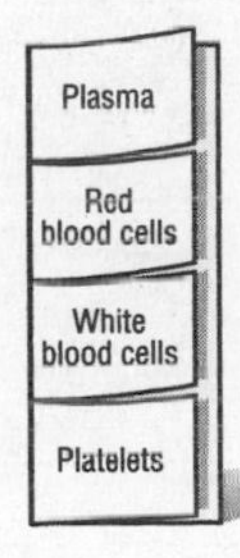

Parts of Blood

Blood is a tissue made of plasma (PLAZ muh), platelets (PLAYT luts), and red and white blood cells. Blood makes up about eight percent of your total body mass. If you weigh 45 kg, you have about 3.6 kg of blood.

What is plasma?

The liquid part of blood is mostly water and is called **plasma.** Nutrients, minerals, and oxygen are dissolved in plasma and carried to cells. Wastes from cells also are carried in plasma.

What is the function of red blood cells?

Red blood cells are different from all other human cells because they have no nuclei. They contain **hemoglobin** (HEE muh gloh bun), which is a molecule that carries oxygen and carbon dioxide. Hemoglobin is made of an iron compound that gives blood its red color. Hemoglobin carries oxygen from your lungs to your body cells. Then it carries some of the carbon dioxide from your body cells back to your lungs. ☑

Reading Check

1. **Determine** What gives blood its red color?

______

______

______

How do white blood cells fight invaders?

White blood cells fight bacteria, viruses, and other invaders of your body. Your body reacts to these invaders by increasing the number of white blood cells. These cells leave the blood through capillary walls and go into the tissues that have been invaded. Here, they destroy bacteria and viruses and absorb dead cells.

What are platelets?

Platelets circulate with red and white blood cells. **Platelets** are irregularly shaped cell fragments that help clot blood.

Blood Clotting

When you cut yourself, platelets stick to the wound and release chemicals. Then materials called clotting factors carry out a series of chemical reactions. These reactions cause threadlike fibers called fibrin (FI brun) to form a sticky net, as shown in the figure below. This net traps escaping blood cells and plasma and forms a clot. The clot becomes hard and skin cells begin the repair process under the scab. After a few days, the scab falls off.

Picture This

2. **Identify** Highlight the name of the blood part that releases chemicals to help with clotting.

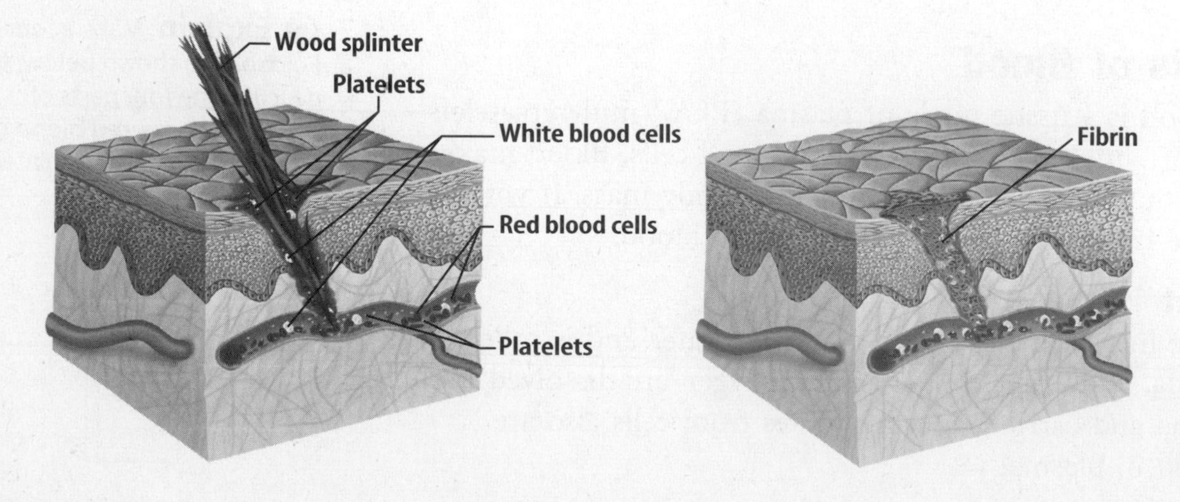

Blood Types

Blood clots stop blood loss quickly in a minor wound. However, a person with a serious wound might lose a lot of blood and need a blood transfusion. During a blood transfusion, a person receives donated blood or parts of blood. The person must get the right type of blood, or the red blood cells will clump together. This causes clots to form in the blood vessels and the person could die.

How are blood types identified?

People can inherit one of four types of blood: A, B, AB, or O. Types A, B, and AB have chemical identification tags called antigens (AN tih junz) on their red blood cells. Type O red blood cells have no antigens, as shown in the table below.

Blood Type	Antigen	Antibody
A	A	Anti-B
B	B	Anti-A
AB	A, B	None
O	None	Anti-A Anti-B

Antibodies and Transfusions Each blood type, except AB, also has specific antibodies in its plasma. Antibodies are proteins that destroy materials that do not belong in or are not part of your body. For example, if type A blood is mixed with type B blood, the type A antibodies cause the type B red blood cells to clump.

Because of these antibodies, certain blood types cannot be mixed. Type AB blood has no antibodies, so people with this blood type can receive blood from A, B, AB, and O types. Type O blood has both A and B antibodies. People with type O blood are sometimes called universal donors because their blood can be transfused into a person with any blood type.

3. **Analyze** Why would it be important for doctors to check your blood type if you were in a serious accident?

____

____

____

Picture This

4. **Identify** Highlight the blood type that produces no antibodies. Circle the blood type that has no antigens.

What is the Rh factor in blood?

The Rh factor is another chemical identification tag in blood. The Rh factor is inherited. If the Rh factor is on red blood cells, the person has Rh-positive (Rh+) blood. If it is not present, the person's blood is Rh-negative (Rh−). If an Rh− person receives a blood transfusion from an Rh+ person, he or she will produce antibodies against the Rh factor. These antibodies can cause Rh+ cells to clump. Clots then form in the blood vessels and the person could die. ☑

Reading Check

5. Define What is the Rh factor?

____

____

____

When an Rh− mother is pregnant with an Rh+ baby, the mother might make antibodies to the child's Rh factor. Close to the time of birth, Rh antibodies from the mother can pass from her blood into the baby's blood. These antibodies can destroy the baby's red blood cells. If this happens, the baby must receive a blood transfusion before or right after birth.

At 28 weeks of pregnancy and immediately after the birth, an Rh− mother can be given an injection that stops the production of antibodies to the Rh+ factor. This keeps the baby from needing a blood transfusion.

Diseases of Blood

Any disease of the blood is a cause for concern, because blood circulates to all parts of your body and performs many important functions. Anemia (uh NEE mee uh) is a common disease of red blood cells. Body tissues cannot get enough oxygen and are not able to carry out their usual activities. Anemia can be caused by the loss of large amounts of blood. It also can be caused by the lack of iron or certain vitamins in the diet. Anemia can be the result of other diseases. Some types of anemia, such as sickle-cell anemia, are inherited. ☑

Reading Check

6. Identify the blood disease that affects red blood cells.

____

Leukemia (lew KEE mee uh) is a disease in which one or more types of white blood cells are made in large numbers. These cells are not able to fight infections well. They crowd out the normal cells. Then not enough red blood cells, normal white blood cells, and platelets can be made. Types of leukemia can affect children or adults. Medicines, blood transfusions, and bone marrow transplants are used to treat this disease. If the treatments are not successful, the person will eventually die from complications related to the disease.

After You Read

Mini Glossary

hemoglobin (HEE muh gloh bun): a molecule that carries oxygen and carbon dioxide

plasma (PLAZ muh): the liquid part of blood that carries nutrients, minerals, and oxygen to cells

platelet (PLAYT lut): an irregularly shaped fragment of a cell that helps clot blood

1. Review the terms and their definitions in the Mini Glossary. Choose one term that describes a part of the blood. Write a sentence that explains the function of this part of the blood.

____

____

____

2. Using the phrases below, fill in the boxes in the correct order to explain how a wound heals.

Clot forms.
Clot hardens.
Clotting factors carry out chemical reaction.
Fibrin forms.
Net traps blood cells.
Platelets stick to wound.
Scab falls off.
Skin cells repair under scab.

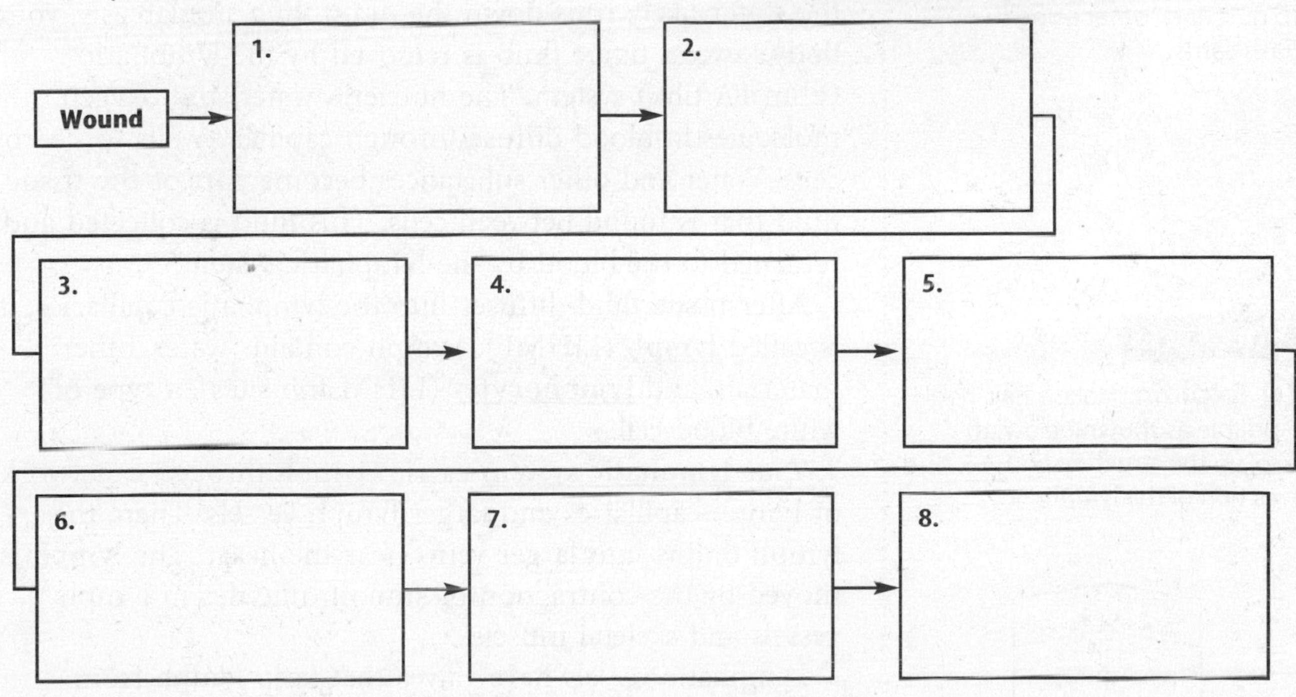

Science Online Visit **life.msscience.com** to access your textbook, interactive games, and projects to help you learn more about blood.

End of Section

Circulation

section 3 The Lymphatic System

What You'll Learn
- the functions of the lymphatic system
- where lymph comes from
- how lymph organs help fight infections

Before You Read

When you fill a glass with water from a water faucet, what happens to the water that does not go into your glass?

____

____

____

Study Coach

Make Flash Cards Write a quiz question on one side of a flash card and the answer on the other side. Work with a partner to quiz each other using the flash cards.

FOLDABLES™

D Explain Make a half-book Foldable as shown below, to explain the structure and function of the lymphatic system.

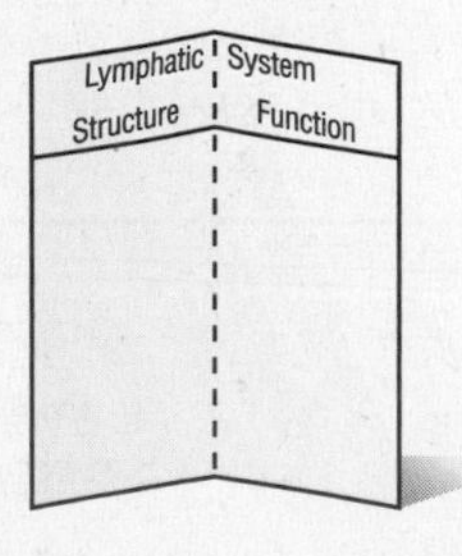

Read to Learn

Functions of the Lymphatic System

When you fill a glass with water from the faucet, some of the water likely runs down the drain. In a similar way, your body's excess tissue fluid is removed by the lymphatic (lihm FA tihk) system. The nutrient, water, and oxygen molecules in blood diffuse through capillary walls to nearby cells. Water and other substances become part of the tissue fluid that is found between cells. This fluid is collected and returned to the blood by the lymphatic system.

After tissue fluid diffuses into the lymphatic capillaries, it is called **lymph** (LIHMF). Lymph contains water, other materials, and **lymphocytes** (LIHM fuh sites), a type of white blood cell.

Your **lymphatic system** carries lymph through a network of lymph capillaries and larger lymph vessels. Then, the lymph drains into larger veins near the heart. The lymph is moved by the contraction of smooth muscles in lymph vessels and skeletal muscles.

Lymphatic vessels have valves that keep lymph from flowing backward. If the lymphatic system is not working properly, swelling occurs because the tissue fluid cannot get back to the blood.

Lymphatic Organs

Before lymph enters the blood, it passes through lymph nodes. **Lymph nodes** are bean-shaped organs found throughout the body, as shown in the figure below. Lymph nodes filter out microorganisms and foreign materials that have been taken up by the lymphocytes. When your body fights an infection, lymphocytes fill the lymph nodes.

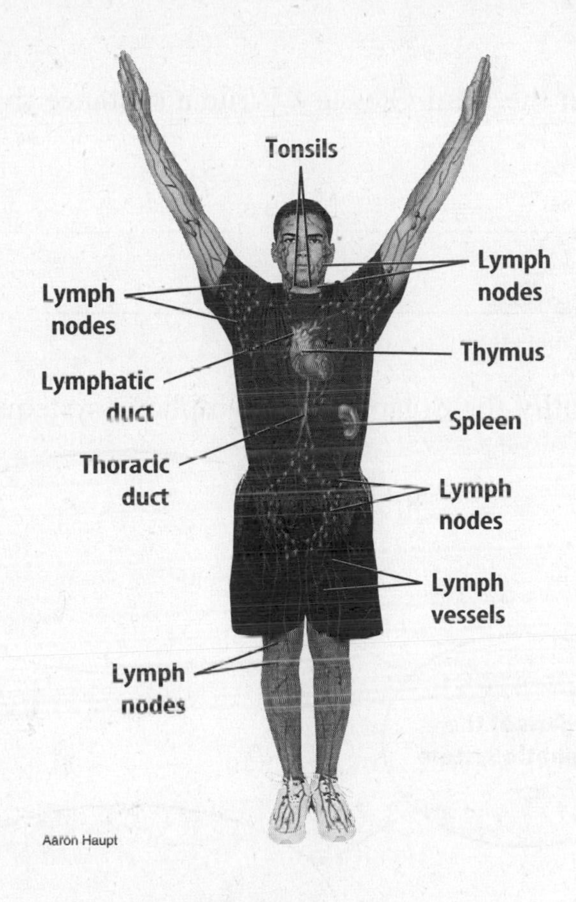

Other important lymphatic organs include the tonsils, the thymus, and the spleen. Tonsils protect you from harmful organisms that enter through your mouth and nose. The thymus makes lymphocytes. The spleen removes worn out and damaged red blood cells from the blood. Cells in the spleen destroy bacteria and other materials that invade your body.

A Disease of the Lymphatic System

HIV is a virus. It destroys lymphocytes called helper T cells that help make antibodies to fight infections. This makes it difficult for a person with HIV to fight some diseases. Usually, the person dies from these diseases, not from the HIV infection.

Picture This

1. Identify Highlight the areas of the body where lymph nodes are found.

FOLDABLES™

E Describe Use a quarter-sheet of notebook paper to take notes about HIV.

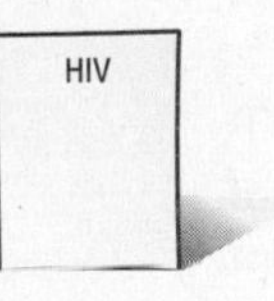

After You Read

Mini Glossary

lymph (LIHMF): tissue fluid that passes into the lymphatic capillaries

lymphatic system: the system that removes lymph that the body does not need through a network of lymph capillaries and larger lymph vessels

lymph nodes: bean-shaped organs found throughout the body, which filter out microorganisms

lymphocyte (LIHM fuh site): a type of white blood cell

1. Review the terms and their definitions in the Mini Glossary. Write a sentence that summarizes what lymph nodes do.

____

____

2. Complete the concept web below to identify the organs of the lymphatic system.

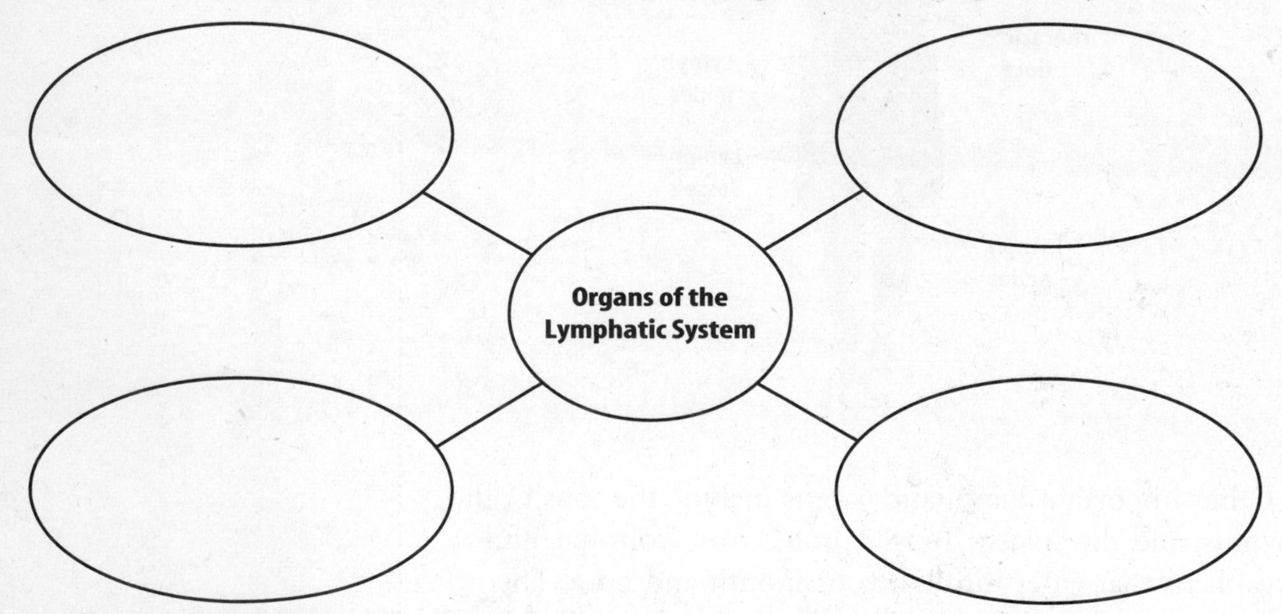

3. Explain how working with a partner helped you learn about the lymphatic system.

____

____

ScienceOnline Visit **life.msscience.com** to access your textbook, interactive games, and projects to help you learn more about the lymphatic system.

Respiration and Excretion

section 1 The Respiratory System

Before You Read

What do people mean when they say they are "out of breath"? What causes them to be out of breath?

____

____

____

What You'll Learn
- the functions of the respiratory system
- how oxygen and carbon dioxide are exchanged in the body
- how air moves in and out of the lungs
- how smoking affects the respiratory system

Read to Learn

Functions of the Respiratory System

You need to breathe air. Earth is surrounded by a layer of gases called the atmosphere (AT muh sfihr). You breathe the gases that are closest to Earth, particularly oxygen.

For thousands of years, people have known that air, food, and water are needed for life. However, people did not know that it is the gas oxygen that is necessary for life until the late 1700s. At that time, a French scientist discovered that animals breathe in oxygen and breathe out carbon dioxide. He then studied the way humans use oxygen. He measured how much oxygen a person uses when resting and when exercising. The measurements showed that the body uses more oxygen during exercise.

Mark the Text

Identify the Main Point After you read each paragraph, put what you have just read into your own words. Then highlight the main idea in each paragraph.

What is breathing?

Breathing is the movement of the chest that brings air into the lungs and removes waste gases. When you breathe in, or inhale, the air that comes into the lungs contains oxygen. The oxygen passes from the lungs into the circulatory system. Blood carries the oxygen to individual cells in the body. This process is shown in the figure on the next page.

FOLDABLES™

A Organize Make a shutterfold Foldable, as shown below. Use the Foldable to organize information about the structure and function of the respiratory system.

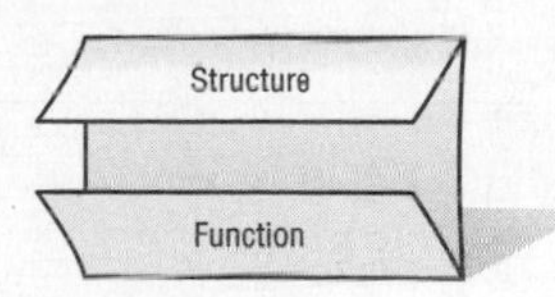

What is respiration?

As blood is carrying oxygen to cells, the digestive system is providing glucose from digested food to the same cells. The oxygen brought to the cells is used to release energy from the glucose. This chemical reaction is called cellular respiration. Oxygen is necessary for respiration to occur. Carbon dioxide and water molecules are the waste products of cellular respiration. The blood carries the waste products back to the lungs. Breathing out, or exhaling, removes the carbon dioxide and some water molecules.

Picture This

1. **Identify** On the two lines labeled A and B, write *inhale* or *exhale*, to show which process is happening.

Obtaining, Transporting, and Using Oxygen

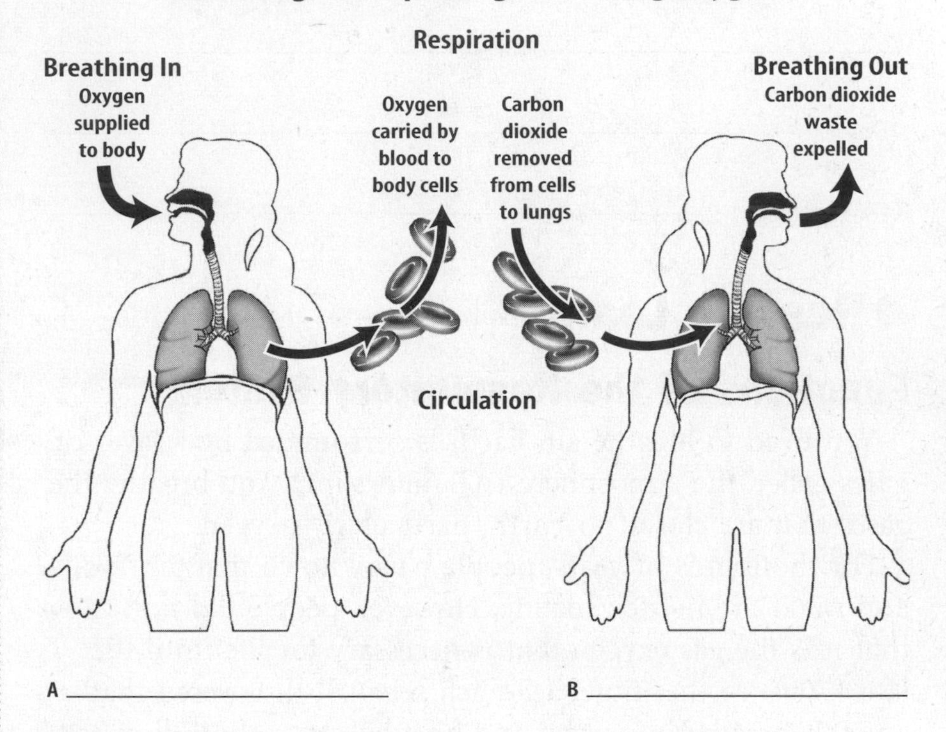

A ______________________ B ______________________

Organs of the Respiratory System

The respiratory system is made up of structures and organs that help move oxygen into the body and take waste gases out of the body. Air comes into the body through the nostrils in your nose or through the mouth. Hairs in the nostrils trap dust from the air. Air then moves through the nasal cavity. There the air is moistened and warmed by the body's heat. Glands that produce sticky mucus line the nasal cavity. The mucus traps dust and other materials that were not trapped by the nasal hairs. This process helps clean the air you breathe. Cilia (SIH lee uh), tiny hairlike structures, move the mucus and trapped material to the back of the throat where it can be swallowed. ☑

☑ Reading Check

2. **Describe** the purpose of nasal hairs and mucus.

____

____

____

What does the pharynx do?

The warmed, moistened air enters a tubelike passageway called the **pharynx** (FER ingks). Food, liquid, and air all use this passage. The epiglottis (eh puh GLAH tus) is a flap of tissue located at the lower end of the pharynx. When you swallow, the epiglottis folds down to prevent food or liquid from entering your airway.

What are the larynx and the trachea?

The air then moves from the pharynx to the larynx (LER ingks). The **larynx** is the airway to which the vocal cords are attached. Forcing air between the vocal cords causes them to vibrate. The vibration produces sounds. When you speak, muscles tighten or loosen your vocal cords, resulting in different sounds.

From the larynx, the air moves into a tube called the **trachea** (TRAY kee uh). The trachea has rings of cartilage, a tough, flexible tissue that prevents the trachea from collapsing. Mucus and cilia line the trachea where they trap dust, bacteria, and pollen. ☑

How do the bronchi and the lungs function?

Air is taken into the lungs by two short tubes called **bronchi** (BRAHN ki). Within the lungs, the bronchi branch into smaller and smaller tubes. The smallest tubes are called bronchioles (BRAHN kee ohlz). At the ends of the bronchioles are clusters of tiny, thin-walled sacs called **alveoli** (al VEE uh li). Air moves into the bronchi, then into the bronchioles, and finally into the alveoli. Lungs are masses of alveoli arranged in grapelike clusters. Capillaries, or small blood vessels, surround the alveoli.

Oxygen and carbon dioxide are exchanged between the alveoli and the capillaries. Oxygen moves through the cell membranes of the alveoli and then through the cell membranes of the capillaries into the blood. This happens easily because the walls of the alveoli and the walls of the capillaries are only one cell thick.

Hemoglobin (HEE muh gloh bun), a molecule in red blood cells, picks up the oxygen and carries it to all the cells of the body. Carbon dioxide and other cellular wastes leave the cells of the body through the membranes of the capillaries into the blood. In the lungs, the waste gases then move through the cell membranes of the capillaries and the alveoli. The waste gases leave the body when you exhale.

Reading Check

3. **Explain** What prevents the trachea from collapsing?

____

____

____

FOLDABLES™

B Describe Make a two-tab book, as shown below. Write notes and draw diagrams to show how oxygen and carbon dioxide are exchanged in the respiratory system.

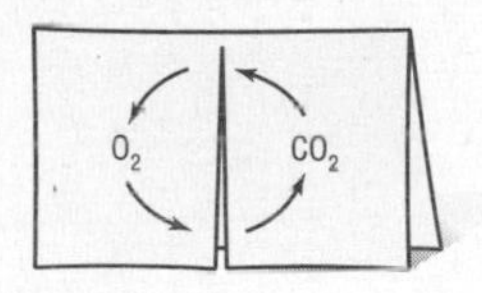

Why do you breathe?

You do not have to think about breathing. Your brain controls your breathing. Signals from your brain tell the muscles in your chest and abdomen to contract and relax. Your brain changes your breathing rate depending on the amount of carbon dioxide that is in your blood. Your breathing rate increases as the amount of carbon dioxide in your blood increases. ☑

Reading Check

4. Determine Why does your breathing rate change?

____

____

____

How do you inhale and exhale?

Breathing is partly the result of changes in air pressure. Generally, a gas moves from a high-pressure area to a low-pressure area. When you squeeze an empty, soft-plastic bottle, air is pushed out. When you stop squeezing the bottle, the air pressure inside the bottle becomes less than the air pressure outside the bottle. Air comes back in and the bottle returns to its original shape.

Your lungs work much like the squeezed bottle. The **diaphragm** (DI uh fram) is a muscle below your lungs that contracts and relaxes to help move gases into and out of your lungs. The figure below shows how your lungs inhale and exhale.

Picture This

5. Identify Circle the diaphragm in each figure. Label the lungs and the trachea in one of the pictures.

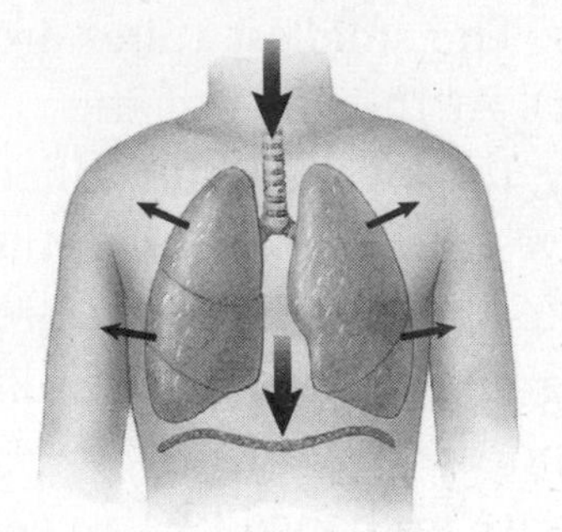

Inhale

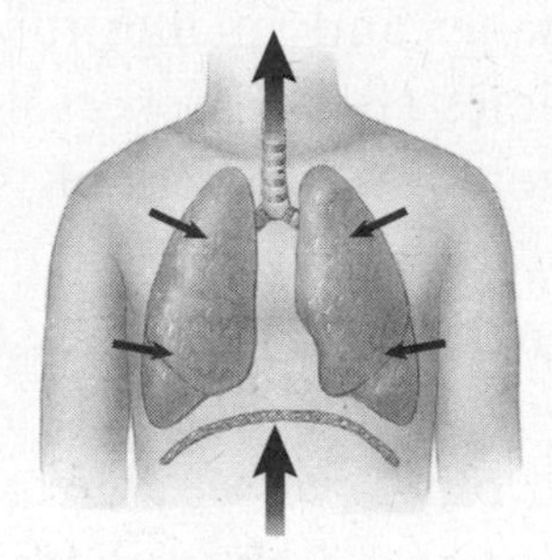
Exhale

Diseases and Disorders of the Respiratory System

Many diseases of the respiratory system are related to smoking. The nicotine and tars in tobacco are poisons that can destroy cells. The high temperatures, smoke, and carbon monoxide produced when tobacco burns also damage a smoker's cells. The respiratory systems of nonsmokers can be harmed by inhaling secondhand smoke from tobacco products. In addition to smoking, polluted air, coal dust, and asbestos (as BES tus) have been linked to various respiratory diseases.

What causes respiratory infections?

Bacteria and viruses can cause infections of the respiratory system. The common cold virus generally affects the upper part of the respiratory system—from the nose to the pharynx. The cold virus also can affect the larynx, trachea, and bronchi. The cilia that line the trachea and bronchi can be damaged. However, cilia usually heal rapidly. ☑

A virus that causes the flu can affect the organs of the respiratory and other body systems. The virus multiplies in the cells lining the alveoli and damages them. Pneumonia is an infection in the alveoli. It can be caused by bacteria, viruses, or other microorganisms. Antibiotics (an ti bi AH tihks) are used to treat bacterial pneumonia. Before antibiotics were available, many people died from bacterial pneumonia.

What is chronic bronchitis?

Bronchitis (brahn KI tus) develops when bronchial tubes are irritated and swell and too much mucus is produced. Sometimes bronchitis is caused by bacteria and can be treated with antibiotics.

Most cases of bronchitis clear up within a few weeks. Sometimes, however, the disease lasts for a long time. This is called chronic (KRAH nihk) bronchitis. A person with this condition must cough often to clear mucus from the airway. But the more a person coughs, the more the cilia and bronchial tubes can be damaged. When cilia are damaged, they do not move mucus, bacteria, and dirt out of the lungs well. Then harmful substances build up in the airways. Sometimes, scar tissue forms and the respiratory system cannot work properly.

What is emphysema?

A disease in which the alveoli in the lungs enlarge is called **emphysema** (em fuh SEE muh). When cells in the alveoli are swollen, an enzyme that causes the walls of the alveoli to break down is released. As a result, alveoli cannot push the air out of the lungs and less oxygen moves into the bloodstream from the alveoli. When the blood has too little oxygen and too much carbon dioxide, shortness of breath occurs.

Some people with emphysema need extra oxygen. People with emphysema may develop heart problems because the heart has to work harder to supply oxygen to body cells.

Reading Check

6. List What are two causes of respiratory infections?

____

____

____

Think It Over

7. Compare What is the difference between bronchitis and chronic bronchitis?

____

____

____

FOLDABLES

C Describe Use a quarter-sheet of notebook paper as shown below, to take notes on the effects of smoking on the respiratory system.

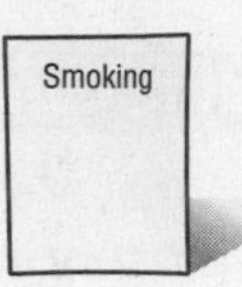

What causes lung cancer?

Cigarette smoking increases the risk of several diseases, as shown in the table below. Cigarette smoking is a major cause of lung cancer. More than 85 percent of all lung cancer is related to smoking.

Tar and other substances found in smoke act as carcinogens (kar SIH nuh junz) in the body. Carcinogens are substances that can cause an uncontrolled growth of cells. In the lungs, this is called lung cancer. Lung cancer is not easy to detect in its early stages. Smoking also has been linked to the development of cancers of the mouth, esophagus, larynx, pancreas, kidneys, and bladder.

Picture This

8. Determine Who has a higher risk of developing lung cancer, males or females?

____

Smokers' Risk of Death from Disease

Disease	Smokers' Risk Compared to Nonsmokers' Risk
Lung cancer	23 times higher for males, 11 times higher for females
Chronic bronchitis and emphysema	5 times higher
Heart disease	2 times higher

What is asthma?

Asthma (AZ muh) is a lung disorder that can cause shortness of breath, wheezing, or coughing. During an asthma attack, the bronchial tubes contract quickly. To treat an asthma attack, a person inhales medicine that relaxes the bronchial tubes.

Asthma often is an allergic reaction. An allergic reaction occurs when the body overreacts to a foreign substance. An asthma attack can happen when a person breathes certain substances such as cigarette smoke, eats certain foods, or experiences stress.

After You Read

Mini Glossary

alveoli: clusters of thin-walled sacs at the end of each bronchiole that are surrounded with capillaries; where the exchange of oxygen and carbon dioxide takes place

asthma: a lung disorder that results in shortness of breath, wheezing, or coughing

bronchi: two short tubes that carry air into the lungs

diaphragm: muscle beneath the lungs that contracts and relaxes to help move gases into and out of the lungs

emphysema: disease in which the alveoli in the lungs enlarge

larynx: airway to which the vocal cords are attached

pharynx: tubelike passageway that is used by food, liquid, and air

trachea: air-conducting tube lined with mucus membranes and cilia

1. Review the terms and their definitions in the Mini Glossary. Choose one of the organs of the respiratory system and write a sentence that describes the organ and explains its function.

2. Choose one of the question headings in the Read to Learn section. Write the question in the space below. Then write your answer to that question on the lines that follow.

Write your question here.

Visit **life.msscience.com** to access your textbook, interactive games, and projects to help you learn more about the respiratory system.

Respiration and Excretion

section 2 The Excretory System

What You'll Learn

- the differences between the excretory and urinary systems
- how the kidneys work
- what happens when urinary organs do not work

Before You Read

On the lines below, write what you know about the job of the body's kidneys.

____

____

____

Study Coach

Create a Quiz After you read this section, create a quiz based on what you have learned. After you have written the quiz questions, be sure to answer them.

Read to Learn

Functions of the Excretory System

It's your turn to take out trash. You carry the bag outside and put it in the trash can. The next day, you bring out another bag of trash, but the trash can is full. When trash isn't collected, it piles up. Just as trash needs to be removed from your home to keep the home livable, your body has to get rid of wastes to stay healthy. The digestive system gets rid of undigested material through the large intestine. The respiratory system and the circulatory system work to rid the body of waste gases. These systems work together as part of the excretory system. If wastes are not removed from the body, toxic substances build up and damage organs. Serious illness or death may occur.

FOLDABLES™

D Organize Make a shutterfold Foldable, as shown below. Use the Foldable to organize information about the structure and function of the excretory system.

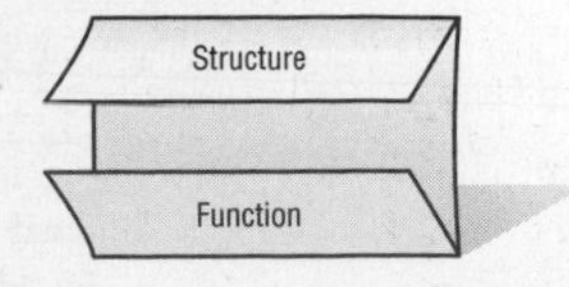

The Urinary System

The **urinary system** rids the blood of wastes produced by the cells. The urinary system controls the volume of blood by removing excess water that the body cells produce during respiration. The urinary system is part of the excretory system. Salts and water that are needed for cell activities are kept in balance by the urinary system.

How are fluid levels in the body regulated?

To stay healthy, the fluid levels in the body have to be balanced. The body also has to keep a normal blood pressure. The hypothalamus (hi poh THA luh mus), an area in the brain, keeps track of the amount of water in the blood. ☑

When the brain detects too much water in the blood, the hypothalamus gives off a lesser amount of a specific hormone. This tells the kidneys to return less water to the blood. It also tells the kidneys to increase the amount of **urine**, or wastewater, that is excreted.

Water in the blood is important for moving gases and getting rid of solid wastes from the body. The urinary system also balances the amounts of certain salts and water that must be present for all cell activities to take place.

1. Describe What area of the brain helps regulate fluid levels in the body?

What organs make up the urinary system?

The organs of the urinary system are also called excretory organs. They are shown in the figure below. The main organs are two **kidneys**. They are located on the back wall of the abdomen at about waist level. The kidneys filter blood that contains wastes collected from cells. Blood enters the kidneys through a large artery and leaves through a large vein.

The Urinary System

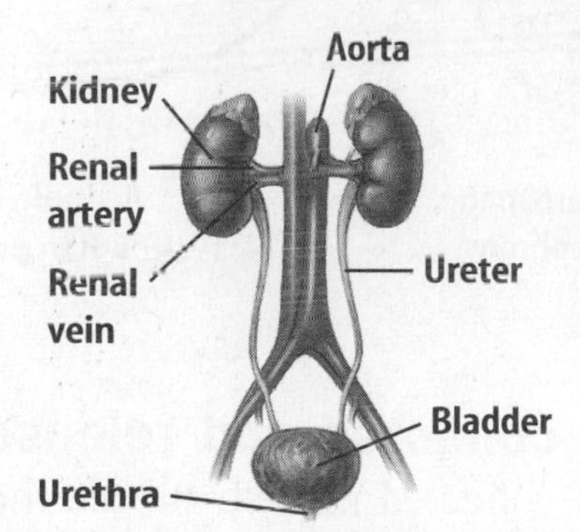

Picture This

2. Identify Circle the main organs of the urinary system.

How does a kidney filter blood?

The kidney is a two-stage filtration system. It is made up of about one million tiny filtering units called **nephrons** (NEF rahnz). Each nephron has a cuplike structure and a tubelike structure called a duct. Blood moves from a renal artery to capillaries in the cuplike structure.

First Stage In the first filtration, water, sugar, salt, and wastes from the blood pass into the cuplike structure. Red blood cells and proteins are left behind in the blood.

FOLDABLES™

E List Make a four-tab Foldable, as shown below. Make notes listing facts about the main structures of the urinary system.

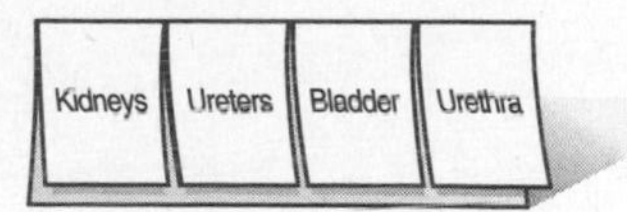

Second Stage Next, liquid in the cuplike structure is squeezed into a narrow tubule. Capillaries that surround the tubule do the second filtration. Most of the water, sugar, and salt are reabsorbed from the tubule and returned to the blood. These capillaries come together to form a renal vein in each kidney. The purified blood is returned to the circulatory system. The liquid left behind flows into collecting tubules in each kidney. This wastewater, or urine, contains excess water, salts, and other wastes that the body did not reabsorb. An average-sized person produces about 1 L of urine per day. Use the figure below to review how the nephrons filter blood.

Picture This

3. **Identify** Work with a partner to explain how the urinary system filters blood and produces urine.

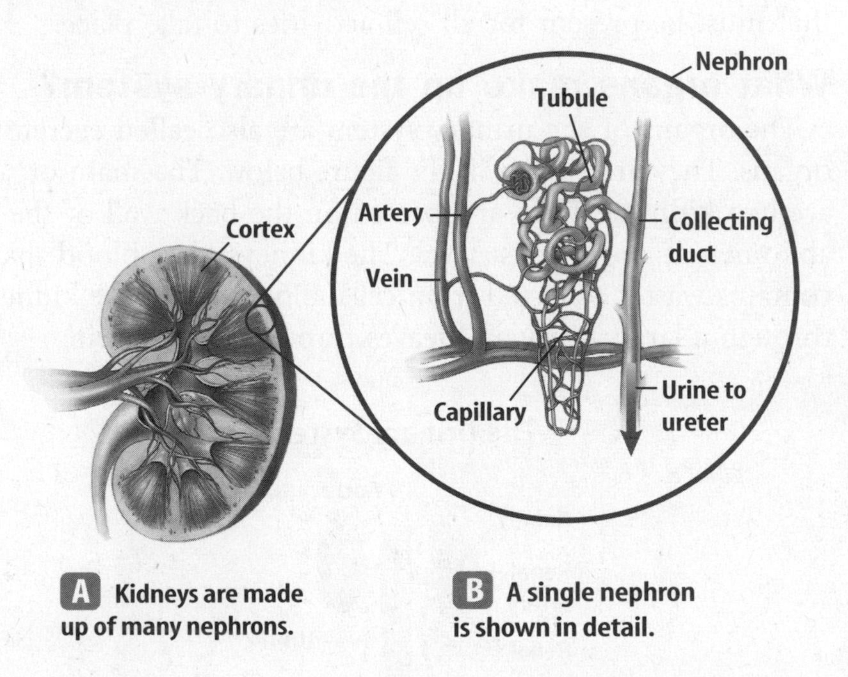

A Kidneys are made up of many nephrons.

B A single nephron is shown in detail.

How is urine collected and released?

The urine that collected in each tubule moves into a funnel-shaped area of each kidney that leads to the ureter (YOO ruh tur). **Ureters** are narrow tubes that lead from each kidney to the bladder. The **bladder** is a muscular organ that holds urine until it leaves the body. The bladder has elastic walls that stretch to hold the urine. A narrow tube called the **urethra** (yoo REE thruh) carries urine from the bladder to the outside of the body.

The body depends on water. Without water, the cells and other body systems could not function. Water is so important to the body that the brain and other body systems are involved in balancing water gain and water loss.

Think it Over

4. **Explain** why the ureter or urethra can become easily blocked.

____

____

Other Organs of Excretion

Your body loses large amounts of liquid wastes in other ways, such as exhaling and perspiring. The amount of fluid lost each day by exhaling and perspiring is about the volume of a soda can. The liver also filters the blood to remove wastes. Some kinds of wastes are changed to other substances. Excess amino acids are changed to a chemical called urea (yoo REE uh) that passes to the urine. Hemoglobin from broken-down red blood cells becomes part of bile, which is digestive fluid from the liver. ☑

Urinary Diseases and Disorders

When kidneys do not work properly, waste products build up and act as poisons in body cells. Water that kidneys normally remove from body tissues builds up and causes swelling of the ankles and feet. Sometimes the fluids build up around the heart. This causes the heart to work harder to move blood to the lungs.

When the urinary system does not work well, there can be an imbalance of salts. If the balance of salts is not restored, the kidneys and other organs can be damaged or fail. This is always a serious problem because the kidney's job is so important to the rest of the body.

Microorganisms can cause infections of the urinary system. Usually, the infection starts in the bladder. Sometimes it spreads to the kidneys. The infection can usually be cured with antibiotics.

If the ureters and urethra become blocked, urine cannot flow out of the body properly. If the blockage is not corrected, the kidneys can be damaged.

How can urinary diseases be discovered?

Urine can be tested for signs of a urinary tract disease. A change in the urine's color suggests kidney or liver problems. High levels of glucose in the urine can be a sign of diabetes. Increased amounts of a protein called albumin (al BYOO mun) can be a sign of kidney disease or heart failure.

What is dialysis?

People can live normally with only one kidney. However, if both kidneys fail, a person will need to have his or her blood filtered by a machine. This process is called dialysis (di AH luh sus). The dialysis machine removes wastes from the blood, just like the kidneys. ☑

Reading Check

5. **List** two additional ways your body gets rid of liquid wastes.

____

____

Think It Over

6. **Identify** two disorders of the urinary system.

____

____

____

Reading Check

7. **Explain** what dialysis does.

____

____

____

After You Read

Mini Glossary

bladder: organ that holds urine until it leaves the body

kidney: urinary system organ made up of about one million nephrons

nephron: tiny filtering unit in the kidney

ureter: one of two tubes that connect a kidney to the bladder

urethra: structure that carries urine from the bladder to the outside of the body

urinary system: body system that rids the blood of wastes made by the body's cells, controls blood volume, and balances salts and water

urine: wastewater that is excreted by the kidneys

1. Review the terms and their definitions in the Mini Glossary. Write a sentence that explains the role of nephrons.

____

____

____

2. Complete the diagram below to show the effects of a urinary system that is not working properly on different parts of the body.

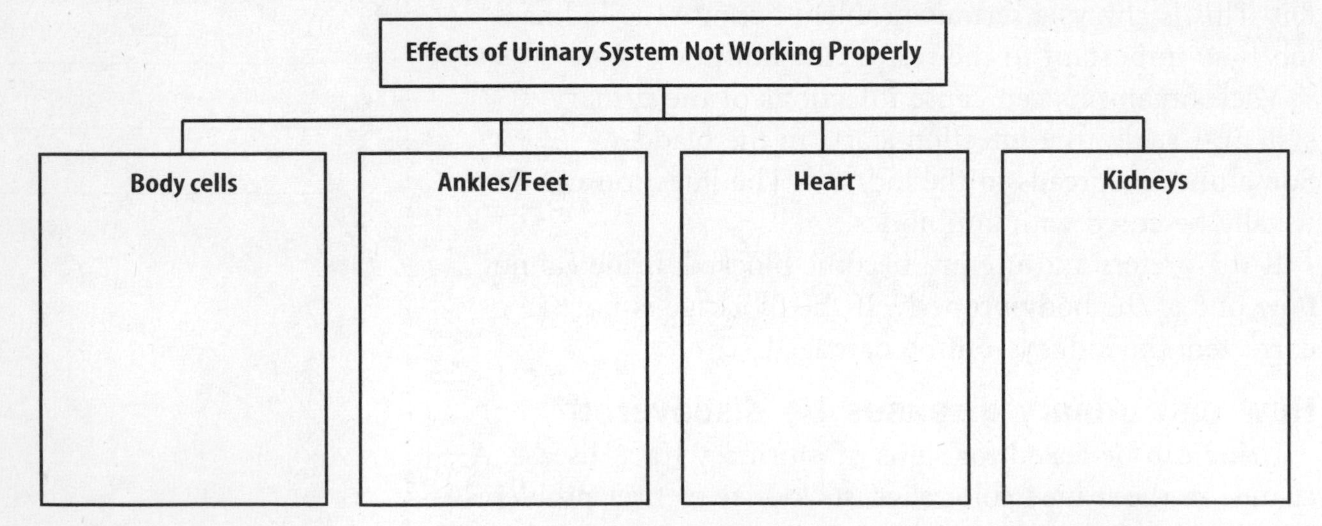

3. How did writing quiz questions help you remember what you read in this section?

____

____

Science Online Visit **life.msscience.com** to access your textbook, interactive games, and projects to help you learn more about the excretory system.

Control and Coordination

section 1 The Nervous System

Before You Read

Do you remember the last time you touched something very hot by mistake? How did you react?

____

____

____

What You'll Learn

- the basic structure of a neuron
- how an impulse moves across a synapse
- differences between central and peripheral nervous systems
- how drugs affect the body

Read to Learn

How the Nervous System Works

What happens when you hear a sudden, loud noise? Your heart may begin to race and your hands may shake. Once the surprise passes, your breathing returns to normal and your heartbeat is back to its regular rate. The way you react to a sudden, unexpected sound or event is one example of how your body responds to changes in your environment.

Mark the Text

Define Words Skim the section before you read it. Circle any words you do not know. As you read the text, underline the words that help you identify the meaning of the words you circled.

What are stimuli?

Any change that brings about a response is called a stimulus (STIHM yuh lus) (plural, *stimuli*). You respond to thousands of stimuli every day. Noise and light are examples of stimuli from outside your body. Hormones are stimuli from inside your body. Your nervous system helps your body adjust to changing stimuli.

What is homeostasis?

Your body has control systems that handle stimuli. The control systems help keep steady internal conditions. The changing of conditions inside an organism to keep it alive, even though the environment changes, is called **homeostasis** (hoh mee oh STAY sus). Your nervous system is one of the control systems that your body uses to maintain homeostasis.

FOLDABLES™

A Organize Make a quarter sheet Foldable, as shown below, to organize information about stimuli.

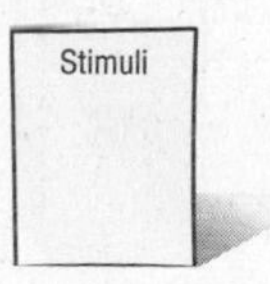

Nerve Cells

The basic units of the nervous system are nerve cells, or **neurons** (NOOR ahnz). The neuron shown in the figure below is made up of a cell body and branches called dendrites and axons. Any message carried by a neuron is called an impulse. **Dendrites** receive impulses from other neurons and send them to the cell body. **Axons** (AK sahns) carry impulses away from the cell body and send them to other neurons. Notice the branching at the end of the axon. This allows impulses to move to many other muscles, neurons, or glands.

Picture This

1. **Identify** In the figure, two dendrites and one axon are labeled. Add labels to identify two more dendrites and one more axon.

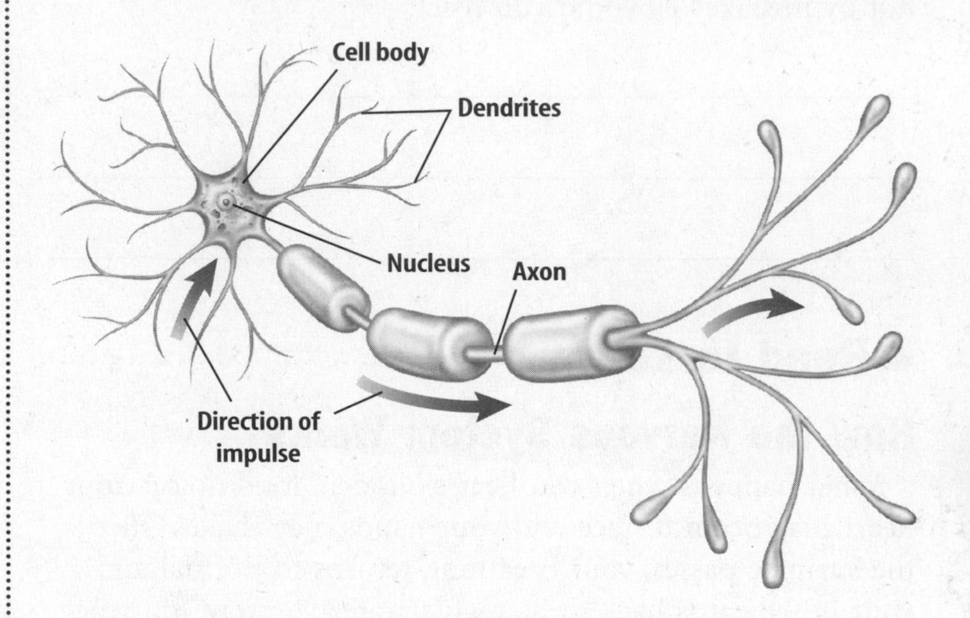

What are the types of nerve cells?

Your body can detect changes in the environment. Sensory receptors detect such things as temperature, sound, pressure, and light. Sensory receptors respond to stimuli by producing electrical impulses that are carried to the brain. Three types of neurons—sensory neurons, motor neurons, and interneurons—carry impulses. Sensory neurons receive information and send impulses to the brain or spinal cord. For example, when you hear a loud noise, sensory receptors in your ears are stimulated. These sensory neurons produce electrical impulses that travel to the brain. In the brain or spinal cord, interneurons send the impulses to motor neurons. The motor neurons move impulses from the brain or spinal cord to muscles or glands throughout your body. In the example of a loud noise, muscles in your arms contract to jerk your arms in response to the noise.

FOLDABLES™

B Describe Make a two-tab Foldable, as shown below, to list the three types of neurons and their functions.

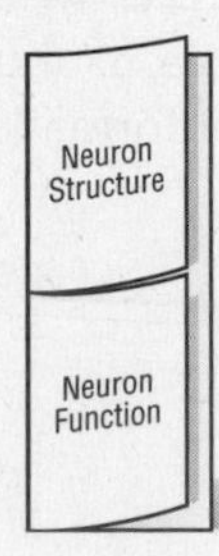

What is a synapse?

Neurons do not touch each other. Neurons are separated by a small space called a **synapse** (SIH naps), as shown in the figure below. To move from one neuron to another, the impulse must cross the synapse. When an impulse reaches the end of an axon, the axon releases a chemical. The chemical flows across the synapse and stimulates an impulse in the dendrite of the next neuron. Neurons allow impulses to move in only one direction.

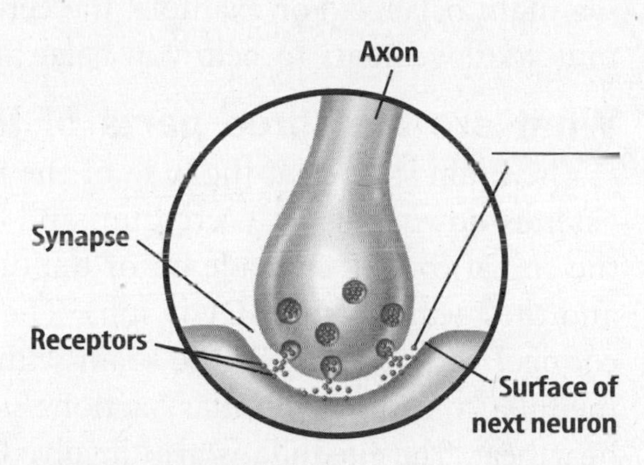

Picture This

2. Draw an arrow to show the direction that chemicals flow across the synapse.

The Central Nervous System

The **central nervous system** (CNS) is made up of the brain and spinal cord. The **peripheral (puh RIH fuh rul) nervous system** (PNS) is made up of all the nerves that are not part of the CNS. The PNS includes the nerves in the head, called cranial nerves. The PNS also includes the spinal nerves, which come from the spinal cord. The PNS connects the brain and spinal cord to other body parts.

What does the brain do?

The brain directs all the activities of the body. If someone tickles your feet, your brain directs your reaction. The brain is made up of about 100 billion neurons. That is about ten percent of all the neurons in the human body. The brain is surrounded and protected by a bony skull, three membranes, and a layer of fluid. The brain is divided into three parts: the brain stem, the cerebellum (ser uh BE lum), and the cerebrum (suh REE brum).

FOLDABLES™

C Describe Make a half book Foldable, like the one shown below, to list things that affect the central and peripheral nervous systems.

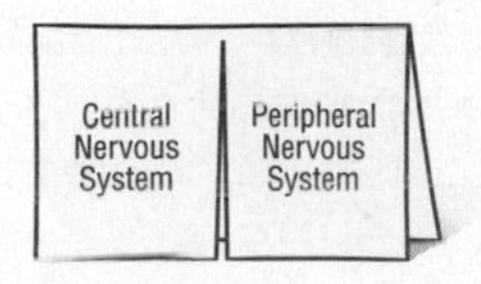

What is the role of the cerebrum?

The largest part of the brain is the **cerebrum**. This is where thinking takes place. The cerebrum interprets the meaning of impulses that come from sensory neurons. The cerebrum also stores memory and controls movements. The outer layer of the cerebrum is the cortex. The cortex has many ridges and grooves that increase its surface area. The greater surface area allows more complex thoughts to be processed.

What does the cerebellum do?

The **cerebellum** is the part of the brain that interprets stimuli from the eyes and ears and from muscles and tendons. Tendons are tissues that connect muscles to bones. The information received by the cerebellum helps it coordinate voluntary muscle movements, maintain muscle tone, and maintain balance. For example, the cerebellum coordinates muscle movements to help you balance while riding a bike.

What are the three parts of the brain stem?

The brain stem is at the base of the brain. The **brain stem** extends down from the cerebrum and connects the brain to the spinal cord. It is made up of the midbrain, the pons, and the medulla (muh DUH luh). The midbrain and pons connect various parts of the brain with each other. The medulla controls involuntary actions such as breathing and heartbeat. The medulla is also involved in actions such as coughing and sneezing. ☑

Reading Check

3. **Explain** What are the three parts of the brain stem?

____

____

____

Why is the spinal cord important?

Your spinal cord is an extension of the brain stem. The spinal cord is made up of neurons that carry impulses from all parts of the body to the brain and from the brain to all parts of the body.

The Peripheral Nervous System

Your brain and spinal cord are connected to the rest of the body by the peripheral nervous system (PNS). The PNS is made up of cranial nerves and spinal nerves. Spinal nerves contain bundles of sensory and motor neurons. Because most spinal nerves have these two kinds of neurons, a single spinal nerve can have impulses going to and from the brain at the same time.

What is the difference between the somatic and autonomic systems?

The peripheral nervous system has two major parts, the autonomic system and the somatic system. The somatic system controls voluntary actions, such as raising your hand. It is made up of cranial and spinal nerves that go from your central nervous system to your skeletal muscles. The autonomic system controls involuntary actions. These are actions that you do without making a choice, such as digestion and the beating of your heart.

Think it Over

4. **Explain** What part of the peripheral nervous system controls blood pressure?

____

____

____

Safety and the Nervous System

The central and peripheral nervous systems are involved with every mental process and physical action of your body. Therefore, an injury to the brain or the spinal cord can be serious. For example, the back of the brain controls vision. A blow to the back of the head could result in a loss of vision.

The spinal cord is surrounded by the bones of the spine, which protect the spinal cord. However, injuries to the spinal cord do occur. A spinal cord injury can damage nerve pathways and lead to paralysis (puh RA luh suhs). Paralysis is the loss of muscle movement.

Types of paralysis Damage to one side of the brain can cause the opposite side of the body to be paralyzed. If the spiral cord is damaged in the neck area, the body can be paralyzed from the neck down. Damage to the middle or lower part of the spiral cord can paralyze the legs and the lower part of the body.

Automobile, motorcycle, and bicycle accidents, as well as sports injuries, are the major causes of head and spinal injuries. You can help protect yourself from serious injury by wearing your seat belt while riding in a car and wearing safety gear while playing sports and riding bicycles.

How do reflexes work?

When you accidentally touch something that is very hot, you experience a reflex. A <u>**reflex**</u> is an involuntary, automatic response to a stimulus. You cannot control a reflex because it occurs before you know what has happened. A reflex involves a simple nerve pathway called a reflex arc.

Reflex Arc

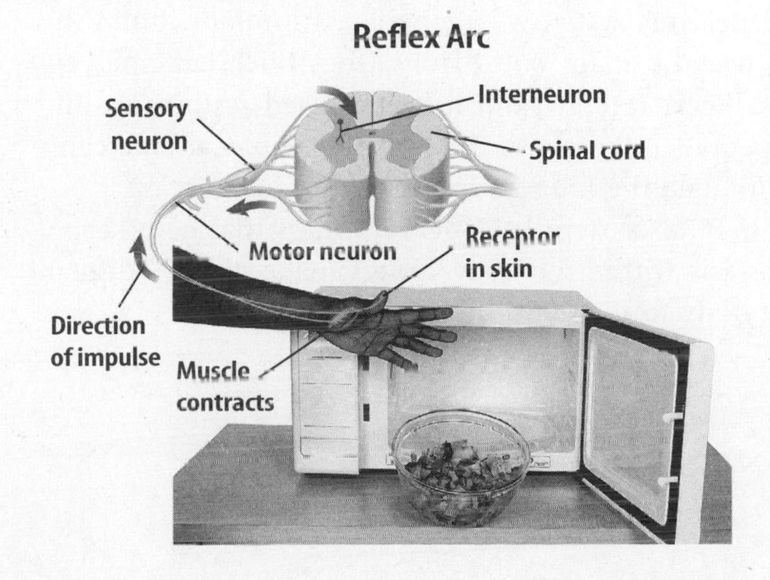

Think it Over

5. **Determine** What safety equipment is used to protect the head or spinal cord?

____

____

____

Picture This

6. **Interpret Scientific Illustrations** Trace the pathway of the reflex arc.

What is an example of a reflex?

A reflex allows the body to respond without having to think about what action to take. If you step on a sharp object, you experience a shooting pain. Sensory receptors in your foot respond to this sharp object, and an impulse is sent to the spinal cord. The impulse then passes to an interneuron in the spinal cord that immediately sends the impulse to motor neurons. Motor neurons send the impulse to muscles in the leg. Instantly, without thinking, you lift your leg away from the sharp object. This is a withdrawal reflex.

Reflex responses are controlled in your spinal cord, not in your brain. Your brain acts after the reflex to help figure out what to do to make the pain stop. Reflexes also happen if you touch something very cold or when you cough or vomit. ☑

7. **Identify** What part of the nervous system controls reflexes?

____

____

____

Drugs and the Nervous System

Many drugs, such as alcohol and caffeine, directly affect the nervous system. When alcohol is swallowed, it passes directly through the walls of the stomach and small intestine into the blood stream. Once in the circulatory system, alcohol can travel throughout the body. When the alcohol reaches neurons, it moves in through their cell membranes and upsets their normal cell functions. Since alcohol slows the activities of the central nervous system, it is called a depressant. Activities such as muscle control, judgment, and memory are reduced, or impaired. Heavy alcohol use destroys brain and liver cells.

A stimulant is a drug that speeds up the activity of the central nervous system. Caffeine is a stimulant found in coffee, tea, and many soft drinks. Too much caffeine can increase heart rate. It can cause increased restlessness in some people and make it difficult to sleep. Caffeine can also stimulate the kidneys to make more urine. ☑

The nervous system controls responses that help keep homeostasis within the body. Drugs make it more difficult for the body to maintain homeostasis.

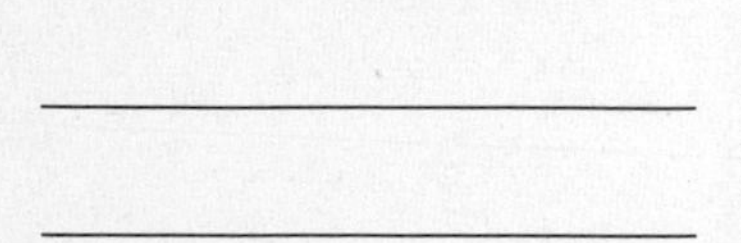

8. **Distinguish** What effect do depressants and stimulants have on the central nervous system?

____

____

____

After You Read

Mini Glossary

axon (AK sahn): the branch of a neuron that carries impulses away from the body of the nerve cell

brain stem: connects the brain to the spinal cord; made up of the midbrain, the pons, and the medulla

central nervous system: the part of the nervous system made up of the brain and spinal cord

cerebellum (ser uh BE lum): the part of the brain that interprets stimuli from the eyes and ears and from muscles

cerebrum (suh REE brum): the largest part of the brain, in which thinking takes place

dendrite: the branch of a neuron that receives impulses from other neurons and sends them to the body of the nerve cell

homeostasis (hoh mee oh STAY sus): the regulation of conditions inside an organism to maintain life, even when the outside environment changes

neuron (NOOR ahn): the basic unit of the nervous system—nerve cell

peripheral (puh RIH fuh rul) nervous system: the part of the nervous system made up of all the nerves outside the central nervous system

reflex: an involuntary, automatic response to a stimulus

synapse (SIH naps): small space between neurons through which an impulse crosses

1. Review the terms and their definitions in the Mini Glossary. Choose one of the parts of the brain and write a sentence that describes the part and explains its function.

____

____

2. Complete the chart below to identify the parts of the central nervous system (CNS) and the peripheral nervous system (PNS) and to explain what each part does.

	Kind of Nervous System (CNS or PNS)	What It Does
Brain		
Spinal cord		
Somatic system		
Autonomic system		

Visit **life.msscience.com** to access your textbook, interactive games, and projects to help you learn more about the nervous system.

chapter 21

Control and Coordination

section 2 The Senses

What You'll Learn

- the sensory receptors in each sense organ
- the type of stimulus each sense organ responds to and how
- why the body needs healthy senses

Before You Read

On the lines below, list which one of your senses you consider the most important. Explain why you think it is the most important.

____

____

____

Study Coach

Create a Quiz As you study the information in this section, create questions about the information you read. After you read, use the questions as a quiz.

Read to Learn

The Body's Alert System

Your sense organs are your body's alert system. They sense stimuli, such as light rays, sound waves, and heat. Your sense organs convert these stimuli into nerve impulses.

Vision

The eye is the sense organ for vision. Your eyes have adaptations that make it possible for you to see objects, shadows, and colors.

How do you see?

Light travels in a straight line unless something causes it to change direction. Your eyes have structures that refract, or bend, light. Two of these structures are the cornea (KOR nee uh) and the lens. The cornea is the transparent part at the front of the eye. As light passes through the cornea, it is refracted. The lens directs the light onto the retina (RET nuh). The **retina** is a tissue at the back of the eye that is sensitive to light energy. Cells called cones and rods are found in the retina. Cones respond to bright light and color. Rods respond to dim light. Rods help you distinguish shapes and movement. ✔

Reading Check

1. **Determine** Which structure of the eye is sensitive to light energy?

____

Structures of the Eye

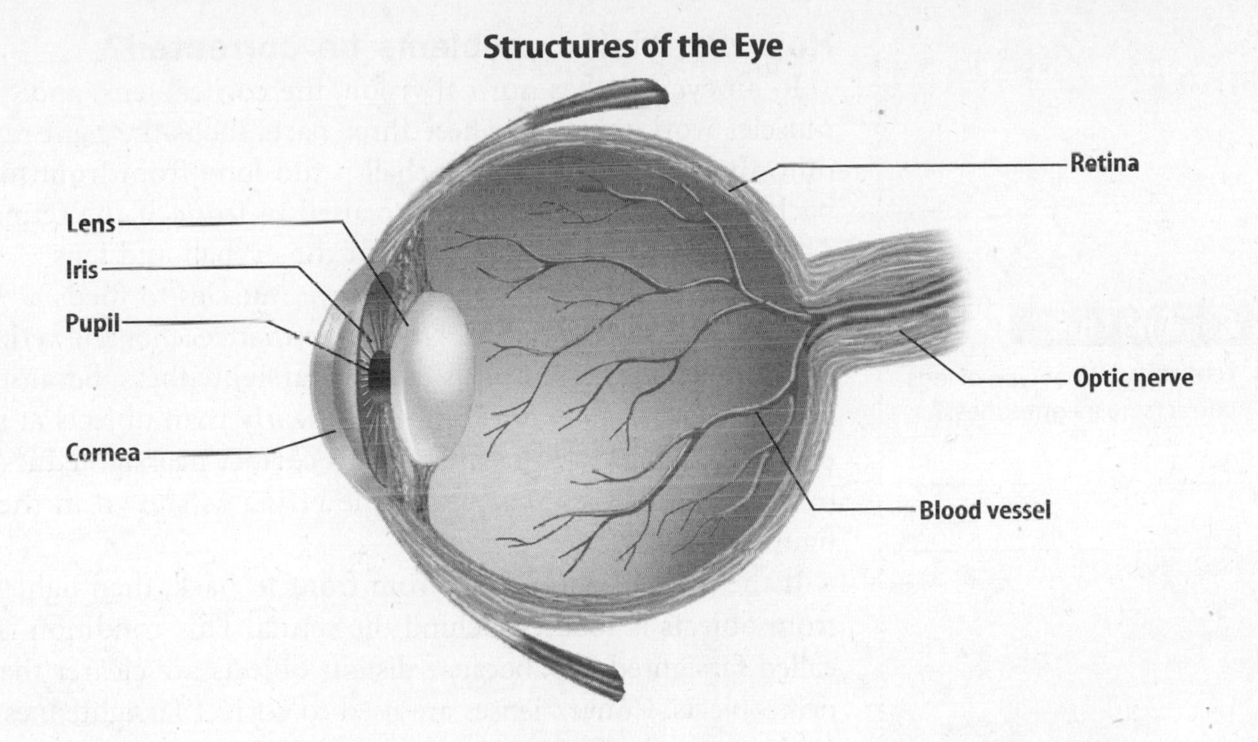

How do your eyes sense distance?

The light energy that reaches the retina stimulates the rods and cones to produce impulses. The impulses pass to the optic nerve, which is shown in the figure above. This nerve carries the impulses to the vision part of the cortex, located in your brain's cerebrum. The image that is passed from the retina to the brain is upside down and reversed. The brain interprets the image correctly, and you see what you are looking at. The brain also interprets the images it receives from both eyes. It blends them into one image that gives you a sense of distance. This helps you to tell how close or how far away something is.

Lenses

Light is refracted when it passes through a lens. The way light refracts depends on the kind of lens it passes through. A lens that is thicker in the middle and thinner on the edges is called a convex lens. The lens in your eye is a convex lens. It bends light so that it passes through a point, called a focal point. Convex lenses can be used to make objects appear larger. The light passes through a convex lens and enters the eye in a way that causes your brain to interpret the image as enlarged. ☑

A lens that is thicker at its edges than in its middle is called a concave lens. A concave lens causes the light to spread out.

Picture This

2. Identify Circle the name of the structure where rods and cones are found.

Reading Check

3. Describe What type of lens is in your eye?

____

How can vision problems be corrected?

In an eye that has normal vision, the cornea, lens, and muscles work together. These three parts focus the light rays onto the retina. But if the eyeball is too long from front to back, the light from objects is focused in front of the retina. This happens because the shape of the eyeball and lens cannot be changed enough by the eye muscles to focus a sharp image on the retina. The image that reaches the retina is blurred. This condition is called nearsightedness, because objects that are near are seen more clearly than objects at a distance. Concave lenses are used to correct nearsightedness by focusing images sharply on the retina as shown in the figure below. ☑

Reading Check

4. **Identify** What type of lens corrects nearsightedness?

____

____

If the eyeball is too short from front to back, then light from objects is focused behind the retina. This condition is called farsightedness, because distant objects are clearer than near objects. Convex lenses are used to correct farsightedness as shown in the figure below.

A nearsighted person cannot see distant objects because the image is focused in front of the retina.

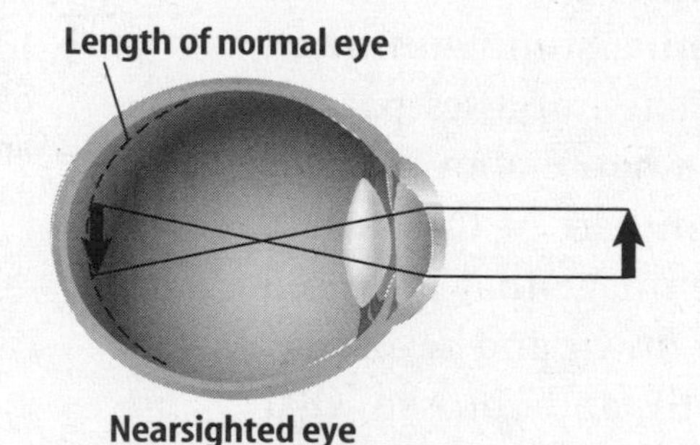

Nearsighted eye

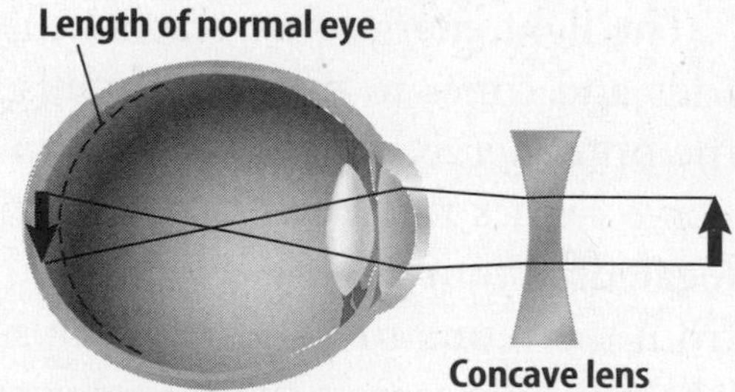

A concave lens corrects nearsightedness.

A farsighted person cannot see close objects because the image is focused behind the retina.

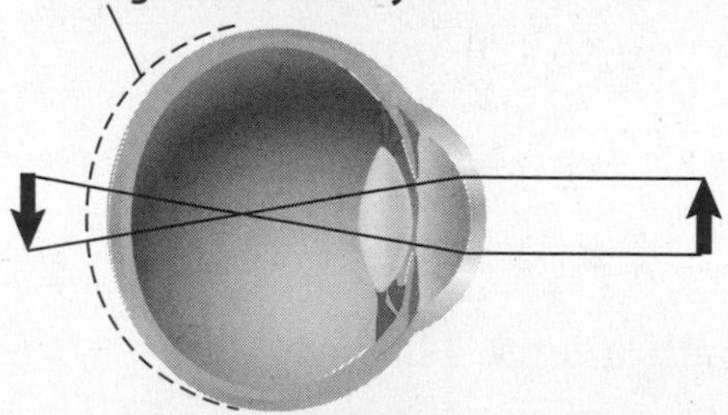

Farsighted eye

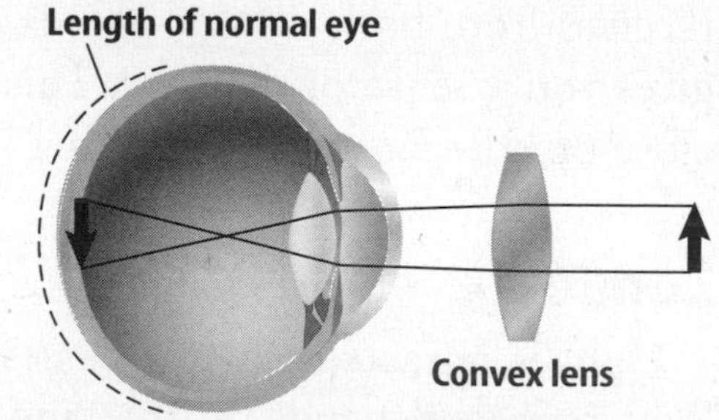

A convex lens corrects farsightedness.

Picture This

5. **Label** the retina and lens in each eye.

____

____

Hearing

You hear a sound when sound waves reach your ears. Sound waves are made when an object vibrates. Sound waves can travel through liquids, solids, and gases. When sound waves reach the ear, they stimulate nerve cells within the ear. Impulses are sent to the hearing area of your brain's cortex. The cortex responds and you hear a sound.

How do the outer ear and middle ear work?

The figure below shows the three sections of the ear: the outer ear, the middle ear, and the inner ear. The outer ear intercepts sound waves and funnels them down the ear canal to the middle ear. The sound waves make the eardrum vibrate much like the membrane on a musical drum vibrates when you tap on it. The vibrations then move through three tiny bones called the hammer, anvil, and stirrup. The stirrup bone rests against a membrane on an opening to the inner ear.

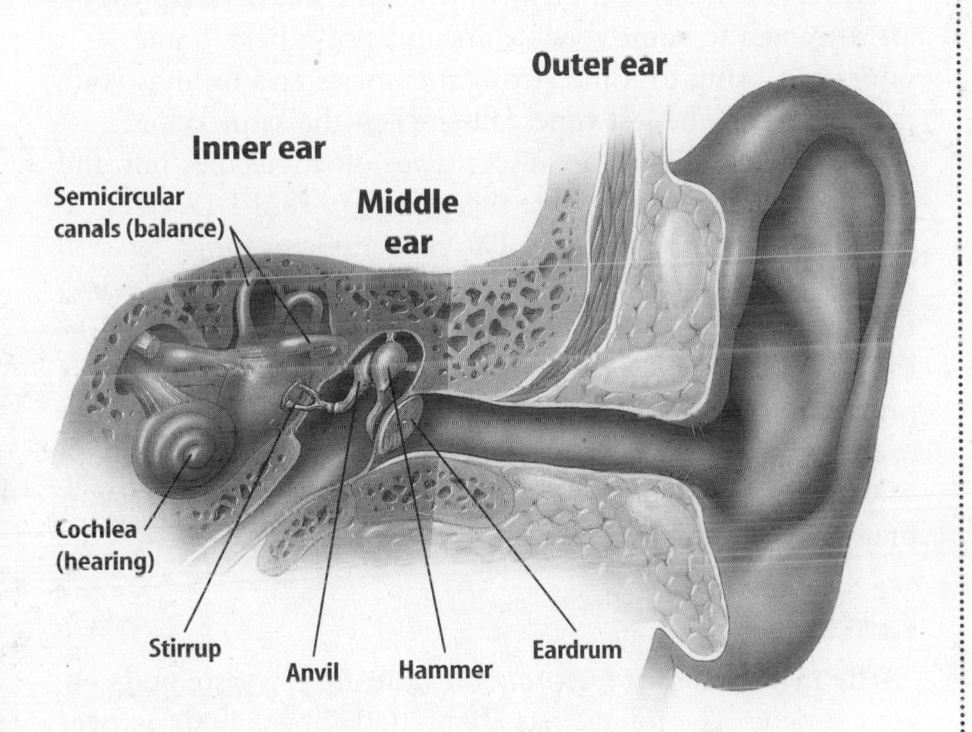

How does the inner ear work?

The <u>cochlea</u> (KOH klee uh) is a fluid-filled structure shaped like a snail's shell. When the stirrup vibrates, fluids in the cochlea begin to vibrate. The vibrations bend hair cells in the cochlea. When a hair cell bends, it sends electrical impulses to the brain. Depending on how the nerves are stimulated, you hear different types of sound. ☑

How does the inner ear control balance?

Some parts of the inner ear also control your balance. These parts are called the cristae ampullaris (KRIHS tee • am pyew LEER ihs) and the maculae (MA kyah lee). These parts sense different types of body movement.

Picture This

6. **Interpret Scientific Illustrations** Use a pencil or pen to trace the path of sound waves from outside the ear to the cochlea.

Reading Check

7. **Explain** What sends the electrical impulses to the brain?

____

____

Think it Over

8. Summarize What are two functions of the ear?

____

____

____

The Muscles Respond to the Inner Ear Both the cristae ampullaris and the maculae have tiny hair cells. As the body moves, the fluid surrounding the hair cells moves. This motion stimulates the nerve cells to send nerve impulses to the brain. The brain interprets the body movements and sends impulses to the skeletal muscles. The impulses cause your muscles to move in a way that helps you keep your balance.

Smell

The sense of smell influences what you eat because you are attracted to some food odors and not others. Some odors can bring to mind strong memories and feelings you may have had the last time you smelled the same odor.

You can smell food because it gives off molecules into the air. The molecules stimulate the **olfactory** (ohl FAK tree) **cells** in your nasal passages. The cells are kept moist by mucus. When molecules in the air dissolve in this moisture, the cells become stimulated and produce impulses. Impulses that start in these cells travel to the brain. The brain interprets the stimulus. If you have smelled the odor before, your brain will recognize it and you can identify the odor. If your brain does not recognize the stimulus, it is remembered and may be identified the next time you smell it. ☑

Reading Check

9. Explain What cells in the nasal passage help you smell food?

____

____

____

Taste

The major sensory receptors for taste are the **taste buds** on your tongue. The tongue has about 10,000 taste buds all over it. The taste buds help you to tell one taste from another. The figure below shows the parts of a taste bud.

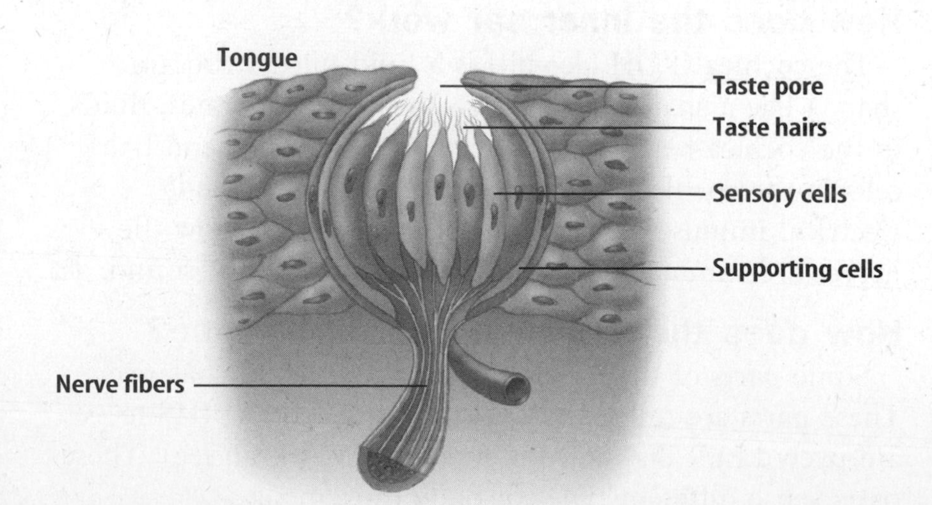

Picture This

10. Circle the name of the structure where the taste buds are found.

____

____

How are you able to taste food?

Most taste buds respond to several kinds of tastes. However, certain areas of the tongue respond more to one taste than another. The five tastes are sweet, salty, sour, bitter, and the taste of MSG (monosodium glutamate). Before you can taste something, it has to be dissolved in water. Saliva begins this process. The saliva and food wash over the taste buds. Taste buds are made up of a group of sensory cells with tiny taste hairs projecting from them.

When food is taken into the mouth, it is dissolved in saliva. This mixture stimulates taste buds to send impulses to the brain. The brain interprets the impulses, and you identify the tastes. ☑

Reading Check

11. Identify the five tastes your brain can detect.

____

____

____

How do smell and taste work together?

Smell and taste are related. You need the sense of smell to identify some foods such as chocolate. When saliva in the mouth mixes with the chocolate, odors travel up the nasal passage in the back of the throat. The olfactory cells are stimulated, and you taste and smell the chocolate.

When you have a stuffy nose, some foods seem tasteless. It may be because the food's molecules are blocked from contacting the olfactory cells in your nasal passage.

Other Sensory Receptors in the Body

Your internal organs have several kinds of sensory receptors. These receptors pick up changes in touch, pressure, pain, and temperature and send impulses to the brain or spinal cord. Your body then responds to the new information.

You also have sensory receptors throughout your skin. Your fingertips have many different types of receptors for touch. These receptors help you tell if an object is rough or smooth, hard or soft, hot or cold. Your lips are very sensitive to heat. They keep you from drinking something so hot that it would burn you.

All of the body's senses work together to maintain homeostasis. Your senses help you enjoy or avoid things around you. You react to your environment because of information that you receive through your senses.

Think it Over

12. Apply Name two things your senses help you avoid.

____

____

After You Read

Mini Glossary

cochlea (KOH klee uh): fluid-filled structure shaped like a snail's shell; located in the inner ear

olfactory (ohl FAK tree) cells: nerve cells in the nasal passages

retina (RET nuh): the tissue at the back of the eye that is sensitive to light energy

taste bud: major sensory receptor for taste

1. Review the terms and their definitions in the Mini Glossary. Using one of the terms, write a sentence explaining how it helps you react to your environment.

____

____

2. Complete the diagram below to identify the structures of the eye and to name a function of each structure.

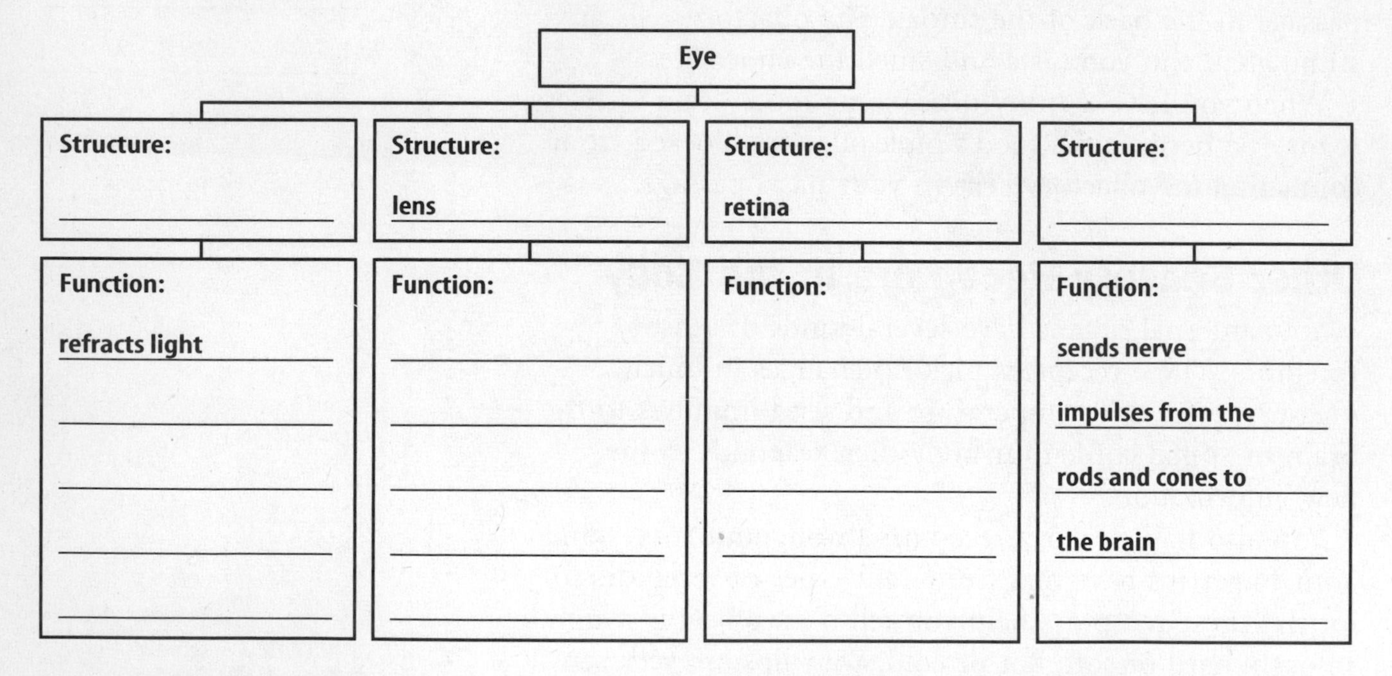

3. How can you use the quiz you created for this section to help you study for a test?

____

____

Science Online Visit **life.msscience.com** to access your textbook, interactive games, and projects to help you learn more about the senses.

Regulation and Reproduction

section 1 The Endocrine System

Before You Read

Have you ever been suddenly frightened? On the lines below, explain how your body reacted.

____

____

What You'll Learn
- how hormones function
- the endocrine glands and the hormones they produce
- how a feedback system works in your body

Read to Learn

Body Controls

Your endocrine system and your nervous system are your body's control systems. The nervous system sends messages to and from the brain to the rest of your body. The endocrine system sends chemical messages to different parts of your body.

Your body reacts very quickly to messages from the nervous system. Your body reacts more slowly to chemical messages from the endocrine system.

Mark the Text

Identify the Main Point Underline the main point of each paragraph. Review the main points after you have finished reading the section.

Endocrine Glands

Endocrine glands are tissues that produce hormones. **Hormones** (HOR mohnz) are chemicals that can speed up or slow down certain cell processes. Each endocrine gland releases its hormones directly into the blood. The blood carries the hormone to other parts of the body.

Endocrine glands produce hormones that control the body in many ways. Some endocrine glands help the body handle stressful situations. Other endocrine glands help the body grow and develop. Endocrine glands coordinate the circulation of the blood and help the body digest and absorb food. The endocrine glands and their functions are listed in the table on the next page.

FOLDABLES™

A Compare Make a three-tab book Foldable, as shown below. Use it to compare the functions and the structure of the endocrine system. Include the names of glands and organs that are part of the endocrine system.

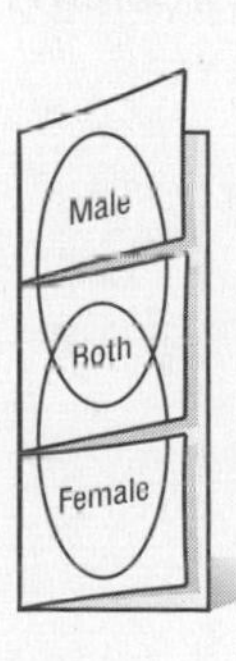

Picture This

1. **Identify** Highlight the names of the endocrine glands located in your brain. Then circle the names of the glands and organs that are involved in reproduction. Which gland is both highlighted and circled?

____

The Endocrine System		
Endocrine Glands and Organs	**Location in the Body**	**Major Function**
Pineal	in the brain	produces the hormone melatonin that may help regulate your body clock
Pituitary	in the brain	produces hormones that regulate various body activities including growth and reproduction
Thymus	upper chest	produces hormones that help the body fight infections
Thyroid	below the larynx	produces hormones that regulate metabolism (the chemical reactions in the body)
Parathyroid	below the larynx	produces hormones that regulate the body's calcium levels
Adrenals	on top of each kidney	produce several hormones that help your body respond to stress and keep your blood sugar levels stable
Pancreas	between the kidneys	produces hormones that help control blood sugar levels in the bloodstream
Testes (male)	in the scrotum	produce testosterone, a male reproductive hormone
Ovaries (female)	in the pelvic cavity	produce estrogen and progesterone, hormones that regulate the female reproductive cycle

A Negative-Feedback System

The organs and glands of the endocrine system control the amount of hormones in your body by sending chemical messages back and forth to each other. This process is called a negative-feedback system. Follow each step in the figure below to learn more about how a negative feedback system works.

Picture This

2. **Identify** Circle the name of the organ that produces insulin.

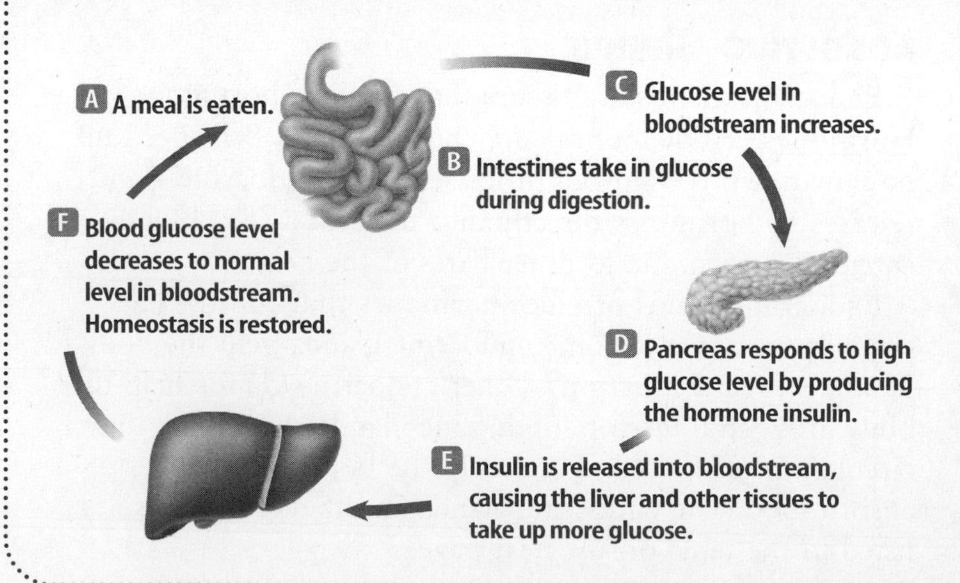

A A meal is eaten.
B Intestines take in glucose during digestion.
C Glucose level in bloodstream increases.
D Pancreas responds to high glucose level by producing the hormone insulin.
E Insulin is released into bloodstream, causing the liver and other tissues to take up more glucose.
F Blood glucose level decreases to normal level in bloodstream. Homeostasis is restored.

After You Read

Mini Glossary

hormones (HOR mohnz): chemical messages in the body that speed up or slow down certain cell processes

1. Review the term and its definition in the Mini Glossary. Write a sentence that explains the purpose of hormones in your body.

____

____

____

2. Use the terms in the box below to complete the sentences that follow.

adrenals	ovaries	parathyroid	pituitary
testes	thymus	thyroid	

a. The ____________ produces a hormone that helps the body fight infection.

b. The ____________ produces testosterone, while the ____________ produce estrogen and progesterone.

c. The glands that help your body react to stress are known as the ____________.

d. The ____________ gland in the brain controls growth.

e. The ____________ and the ____________ are located below the larynx.

Visit **life.msscience.com** to access your textbook, interactive games, and projects to help you learn more about the endocrine system.

Regulation and Reproduction

chapter 22

section 2 The Reproductive System

What You'll Learn
- the function of the reproductive system
- the major structures of the male and female reproductive systems
- the stages of the menstrual cycle

Before You Read

On the lines below, describe one way in which a male body differs from a female body.

____

____

____

Study Coach

Create a Quiz As you study the information in this section, create questions about the information you read. The questions can be used to review the section's content.

Read to Learn

Reproduction and the Endocrine System

Most human body systems are the same in males and females, but the reproductive systems are different. As you can see in the figure below, the pituitary gland makes the sex hormones that control the male and female reproductive systems. Sex hormones are needed to develop sexual characteristics. Sex hormones from the pituitary gland begin the process of making eggs in females and sperm in males. Eggs and sperm pass hereditary information from one generation to the next.

Picture This

1. **Explain** Use the diagram to explain to a classmate what the pituitary gland does in females and then have the classmate explain what the pituitary gland does in males.

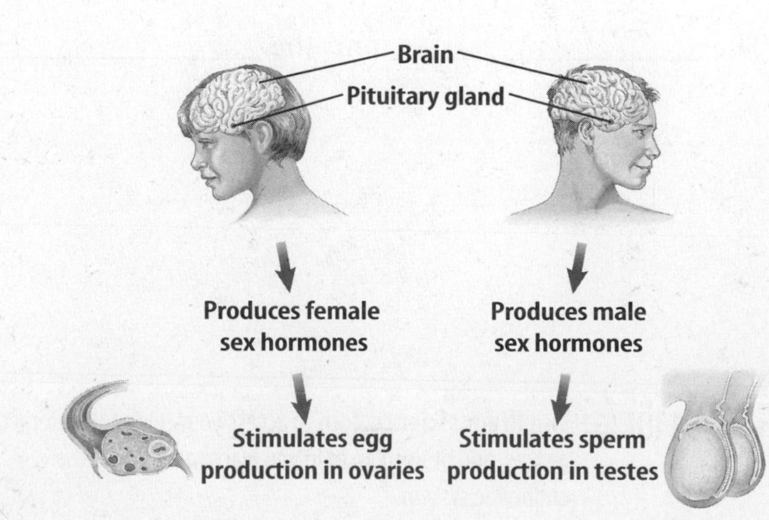

The Male Reproductive System

The male reproductive organs are inside and outside the body. As shown in the figure below, the organs outside the body are the penis and the scrotum (SKROH tum). The scrotum contains two organs called testes (TES teez) (singular, *testis*). The **testes** make the male hormone, testosterone (tes TAHS tuh rohn). They also make male reproductive cells, called **sperm**.

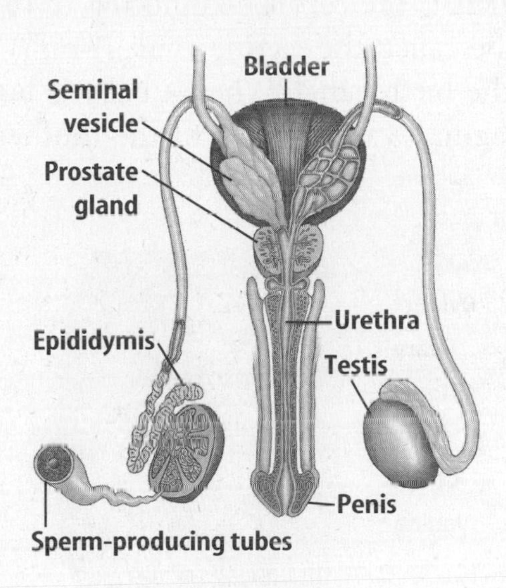

Picture This

2. Identify Underline the name of the structure that produces testosterone.

What happens to sperm?

Each sperm cell has a head and tail. The head contains hereditary information. The tail moves back and forth to push the sperm through fluid. Sperm travel out of the testes through sperm ducts that circle the bladder. The seminal vesicle (VEH sih cuhl) provides the sperm with a fluid. The fluid provides energy to the sperm and helps them move. The mixture of sperm and fluid is called **semen** (SEE mun). Semen leaves the body through the urethra. The urethra is the same tube that carries urine from the body.

FOLDABLES™

B Explain Make a two-tab book Foldable, as shown below, to record notes about the structures and functions of the male and female reproductive systems.

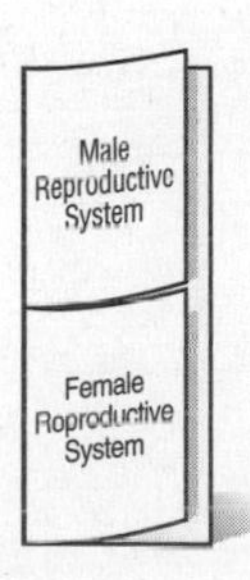

The Female Reproductive System

Most of the female reproductive organs are inside the body. The female sex organs are called the **ovaries.** The ovaries produce eggs. Eggs are the female reproductive cells.

What happens to the eggs?

About once a month, hormones cause one of the ovaries to release an egg. The release of an egg from an ovary is called **ovulation** (ahv yuh LAY shun). After the egg is released, it enters the oviduct. Short, hairlike structures called cilia (SIH lee uh) help move the egg through the oviduct to the uterus (YEW tuh rus). The **uterus** is a muscular organ with thick walls. The fertilized egg develops in the uterus. ☑

Reading Check

3. **Identify** the structure the egg enters when it is released from the ovary.

As you can see in the figure below, at the lower hollow end of the uterus is the cervix. Connected to the cervix is a muscular tube called the **vagina** (vuh JI nuh). The vagina is also called the birth canal. When a baby is born, it travels through the vagina to the outside of the mother's body.

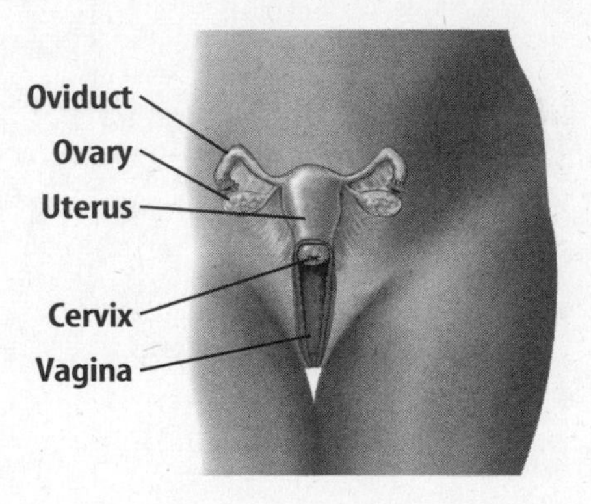

Picture This

4. **Explain** Trace the path of an egg after ovulation.

The Menstrual Cycle

The **menstrual** (MEN strul) **cycle** is the monthly cycle of changes in the female reproductive system. The menstrual cycle lasts about 28 days. During each cycle, an egg matures, female sex hormones are produced, the uterus prepares to receive a fertilized egg, and menstrual flow occurs. The first menstrual period happens between ages nine and 13 for most females.

What controls the menstrual cycle?

The pituitary gland releases several hormones that control the menstrual cycle. These hormones begin the process that results in the release of the egg from the ovary. They also stimulate the production of two other hormones, estrogen (ES truh jun) and progesterone (proh JES tuh rohn). The interaction of all these hormones causes the menstrual cycle. The menstrual cycle has three parts, or phases.

Phase One of the Menstrual Cycle Phase 1 starts with the menstrual flow, called **menstruation** (men STRAY shun). This flow is made up of blood and tissue cells released from the thickened lining of the uterus. Menstruation lasts up to six days.

Phase Two of the Menstrual Cycle During phase 2 of the menstrual cycle, hormones cause the lining of the uterus to thicken. During phase 2, an egg develops in the ovary. The release of the egg, or ovulation, occurs about 14 days before menstruation begins. The egg must be fertilized within 24 hours or it begins to break down. Sperm can live in a female's body for up to three days, so fertilization can happen soon after ovulation. ☑

Phase Three of the Menstrual Cycle During phase 3, the lining of the uterus continues to thicken. If a fertilized egg arrives, the thickened lining of the uterus begins to support and feed the developing embryo. If the egg is not fertilized, the lining of the uterus breaks down and the menstrual cycle starts over. The changes to the uterus during the phases of the menstrual cycle are shown in the figure below.

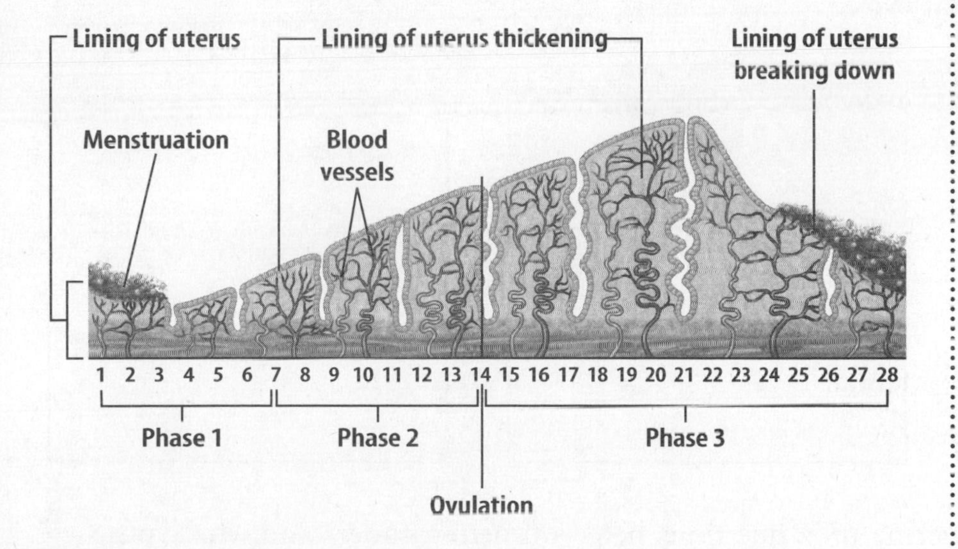

What is menopause?

For most females, the menstrual cycle ends between ages 45 and 60. Menopause occurs when the menstrual cycle ends. During menopause, the ovaries produce fewer and fewer sex hormones. The completion of menopause may take several years.

Reading Check

5. **Explain** For how long after ovulation can an egg be fertilized?

____

Picture This

6. **Identify** During which phase is the lining of the uterus the thickest?

____

After You Read

Mini Glossary

menstrual (MEN strul) cycle: the monthly cycle of changes in the female reproductive system

menstruation (men STRAY shun): phase 1 of the menstrual cycle, when blood and tissue cells are released from the thickened lining of the uterus

ovaries: the female sex organs that produce eggs

ovulation (ahv yuh LAY shun): the process that releases an egg from an ovary

semen (SEE mun): a mixture of sperm and fluid

sperm: male reproductive cells

testes (TES teez): male reproductive organs that produce sperm and the male hormone, testosterone

uterus (YEW tuh rus): the female organ in which a fertilized egg develops

vagina (vuh JI nuh): part of the female reproductive system, a muscular tube connected to the cervix

1. Review the terms and their definitions in the Mini Glossary. Use at least two of the terms in a sentence to describe either the male or female reproductive system.

____

____

2. Complete the flow chart below by writing a phrase that describes what happens during each phase of the menstrual cycle.

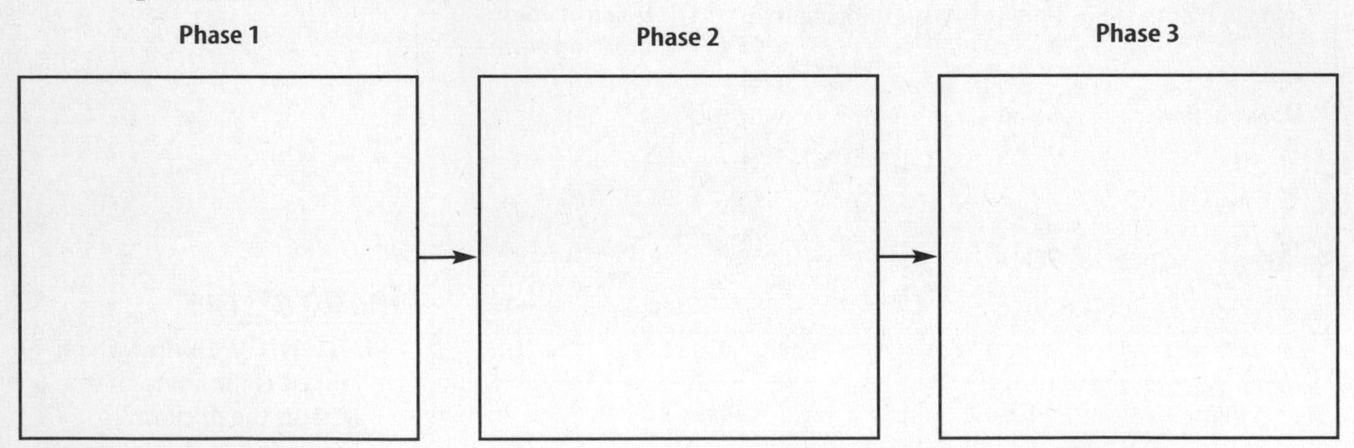

3. How did writing and answering quiz questions help you better understand what you have read?

____

____

Visit **life.msscience.com** to access your textbook, interactive games, and projects to help you learn more about the reproductive system.

Regulation and Reproduction

section 3 Human Life Stages

Before You Read

Describe the changes that you have seen happen in a young child over a year's time.

____

____

What You'll Learn

- how a human egg is fertilized
- how the embryo and fetus develop
- the life stages of infancy, childhood, adolescence, and adulthood

Read to Learn

Mark the Text

Locate Information As you read this section, highlight the portions of the text that describe the changes to an embryo and fetus during pregnancy.

Fertilization

A human develops from an egg that has been fertilized by a sperm. As sperm enter the vagina, they come in contact with chemicals given off in the vagina. These chemicals cause changes in the sperm that make it possible for the sperm to fertilize the egg. A sperm that touches the egg releases an enzyme. This enzyme helps the sperm enter the egg. Fertilization takes place when sperm and egg unite.

How does a zygote form?

Once a sperm enters an egg, the nucleus of the sperm joins with the nucleus of the egg. This joining creates a fertilized cell called the zygote (ZI goht).

Multiple Births

Mothers sometimes give birth to two or more babies at once. These are called multiple births. Multiple births can happen when an ovary releases more than one egg at a time or when a zygote divides into two or more zygotes.

Sometimes an ovary releases two eggs at the same time. If both eggs are fertilized, fraternal twins are born. Fraternal twins do not have the same hereditary information because they came from two different eggs. Fraternal twins can be the same or different sexes. ☑

Reading Check

1. **Explain** How many eggs must be fertilized for fraternal twins to be born?

____

FOLDABLES™

C Describe Make a five-tab book Foldable on notebook paper, as shown below, to record notes about the changes that occur during each stage of human development.

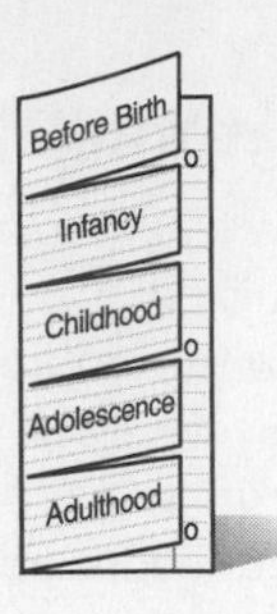

Picture This

2. **Explain** Place the numbers 1, 2, or 3 beside the following words in the figure to show the order in which they happen: Fertilization, Implantation, and Ovulation.

Think it Over

3. **Conclude** Why is the mother's good nutrition important to the embryo?

____

____

When are twins identical?

Identical twins develop from one egg that has been fertilized by one sperm. The zygote divides into two separate zygotes. Identical twins have the same hereditary information because they come from the same fertilized egg. Identical twins are always the same sex.

Development Before Birth

As you can see in the figure below, the zygote moves along the oviduct to the uterus. During this time, the zygote goes through many cell divisions. After about seven days, the zygote attaches to the wall of the uterus. This is called implantation. A zygote that attaches to the wall of the uterus will develop into a baby in about nine months. The period of development from fertilized egg to birth is called **pregnancy**.

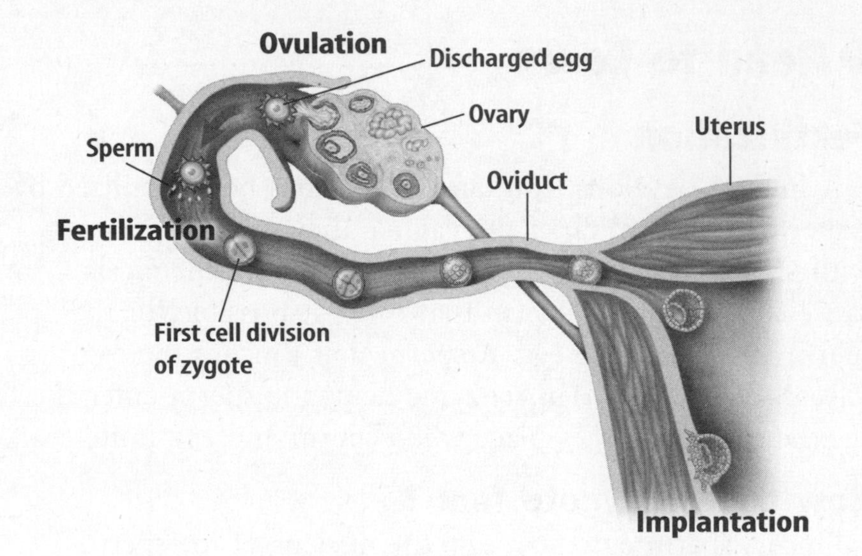

When does a zygote become an embryo?

After the zygote attaches to the wall of the uterus, it is called an **embryo** (EM bree oh).

How does an embryo get food and oxygen?

After an embryo attaches to the uterus, a placenta (pluh SEN tuh) develops from tissues of the uterus and the embryo. An umbilical (um BIH lih kul) cord connects the embryo to the placenta. Blood vessels in the umbilical cord carry nutrients and oxygen from the mother's blood through the placenta to the embryo. Other blood vessels in the umbilical cord carry wastes from the embryo to the mother's blood.

What protects the embryo?

During the third week of pregnancy, a thin membrane called the **amniotic** (am nee AH tihk) **sac** forms around the embryo. The amniotic sac is filled with a clear fluid called amniotic fluid. The amniotic fluid acts as a cushion to protect the embryo. Amniotic fluid also stores nutrients and wastes. ☑

When does the embryo develop body parts?

During the first two months of development, the embryo's major organs form and the heart begins to beat. At five weeks, the embryo has a head with eyes, nose, and mouth. During the sixth and seventh weeks, fingers and toes develop.

How does a fetus develop?

Pregnancy in humans lasts about 38 to 39 weeks. After the first two months of pregnancy, the developing embryo is called a **fetus** (FEE tus). The fetus has all its body organs and is about 8 cm to 9 cm long. By the end of the seventh month of pregnancy, the fetus is 30 cm to 38 cm long. By the ninth month, the fetus is about 50 cm long. It weighs from 2.5 kg to 3.5 kg. During the ninth month, the fetus moves to a head-down position within the uterus. This is the best position for delivery.

Reading Check

4. **Explain** What are the functions of the amniotic fluid?

______

______

______

The Birthing Process

The process of childbirth begins when the muscles of the uterus start to contract. This is called labor. As the contractions increase, the amniotic sac breaks and the fluid comes out. Over a period of hours, the contractions cause the opening of the uterus to get wider. More powerful and more frequent contractions push the baby out through the vagina into the world. After the baby is born, more contractions push the placenta out of the mother's body. ☑

When are babies delivered through surgery?

Sometimes babies cannot be born through the birth canal. In these cases, a baby is delivered through surgery called a cesarean (suh SEER ee uhn) section. In this surgery, a cut is made in the abdominal wall of the mother, then through the wall of the uterus. The baby is delivered through this opening.

Reading Check

5. **Identify** two things contractions help push from the mother's body.

______

______

______

What happens after birth?

After birth, the baby is still attached to the umbilical cord. Two clamps are placed on the umbilical cord and it is cut between the clamps. The scar where the cord was attached is called the navel.

The experiences that a fetus goes through during childbirth can cause **fetal stress.** After it is born, the fetus must adapt from a dark, watery environment with a constant temperature to an environment with more light, less water, and changes in temperature. The first four weeks after birth are known as the neonatal (nee oh NAY tul) period. Neonatal means "newborn." During this time the baby's body begins to function normally.

Stages After Birth

After birth, four stages of development occur: infancy, childhood, adolescence, and adulthood. Infancy lasts from birth to around 18 months of age. Childhood lasts from the end of infancy to puberty (PYEW bur tee), the time of development when a person becomes physically able to reproduce. Adolescence is the teen years. Adulthood lasts from about the early 20s until death. ☑

Reading Check

6. Explain When do humans become physically able to reproduce?

____

How does a baby develop during infancy?

Human babies depend on other humans for their survival. During infancy a baby learns how to coordinate the movements of its body, as shown in the figure below. Its mental abilities increase, and it grows rapidly. Many infants triple their weight in the first year of life.

Picture This

7. Identify Study the table to answer the following questions.

a. At what age can most infants sit alone?

____

b. At what age do infants learn to walk?

____

Infant Development

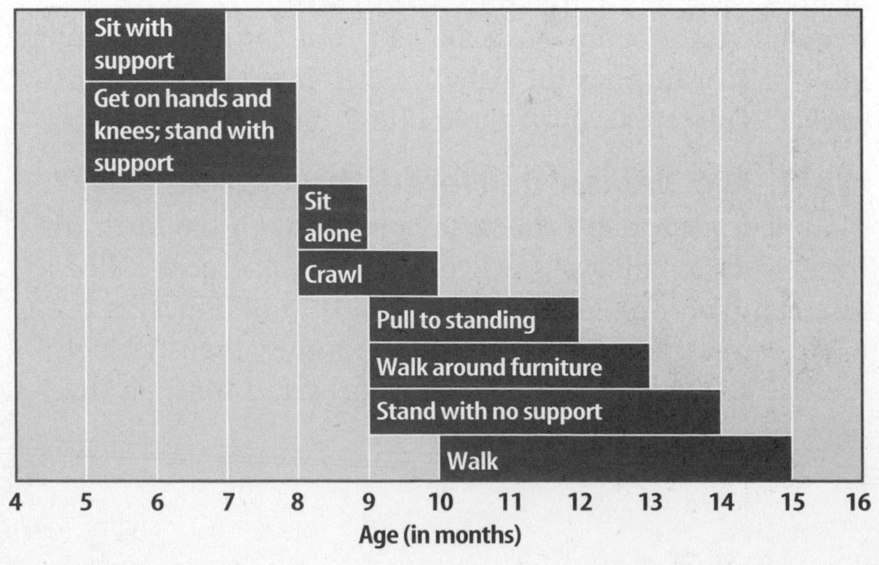

What developments take place in childhood?

Childhood lasts from the age of about 18 months to about 12 years. Growth during childhood is rapid. Between two and three years of age, the child learns to control his or her bladder and bowels. Most children also can speak in simple sentences at age two or three. Around age four, the child can get dressed and undressed with some help. By age five, many children can read some words. Throughout childhood, children develop their abilities to speak, read, write, and reason.

What happens during adolescence?

Adolescence begins at about age 12 or 13 and ends at about age 20. Puberty is a part of adolescence. For girls, puberty happens between ages nine and 13. For boys, puberty occurs between ages 13 and 16. During puberty, hormones produced in the pituitary gland cause changes in the body. Females develop breasts, pubic and underarm hair, and fatty tissue around the thighs and buttocks. Males develop deeper voices, increased muscle size, and facial, pubic, and underarm hair. ☑

Adolescence is usually when a final growth spurt occurs. Most girls begin this final growth phase around age 11 and end around age 16. For boys, the final growth spurt begins around age 13 and ends around age 18. However, different people have different growth rates.

What happens during adulthood?

Adulthood begins when adolescence ends, at about age 20, and continues through old age. From about age 45 to age 60, middle-aged adults begin to lose physical strength. Their blood circulation and breathing become less efficient. Bones break more easily, and skin becomes wrinkled. ☑

What changes occur in older adults?

After age 60, adults may have an overall decline in their health. Their body systems do not work as well as they once did. Muscles and joints become less flexible. Bones become thinner and break more easily. Older adults may lose some of their ability to hear and see. Their lungs and heart do not work as well as they used to. Eating well and exercising throughout life can help delay these changes.

Reading Check

8. Identify When do most girls experience puberty?

____

____

____

Reading Check

9. Explain What physical changes occur during middle age?

____

____

____

● After You Read

Mini Glossary

amniotic (am nee AH tihk) sac: a thin membrane that forms around the embryo, acting as a cushion and a place to store nutrients and wastes

embryo (EHM bree oh): a fertilized egg or zygote after it attaches to the wall of the uterus

fetal stress: the experiences that a fetus goes through during childbirth

fetus (FEE tus): a developing human embryo after two months of pregnancy

pregnancy: the period of development from fertilized egg to birth

1. Review the terms and their definitions in the Mini Glossary. Write one or two sentences that explain the relationship of a zygote, an embryo, and a fetus.

____

____

____

2. Fill in the table below to identify and describe the stages of development after birth.

Stage of Development	Period of Time	Development Changes
	Birth to 18 months	
	18 months to 12 years	
	12 years to 20 years	
	20 years to 60 years	
	After age 60	

Visit **life.msscience.com** to access your textbook, interactive games, and projects to help you learn more about human life stages.

Immunity and Disease

section 1 The Immune System

Before You Read

Think about the last time that you had a cold. On the lines below, describe three ways your body reacted to the cold.

____

____

____

What You'll Learn

- the body's natural defenses
- the difference between an antigen and an antibody
- the differences between active and passive immunity

Read to Learn

Lines of Defense

Your body has many ways to defend itself from illness. Your first-line defenses are general. First-line defenses work against harmful substances and all types of disease-causing organisms, called **pathogens** (PA thuh junz). Your second-line defenses are specific. They work against specific pathogens. The combination of first-line and second-line defenses is called your **immune system**.

Mark the Text

Locate Information Read all the headings for this section and circle any word you cannot define. Then review the circled words and underline the part of the text that helps you define the words.

What are your body's first-line defenses?

Your skin and your respiratory, digestive, and circulatory systems are your first-line defenses against pathogens. Your skin stops many pathogens from entering the body. Sweat and oils produced by your skin cells can slow the growth of some pathogens.

Respiratory System Defenses The respiratory system traps pathogens with hairlike structures, called cilia (SIH lee uh), and mucus. Mucus has enzymes (EN zimez) that weaken the cell walls of some pathogens. Coughs and sneezes help get rid of pathogens from your lungs and nasal passages. ☑

Reading Check

1. Explain What do cilia do?

____

____

Digestive System Defenses Your digestive system has four defenses against pathogens—saliva, enzymes, hydrochloric acid, and mucus. Saliva contains substances that kill bacteria. Enzymes in your stomach, pancreas, and liver help destroy pathogens. Hydrochloric acid in your stomach kills some bacteria and stops some viruses that enter your body on the food you eat. The mucus in your digestive tract has a chemical that prevents bacteria from attaching to the inner lining of your digestive organs.

Circulatory System Defenses Your circulatory system contains white blood cells that surround and destroy foreign organisms and chemicals. White blood cells constantly patrol your body, destroying harmful bacteria. If the white blood cells cannot destroy the bacteria fast enough, you may develop a fever. A fever is a slight increase in body temperature that slows the growth of pathogens. A fever speeds up your body's defenses. ☑

Reading Check

2. **Determine** What does a fever do?

____

____

____

How do you know when tissue is damaged?

When tissue is damaged by injury or infected by pathogens, it becomes inflamed. Signs that tissue is inflamed include redness, an increase in temperature, swelling, and pain. Damaged cells release chemicals that cause nearby blood vessels to widen, allowing more blood to flow into the inflamed area. Other chemicals released by damaged cells attract white blood cells that surround and destroy the pathogens. If pathogens get past these first-line defenses, your body uses its second-line defenses. Second-line defenses work against specific pathogens.

What are antigens?

Molecules that are foreign to your body are called **antigens** (AN tih junz). Antigens can be separate molecules, or they can be attached to the surface of pathogens. When your immune system recognizes antigens in your body, it releases special kinds of white blood cells that fight infection. White blood cells that fight infections are called lymphocytes.

The first lymphocytes to respond to an antigen are the T cells. There are two kinds of T cells, killer T cells and helper T cells. Killer T cells release enzymes that help destroy foreign matter. Helper T cells cause the body to produce another kind of lymphocyte, called a B cell.

FOLDABLES™

A Explain Use quarter-sheets of notebook paper to make Foldables to define and explain antigens and antibodies.

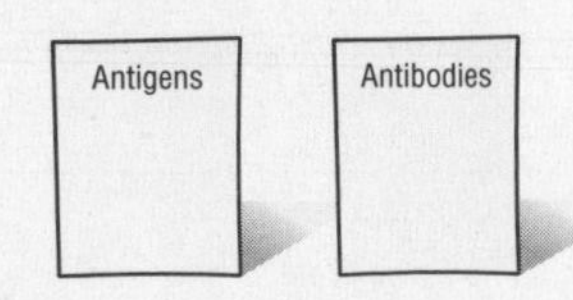

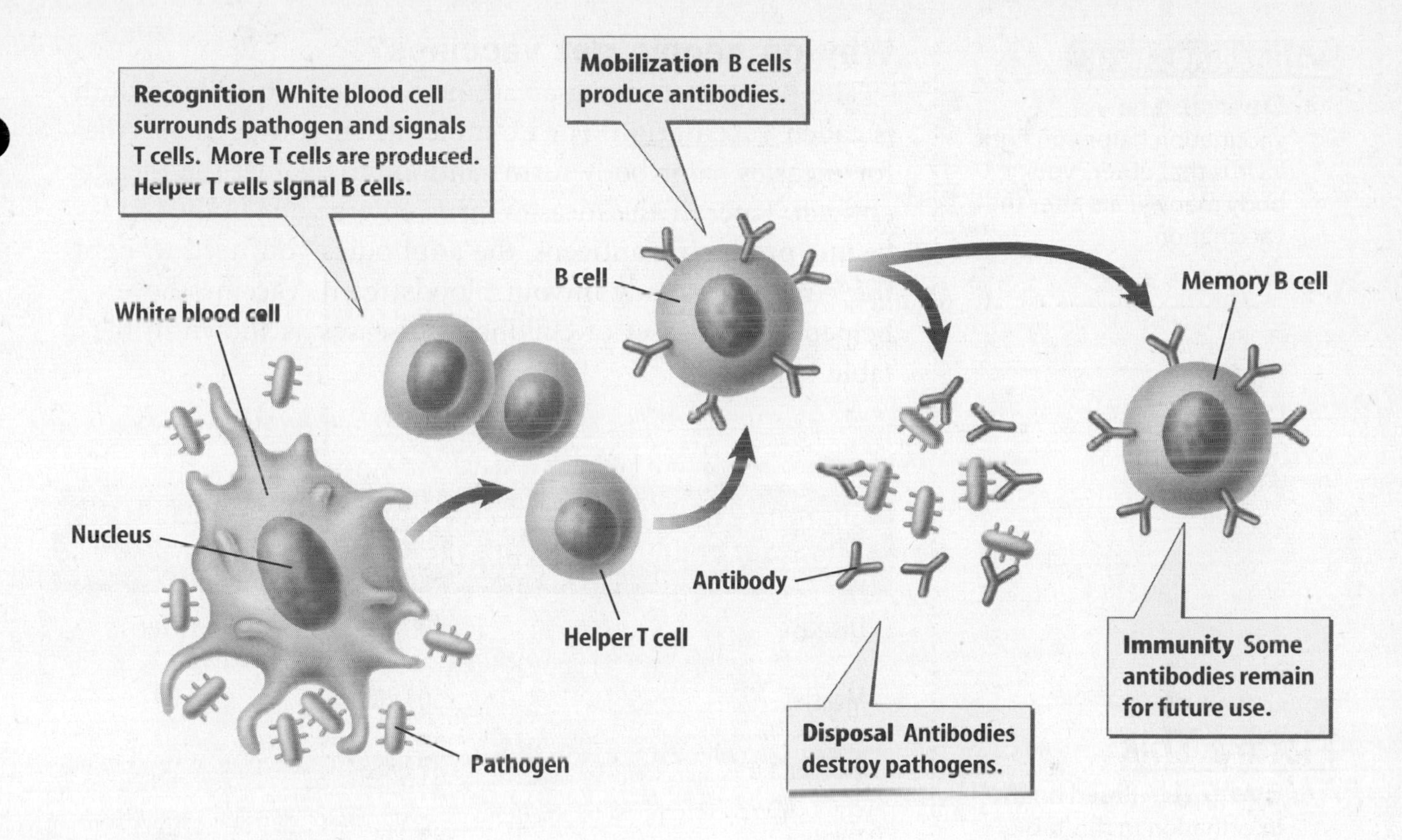
Recognition White blood cell surrounds pathogen and signals T cells. More T cells are produced. Helper T cells signal B cells.
Mobilization B cells produce antibodies.
B cell
Memory B cell
White blood cell
Nucleus
Antibody
Helper T cell
Pathogen
Disposal Antibodies destroy pathogens.
Immunity Some antibodies remain for future use.

What are antibodies?

B cells form antibodies to specific antigens. An **antibody** is a protein your body makes to fight a specific antigen. The antibody can attach to the antigen and make the antigen harmless. The antibody can also make it easier for a killer T cell to destroy the antigen.

Other lymphocytes, called memory B cells, also have antibodies against specific pathogens. Memory B cells stay in the blood ready to destroy that same pathogen if it invades your body again. The response of your immune system to a pathogen is summarized in the figure above.

What are active and passive immunity?

Antibodies help your body build defenses in two ways—actively and passively. In **active immunity**, your body makes its own antibodies in response to an antigen. In **passive immunity**, the antibodies have been produced in another animal and put into your body. Vaccines are antigens produced in another organism and then placed in your body to build immunity against a disease. Passive immunity does not last as long as active immunity does.

Picture This

3. **Identify** Circle the name of the step in which antibodies are produced. Highlight the name of the step in which pathogens are destroyed.

FOLDABLES™

B Describe Make a three-tab Foldable, as shown below, to compare and contrast active immunity and passive immunity.

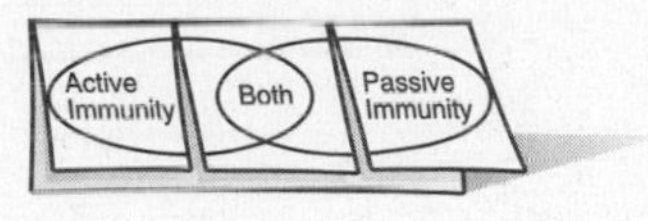
Active Immunity
Both
Passive Immunity

Reading Check

4. **Describe** how a vaccination helps you fight a virus that enters your body many years after the vaccination.

____

____

____

Why do people get vaccines?

The process of giving a vaccine by injection or by mouth is called **vaccination**. For example, when you get a vaccine for measles, your body forms antibodies against the measles antigen. Later, if the measles virus enters your body and begins producing antigens, the antibodies you need to fight the virus are already in your bloodstream. Vaccines have helped reduce cases of childhood diseases as shown in the table below.

Picture This

5. **Evaluate** Based on the information in this table, have vaccines been effective? Explain.

____

____

____

Annual Cases of Disease Before and After Vaccine Availability in the U.S.

Disease	Before	After
Measles	503,282	89
Diptheria	175,885	1
Tetanus	1,314	34
Mumps	152,209	606
Rubella	47,745	345
Pertussis (whooping cough)	147,271	6,279

Antibodies that protect you from one virus may not help you fight another virus. Each year a different set of flu viruses causes the flu. As a result, people get a new flu shot each year.

What is tetanus?

Tetanus is a disease caused by bacteria in the soil. Bacteria can enter the body through an open wound. The bacteria that causes tetanus produces a chemical that makes muscles unable to move. In early childhood, you received several tetanus vaccines to help you develop immunity to this disease. You need to continue to get tetanus vaccines every 10 years to stay protected.

After You Read

Mini Glossary

active immunity: long-lasting immunity that results when the body makes its own antibodies in response to an antigen

antibody: a protein made in response to a specific antigen

antigen (AN tih jun): any molecule that is foreign to your body

immune system: the complex group of defenses against harmful substances and disease-causing organisms

passive immunity: immunity that results when antibodies produced in another animal are introduced into your body

pathogen (PA thuh jun): a disease-causing organism

vaccination: the process of giving a vaccine by injection or mouth to provide active immunity

1. Review the terms and their definitions in the Mini Glossary. Write a sentence or two that explains the difference between an antigen and an antibody.

____

____

____

2. Complete the concept web below to identify four first-line defenses your body has against disease.

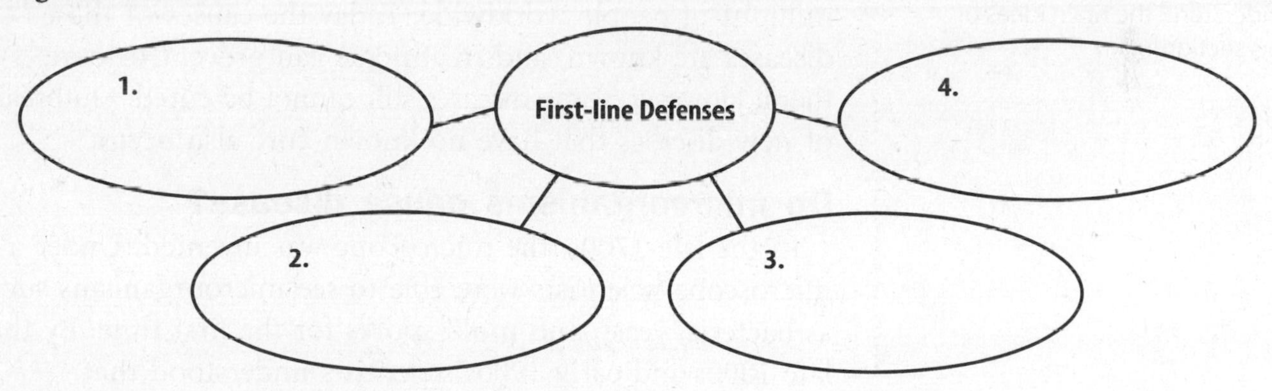

3. How did finding definitions of unfamiliar words help you understand the immune system?

____

____

Visit **life.msscience.com** to access your textbook, interactive games, and projects to help you learn more about the immune system.

chapter 23 Immunity and Disease

section 2 Infectious Diseases

What You'll Learn

- the work done by scientists to discover and prevent disease
- diseases caused by viruses and bacteria
- the causes of sexually transmitted diseases

Before You Read

How do you think washing hands helps prevent disease?

____

____

____

Study Coach

Read-and-Say Work with a partner. Read the information under a heading to yourselves. Then discuss together what you learned. Continue until you both understand the main ideas of this section.

Read to Learn

Disease in History

In the past, there were no treatments for diseases such as the plague, smallpox, and influenza. These diseases killed millions of people worldwide. Today the causes of these diseases are known, and treatments can prevent or cure them. However, some diseases still cannot be cured. Outbreaks of new diseases that have no known cure also occur.

Do microorganisms cause disease?

In the late 1700s, the microscope was invented. Under a microscope, scientists were able to see microorganisms such as bacteria, yeast, and mold spores for the first time. By the late 1800s and early 1900s, scientists understood that microorganisms could cause diseases and carry them from one person to another.

1. Infer What liquids do you drink that you think have undergone pasteurization?

____

____

____

What did Louis Pasteur discover?

The French chemist Louis Pasteur discovered that micro-organisms could spoil wine and milk. He then realized that microorganisms could attack the human body in the same way, causing diseases. Pasteur invented **pasteurization** (pas chuh ruh ZAY shun), which is the process of heating liquid to a specific temperature that kills most bacteria.

Which microorganisms cause diseases?

Many diseases are caused by bacteria, viruses, protists (PROH tihsts), or fungi. Bacteria can slow the normal growth and activities of body cells and tissues. Some bacteria produce toxins, or poisons, that kill body cells on contact. The table below lists some of the diseases caused by different groups of pathogens.

Human Diseases and the Pathogens that Cause Them

Pathogens	Diseases Caused
Bacteria	Tetanus, tuberculosis, typhoid fever, strep throat, bacterial pneumonia, plague
Protists	Malaria, sleeping sickness
Fungi	Athlete's foot, ringworm
Viruses	Colds, influenza, AIDS, measles, mumps, polio, smallpox

Picture This

2. Identify What type of pathogen causes strep throat?

____

Viruses A <u>virus</u> is a tiny piece of genetic material surrounded by a protein coating that infects host cells and multiplies inside them. The host cells die when the viruses break out of them. These new viruses infect other cells. Viruses destroy tissues or interrupt important body activities.

Other Pathogens Protists can destroy tissues and blood cells. They also can interfere with normal body functions. Fungus infections work in a similar way and can cause athlete's foot, nonhealing wounds, and chronic lung disease.

What did Robert Koch develop?

In the 1880s, Robert Koch developed a way to isolate and grow one type of bacterium at a time. Koch developed rules for identifying which organism causes a particular disease. Koch's rules are still used by doctors today.

What did Joseph Lister discover?

Today we know that washing hands kills bacteria and other organisms that spread disease. But until the late 1800s, people, including doctors, did not know this. Joseph Lister, an English surgeon, saw that infection and cleanliness were related. Lister learned that carbolic (kar BAH lik) acid kills pathogens. He greatly reduced the number of deaths among his patients by washing their skin, his hands, and his surgical instruments with carbolic acid. ☑

FOLDABLES™

C Organize Make a folded table with three columns and three rows, as shown below. Use the Foldable to record facts about types of diseases.

	Diseases Caused By	How Caused
Bacteria		
Viruses		

What operating procedures are followed today?

Today special soaps are used to kill pathogens on skin. Every person who helps perform surgery must wash his or her hands thoroughly and wear sterile gloves and a covering gown. The patient's skin is cleaned around the area of the body to be operated on and then covered with sterile cloths. Surgery instruments and all operating equipment are sterilized. The air in the operating room is filtered to keep out pathogens. ☑

Reading Check

3. **List** two operating procedures followed today.

____

____

____

How Diseases Are Spread

An **infectious disease** is a disease that is spread from an infected organism or the environment to another organism. An infectious disease can be caused by a virus, bacterium, protist, or fungus. Infectious diseases are spread in many ways. They can be spread by direct contact with the infected organism, through water and air, on food, or by contact with contaminated objects. They can also be spread by disease-carrying organisms called **biological vectors**. Rats, birds, and flies are examples of biological vectors.

People also can be carriers of diseases. When you have the flu and sneeze, you send thousands of virus particles into the air. These particles can spread the virus to others. Colds and many other diseases also can be spread by contact. Everything you touch may have disease-causing bacteria or viruses on it. Washing your hands regularly is an important way to avoid disease.

FOLDABLES™

D Explain Use a quarter-sheet of notebook paper to define, list the types of, and explain STDs.

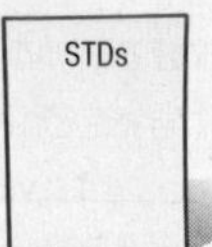

Sexually Transmitted Diseases

Infectious diseases that are passed from person to person during sexual contact are called **sexually transmitted diseases (STDs)**. STDs are caused by bacteria or viruses.

What are bacterial STDs?

STDs caused by bacteria are gonorrhea (gah nuh REE uh), chlamydia (kluh MIH dee uh), and syphilis (SIH fuh lus). The symptoms for gonorrhea and chlamydia may not appear right away, so a person may not know that he or she is infected. The symptoms for these STDs are pain when urinating, genital discharge, and genital sores. Bacterial STDs can be treated with antibiotics. If left untreated, gonorrhea and chlamydia can damage the reproductive system, leaving the person unable to have children.

What are the symptoms for syphilis?

Syphilis has three stages. In stage 1, a sore that lasts 10 to 14 days appears on the mouth or sex organ. Stage 2 may involve a rash, fever, and swollen lymph glands. In stage 3, syphilis may infect the cardiovascular and nervous systems. Syphilis can be treated with antibiotics in all stages. However damage to body organs in stage 3 cannot be reversed and may lead to death.

What is genital herpes?

Genital herpes is a lifelong STD caused by a virus. The symptoms include painful blisters on the sex organs. Genital herpes can be passed from one person to another during sexual contact or from an infected mother to her child during birth. The herpes virus hides in the body for long periods of time without causing symptoms and then reappears suddenly. The symptoms for genital herpes can be treated with medicine, but there is no cure or vaccine for the disease.

HIV and Your Immune System

Human immunodeficiency virus (HIV) can exist in blood and body fluids. This virus can hide in body cells, sometimes for years. HIV can be passed on by an infected person through sexual contact. A person can also be infected by reusing an HIV-contaminated needle for an injection. A sterile needle, however, cannot pass on HIV. The risk of getting HIV through blood transfusion is small because all donated blood is tested for HIV. An HIV-infected pregnant woman can infect her unborn child. A baby can get HIV after birth when nursing from an HIV-infected mother. ☑

What is AIDS?

An HIV infection can lead to Acquired Immune Deficiency Syndrome (AIDS). AIDS is a disease that attacks the body's immune system.

HIV is different from other viruses. It attacks the helper T cells in the immune system. HIV enters the T cell and multiplies. When the infected T cell bursts open, it releases more HIV that infects more T cells. Soon, so many T cells are destroyed that not enough B cells are formed to produce antibodies. Once HIV has reached this stage, the infected person has AIDS. The immune system can no longer fight HIV or any other pathogen. There is no cure for AIDS, but several kinds of medicines help treat AIDS in some patients.

Think it Over

4. **Determine** At which stage does syphilis cause permanent damage?

____

____

____

✔ Reading Check

5. **Identify** What are two ways that a teenager or adult can get HIV?

____

____

____

Fighting Disease

The first step to preventing infections is to wash small wounds with soap and water. Cleaning the wound with an antiseptic and covering it with a bandage also help fight infection.

Washing your hands and body helps prevent body odor. Washing also removes and destroys microorganisms on your skin. Health-care workers, such as the one shown below, wash their hands between patients. This reduces the spread of pathogens from one person to another.

Picture This

6. Identify other types of jobs where workers should wash their hands often.

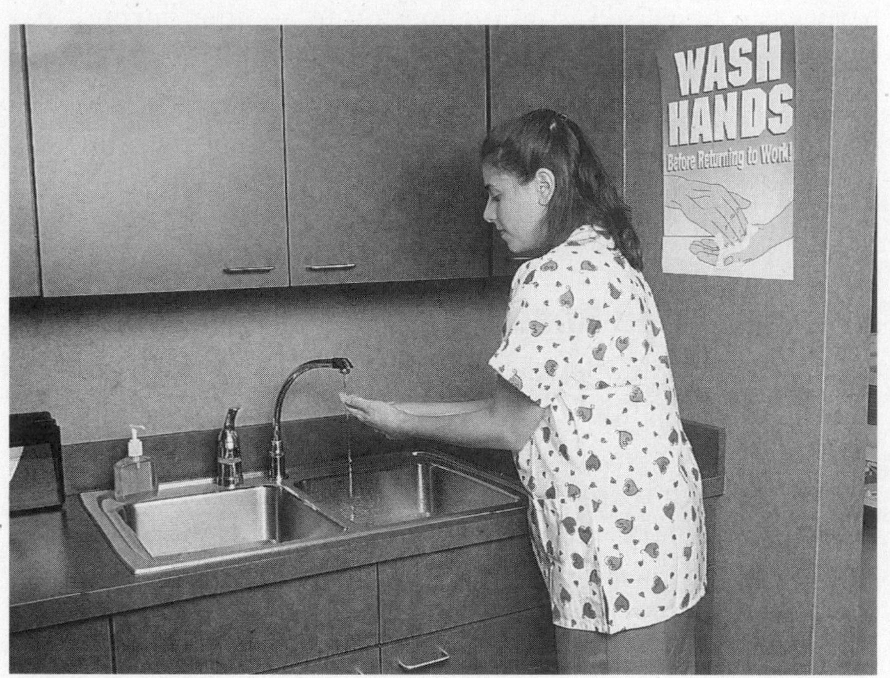

Mark Burnett

Microorganisms in your mouth cause mouth odor and tooth decay. Brushing and flossing your teeth every day keep these microorganisms under control.

Exercising, eating healthy foods, and getting plenty of rest help keep you healthy. You are less likely to get a cold or the flu if you have good health habits. Having checkups every year and getting the recommended vaccinations also help you stay healthy.

Think it Over

7. Apply List two things you do every day to keep healthy.

After You Read

Mini Glossary

biological vector: a disease-carrying organism

infectious disease: a disease that is caused by a virus, bacterium, protist, or fungus and is spread by an organism or the environment to another organism

pasteurization (pas chuh ruh ZAY shun): the process of heating a liquid to a specific temperature that kills most bacteria

sexually transmitted disease (STD): an infectious disease that is passed from person to person during sexual contact

virus: a small piece of genetic material surrounded by a protein coating that infects and multiplies in host cells

1. Review the terms and their definitions in the Mini Glossary. Choose one term that identifies a way a person gets a disease. Write a sentence about how the term you selected causes infection.

____

____

____

2. Complete the table below to identify the causes, symptoms, and treatments of STDs.

Kinds of STDs	Causes (Bacteria or Virus)	Symptoms	Treatment
Gonorrhea			
Chlamydia			
Syphilis			
Genital herpes			

Visit **life.msscience.com** to access your textbook, interactive games, and projects to help you learn more about infectious diseases.

chapter 23

Immunity and Disease

section 3 Noninfectious Diseases

What You'll Learn

- the causes of noninfectious diseases
- what happens during an allergic reaction
- the characteristics of cancer
- how chemicals in the environment can harm humans

Before You Read

Explain on the lines below why it is important to read labels and follow directions when using household products.

____

____

____

Mark the Text

Identify the Main Point Underline the main point of each paragraph. Review these ideas as you study the section.

Read to Learn

Chronic Disease

Diseases and disorders that are not caused by pathogens are called **noninfectious diseases**. Allergies, diabetes, asthma, cancer, and heart disease are noninfectious diseases. Many are chronic (KRAH nihk) diseases, or can become chronic diseases if not treated. A chronic disease is an illness that can last a long time. Some chronic diseases can be cured, but others cannot be cured.

Allergies

An **allergy** is an overly strong reaction of the immune system to a foreign substance. Allergic reactions include itchy rashes, sneezes, and hives. Most allergic reactions do not cause major problems. However, some allergic reactions can cause shock and even death if not treated right away.

What causes allergies?

A substance that causes an allergic reaction is called an **allergen**. Examples of allergens include dust, chemicals, certain foods, pollen, and some antibiotics. Asthma (AZ muh) is a lung disorder that is caused by allergens. The symptoms of asthma include shortness of breath, wheezing, and coughing. ☑

Reading Check

1. **Explain** How are allergens and asthma related?

____

____

____

How does the body react to allergens?

When you come in contact with an allergen, your immune system usually forms antibodies. Your body also reacts to allergens by releasing chemicals called histamines (HIHS tuh meenz) that cause red, swollen tissues. Antihistamines are medications that can be used to treat allergic reactions and asthma.

Diabetes

Diabetes is a chronic disease that has to do with the levels of insulin made by the pancreas. Insulin is a hormone that helps glucose, a form of sugar, pass from the bloodstream into your cells. There are two types of diabetes. Type I diabetes is the result of too little or no insulin production. Type II diabetes happens when your body does not properly use the insulin it produces. Symptoms of diabetes include tiredness, great thirst, the need to urinate often, and tingling feelings in the hands and feet. ☑

People with Type I diabetes often need daily injections of insulin to control their glucose levels. People with Type II diabetes usually can control the disease by watching their diet and their weight.

If diabetes is not treated, health problems can develop. These problems include blurred vision, kidney failure, heart attack, stroke, loss of feeling in the feet, and the loss of consciousness, or a diabetic coma.

Chemicals and Disease

Chemicals are everywhere—in your body, the foods you eat, cosmetics, and cleaning products. Most chemicals used by consumers are safe, but a few are harmful. A chemical that is harmful to living things is called a toxin. Toxins can cause a variety of diseases, as well as birth defects, tissue damage, and death. Some toxins and the damage they cause are shown in the table below.

Toxin	Effect
asbestos	lung disease
lead-based paints	damage to central nervous system
alcohol (consumed during pregnancy)	birth defects

Reading Check

2. **List** four symptoms of diabetes.

____

____

____

____

FOLDABLES™

E Explain Make a four-tab Foldable to explain the causes of noninfectious diseases.

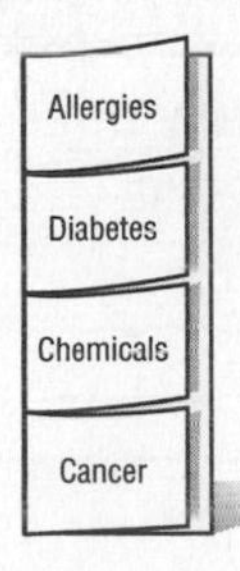

Cancer

Cancer is a group of closely related diseases that are caused by uncontrolled cell growth. The table below shows characteristics of cancer cells.

Picture This

3. **Explain** How can cancer cells spread through the body?

Characteristics of Cancer Cells
Cell growth is out of control.
Cells do not function as part of the body.
Cells take up space and cause problems with normal body functions.
Cells travel throughout the body by way of blood and lymph vessels.
Cells produce tumors and unusual growths anywhere in the body.

What are some types of cancers?

Leukemia (lew KEE mee uh) is a cancer of white blood cells. The cancerous white blood cells cannot fight diseases. These cancer cells multiply and crowd out normal blood cells. Cancer of the lungs makes breathing difficult. Cancer of the large intestine is a leading cause of death in men and women. Breast cancer causes tumors to grow in the breast. Cancer of the prostate gland, an organ that surrounds the urethra, is the second most common cancer in men.

What are some causes of cancer?

Carcinogens (kar SIH nuh junz) are substances that can cause cancer. Some of these substances are shown in the photograph below. Coming in contact with carcinogens increases your chance of getting cancer. Carcinogens include asbestos, some cleaning products, heavy metals, tobacco, alcohol, and some home and garden products. Smoking has been linked to lung cancer. Exposure to X rays and radiation increase your chances of getting cancer. Some foods, such as smoked or barbecued meats, can give rise to cancers. ☑

Reading Check

4. **Identify** three carcinogens.

KS Studios

Genetics and Cancer The genetic makeup of some people increases their risk of developing cancer. That does not mean they will definitely get cancer, but it increases their chances of developing cancer.

How is cancer treated?

Finding cancer in its early stages is important for successful treatment. The early warning signs of cancer are listed in the table below.

Early Warning Signs of Cancer
Changes in bowel movements or urination
A sore that does not heal
Unusual bleeding or discharge
Thickening or lump in the breast or elsewhere
Difficulty in digesting or swallowing food
Changes in a wart or mole
Cough or hoarseness that will not go away

Surgery to remove cancerous tissue is one treatment for cancer. Radiation with X rays may be used to kill cancer cells. In **chemotherapy** (kee moh THER uh pee), chemicals are used to kill cancer cells.

What can you do to help prevent cancer?

Knowing the causes of cancer can help you prevent it. One way to help prevent cancer is to follow a healthy lifestyle. Avoiding tobacco and alcohol products can help prevent mouth and lung cancers. Eating a healthy diet that is low in fats, salt, and sugar can help prevent cancer. Using sunscreen and limiting the amount of time you spend in the sunlight are ways to prevent skin cancer. Avoid harmful home and garden chemicals. If you choose to use them, read all the labels and carefully follow the directions for their use.

Picture This

5. **Conclude** What should a person do if they notice one of these early warning signs?

6. **Identify** one thing that you need to avoid or that you need to start doing to help prevent cancer.

After You Read

Mini Glossary

allergen: a substance that causes an allergic response

allergy: an overly strong reaction of the immune system to a foreign substance

chemotherapy (kee moh THER uh pee): the use of chemicals to destroy cancer cells

noninfectious disease: a disease not caused by pathogens

1. Review the terms and their definitions in the Mini Glossary. Write a sentence that describes an allergy that you have or that someone you know has.

____

____

____

2. Fill in the table below to identify the causes of some noninfectious diseases.

Noninfectious Diseases	Causes
Asthma	
Diabetes Type I	
Diabetes Type II	
Lung cancer	
Skin cancer	

3. How did reviewing the main ideas help you study this section?

____

____

____

Visit **life.msscience.com** to access your textbook, interactive games, and projects to help you learn more about noninfectious diseases.

Interactions of Life

section 1 Living Earth

Before You Read

On the lines below, list the living things that are part of your neighborhood.

___

___

___

What You'll Learn

- places where life is found on Earth
- what ecology is
- how the environment influences life

Read to Learn

Study Coach

Two-Column Notes Organize notes into two columns. On the left, list a main idea about the material in each subheading. On the right, list the details that support the main idea.

The Biosphere

Earth has many living organisms. The part of Earth that supports life is the **biosphere** (BI uh sfihr). The biosphere includes the top part of Earth's crust, the waters that cover Earth's surface, and the atmosphere that surrounds Earth.

The biosphere is made up of different environments. Different kinds of organisms live in each environment. For example, a desert environment gets little rain. Organisms that live in a desert environment include cactus plants, coyotes, and lizards. Tropical rain forest environments get a lot of rain and warm weather. Parrots, monkeys, and tens of thousands of other organisms live in tropical rain forests. Arctic regions near the north pole are covered with ice and snow. Polar bears and walruses are two organisms that live in an arctic environment.

Why is life on Earth possible?

In our solar system, Earth is the third planet from the Sun. The amount of energy that reaches Earth from the Sun helps make the temperature just right for life. Other planets are either too close or too far from the Sun to have the right conditions for life. ☑

Reading Check

1. **Describe** how the energy from the Sun helps make life on Earth possible.

___

___

___

Ecosystems

An **ecosystem** is all the organisms living in an area and the nonliving parts of that environment. In a prairie ecosystem, the living organisms include bison, grass, and birds. Water, sunlight, and soil are nonliving parts of the ecosystem. **Ecology** is the study of interactions that occur among organisms and their environments. Scientists who study these interactions are ecologists.

FOLDABLES™

A Organize Make a two-tab concept map Foldable, as shown below. List facts about the living factors and the nonliving factors that help an organism survive in its habitat.

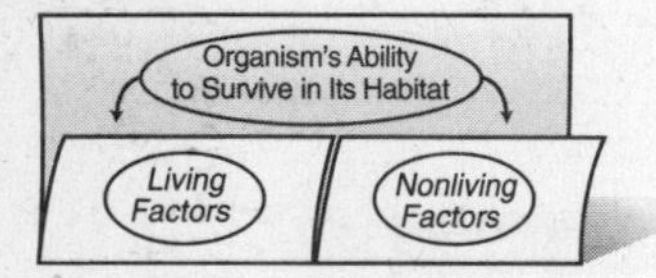

Populations

A **population** is all organisms of the same species that live in an area at the same time. For example, all the bison in a prairie ecosystem make up one population.

Ecologists often study how populations in an ecosystem interact. For example, they might study a prairie ecosystem. How does grazing by bison affect prairie grasses and the insects that live in the grass? By studying the interactions of organisms in a place, ecologists are studying a community. A **community** is all the populations of all species living in an ecosystem, as shown in the figure below.

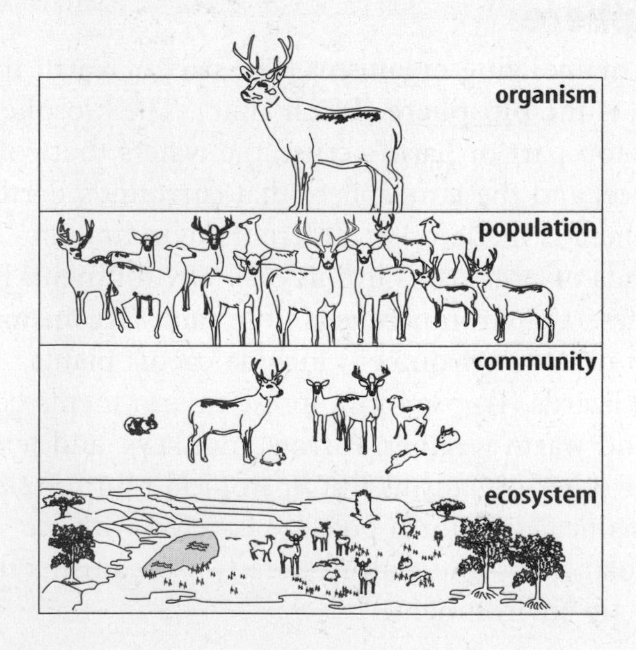

Picture This

2. **Identify** the two different populations shown in the community in the figure.

Habitats

The place in which an organism lives is called its **habitat.** In a forest ecosystem, trees are the habitat of the woodpecker. The forest floor is the habitat of the salamander. An organism's habitat provides the food, shelter, temperature, and the amount of moisture the organism needs to survive.

After You Read

Mini Glossary

biosphere (BI uh sfihr): the part of Earth that supports life

community: the populations of all species living in an ecosystem

ecology: the study of interactions that occur among organisms and their environments

ecosystem: all the organisms living in an area and the nonliving parts of that environment

habitat: the place in which an organism lives

population: all organisms of the same species that live in an area at the same time

1. Review the terms and their definitions in the Mini Glossary. Write a sentence that explains how a community is different from an ecosystem.

____

____

____

2. Complete the illustration below to help you understand how scientists organize the living organisms on Earth.

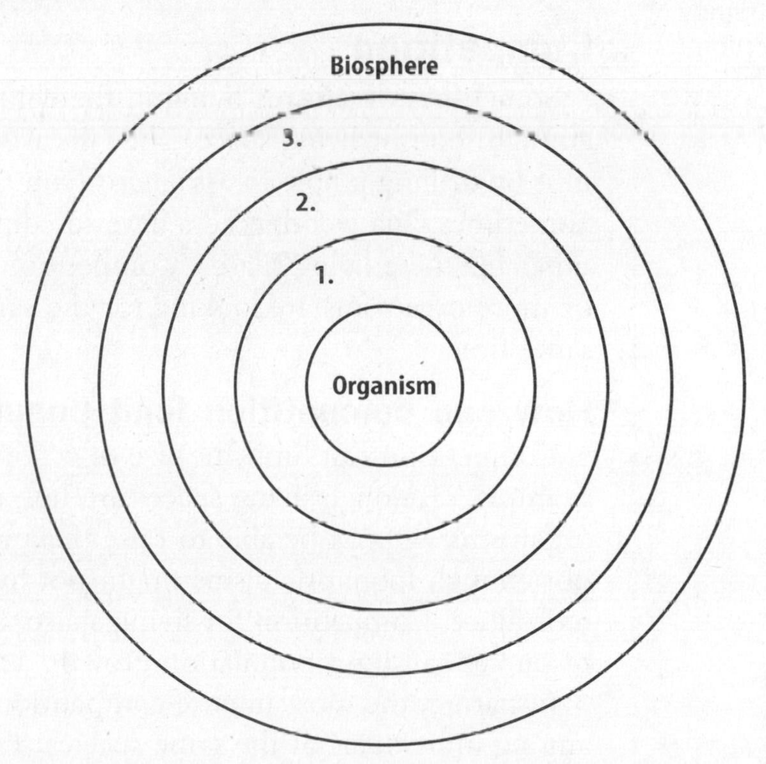

Visit **life.msscience.com** to access your textbook, interactive games, and projects to help you learn more about the living Earth.

chapter 24

Interactions of Life

section 2 Populations

What You'll Learn

- how population sizes are estimated
- how competition limits population growth
- what factors influence changes in population size

Before You Read

Is the human population in your area getting larger or smaller? What is causing the increase or decrease?

____

____

____

Study Coach

Create a Quiz As you study the information in this section, create questions about the information you read. The questions can be used to review the section's content.

Read to Learn

Competition

Sometimes organisms living in the wild do not have enough food or living space. The Gila woodpecker makes its nest by drilling a hole in a saguaro (suh GWAR oh) cactus. Sometimes Gila woodpeckers have to compete with each other for these living spaces. Competition occurs when two or more organisms are looking for the same resource at the same time.

How can competition limit population growth?

Competition can limit the size of a population. For example, if enough living spaces are not available, some organisms will not be able to raise their young. If there is not enough food, organisms might not live long enough to reproduce. Competition for living space, food, and other resources can limit population growth. ☑

In nature, the most intense competition usually occurs among individuals of the same species. This is because they need the same kinds of food and shelter. Competition also takes place among different species. For example, after a Gila woodpecker has moved from its nest, owls, snakes, and lizards might compete for the empty hole.

☑ Reading Check

1. Explain What do organisms compete for?

____

____

____

Population Size

Ecologists often need to measure the size of a population to find out whether or not the population is healthy and growing. Measuring the size of the population can help ecologists know if a population is in danger of disappearing. One measurement ecologists use is population density. Population density is the number of individuals of one species in a specific area. ☑

Reading Check

2. **Explain** Why do ecologists measure the size of a population?

___

___

___

How are populations measured?

Imagine having to count all the crickets in an area. They look alike, move a lot, and hide. You might count a cricket more than once. Or you might miss other crickets completely. One method ecologists use to count populations is called trap-mark-release. When ecologists want to count wild rabbits, for example, they set traps that catch the rabbits without hurting them. Each captured rabbit is then marked and let go. Later, another set of rabbits is caught. Some of these rabbits will have marks, but others will not. The ecologists compare the number of marked and unmarked rabbits in the second sample. By doing this, they can estimate the size of the rabbit population.

How are sample counts used?

To estimate the size of large populations, ecologists use sample counts. For example, pretend you wanted to estimate the number of rabbits in an area of 100 acres. You might count the rabbits in one acre and then multiply by 100.

How does a limiting factor affect population?

In an ecosystem, food, water, space, and other resources are limited. A **limiting factor** is anything that restricts the number of individuals in a population.

A limiting factor can affect more than one population. For example, when the plants in a meadow do not get enough rain, fewer plants survive. Because there are fewer plants, fewer seeds are produced. The seeds are a source of food for the seed-eating mice that live in the meadow. The smaller food supply could become a limiting factor for mice. In turn, a smaller mouse population could be a limiting factor for the hawks and owls that eat the mice. Limiting factors include living and nonliving parts in a community of an ecosystem.

FOLDABLES™

B Describe Make a three-tab Foldable, as shown below. Use the Foldable to describe factors that affect population size.

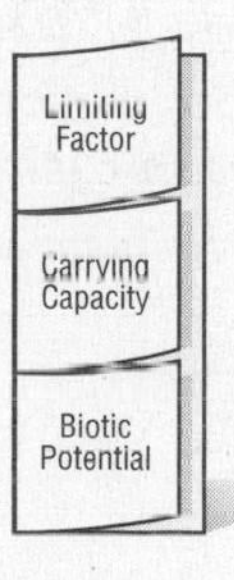

How does carrying capacity affect population?

The largest number of individuals of one species that an ecosystem can support over time is the **carrying capacity.** For example, if the number of robins living in a park increases, nesting space might become difficult to find. Available nesting space limits the robin population. If the population gets larger than its carrying capacity, some individuals of a species will not have enough resources. They could die or have to move somewhere else. ☑

Reading Check

3. **Explain** How is a population affected when it goes beyond the carrying capacity of the ecosystem?

____

____

____

What is biotic potential?

If a population had an unlimited supply of food, water, and living space, and was not limited by disease, predators, or competition with other species, the population would continue to grow. The highest rate of reproduction under ideal conditions is a population's biotic potential. The more offspring organisms produce, the higher the species' biotic potential. Tangerines have a higher biotic potential than avocados because tangerines have many seeds in each fruit, while an avocado has only one seed in each fruit.

Changes in Populations

A population's birthrate and death rate also influence the size of the population and its rate of growth. A population gets larger when the number of individuals born is greater than the number of individuals that die. A population gets smaller when the number of deaths is greater than the number of births. As the table below shows, countries with a faster population growth have birthrates much higher than death rates. Countries with a slower population growth have only slightly higher birthrates than death rates.

Picture This

4. **Identify** Highlight the country listed in the table that has the fastest growing population. Circle the country that has a population that is getting smaller.

Population Growth			
	Birthrate*	Death Rate*	Population Increase
Countries with Rapid Growth			
Jordan	38.8	5.5	3.3%
Uganda	50.8	21.8	2.9%
Zimbabwe	34.3	9.4	5.2%
Countries with Slow Growth			
Germany	9.4	10.8	-1.5%
Sweden	10.8	10.6	0.1%
United States	14.8	8.8	0.6%

*Number per 1,000 people

How does moving around affect population size?

When animals move from place to place, the movements can affect population size. For example, a male animal may move many miles to find a mate. After he finds a mate, their offspring might start a completely new population, far from the male's original population.

Plants and microscopic organisms also can move from place to place. The seeds of dandelions, for example, have special parts that allow them to be carried to other places by the wind. Many kinds of seeds can be moved to new places by water currents. Animals also spread seeds from place to place. ☑

What is exponential growth?

When a species moves to a new area that has plenty of food, living space, and other resources, the population can grow quickly. This pattern of growth is called exponential growth. Exponential growth means that the larger a population gets, the faster it grows. Over time, the population will reach the carrying capacity of the ecosystem for that species.

The figure below shows the exponential growth of the human population. By the year 2050, the population could reach 9 billion people.

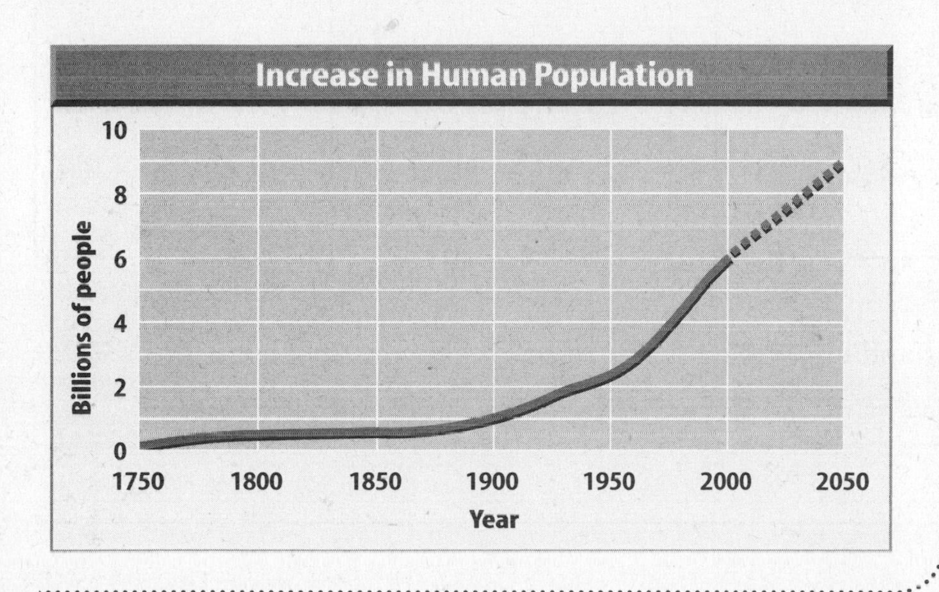

Reading Check

5. **Identify** three things that move plant seeds from place to place.

____

____

____

Picture This

6. **Estimate** Use the graph to estimate the increase in the human population from 1950 to 2000.

____

After You Read

Mini Glossary

carrying capacity: the largest number of individuals of one species that an ecosystem can support over time

limiting factor: anything that restricts the number of individuals in a population

1. Review the terms and their definitions in the Mini Glossary. Choose one of the terms and explain how it can affect the population size of a species.

2. Complete the diagram below to help you describe the things that affect changes in population size.

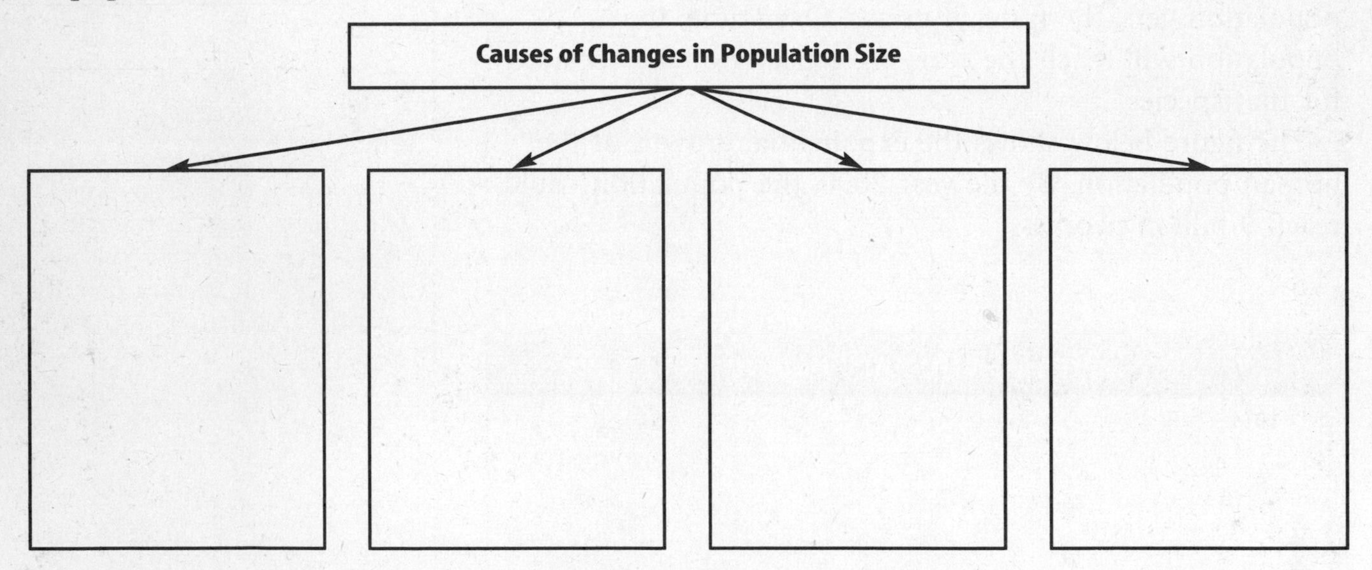

3. How do the quiz questions and answers help you review what you have learned?

Visit **life.msscience.com** to access your textbook, interactive games, and projects to help you learn more about populations.

Interactions of Life

section ❸ Interactions Within Communities

Before You Read

How do you get the energy you need to do the things you want to do?

____

____

____

What You'll Learn

- how organisms get energy for life
- how organisms interact
- that every organism occupies a niche

Read to Learn

Obtaining Energy

Living organisms need a constant supply of energy. The Sun provides the energy for most of life on Earth. Some organisms use this energy to make energy-rich molecules through photosynthesis. The energy-rich molecules are food for the organism. They are made up of different combinations of carbon, hydrogen, and oxygen atoms. Chemical bonds hold the atoms of these molecules together. Energy is stored in the chemical bonds. During digestion, the molecules break apart and release energy. The organism uses the energy to grow, develop, and stay alive.

What are producers?

Organisms that use an outside energy source like the Sun to make energy-rich molecules are called **producers**. Most producers have chlorophyll (KLOR uh fihl). Chlorophyll is needed for photosynthesis. Green plants are producers.

Not all producers have chlorophyll or use energy from the Sun. Some use chemosynthesis (kee moh SIHN thuh sus) to make energy-rich molecules. These organisms live near volcanic vents on the ocean floor. Inorganic molecules in the water provide the energy for chemosynthesis.

Mark the Text

Defining Important Words Read the headings in this section and circle any word you cannot define. At the end of each section, review the circled words and underline the part of the text that helps you define the words.

FOLDABLES™

Ⓒ Organize Make Foldables, as shown below, to organize information as you read. Take notes to define the two types of living organisms.

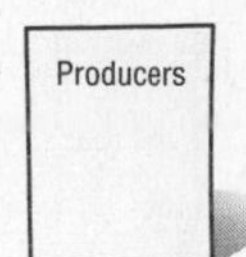

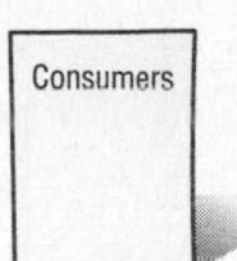

Think it Over

1. **Draw Conclusions** To which consumer group do humans belong? Why?

____

____

What are consumers?

Organisms that get energy by eating other organisms are called **consumers**. There are four kinds of consumers. Herbivores, such as rabbits, eat plants. Carnivores, such as frogs, eat other animals. Omnivores, such as pigs, eat both plants and animals. Decomposers, such as earthworms, consume wastes and dead organisms. Decomposers help recycle once-living matter.

What are food chains?

Ecology includes the study of how organisms depend on each other for food. A food chain is a simple model of the feeding relationships and energy flow in an ecosystem. For example, shrubs are food for deer. Deer are food for mountain lions, as shown in the figure below.

Picture This

2. **Identify** Beneath the figure, write the food chain in words, using arrows to show relationships.

______________ ______________ ______________

Symbiotic Relationships

Organisms may share food and other resources. Any close relationship between species is called **symbiosis**.

What is mutualism?

A symbiotic relationship in which both species benefit is called **mutualism** (MYEW chuh wuh lih zum). Ants and acacia trees illustrate mutualism. The ants protect the tree by attacking any animal that tries to feed on it. The tree provides food and a home for ants.

What is commensalism?

A symbiotic relationship in which one organism benefits and the other one is not affected is called **commensalism** (kuh MEN suh lih zum). For example, a sea anemone has tentacles that have a mild poison. The clown fish is not harmed by the poison. It swims among the tentacles and is protected from predators. The clown fish benefits, but the sea anemone is not helped or hurt.

FOLDABLES

D Contrast Make a three-tab Foldable, as shown below. Use the Foldable to describe the three kinds of symbiosis.

What is parasitism?

A symbiotic relationship in which one organism benefits and one is harmed is called **parasitism** (PER uh suh tih zum). An example of this relationship is a pet dog and roundworms. A roundworm sometimes attaches itself to the inside of the dog's intestine. It feeds on the nutrients in the dog's blood. The dog may have abdominal pain and diarrhea. Sometimes the dog may die. In this relationship, the roundworm benefits, but the dog is harmed.

Niches

Hundreds of species might live in one habitat. For example, a rotting log is home to many species. Spiders, ants, termites, and worms are some species that live on or under the rotting log. Although many species use the log as their habitat, the species do not compete for resources. This is because each species needs different things to survive. So, each species has its own niche (NICH). An organism's **niche** is its role in its environment—how it obtains food and shelter, finds a mate, cares for its young, and avoids danger. ☑

Special adaptations that improve survival are often part of an organism's niche. For example, a poison in milkweed plants stops many insects from eating them. Monarch butterfly caterpillars have an adaptation that lets them eat milkweed. When they eat milkweed, the caterpillars become slightly poisonous. Birds avoid eating these caterpillars because they know that the caterpillars and adult butterflies have an awful taste and can make them sick.

How do predator and prey fit in a niche?

An organism's niche includes how it avoids being eaten and how it gets its food. Predators are consumers that capture and eat other consumers. The prey is the organism that is captured by the predator. Having predators in an ecosystem usually increases the number of species that can live in the ecosystem. Predators limit the size of the prey population. So, food and other resources are less likely to become difficult to find. Competition between species is reduced. ☑

How do species in a niche cooperate?

Individual organisms often cooperate, or work together, in ways that improve survival. For example, a white-tailed deer that detects the presence of a wolf will warn other deer in the herd. These cooperative actions are part of the species' niche.

Reading Check

3. **Explain** What is a niche?

____

____

____

Reading Check

4. **Explain** How do predators increase the number of different species that can live in an ecosystem?

____

____

____

● After You Read

Mini Glossary

commensalism (kuh MEN suh lih zum): a symbiotic relationship in which one organism benefits and the other is not affected

consumer: an organism that gets energy by eating other organisms

mutualism (MYEW chuh wuh lih zum): a symbiotic relationship in which both species benefit

niche (NICH): an organism's role in its environment

parasitism (PER uh suh tih zum): a symbiotic relationship in which one organism benefits but the other is harmed

producer: an organism that uses an outside energy source like the Sun to make energy-rich molecules

symbiosis: any close relationship between species

1. Review the terms and their definitions in the Mini Glossary. Write a sentence that explains the difference between consumers and producers.

____

____

____

2. Choose one of the question headings in the Read to Learn section. Write the question in the space below. Then write your answer to that question on the lines that follow.

Write your question here.

____

____

____

____

Visit **life.msscience.com** to access your textbook, interactive games, and projects to help you learn more about interactions within communities.

The Nonliving Environment

section 1 Abiotic Factors

Before You Read

How would you describe the climate where you live? How does it affect the plant and animal life around you?

____

____

____

What You'll Learn

- the common abiotic factors in most ecosystems
- the components of air that are needed for life
- how climate influences life in an ecosystem

Read to Learn

Mark the Text

Summarize Write a phrase beside each main heading that summarizes the main point of the section.

Environmental Factors

Living things depend on one another for food and shelter. The features of the environment that are alive, or were once alive, are called **biotic** (bi AH tihk) factors.

Biotic factors are not the only things needed for life. Plants and animals cannot survive without the nonliving environment. The nonliving, physical features of the environment are called **abiotic** (ay bi AH tihk) factors. Abiotic factors include air, water, sunlight, soil, temperature, and climate. These factors often determine the kinds of organisms that live there.

Air

The air that surrounds Earth is called the **atmosphere**. Air is made up of 78 percent nitrogen, 21 percent oxygen, 0.94 percent argon, 0.03 percent carbon dioxide, and trace amounts of other gases. Some of these gases are important in supporting life.

Carbon dioxide (CO_2) is necessary for photosynthesis. Photosynthesis uses CO_2, water, and energy from sunlight to make sugar molecules. Organisms such as plants use photosynthesis to produce their own food. ☑

Reading Check

1. **List** What three things are needed for photosynthesis?

____

____

____

Respiration Oxygen is released into the atmosphere during photosynthesis. Cells use oxygen to release the chemical energy stored in sugar molecules. This process, called respiration, provides cells with the energy needed for all life processes.

Water

Water is necessary to life on Earth. It is a major part of the fluid inside the cells of all organisms. Most organisms are 50 percent to 95 percent water. Processes such as respiration, digestion, and photosynthesis occur only if water is present. Environments that have plenty of water usually have a greater variety of and a larger number of organisms than environments that have little water.

2. Recognizing Cause and Effect How does soil affect plant life in an area?

____

____

____

Soil

Soil is a mixture of mineral and rock particles, the remains of dead organisms, water, and air. Soil is the top layer of Earth's crust where plants grow. It is formed partly of rock that has been broken down into tiny particles.

Soil is considered an abiotic factor because most of it is made up of nonliving rock and mineral particles. But soil also contains living organisms and the remains of dead organisms. The decaying matter in soil is called humus. Soils contain different combinations of sand, clay, and humus. The kind of soil in a region affects the kinds of plant life that grow there.

Sunlight

Sunlight is the energy source for almost all life on Earth. Plants and other organisms that use photosynthesis are called producers. They use light energy from the Sun to produce their own food. Organisms that cannot make their own food are called consumers. Energy is passed to consumers when they eat producers or other consumers.

3. Identify What are two types of energy the Sun provides?

____

____

Temperature

Sunlight provides the light energy for photosynthesis and the heat energy for warmth. Most organisms can live only if their body temperatures are between the freezing point of water, 0°C, and 50°C. The temperature of a region depends partly on the amount of sunlight it gets. The amount of sunlight depends on the area's latitude and elevation. ☑

How does latitude affect temperature?

The temperature of a region is affected by its latitude. Places farther from the equator generally have colder temperatures than places at latitudes nearer to the equator. Look at the figure below. Near the equator, sunlight directly hits Earth. Sunlight hits Earth at an angle near the poles. This spreads the energy over a larger area.

Picture This

4. Identify Use one color to highlight the sunlight directly hitting Earth at the equator. Use another color to highlight the sunlight hitting Earth at the poles.

How does elevation affect temperature?

A region's elevation, or distance above sea level, affects its temperature. Earth's atmosphere traps the Sun's heat. At higher elevations, the atmosphere is thinner than at lower elevations. Air becomes warmer when sunlight heats the air molecules. Because there are fewer air molecules at higher elevations, the air temperature at higher elevations tends to be cooler. ☑

Trees at higher elevations are usually shorter. The timberline is the elevation above which trees do not grow. Only low-growing plants exist above the timberline. The tops of some mountains are so cold that no plants grow there.

Reading Check

5. Explain Why is the air temperature at higher elevations usually cooler than the air temperature at lower elevations?

____

____

Climate

In Fairbanks, Alaska, winter temperatures may be as low as −52°C. More than one meter of snow might fall in one month. In Key West, Florida, winter temperatures rarely go below 5°C. Snow never falls. These two cities have different climates. The **climate** of an area is its average weather conditions over time. Climate includes temperature, rainfall or other precipitation, and wind.

6. Determine What are the most important parts of climate for most living things?

____

____

How does climate affect life in an area?

Temperature and precipitation are the two most important parts of climate for most living things. They affect the kinds of organisms that live in an area. For example, an area that has an average temperature of 25°C and gets less than 25 cm of rain per year probably has cactus plants growing there. An area with the same average temperature and more than 300 cm of rain every year is probably a tropical rain forest. ☑

How are winds created?

In addition to affecting the temperature of an area, the heat energy from the Sun causes wind. Air is made up of gas molecules. As the temperature increases, the molecules spread farther apart. So, warm air is lighter than cold air. Colder air sinks below warmer air and pushes it upward. This movement creates air currents that are called wind.

What is the rain shadow effect?

Mountains can affect rainfall patterns. As the figure below shows, moist air is carried toward land by the wind. The wind is forced upward by the slope of the mountain. As the air moves to the top, it cools. When air cools, the moisture in it falls as rain or snow. By the time the air crosses over the top of the mountain, it has lost most of its moisture. The drier air warms as it flows down the mountain. The other side of the mountain is in a rain shadow and receives much less precipitation. As a result, one side of the mountain could be covered with forests, while the other side is a desert.

Picture This

7. Explain On the figure below, label the first and fourth arrows to complete the explanation of the rain shadow effect.

After You Read

Mini Glossary

abiotic (ay bi AH tihk): nonliving, physical features of the environment, including air, water, sunlight, soil, temperature, and climate

atmosphere: the air that surrounds Earth

biotic (bi AH tihk): features of the environment that are alive or were once alive

climate: an area's average weather conditions over time, including temperature, rainfall or other precipitation, and wind

soil: a mixture of mineral and rock particles, the remains of dead organisms, water, and air

1. Review the terms and their definitions in the Mini Glossary. Write a sentence that explains the difference between abiotic and biotic factors.

____

____

____

2. Complete the chart below to identify a way that each abiotic factor is important to life.

Abiotic Factor	Importance to Life
Air	
Water	
Soil	
Sunlight	
Temperature	
Climate	

ScienceOnline Visit **life.msscience.com** to access your textbook, interactive games, and projects to help you learn more about abiotic factors.

End of Section

chapter 25 The Nonliving Environment

section 2 Cycles in Nature

What You'll Learn

- why Earth's water cycle is important
- about the carbon cycle
- how nitrogen affects life on Earth

Before You Read

What happens when you boil water in a covered pot? What do you see on the lid of the pot when you remove it?

____

____

Study Coach

Outline As you read, make an outline to summarize the information in the section. Use the main headings in the section as the main headings in the outline. Complete the outline with the information under each heading in the section.

Read to Learn

The Cycles of Matter

Imagine an aquarium with water, fish, snails, plants, algae, and bacteria. The tank is sealed so that only light can enter. How can the organisms survive without adding food, water, and air? The plants and algae produce their own food through photosynthesis. They also supply oxygen to the tank. The fish and snails eat the plants and algae and take in the oxygen. The wastes from the fish and snails fertilize the plants and algae. Bacteria decompose those organisms that die. The organisms in this closed environment can survive because the materials are recycled.

The environment in the aquarium is similar to Earth's biosphere. Earth only has a certain amount of water, carbon, nitrogen, oxygen, and other materials needed for life. These materials are constantly being recycled.

Picture This

1. **Explain** to a partner how the fish in the tank survive without anyone adding food, water, and air.

The Water Cycle

When you leave a glass of water on a sunny windowsill, the water evaporates. **Evaporation** takes place when liquid water changes into a gas, called water vapor, and enters the atmosphere. Water evaporates from the surfaces of lakes, streams, and oceans. It enters the atmosphere from plants in a process known as transpiration (trans puh RAY shun). Animals release water vapor as they exhale. Water is returned to the environment from animal wastes.

What Is condensation?

After water vapor enters the atmosphere, eventually it will come into contact with colder air. The temperature of the water vapor drops. Over time, the water vapor becomes cool enough to change back into liquid water. The process of changing from a gas to a liquid is called **condensation**.

The water vapor condenses on particles of dust in the air and forms tiny droplets. The droplets join together to form clouds. When the droplets become large and heavy enough, they fall to the ground as rain or other precipitation.

As the figure below shows, the **water cycle** is a model that describes how water moves from the surface of Earth to the atmosphere and back to the surface again.

FOLDABLES™

A Describe Make a three-tab book Foldable, as shown below. Use the Foldable to describe the water, carbon, and nitrogen cycles.

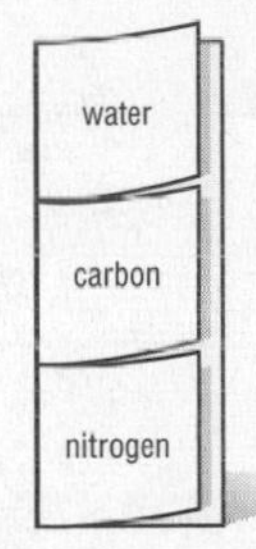

Picture This

2. **Identify** Complete the figure by labeling the missing steps in the water cycle.

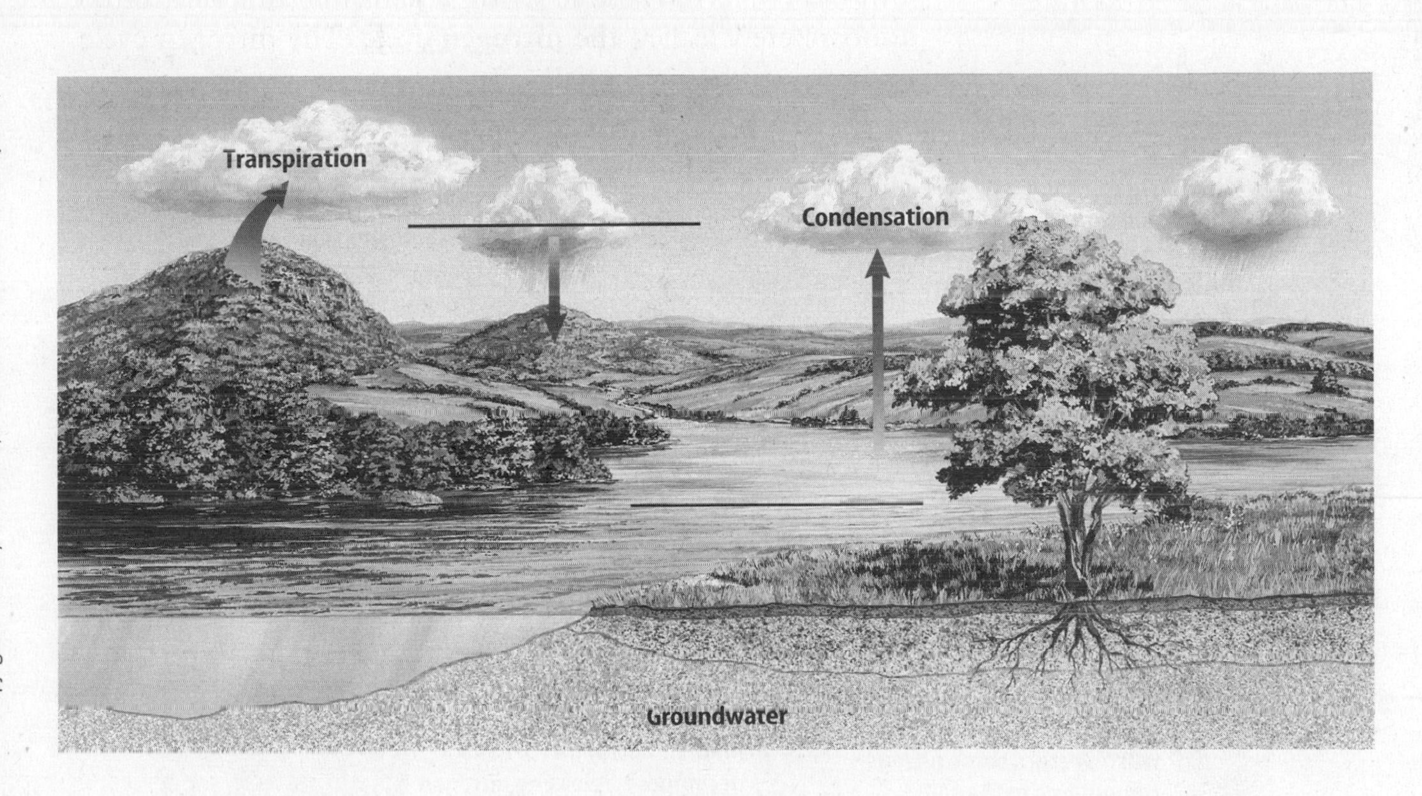

Think it Over

3. Analyze List some of the ways you use water.

____

____

____

Picture This

4. Discuss What is one role animals play in the nitrogen cycle?

____

____

____

How do humans affect the water cycle?

Humans take water from reservoirs, rivers, and lakes to use in their homes, businesses, and farms. Using this water can reduce the amount of water that evaporates into the atmosphere. Humans also influence how much water returns to the atmosphere by limiting the amount of water available to plants and animals.

The Nitrogen Cycle

Nitrogen is important to all living things. It is a necessary part of proteins. Proteins are needed for the life processes that take place in the cells of all organisms. Nitrogen is the most plentiful gas in the atmosphere. However, most organisms cannot use nitrogen directly from the air.

Plants need nitrogen that has been combined with other elements to form nitrogen compounds. Through a process called **nitrogen fixation**, some types of soil bacteria form the nitrogen compounds that plants need. Plants take in these nitrogen compounds through their roots. Animals get the nitrogen they need by eating plants or other animals. When dead organisms decay, the nitrogen in their bodies returns to the soil or the atmosphere. This transfer of nitrogen from the atmosphere to the soil, to living organisms, and back to the atmosphere is called the **nitrogen cycle**. The nitrogen cycle is shown in the figure below.

Nitrogen gas is changed into usable compounds by lightning or by nitrogen-fixing bacteria that live on the roots of certain plants.
Plants use nitrogen compounds to build cells.
Animals eat plants. Animal wastes return some nitrogen compounds to the soil.
Animals and plants die and decompose, releasing nitrogen compounds back into the soil.

How do human activities affect soil nitrogen?

Humans can affect the part of the nitrogen cycle that takes place in the soil. After crops are harvested, farmers often remove the rest of the plant material. The plants are not left in the field to decay and return their nitrogen compounds to the soil. If the nitrogen compounds are not replaced, the soil could become infertile. Fertilizers can be used to replace soil nitrogen. Compost and animal manure also contain nitrogen compounds that plants can use. They can be added to soil to make it more fertile. ☑

Another way to replace soil nitrogen is by growing nitrogen-fixing crops. Most nitrogen-fixing bacteria live on or in the roots of certain plants. Some plants, such as peas, have roots with nodules that contain nitrogen-fixing bacteria. These bacteria supply nitrogen compounds to the plants and add nitrogen compounds to the soil.

Reading Check

5. **Identify** two ways to add nitrogen to soil.

____

____

____

The Carbon Cycle

Carbon atoms are found in the molecules of living organisms. Carbon is part of soil humus and is found in the atmosphere as carbon dioxide gas (CO_2). The **carbon cycle** describes how carbon molecules move between the living and nonliving world.

The cycle begins when producers take CO_2 from the air during photosynthesis. They use CO_2, water, and sunlight to make energy-rich sugar molecules. Energy is released from these molecules during respiration—the chemical process that provides energy for cells. Respiration uses oxygen and releases CO_2. Photosynthesis uses CO_2 and releases oxygen. The two processes help recycle carbon on Earth. ☑

Human activities also release CO_2 into the atmosphere. For example, when fossil fuels are burned, CO_2 is released into the atmosphere as a waste product. People also use wood for building and for fuel. Trees that are cut down for these purposes cannot remove CO_2 from the atmosphere during photosynthesis. The amount of CO_2 in the atmosphere is increasing. The extra CO_2 could trap more heat from the Sun and cause average temperatures on Earth to rise.

Reading Check

6. **Explain** What two processes recycle carbon on Earth?

____

____

____

● After You Read

Mini Glossary

carbon cycle: a model that describes how carbon molecules move between the living and nonliving world

condensation: process that occurs when a gas changes to a liquid

evaporation: process that occurs when liquid water changes into water vapor and enters the atmosphere

nitrogen cycle: the transfer of nitrogen from the atmosphere to the soil, to living organisms, and back to the atmosphere

nitrogen fixation: process in which some types of soil bacteria form the nitrogen compounds that plants need

water cycle: a model that describes how water moves from the surface of Earth to the atmosphere and back to the surface again

1. Review the terms and their definitions in the Mini Glossary. Write a sentence that explains the difference between condensation and evaporation.

____

____

____

2. In the chart, list the steps in the nitrogen cycle.

Steps in the Nitrogen Cycle
1. ______________________________
____
2. ______________________________
____
3. ______________________________
____
4. ______________________________
____

Visit **life.msscience.com** to access your textbook, interactive games, and projects to help you learn more about the cycles in nature.

The Nonliving Environment

section 3 Energy Flow

Before You Read

Why do you need energy? What is your source of energy?

____

____

____

____

What You'll Learn

- how organisms make energy-rich compounds
- how energy flows through ecosystems
- how much energy is available at different levels in a food chain

Read to Learn

Converting Energy

All living things are made up of matter, and all living things need energy. Matter can be recycled over and over. Energy is not recycled, but it is converted from one form to another. This conversion is important to all life on Earth.

Mark the Text

Locate Information Read all the headings for this section and circle any word you cannot define. At the end of each section, review the circled words and underline the part of the text that helps you define the words.

How is energy converted during photosynthesis?

During photosynthesis, producers convert light energy into the chemical energy in sugar molecules. Some of these sugar molecules are broken down as energy. Some are used to build complex carbohydrate molecules that become part of the producer's body. Fats and proteins also contain stored energy.

What are hydrothermal vents?

Some producers do not rely on light for energy. These producers live deep underwater in total darkness. They live near powerful hydrothermal vents. Hydrothermal vents are deep cracks in the ocean floor. The water from these vents is very hot from contact with molten rock deep in the Earth's crust.

FOLDABLES™

B Compare Make a two-tab Foldable, as shown below, to compare how producers use photosynthesis and chemosynthesis to convert energy.

Photosynthesis | Chemosynthesis

What is chemosynthesis?

Because sunlight does not reach deep ocean regions, the organisms that live there cannot get energy from sunlight. Scientists have learned that the hot water has nutrients that bacteria use to make their own food. The production of energy-rich nutrient molecules from chemicals is called **chemosynthesis** (kee moh SIN thuh sus). Consumers that live in hydrothermal vent communities rely on chemosynthetic bacteria for nutrients and energy. ☑

Reading Check

1. **Explain** What type of organisms use chemosynthesis?

____

____

____

Energy Transfer

Energy can be converted from one form to another. It also can be transferred from one organism to another. Consumers cannot make their own food. Instead, they obtain energy by eating producers or other consumers. The energy that is stored in the molecules of one organism is transferred to another organism. That organism can release the energy stored in the food. It can use the energy for growth, or it can transform the energy into heat. At the same time, the matter that makes up those molecules is transferred from one organism to another. Throughout nature, energy and matter are transferred from organism to organism.

How does energy flow in food chains?

The food chain in the figure below shows how matter and energy pass from one organism to another. Producers, such as plants, are the first step in a food chain. All producers make their own food using either photosynthesis or chemosynthesis. Animals, such as herbivores, that eat producers are the second step. Animals that eat other consumers are the third and higher steps of food chains.

Picture This

2. **Identify** Write Producer or Consumer below each organism on the food chain.

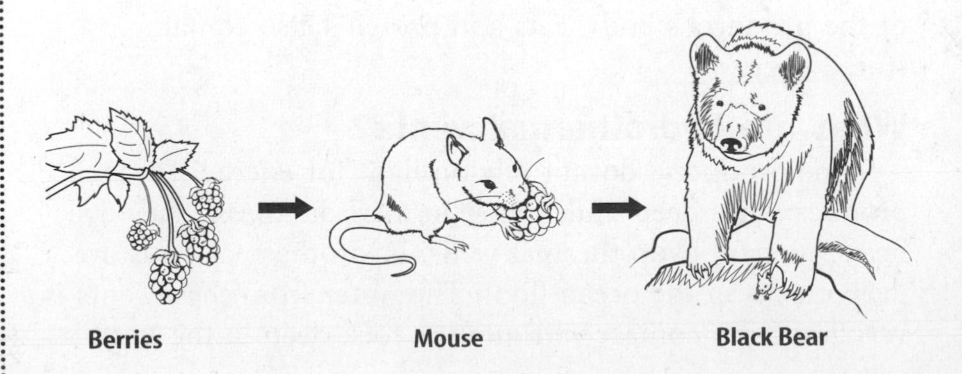

______________ ______________ ______________

What are food webs?

There are many feeding relationships in a forest community. For example, bears eat berries, insects, and fish. Berries are eaten by many different organisms. A **food web** is a model that shows all the possible feeding relationships among the organisms in a community. A food web is made up of many different food chains.

Energy Pyramids

Most food chains have three to five links. The number of links is limited because the amount of available energy is reduced as you move from one level to the next.

How does available energy decline?

When a mouse eats seeds, energy stored in the seeds transfers to the mouse. But most of the energy the plant took in from the Sun was used to help the plant grow. The mouse uses energy from the seed for its own processes, such as digestion and growth. Some of the energy is given off as heat. A hawk that eats the mouse gets even less energy. The amount of available energy is reduced from one level of a food chain to another.

An **energy pyramid** shows the amount of energy available at each feeding level in an ecosystem. The bottom of the pyramid below includes all producers. It is the first and largest level because it contains the most energy and the largest number of organisms. As the energy is reduced from one level to another, each level becomes smaller. In fact, only about 10 percent of the energy available at each feeding level is transferred to the next higher level.

FOLDABLES™

C Identify Make a pyramid Foldable, as shown below, to identify the flow of energy from producers, to herbivores, to carnivores.

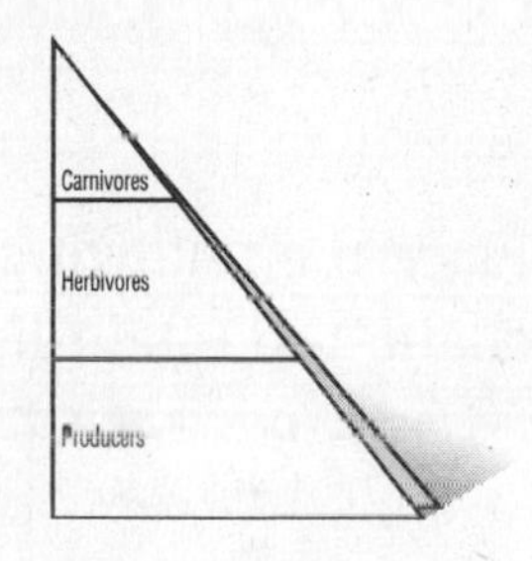

Think it Over

3. **Synthesize** Why are there more producers than consumers?

____

____

____

After You Read

Mini Glossary

chemosynthesis (kee moh SIN thuh sus): the production of energy-rich nutrient molecules from chemicals

energy pyramid: a model that shows the amount of energy available at each feeding level in an ecosystem

food web: a model that shows all the possible feeding relationships among the organisms in a community

1. Review the terms and their definitions in the Mini Glossary. Choose the term that explains how energy-rich molecules are produced and write a sentence explaining how the process works.

____

____

____

____

2. Place the following organisms in the order of steps in which they would appear in a food chain: mountain lion, plant, bird, insect.

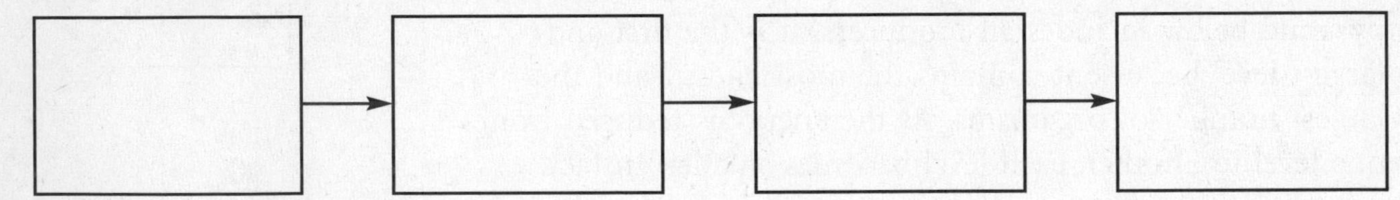

3. How did finding definitions of words you did not know help you understand energy flow?

____

____

____

____

Visit **life.msscience.com** to access your textbook, interactive games, and projects to help you learn more about energy flow.

Ecosystems

section 1 How Ecosystems Change

Before You Read

List two ways the area you live in has changed over time.

____

____

____

What You'll Learn

- how ecosystems change over time
- how new communities begin in areas
- how pioneer species and climax communities differ

Read to Learn

Study Coach

Create a Quiz After you read this section, create a quiz of five to seven questions that you think might be on a test. Be sure to answer your questions.

Ecological Succession

What would happen if the lawn at your home were never cut? The grass would get longer, and it would look like a meadow. Later, larger plants would grow from seeds brought to the area by animals or the wind. Then trees might sprout. In 20 years or less you wouldn't be able to tell that the land was once a mowed lawn.

An ecologist can tell you what type of ecosystem your lawn would become. Ecosystems are all the organisms that live in an area and the nonliving parts of that environment. **Succession** is the normal, gradual changes that occur in the types of species that live in an area. Succession occurs differently in different places around the world.

What is primary succession?

The process of succession that begins in a place where no plants grew before is called primary succession. It begins with the arrival of living things such as lichens (LI kunz). The first living things to inhabit an area are called **pioneer species**. They can survive the harsh conditions of the area, such as drought and extreme heat and cold.

FOLDABLES™

A Compare Make a Foldable as shown below to compare and contrast pioneer species and climax communities.

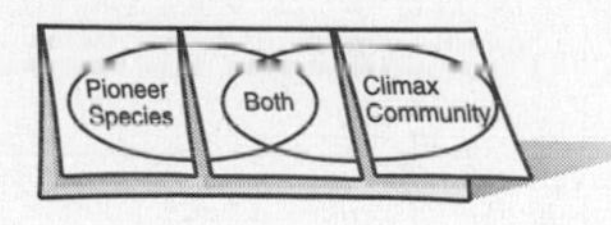

How does soil form?

Pioneer species often start the soil-building process in an area that is made up of rock. Soil begins to form as lichens and the forces of weather and erosion help to break down rocks into smaller pieces. When lichens die, they decay, adding organic matter to the rocks. Moss and ferns can grow in this new soil as shown in the photo. When these plants die, they add more organic material to the soil. Soon there is enough soil for grasses, wildflowers, and other plants to grow. When these plants die, they make the soil richer and deep enough for shrubs and trees to grow. During these changes, insects, small birds, and mammals have begun to move into the area.

David Wrobel/Visuals Unlimited

Where does secondary succession occur?

Succession that begins in a place that already has soil and was once home to living organisms is called secondary succession. Since the area already has soil, secondary succession is much faster than primary succession. The soil in an area that had a forest fire or a building torn down will not remain lifeless for long. The soil already contains seeds. Wind and birds will carry more seeds to the area. Wildlife will move in. ☑

Reading Check

1. **Describe** Where does secondary succession occur?

____

____

____

What are climax communities?

A **climax community** is a community of plants that is mostly stable and has reached the end stage of succession. New trees grow when larger, older trees die. The individual trees change, but the species does not. For example, a climax community that is a forest of beeches and maples will stay a forest of beeches and maples even though some older trees will die and new trees grow. It can take hundreds or thousands of years for a climax community to develop.

After You Read

Mini Glossary

climax community: a community of plants that is mostly stable and has reached the end stage of succession

pioneer species: the first living things to inhabit an area

succession: the normal, gradual changes that occur in the types of species that live in an area

1. Review the terms and their definitions in the Mini Glossary. Write two or three sentences that explain the difference between pioneer species and climax communities.

____

____

____

2. Fill in the blanks in the graphic organizer below to show how soil is formed.

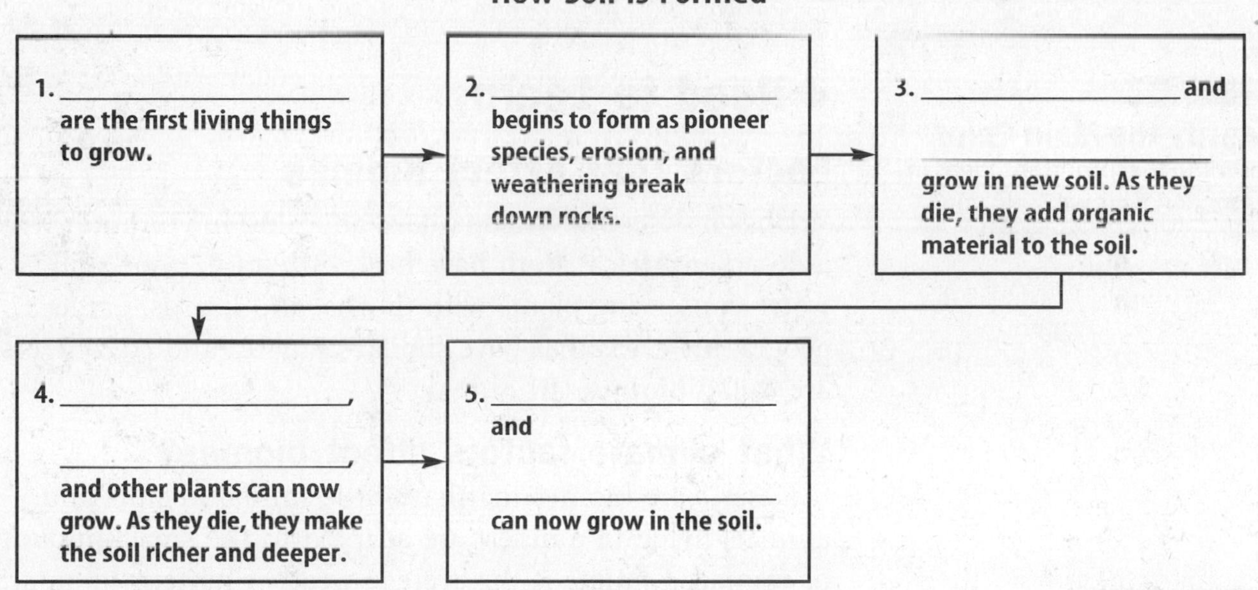

How Soil Is Formed
1. ____________ are the first living things to grow.
2. ____________ begins to form as pioneer species, erosion, and weathering break down rocks.
3. ____________ and ____________ grow in new soil. As they die, they add organic material to the soil.
4. ____________, ____________, and other plants can now grow. As they die, they make the soil richer and deeper.
5. ____________ and ____________ can now grow in the soil.

3. How does the quiz you created help you prepare for a test?

____

____

Science Online Visit **life.msscience.com** to access your textbook, interactive games, and projects to help you learn more about how ecosystems change.

Ecosystems

section 2 Biomes

What You'll Learn
- how climate affects land environments
- the seven biomes of Earth
- how organisms adapt to different biomes

Before You Read

On the lines below, describe the geographic area where you live. Include information about the climate, the landforms, and the kinds of plants and animals that live there.

____

____

____

Mark the Text

Identify the Main Point Underline the important ideas in this section. This will help you remember what you read.

Read to Learn

Factors That Affect Biomes

Does a desert in Arizona have anything in common with a desert in Africa? Both have heat, little rain, poor soil, water-conserving plants with thorns, and lizards. Large geographic areas that have similar climates and ecosystems are called **biomes** (BI ohmz).

What climate factors affect biomes?

Deserts are biomes that have little rainfall. Plants and animals living in a desert are adapted to the small amount of rainfall. Climate is the average weather pattern in an area over many years. The two most important factors of climate that affect life are temperature and precipitation.

FOLDABLES™

B Describe Make a pocket Foldable, as shown below. Use quarter sheets of notebook paper to describe the characteristics of each type of biome.

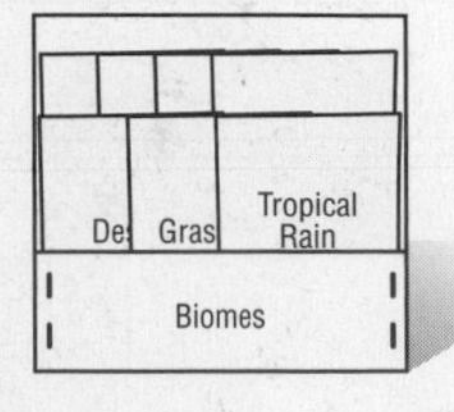

Major Biomes

The seven types of land biomes are shown on the map on the next page. The major land biomes are tundra, taiga, temperate deciduous forest, temperate rain forest, tropical rain forest, desert, and grassland. Areas with similar climates have similar plants and animals.

What kind of climate does tundra have?

The <u>tundra</u> is a cold, dry, treeless area. The tundra is found in latitudes just south of the North Pole or on high mountains.

Locate the tundra areas on the map below. Notice how far these areas are from the equator. The average amount of precipitation in the tundra is less than 25 cm per year. The average daily temperature is −12°C. The tundra is covered with ice most of the year. Summers are short and cold. The top part of the soil thaws in summer. Below this thawed surface is a layer of soil called permafrost that is always frozen.

What plants and animals live on the tundra?

Tundra plants include mosses, grasses, small shrubs, and lichens. Since the growing season is so short, it can take many years for the plant life to recover when damaged. During the summer, insects and migratory birds such as ducks and geese live on the tundra. Other animals that live on the tundra include hawks, owls, mice, reindeer, and musk oxen.

1. Infer Would you expect to find few or many species of plants and animals in the tundra? Explain.

___

___

___

Picture This

2. Locate Circle the names of continents on which deserts are found.

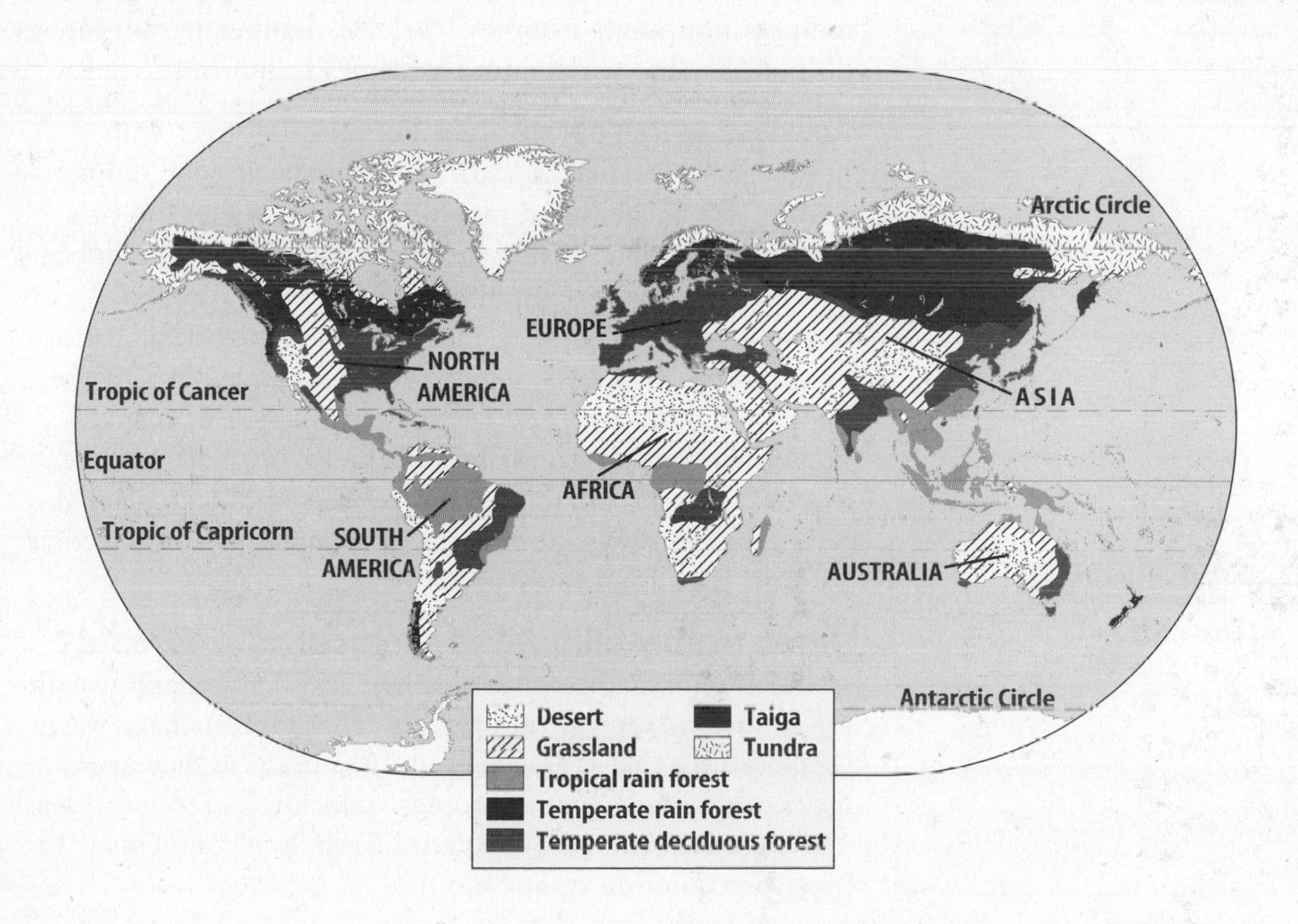

What is the world's largest biome?

The **taiga** (TI guh) is the world's largest biome. The taiga is located between latitudes 50°N and 60°N and stretches across North America, northern Europe, and Asia. The taiga is a cold, forest region. Its climate is warmer and wetter than the tundra's. Precipitation is mostly snow and averages 35 cm to 100 cm a year. Cone-bearing evergreen trees grow in the taiga.

What are temperate deciduous forests like?

The **temperate deciduous forests** are climax communities of deciduous trees, which lose their leaves every autumn. The yearly precipitation is between 75 cm and 150 cm. Precipitation is received evenly throughout the year. Temperatures range from below freezing during the winter to 30°C or more during the summer. White-tailed deer are one of the many species found in temperate deciduous forests. ☑

Reading Check

3. **Explain** What happens to the leaves of deciduous trees in autumn?

____

____

____

Where are temperate rain forests located?

Temperate rain forests are found in places such as New Zealand, southern Chile, and the Pacific Northwest of the United States. This biome receives precipitation ranging from 200 cm to 400 cm throughout the year. The average temperature ranges from 9°C to 12°C. Temperate rain forests do not have the temperature extremes found in the taiga.

Activities in Temperate Rain Forests Tall trees with needlelike leaves, like fir, cedar, and spruce, grow in temperate rain forests. Lichens and mosses also grow there. Animals that live in temperate rain forests include black bear, bobcats, and many species of amphibians.

The logging industry in the Northwest provides jobs for many people. However, logging removes large parts of the temperate rain forest and destroys the habitat of many organisms. Logging companies in the Pacific Northwest of the United States are required to replant trees to replace the ones they cut down. Some rain forest areas are protected as national parks and forests. ☑

Reading Check

4. **Describe** why logging can be harmful to temperate rain forests.

____

____

____

What is the climate in tropical rain forests?

Warm temperatures, wet weather, and dense plant growth are found in **tropical rain forests.** These forests have warm temperatures that average about 25°C because they are located near the equator. Tropical rain forests receive at least 200 cm and as much as 600 cm of rain per year. This is the most precipitation of any biome.

Zones in Tropical Rain Forests More species of animals are found in tropical rain forests than in any other biome. The variety of species is so large that many have not been discovered. Scientists divide the rain forests into zones based on the types of plants and animals that live there. As shown in the figure below, the zones include forest floor, understory, canopy, and emergents (e MERH gentz).

Picture This

5. Identify Highlight the zones in which birds live. Circle the zones in which insects live.

Emergents These giant trees are much higher than the average canopy tree. Birds, such as the macaw, and insects are found here.
Canopy The canopy includes the upper parts of the trees. It's full of life—insects, birds, reptiles, and mammals.
Understory This dark, cool environment is under the canopy leaves but above the ground. Many insects, reptiles, and amphibians live in the understory.
Forest Floor The forest floor is home to many insects and the largest mammals in the rain forest generally live here.

Human Impact Farmers clear the land in tropical areas to farm and to sell the wood. After a few years, the crops use up the nutrients in the soil, and more land is cleared. This process destroys the rain forests. Through education, people are learning the value of preserving the species of the rain forest. Logging is not allowed in some areas. In other areas, farmers use new farming methods so they do not need to clear as much rain forest land.

What is the driest biome?

The **desert** is the driest biome. Deserts receive less than 25 cm of rain each year. The temperatures are extreme heat and cold. Few plants live in desert areas and much of the ground is bare. Most deserts are covered with a thin, sandy, or rocky soil that contains little organic matter. The driest deserts have windblown sand dunes.

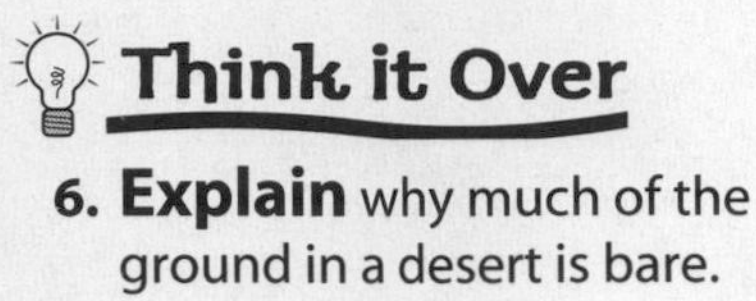

6. Explain why much of the ground in a desert is bare.

____

____

____

Desert Plants and Animals Most desert plants, like cactus, survive the extreme dryness because they are able to store water. Desert plants and animals also are adapted to hot and cold temperatures. Some animals, like the kangaroo rat, never need to drink water. They get the moisture they need from the food they eat. Most animals are active only during the night, late afternoon, or early morning when the temperatures are less extreme. Most animals in the desert are small.

What are grasslands like?

Temperate and tropical regions that receive between 25 cm and 75 cm of precipitation each year and are made up of climax communities of grasses are called **grasslands**. Most grasslands have a dry season, with little or no rain. This lack of rain prevents the development of forests. ☑

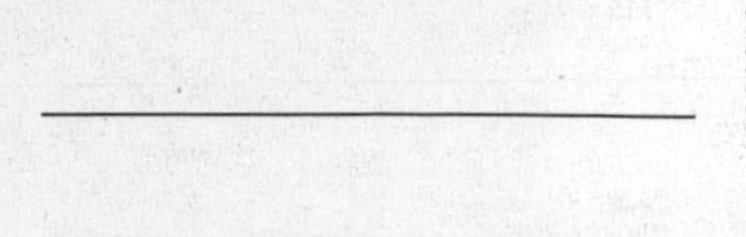

7. Determine What keeps forests from developing on grasslands?

____

____

____

Grassland Plants and Animals The animals in grasslands are mostly mammals that eat the stems, leaves, and seeds of grass plants. Kangaroos are found in the grasslands of Australia. Zebras live in the grasslands of Africa. Many crops, such as wheat, rye, and corn are grown in grasslands. Sheep and cattle are raised on grasslands.

After You Read

Mini Glossary

biomes (BI ohmz): large geographic areas that have similar climates and ecosystems

desert: dry biome with extreme hot and cold temperatures

grasslands: biome of temperate and tropical regions that receive little precipitation and are made up of climax communities of grasses

taiga (TI guh): biome with long, cold winters, moderate precipitation, and forests of evergreen trees

temperate deciduous forests: biome with four seasons and climax forests of deciduous trees, which lose their leaves every autumn

temperate rain forests: biome with warm temperatures, much precipitation, and forests of tall trees that have needlelike leaves

tropical rain forests: biome of warm temperatures, wet weather, and dense plant growth

tundra: a cold, dry, treeless biome that gets little precipitation and is covered with ice most of the year

1. Review the terms and their definitions in the Mini Glossary. Write two sentences that explain the difference between temperate deciduous forests and temperate rain forests.

2. How did underlining the important ideas in this section help you understand biomes?

3. Complete the chart below to help you compare and contrast the seven biomes of the world.

Biomes	Climate	Plants and Animals	Locations
Tundra			
Taiga			
Temperate deciduous forests			
Temperate rain forests			
Tropical rain forests			
Deserts			
Grasslands			

End of Section

ScienceOnline Visit **life.msscience.com** to access your textbook, interactive games, and projects to help you learn more about the biomes.

Ecosystems

section 3 Aquatic Ecosystems

Before You Read

On each line below, name a different body of water. Next to each body of water, classify it as freshwater or salt water.

____

____

____

____

What You'll Learn

- the differences between flowing freshwater and standing freshwater ecosystems
- the importance of saltwater ecosystems
- problems that affect aquatic ecosystems

Read to Learn

Factors That Affect Aquatic Ecosystems

Aquatic ecosystems are places where organisms grow or live in water. There are four factors that affect aquatic ecosystems—water temperature, the amount of sunlight present, dissolved oxygen, and salt in the water.

Freshwater Ecosystems

Earth's freshwater ecosystems include flowing water such as rivers and streams. Freshwater ecosystems also include standing water such as lakes, ponds, and wetlands. Freshwater ecosystems contain very low amounts of salt.

How are river and stream environments alike?

Rivers and streams that flow fast have clearer water and higher levels of oxygen than slow-flowing rivers and streams. This is because the faster the water moves, the more air mixes in. In flowing-water ecosystems, nutrients that support life are washed in from the land. Plants and animals that live in rivers and streams are adapted to the flowing water.

Study Coach

Sticky-Note Discussions As you read the section, use sticky-note paper to mark at least four paragraphs that you find interesting or that you have a question about. Your teacher can help you better understand what you have read.

FOLDABLES™

C Describe Make a layered-look book, as shown below. Under each flap, write descriptions of the aquatic ecosystems.

Aquatic Ecosystems
Freshwater
Wetlands
Salt Water

How are lake and pond environments alike?

The water in lakes and ponds hardly moves. These environments have more plants than flowing-water environments. Lakes and ponds contain organisms that are not well adapted to flowing-water environments.

Lakes are larger and deeper than ponds. They have more open water because plant growth is limited to shallow areas along the shoreline. Colder temperatures and lower light levels limit the types of organisms that can live in deep lake waters. Microscopic algae, plants, and other organisms known as plankton live near the surface and the shoreline of freshwater lakes and in ponds where the water is warm and sunlit. Many ponds are filled almost completely with plant material, which make them high in nutrients. ☑

✔ Reading Check

1. Compare What is the difference between a lake and a pond?

____

____

____

____

What are wetlands?

Regions that are wet for all or most of the year are called **wetlands**. These regions, also known as swamps, bogs, and fens, are located between land areas and water. Wetlands are filled with plants and animals that are adapted to water-logged soil. Fish, shellfish, and cranberries are some products that come from wetlands. Wetland animals include beavers, muskrats, alligators, and some species of turtles. Many birds use wetland areas to have their young.

How do humans affect freshwater ecosystems?

Sometimes freshwater ecosystems are used as places to dump waste and other pollutants. Fertilizer from farms and lawns runs off into freshwater. Wetlands were once drained and destroyed because people thought they were useless and full of diseases. The drained land was used for shopping centers and houses.

People are being educated about the damage caused by polluting freshwater ecosystems. Sewage is treated before it is released into the water to prevent problems. People who pollute waterways may be fined. Many developers now are working to restore wetlands.

Applying Math

2. Graph In the circle below, make and label a circle graph that shows the percent of Earth's water that is salt water and the percent that is freshwater.

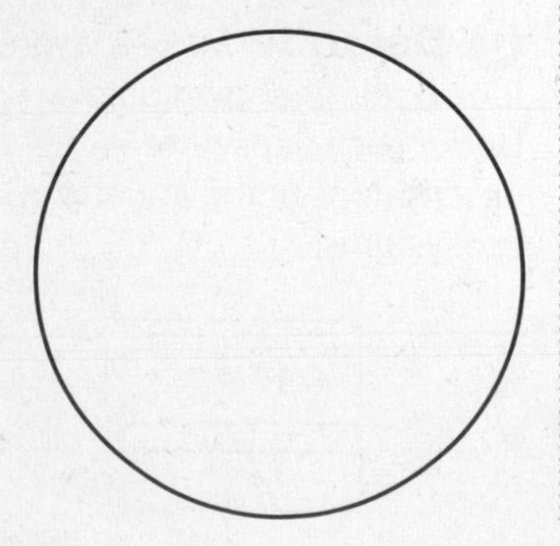

Saltwater Ecosystems

About 95 percent of Earth's water contains high amounts of salts. Saltwater ecosystems include oceans, seas, a few inland lakes such as the Great Salt Lake in Utah, coastal inlets, and estuaries.

What are ocean life zones?

Scientists divide the ocean into life zones. There are two zones based on the depth to which sunlight penetrates the water—the lighted zone and the dark zone. The lighted zone of the ocean is about the upper 200 m. Plankton make up the base of the food chain in this zone. Below about 200 m is the dark zone. Animals living in this zone feed on each other or on material that floats down from the lighted zone. A few organisms produce their own food. ☑

How do coral reefs form?

Coral reefs are one of the most varied ecosystems in the world. **Coral reefs** form in oceans over long periods of time from the calcium carbonate shells of ocean animals called corals. When corals die, their shells remain. Over time the shell deposits form coral reefs. Coral reefs contain colorful fish and many other organisms.

Waste materials easily damage coral reefs. World organizations are helping protect coral reefs from harm.

What are the characteristics of seashores?

The shallow waters along the world's coastlines have many kinds of saltwater ecosystems. These waters are affected by the tides and by the action of the waves. The height of the tides changes based on the phases of the Moon, the season, and the slope of the shoreline. The part of the shoreline that is covered with water at high tide and exposed to air during low tide is called the **intertidal zone**. Organisms that live in the intertidal zone must withstand the force of the waves. They must also be adapted to changes in temperature, moisture, and the amount of salt in the water.

What is an estuary?

Almost every river eventually flows into an ocean. The area where they meet contains a mixture of freshwater and salt water and is called an **estuary** (ES chuh wer ee). Estuaries are located near coastlines and border the land. Other names for estuaries include bays, lagoons, and sounds. An estuary is a very fertile environment. Freshwater streams bring in great amounts of nutrients washed from inland soils. An estuary is an important aquatic ecosystem because many kinds of organisms live there, including algae, grasses, shrimp, crabs, clams, and fish. Estuaries are places where the young of many species of ocean fish grow and develop.

Reading Check

3. **Identify** What are the two ocean life zones?

____

____

Think it Over

4. **List** three things that organisms that live in intertidal zones must be adapted to in order to survive.

____

____

____

____

After You Read

Mini Glossary

coral reefs: ecosystems in oceans that formed over long time periods from the calcium carbonate shells of corals

estuary (ES chuh wer ee): the area where a river meets an ocean and contains a mixture of freshwater and salt water

intertidal zone: the part of the shoreline that is covered with water at high tide and exposed to air during low tide

wetlands: regions that are wet for all or most of the year

1. Review the terms and their definitions in the Mini Glossary. Write a sentence that explains the difference between a wetland and an estuary.

____

____

____

2. Complete the graphic organizer below to identify the kinds of aquatic ecosystems.

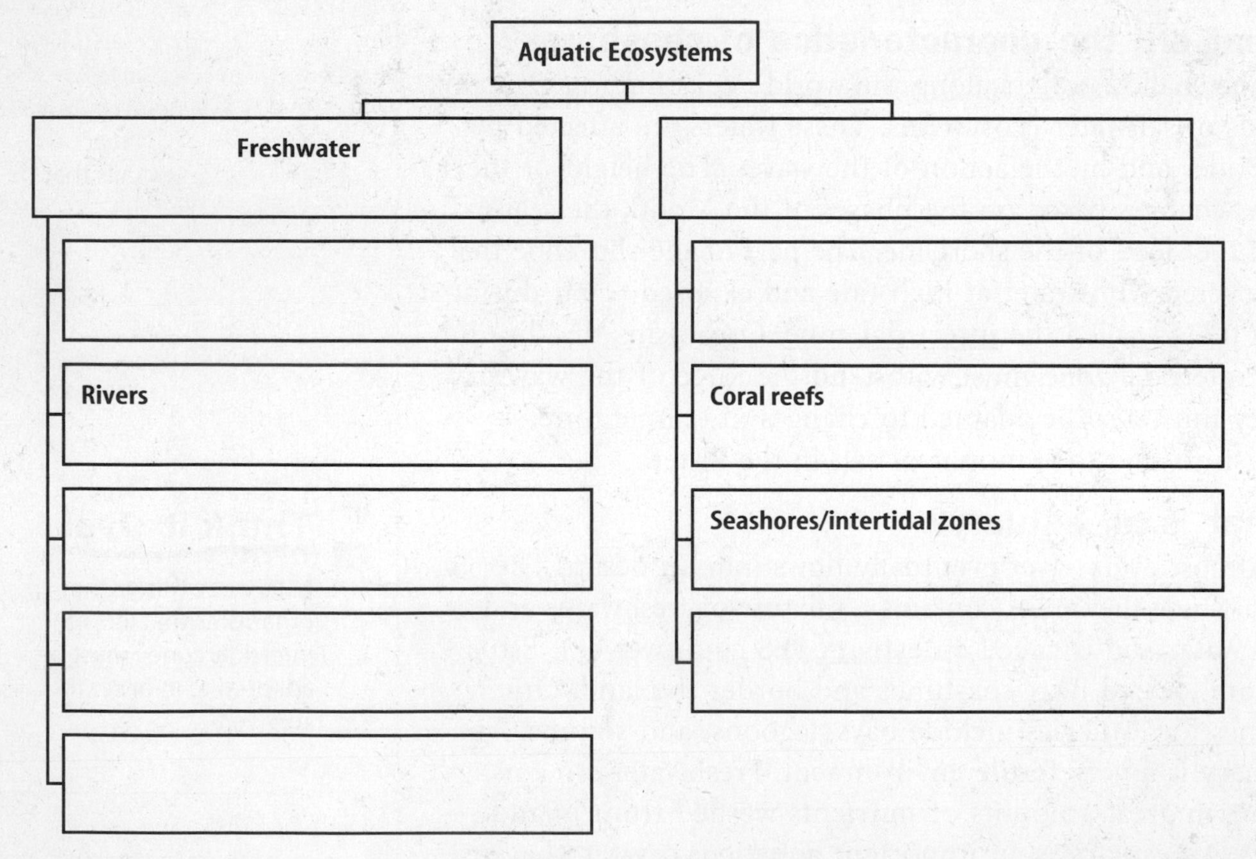

End of Section

Science Online Visit **life.msscience.com** to access your textbook, interactive games, and projects to help you learn more about aquatic ecosystems.

Conserving Resources

section 1 Resources

Before You Read

Identify two objects in the room you are in. What products from the environment were used to make them?

What You'll Learn

- the difference between renewable and nonrenewable resources
- how fossil fuels are used
- alternatives to using fossil fuels

Read to Learn

Natural Resources

An earthworm eats decaying plant material. A robin catches the worm and flies to a tree. The leaves of the tree use sunlight during photosynthesis. Leaves fall to the ground and decay. What do these living things have in common? They rely on Earth's natural resources. **Natural resources** are the parts of the environment that are useful or necessary for the survival of living organisms. Like other organisms, humans need food, air, and water. Humans also use resources to make everything from clothes to cars.

What are renewable resources?

A **renewable resource** is any natural resource that is recycled or replaced constantly by nature. For example, the Sun provides a constant supply of heat and light. Plants add oxygen to the air when they carry out photosynthesis. Rain fills lakes and streams with water.

Why are some resources in short supply?

Although renewable resources are recycled or replaced, they are sometimes in short supply. Sometimes there may not be enough rain or water provided from melting snow to supply water to people, plants, and animals. In desert regions, water and other resources are often scarce.

Study Coach

Identify the Main Idea As you read this section, organize notes into two columns. On the left, list a main idea about the material in each subhead. On the right, list the details that support the main idea.

FOLDABLES™

A Identify Make a vocabulary book using notebook paper. As you read the section, add each boldface underlined term. Write the definitions under the tabs.

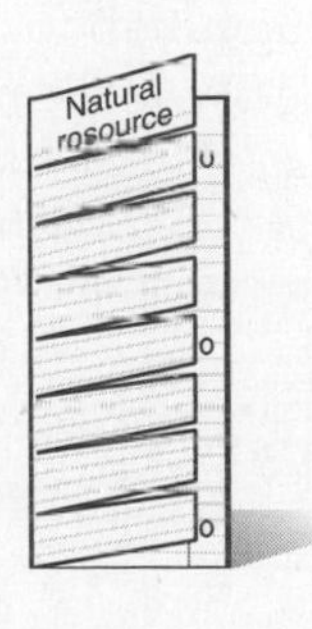

What are nonrenewable resources?

Natural resources that are used up more quickly than they can be replaced by natural processes are **nonrenewable resources**. Earth's supply of nonrenewable resources is limited. For example, plastics and gasoline are made from a nonrenewable resource called petroleum, or oil. **Petroleum** is formed mostly from the remains of microscopic marine organisms buried in Earth's crust. Petroleum is nonrenewable because it takes hundreds of millions of years for it to form. ☑

Reading Check

1. **Explain** why petroleum is a nonrenewable resource.

____

____

____

Fossil Fuels

Coal, oil, and natural gas are nonrenewable resources that supply energy. Most of the energy you use comes from these fossil fuels, as you can see in the figure below. **Fossil fuels** are fuels formed in Earth's crust over hundreds of millions of years. Cars are powered by gasoline, which is made from oil. Many power plants use coal to produce electricity. Natural gas is used for heating and cooking.

Sources of Energy in the United States

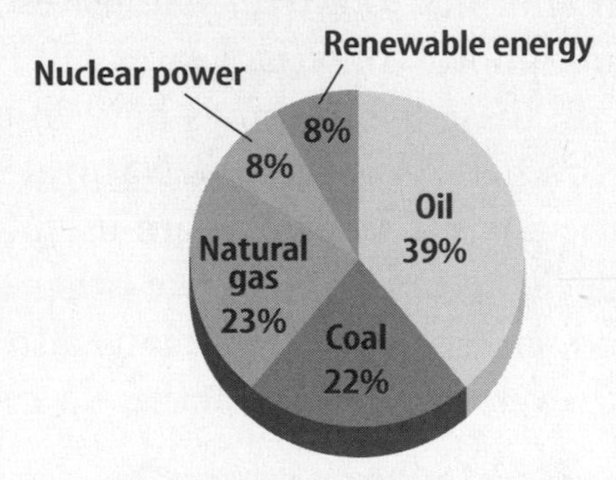

Picture This

2. **Identify** On the circle graph, outline the sections that represent fossil fuels. On the line below, write the percentage of U.S. energy that comes from sources other than fossil fuels.

____

Why should fossil fuels be conserved?

People all over the world use fossil fuels every day. Earth's supply of these fuels is limited. In the future, fossil fuels may become more expensive and harder to get.

The use of fossil fuels can cause environmental problems. Layers of soil and rock are often stripped away when mining for coal. This destroys ecosystems. Another problem with fossil fuels is that they have to be burned to release energy. The burning results in waste gases that cause air pollution. Two forms of air pollution are smog and acid rain. To reduce the problems caused by fossil fuels, many people suggest using fossil fuels less and finding other sources of energy.

FOLDABLES™

B Explain Make a two-tab book using notebook paper, as shown below. Make notes about the effects of fossil fuels and alternatives to fossil fuels.

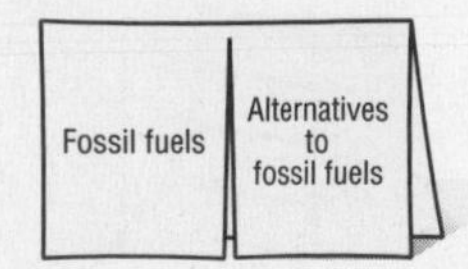

Reducing the Use of Fossil Fuels You can turn off the television when you are not watching it. This will reduce the use of electricity. You can ride in a car pool or use public transportation to reduce the use of gasoline. Walking or riding a bicycle also can reduce the use of fossil fuels.

Alternatives to Fossil Fuels

Another way of reducing the use of fossil fuels is to find other sources of energy. Power plants use fossil fuels to power the turbines that produce electricity. Alternative energy sources such as water, wind, and nuclear energy can be used instead of the fossil fuels to turn the turbines. Another alternative is solar cells that use only sunlight to produce electricity.

How can water generate electricity?

Water is a renewable resource that can be used to produce electricity. **Hydroelectric power** is electricity that is made when the energy of falling water is used to turn the turbines of an electric generator. Hydroelectric power does not burn fuel, so it does not cause air pollution. However, this type of power can cause environmental problems. To build a hydroelectric plant, usually a dam needs to be constructed across a river. The dam raises the water level to produce the energy that is needed to make electricity. Many acres of land behind the dam are flooded, destroying land habitats and turning part of the river into a lake.

How can wind be used to produce energy?

Wind power is another renewable energy source that can be used to make electricity. Wind turns the blades of a turbine, which powers an electric generator. Wind power does not cause air pollution. However, electricity can be produced only when the wind is blowing.

Where does geothermal energy come from?

The hot, molten rock that lies beneath Earth's surface is another energy source. You can see the effects of this energy when a volcano erupts. **Geothermal energy** is the heat energy contained in Earth's crust. Geothermal power plants use this energy to produce steam to produce electricity. Geothermal energy is available only where there are natural geysers or volcanoes. Iceland, an island nation, was formed by volcanoes. Geothermal energy supplies most of Iceland's power. ☑

Think it Over

3. **Identify** one advantage and one disadvantage of hydroelectric power.

______

______

______

Reading Check

4. **Explain** the source of geothermal energy.

______

______

______

What is nuclear power?

Another alternative to fossil fuels is nuclear energy. **Nuclear energy** is released when billions of atomic nuclei from uranium, a radioactive element, are split apart in a nuclear fission reaction as shown below. This energy is used to make the steam that turns the turbines of an electric generator.

Nuclear power does not cause air pollution, but it does cause other problems. Mining uranium can harm ecosystems. Nuclear power plants produce radioactive wastes that can harm living organisms. Disposing of these wastes can be a problem. Accidents also are a danger.

Picture This

5. **Describe** Use the figure to explain to a partner how heat is produced from uranium.

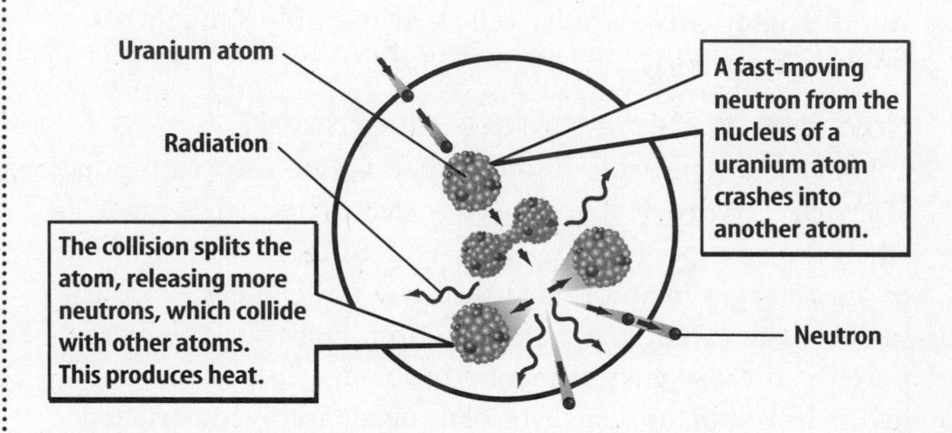

Uranium atom
Radiation
A fast-moving neutron from the nucleus of a uranium atom crashes into another atom.
The collision splits the atom, releasing more neutrons, which collide with other atoms. This produces heat.
Neutron

What is solar energy?

Solar energy is another alternative to fossil fuels. Solar energy comes from the Sun. It is an inexhaustible source of energy—it cannot be used up. One use of solar energy is to heat buildings. During winter in the northern hemisphere, the parts of a building that face south receive the most sunlight. Large windows on the south side of the building let in warm sunshine during the day. The floors and walls of solar-heated buildings are made of materials that absorb heat during the day. At night, the heat is slowly released, keeping the building warm.

6. **Determine** What do PV cells use to produce electricity?

____

____

____

What are solar cells?

A solar-powered calculator uses photovoltaic (foh toh vohl TAY ihk) cells to turn sunlight into electric current. Photovoltaic (PV) cells, also known as solar cells, are small and easy to use. But they can produce electricity only in sunlight. Batteries are needed to store electricity for use at night or on cloudy days. PV cells are considered too expensive to use to make large amounts of electricity. ☑

After You Read

Mini Glossary

fossil fuel: fuel formed in Earth's crust over hundreds of millions of years

geothermal energy: heat energy within Earth's crust that is available only where geysers and volcanoes are found

hydroelectric power: electricity produced when the energy of falling water turns the blades of a turbine that generates electricity

natural resource: part of the environment that is useful or necessary for the survival of living organisms

nonrenewable resource: natural resource that is used up more quickly than it can be replaced by natural processes

nuclear energy: energy released when billions of uranium nuclei are split apart in a nuclear fission reaction

petroleum: nonrenewable resource formed from the remains of microscopic marine organisms buried in Earth's crust

renewable resource: natural resource that is recycled or replaced constantly by nature

1. Review the terms and their definitions in the Mini Glossary. Write a sentence that explains the difference between renewable and nonrenewable resources.

____

____

2. Complete the chart below to compare the advantages and disadvantages of using each of the following forms of energy.

Energy Source	Advantages	Disadvantages
fossil fuels		
hydroelectric power		
wind power		
nuclear power		
geothermal power		
solar power		

Visit **life.msscience.com** to access your textbook, interactive games, and projects to help you learn more about resources.

chapter 27 Conserving Resources

section 2 Pollution

What You'll Learn
- the types of air pollution
- the causes of water pollution
- how erosion can be prevented

Before You Read

What do you think are the major causes of pollution in your community?

____

____

____

Study Coach

Make Flash Cards to help you learn more about the section. Write a quiz question for each paragraph on one side of the flash card and the answer on the other side. Keep quizzing yourself until you know all of the answers.

FOLDABLES

C Describe Make a trifold book using notebook paper, as shown below. Use the Foldable to describe the three types of pollution.

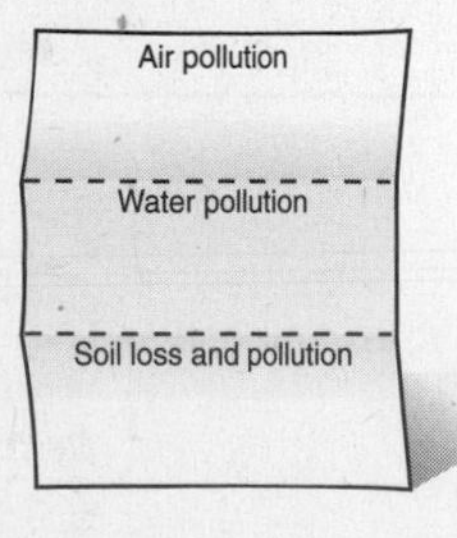

Read to Learn

Keeping the Environment Healthy

More than six billion people live on Earth. This puts a strain on the environment. You can help protect the environment by paying attention to how your use of natural resources affects air, water, and land.

Air Pollution

On a still, sunny day in most large cities, you might see a dark haze in the air. The haze comes from pollutants that form when wood or fuels are burned. A **pollutant** is a substance that contaminates the environment. Air pollution is likely wherever there are cars, airplanes, factories, homes, and power plants. Volcanic eruptions and forest fires also can cause air pollution.

What is smog?

Smog is a form of pollution that is created when sunlight reacts with pollutants produced by burning fuels. Smog can irritate the eyes and make it difficult for people who have lung diseases to breathe. Smog can be reduced if more people take buses or trains instead of driving. Other vehicles, such as electric cars, that produce fewer pollutants also can help reduce smog.

Acid Precipitation

Water vapor condenses on dust particles in the air to form droplets. The droplets create clouds. Eventually, the droplets become large enough to fall as precipitation—mist, rain, snow, sleet, or hail. Air pollutants from the burning of fossil fuels can react with water in the atmosphere to form strong acids. Acidity is measured by a value called pH. **Acid precipitation** has a pH below 5.6, as shown in the figure below. ☑

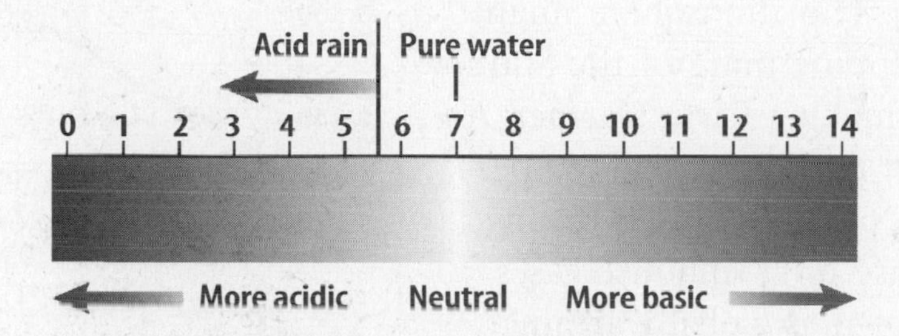

What are the effects of acid rain?

Acid precipitation, or acid rain, washes nutrients from the soil. This can cause trees and plants to die. Acid rain runs off into lakes and ponds, lowering the pH of the water. If the water is too acidic, it can kill the algae and microscopic organisms in the water. This means that fish and other organisms that depend on them for food also die.

How can acid rain be prevented?

When factories burn coal, sulfur is released into the air. Vehicle exhaust contains nitrogen oxide. Sulfur and nitrogen oxide are the main pollutants that cause acid rain. Using low-sulfur fuels, such as low-sulfur coal or natural gas, can reduce acid rain. However, these fuels are more expensive than high-sulfur coal. Smokestacks that remove sulfur dioxide before it enters the air can also help reduce acid rain. Reducing automobile use or using electric cars can help reduce acid rain caused by nitrogen oxide pollution.

Greenhouse Effect

When sunlight reaches Earth's surface, some of it is reflected back into space. The rest is trapped by atmospheric gases. This heat-trapping feature of the atmosphere is the **greenhouse effect**. Without it, temperatures on Earth would probably be too cold to support life.

Reading Check

1. **Explain** the causes of acid precipitation.

____

____

____

Picture This

2. **Identify** You measure the pH of rainwater several times. For each reading below, use the scale to determine if it is acid precipitation. Write *Yes* or *No* beside each measurement.

pH of 3.2 ______________

pH of 8.5 ______________

pH of 6.0 ______________

What are greenhouse gases?

The gases in the atmosphere that trap heat are called greenhouse gases. Carbon dioxide (CO_2) is one of the most important greenhouse gases. CO_2 is a normal part of the atmosphere. It is also a by-product of burning fossil fuels. Over the past century, more fossil fuels have been burned than ever before. This is increasing the percentage of CO_2 in the atmosphere, as you can see in the graph above. The atmosphere might be trapping more of the Sun's heat, making Earth warmer. A rise in Earth's average temperature, possibly caused by an increase in greenhouse gases, is known as global warming.

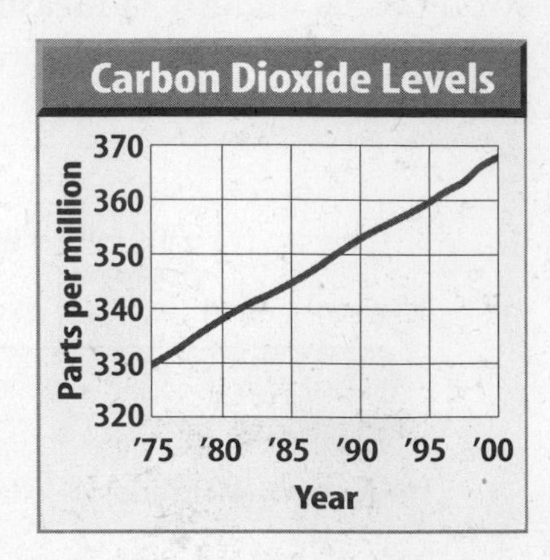

Picture This

3. **Describe** the trend shown on this graph.

____

____

____

Is Earth's average temperature changing?

Between 1895 and 1995, Earth's average temperature increased 1°C. No one is certain whether the rise in temperature was caused by human activities or is a natural part of Earth's weather cycle.

Global warming might have several effects. It might cause a change in rainfall patterns, which can affect ecosystems. The rate of plant growth and the plants that can be grown in different parts of the world may change. The number of storms might increase. The polar ice caps might begin to melt, raising sea levels and flooding coastal areas. Many people think that the possibility of global warming is a good reason to reduce the use of fossil fuels. ☑

Reading Check

4. **Identify** What are two possible effects of global warming?

____

____

____

Ozone Depletion

Ozone (OH zohn) is a form of oxygen in the atmosphere. Ozone molecules are made of three oxygen atoms. They are formed in a chemical reaction between sunlight and oxygen. The oxygen you breathe has two oxygen atoms in each molecule.

The ozone layer is found about 20 km above Earth's surface, as shown in the figure at the top of the next page. The ozone layer in Earth's atmosphere absorbs some of the Sun's harmful ultraviolet (UV) radiation. This radiation can damage living cells.

CFCs The ozone layer becomes thinner over each polar region during the spring. This thinning of the ozone layer is called **ozone depletion**. It is caused by pollutant gases, especially chlorofluorocarbons (klor oh FLOR oh kar bunz) (CFCs). These gases are sometimes used in the cooling systems of refrigerators and air conditioners. When CFCs leak into the air, they rise in the atmosphere until they reach the ozone layer. CFCs react chemically with ozone, breaking apart the ozone molecules.

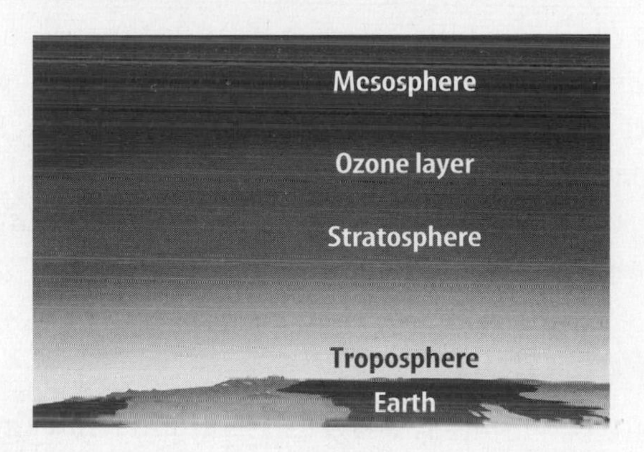

Why is ozone depletion a problem?

Because of ozone depletion, the amount of UV radiation that reaches Earth could be increasing. This radiation may be causing an increase in the number of skin cancer cases in humans. The ozone layer is important to the survival of life on Earth. For this reason, many countries and industries have agreed to stop making and using CFCs.

The ozone that is high in the atmosphere protects life on Earth. However, ozone that is near Earth's surface can be harmful. Ozone is produced when fossil fuels are burned. This ozone stays lower in the atmosphere and pollutes the air. Ozone damages lungs and other tissues of animals and plants.

Indoor Air Pollution

Air pollution also can occur indoors. Buildings today are better insulated to conserve energy. The insulation reduces the flow of air into and out of a building, so air pollutants can build up indoors. Burning cigarettes release hazardous particles and gases into the air. Even people who do not smoke can be affected by this secondhand cigarette smoke. For this reason, smoking is not allowed in many buildings. Other dangerous gases in buildings are released by paints, carpets, and photocopiers.

Picture This

5. **Identify** Add the approximate height at which the ozone layer can be found to the figure.

Reading Check

6. Explain why carbon monoxide is hard to detect.

Carbon Monoxide Carbon monoxide (CO) is a poisonous gas. It is produced when fuels such as charcoal and natural gas are burned. CO is colorless and odorless, so it is difficult to detect. CO poisoning can cause illness or even death. Today, fuelburning stoves and heaters have to be designed to prevent CO from building up indoors. Many buildings today have alarms that warn of buildups of CO.

Radon Radon is a naturally occurring, radioactive gas that is given off by some types of rock and soil. It has no color or odor. It can seep into basements and lower floors in buildings. Radon exposure is the second leading cause of lung cancer in the United States. Radon detectors sound an alarm if the levels of radon in a building are too high. If radon is present, increasing a building's ventilation can eliminate any damaging effects.

Water Pollution

Pollutants enter water, too. Air pollutants can drift into water or be washed out of the sky by rain. Wastewater from factories and sewage-treatment plants is often released into waterways. Pollution also occurs when people dump litter and waste into rivers, lakes, and oceans.

Reading Check

7. Identify What are two causes of water pollution?

What happens when surface water is polluted?

Some water pollutants can poison fish and other animals. People who swim in or drink the polluted water can be harmed. Pesticides used on farms can wash into lakes and streams. The chemicals can harm the insects that fish eat. The fish may die from a lack of food.

Another effect of water pollution is algal blooms. Fertilizers and raw sewage contain large amounts of nitrogen. If they are washed into a lake or pond, they can cause algae to grow quickly. When the algae die, bacteria decompose them. The bacteria use up much of the oxygen in the water during this process. Fish and other organisms can die from a lack of oxygen in the water.

Think it Over

8. Explain how a polluted river will eventually affect an ocean.

How is ocean water polluted?

Rivers and streams flow into oceans, bringing their pollutants along. Ocean water can be polluted by the wastewater from factories and sewage-treatment plants along the coast. Oil spills also cause pollution. About 4 billion kg of oil are spilled into ocean waters every year.

How is groundwater polluted?

Groundwater comes from precipitation and runoff that soaks into the soil. This water moves slowly through layers of rock called aquifers. If the water comes in contact with pollutants as it moves through the soil, the aquifer could become polluted. Polluted groundwater is difficult to clean.

Soil Loss

Most plants need fertile topsoil in order to grow. New topsoil takes hundreds or thousands of years to form. Topsoil can be blown away by wind and washed away by rain. The movement of soil from one place to another is called **erosion** (ih ROH zhun). Eroded soil that washes into a river or stream can block sunlight and slow photosynthesis. It also can harm fish and other organisms. Erosion happens naturally, but human activities increase the rate of erosion. For example, when a farmer plows a field, soil is left bare. Bare soil is more easily carried away by rain and wind. Some methods of farming can help reduce soil erosion. ☑

Reading Check

9. **Identify** What is erosion?

Soil Pollution

Soil becomes polluted when air pollutants fall to the ground or when water leaves pollutants behind as it flows through the soil. Soil also becomes polluted when people throw litter on the ground or dump trash in landfills.

What happens to solid wastes?

Most of the trash that people throw away every week is dumped in landfills. Most landfills are designed to seal out air and water to keep pollutants from seeping into surrounding soil. However, this also slows normal decay processes. Food scraps and paper, which usually break down quickly, can last for many years in landfills. By reducing the amount of trash that people produce, the need for new landfills can also be reduced.

Think it Over

10. **Explain** why hazardous wastes should not be dumped into landfills.

What happens to hazardous wastes?

Waste materials that are harmful to human health or poisonous to living organisms are **hazardous wastes**. Pesticides and oil are hazardous wastes. Many household items such as leftover paint and batteries also are hazardous wastes. Hazardous wastes should be treated separately from regular trash to prevent them from polluting the environment.

After You Read

Mini Glossary

acid precipitation: precipitation that has a pH below 5.6

erosion: the movement of soil from one place to another

greenhouse effect: the heat-trapping feature of the atmosphere that keeps Earth warm enough to support life

hazardous waste: waste materials that are harmful to human health or poisonous to living organisms

ozone depletion: the thinning of the ozone layer

pollutant: a substance that contaminates the environment

1. Review the terms and their definitions in the Mini Glossary. Choose one of the terms and write a sentence explaining how it can harm the environment.

____

____

____

2. Choose one of the question headings in the Read to Learn section. Write the question in the space below. Then write your answer to that question on the lines that follow.

Write your question here.

____

____

____

3. How do flash cards help you remember what you have read?

____

____

____

Visit **life.msscience.com** to access your textbook, interactive games, and projects to help you learn more about pollution.

Conserving Resources

section 3 The Three Rs of Conservation

Before You Read

In what ways do you and your family help to conserve natural resources?

What You'll Learn

- how use of natural resources can be reduced
- how resources can be reused
- that many materials can be recycled

Read to Learn

Conservation

Conserving resources can help prevent shortages of natural resources. It also can slow the growth of landfills and lower levels of pollution. You can conserve resources in several ways. The three Rs of conservation are reduce, reuse, and recycle.

Mark the Text

Identify Main Ideas Highlight the main idea of each paragraph. Then underline the details that support the main idea.

Reduce

You help conserve natural resources when you reduce your use of them. For example, you use less fossil fuel when you walk instead of ride in a car. You also can reduce your use of natural resources by buying only the things that you need. You can buy products that use less packaging or that use packaging made from recycled materials.

Reuse

Another way to conserve natural resources is to use items more than once. Reusing an item means that it can be used again without changing it or reprocessing it. Bring reusable canvas bags to the grocery store to carry home your purchases. Donate outgrown clothes to charity so that others can reuse them.

FOLDABLES™

D Describe Make a layered-look Foldable using notebook paper, as shown below. Make notes describing the three Rs of conservation.

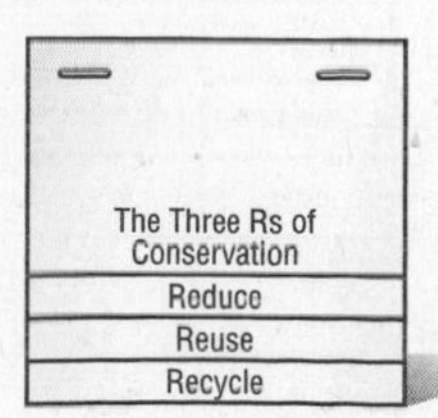

Recycle

If you cannot avoid using an item, and if you cannot reuse it, then you may be able to recycle it. **Recycling** is a form of reuse that requires changing or reprocessing an item or natural resource. Many communities have a curbside recycling program. Items that can be recycled include glass, paper, and plastics. The figure below shows the rates at which some household items are recycled in the United States.

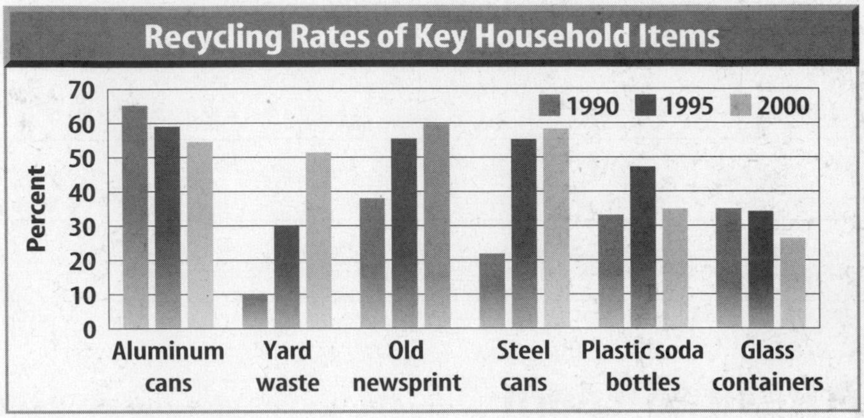

Recycling Rates of Key Household Items

1990 1995 2000

Percent

70
60
50
40
30
20
10
0

Aluminum cans
Yard waste
Old newsprint
Steel cans
Plastic soda bottles
Glass containers

Source: U.S. EPA, 2003

Picture This

1. Identify Circle the products on the chart that show increasing recycling rates. Place an X through products that show declining recycling rates.

What makes plastic difficult to recycle?

Plastic is more difficult to recycle than other items because there are several types of plastic. Every plastic container is marked with a code that tells the type of plastic it is made of. Plastic soft-drink bottles are the type of plastic easiest to recycle. Some types of plastics cannot be recycled at all because they are made of a mixture of different plastics. Before plastic can be recycled, it has to be separated carefully. One piece of a different type of plastic can ruin an entire batch.

How are metals recycled?

About one quarter of steel used in cans, appliances, and automobiles is recycled steel. Using recycled steel saves iron ore and coal, the resources needed to make steel. Metals such as iron, copper, and aluminum also can be recycled.

You can conserve metals by recycling food cans, which are mostly steel, and aluminum cans. It takes less energy to make a can from recycled aluminum than from raw materials. Also, a can that is recycled is not taking up space in landfills.

FOLDABLES™

E Identify Make a folded table using notebook paper, as shown below. For each material listed in the first column, identify a recycled product that is made from it. Write the product in the second column.

Material	Recycled Products
plastic	
metals	
glass	
paper	
compost	

How is glass recycled?

Glass bottles and jars can be sterilized and then reused. They also can be melted and made into new bottles. Glass can be recycled again and again. Most glass bottles today already contain at least 25 percent recycled glass. Recycling glass saves the mineral resources needed to make glass. Recycling glass requires less energy than making new glass.

What are some uses of recycled paper?

Used paper can be recycled to make paper towels, newsprint, and cardboard. Ranchers and farmers sometimes use shredded paper instead of straw for bedding in barns and stables. Used paper can be made into compost. Recycling one metric ton of paper saves 17 trees. It also saves water, oil, and electric energy. You can help by recycling newspapers, notebook paper, and junk mail. ☑

Why is composting useful?

When grass clippings, leaves, and fruit and vegetable scraps are dumped in landfills, they stay there for many years without breaking down. Instead, these items can be turned into compost, which can help to enrich the soil. Many communities distribute compost bins to encourage residents to recycle fruit and vegetable scraps and yard waste.

How are recycled materials used?

Many people have learned to recycle. As a result, many recyclable materials are piling up just waiting to be put to use. When you shop, check labels and buy products that contain recycled materials. Buying products made of recycled material will reduce the backlog of recyclable material.

Reading Check

2. **Explain** In addition to trees, what resources are saved when paper is recycled?

____

____

____

After You Read

Mini Glossary

recycling: a form of reuse that requires changing or reprocessing an item or natural resource

1. Review the term and its definition in the Mini Glossary. Write a sentence explaining how you can participate in recycling.

____

____

____

2. Use the web diagram below to explain the three Rs of conservation. In the ovals, identify the three Rs and include an example of each.

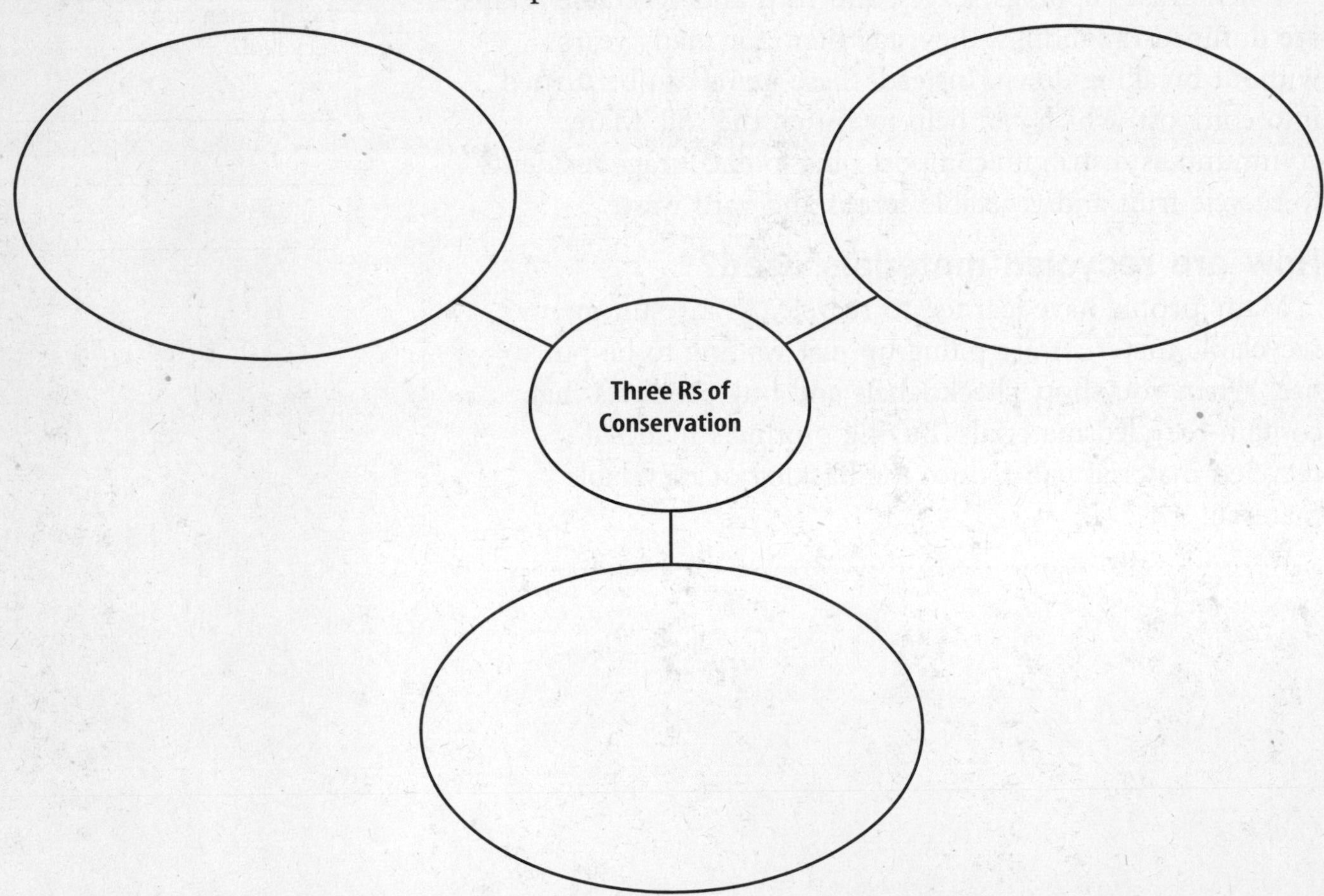

Science Online Visit **life.msscience.com** to access your textbook, interactive games, and projects to help you learn more about the three Rs of conservation.

PERIODIC TABLE OF THE ELEMENTS

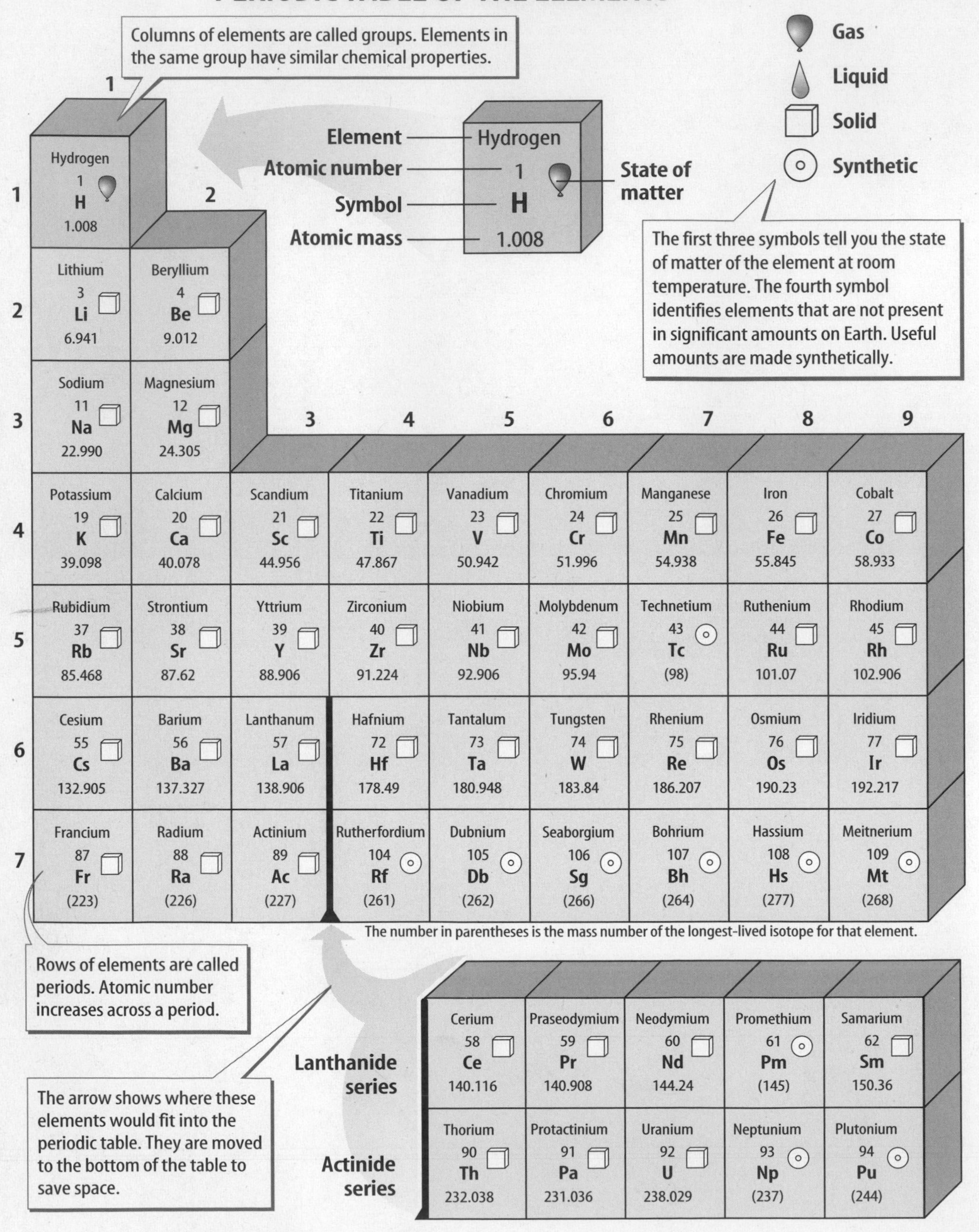

Columns of elements are called groups. Elements in the same group have similar chemical properties.
Gas
Liquid
Solid
Synthetic
Element
Atomic number
Symbol
Atomic mass
Hydrogen
1
H
1.008
State of matter
The first three symbols tell you the state of matter of the element at room temperature. The fourth symbol identifies elements that are not present in significant amounts on Earth. Useful amounts are made synthetically.
1
2
3
4
5
6
7
8
9
Hydrogen 1 H 1.008
Lithium 3 Li 6.941
Beryllium 4 Be 9.012
Sodium 11 Na 22.990
Magnesium 12 Mg 24.305
Potassium 19 K 39.098
Calcium 20 Ca 40.078
Scandium 21 Sc 44.956
Titanium 22 Ti 47.867
Vanadium 23 V 50.942
Chromium 24 Cr 51.996
Manganese 25 Mn 54.938
Iron 26 Fe 55.845
Cobalt 27 Co 58.933
Rubidium 37 Rb 85.468
Strontium 38 Sr 87.62
Yttrium 39 Y 88.906
Zirconium 40 Zr 91.224
Niobium 41 Nb 92.906
Molybdenum 42 Mo 95.94
Technetium 43 Tc (98)
Ruthenium 44 Ru 101.07
Rhodium 45 Rh 102.906
Cesium 55 Cs 132.905
Barium 56 Ba 137.327
Lanthanum 57 La 138.906
Hafnium 72 Hf 178.49
Tantalum 73 Ta 180.948
Tungsten 74 W 183.84
Rhenium 75 Re 186.207
Osmium 76 Os 190.23
Iridium 77 Ir 192.217
Francium 87 Fr (223)
Radium 88 Ra (226)
Actinium 89 Ac (227)
Rutherfordium 104 Rf (261)
Dubnium 105 Db (262)
Seaborgium 106 Sg (266)
Bohrium 107 Bh (264)
Hassium 108 Hs (277)
Meitnerium 109 Mt (268)
The number in parentheses is the mass number of the longest-lived isotope for that element.
Rows of elements are called periods. Atomic number increases across a period.
The arrow shows where these elements would fit into the periodic table. They are moved to the bottom of the table to save space.
Lanthanide series
Cerium 58 Ce 140.116
Praseodymium 59 Pr 140.908
Neodymium 60 Nd 144.24
Promethium 61 Pm (145)
Samarium 62 Sm 150.36
Actinide series
Thorium 90 Th 232.038
Protactinium 91 Pa 231.036
Uranium 92 U 238.029
Neptunium 93 Np (237)
Plutonium 94 Pu (244)

T C

chef'testant • \shef-'test-ent\

(*noun*)

:one of the contestants appearing on Bravo's *Top Chef*. The genesis of the term is hazily credited to multiple television bloggers.

Page 223 constitutes a continuation of the copyright page.

Library of Congress Cataloging-in-Publication Data:

Miller, Emily Wise.
Top chef : the Quickfire cookbook / foreword by Padma Lakshmi ; text by Emily Miller ; food photographs by Antonis Achilleos.
p. cm.
Includes index.
ISBN 978-0-8118-7082-5
1. Cookery. 2. Quickfire (Television program) I. Title.

TX714.M5494 2009
641.5—dc22

2009021466

Manufactured in China

Design by Vanessa Dina, Anne Donnard, and Catherine Grishaver.
Full-page food styling by Jamie Kimm and Alison Attenborough.
Full-page prop styling by Marina Malchin.

The recipes included in this book have been re-created from live cooking events on the *Top Chef* television series with some modifications for the home cook. The information in this book has been researched and tested, and all efforts have been made to ensure accuracy. Neither the publisher nor the creators can assume responsibility for any accident, injuries, losses, or other damages resulting from the use of this book.

10 9 8 7 6 5 4 3 2 1

Chronicle Books LLC
680 Second Street
San Francisco, California 94107
www.chroniclebooks.com

TOP CHEF

THE QUICKFIRE COOKBOOK

FOREWORD BY **PADMA LAKSHMI**

TEXT BY **EMILY MILLER**

FOOD PHOTOGRAPHS BY **ANTONIS ACHILLEOS**

TABLE OF CONTENTS

FROM FRIES TO FOIE GRAS: INGREDIENT CHALLENGE

UTENSILS DOWN HANDS UP: TIME CHALLENGE

OUTSIDE THE BREAD BOX: CREATIVITY CHALLENGE

SHOW YOUR MAD SKILLS: TECHNIQUE CHALLENGE

FACE THE FIRING SQUAD: JUDGES' CHALLENGE

"ROCK IT OUT.
COOK YOUR OWN FLAVORS.
BE TOP CHEFS."

JENNIFER BIESTY, SEASON 4

FOREWORD BY PADMA LAKSHMI

No other exercise, no other endeavor, brings out a chef's culinary voice more so than the Quickfire Challenge. Each morning before filming, I wonder, "What will they pull off today?" "Who will do well, who won't?" It is the most suspenseful and nerve-racking part of the show, and also the highlight of my job as an eater and as a thinker on food.

Why is the Quickfire Challenge the best part of *Top Chef*? Why is it my favorite time in the *Top Chef* kitchen? Why is it the part of my day at work I most look forward to? After all, no one goes home for a bad Quickfire, right? It is because the Quickfire begs the most visceral, instinctive, immediate, knee-jerk culinary response from our chef'testants. It comes from that quiet place in each chef that is shaken awake with the adrenaline-fueled intensity of the time-pressured challenge, that place that is the inner core of what a chef is really about.

Since we usually film the Quickfire in the morning, I tend to skip breakfast—although there can sometimes be a two- or three-hour delay before I eat. But I always try and come in hungry, to give our chef'testants the full audience of my appetite. The thing I love about our Quickfires is that it is the most succinct way to get to know how a chef thinks and acts when they have no time to second-guess themselves.

Anyone can cook when you've researched, planned, and plotted about ideas, spices, cooking methods, ingredients, and flavor profiles for hours on end. But what will a chef make when he or she is asked to work with instant rice, just thirty minutes of time, or only $10 worth of groceries? How will he do when he is asked to immediately translate a childhood fantasy with only ingredients found in one aisle of the store? Or how will she fare when she has to make an omelet with one hand literally tied behind her back? Most of us do this in our daily lives, probably not with one hand literally tied, but perhaps with one hand hanging on to a child, or a BlackBerry, or both. We have six friends coming over unannounced and only some sad zucchini and canned tomatoes with a half a box of dried couscous and . . . voilà! After a little Quickfiresque scrambling, Tuscan Papa al Pomodoro reinvented with a Moroccan twist.

I've actually written many of my recipes this way. I call it MacGyvering in the kitchen after that TV show where the title character can make a bomb out of a rubber band and an old wine cork. But our heroes and heroines mostly come from a professional environment, where creativity is cushioned by a well-stocked pantry, a vast array of mise en place, and restaurant-caliber imported ingredients at hand.

THE QUICKFIRE FORCES OUR CHEF'TESTANTS TO DO, IN A SENSE, WHAT WE DO AT HOME ON MOST NIGHTS. AND THAT IS WHAT THIS BOOK IS ABOUT: TAKING THE FRUITS OF THE CREATIVITY AND TALENT OF OUR CHEFS AND BRINGING THEM HOME TO YOU.

It shows you how to make some of the most beloved and wackiest recipes we've all salivated over (vending-machine treats anyone?), and deconstructs them for us to reproduce in the home kitchen without the enormous pressure.

Many chef'testants are stymied by the Quickfire Challenge and for good reason. No matter how hard it looks on television, as the one person who is there from start to finish, I can tell you it is much, much harder in real life. There's not much room for meditation, and the palate's subconscious must take over. You have to be "in the zone" and let the gastronomic spirit move you. And move you fast because I will usually be scooting in to tell the chefs they have five minutes before I say "put your utensils down and your hands up!" Which, by the way, came about quite naturally around the last few Quickfires of the Los Angeles season.

I am the only judge to taste every single thing cooked on the show, and I often find myself rooting for all the chef'testants to perform well in front of whatever culinary heavyweight is before them. In fact, at times, my nerves are worse than theirs, something akin to an overzealous parent at the national little league finals.

Since filming my first season of *Top Chef*, it's been a joy to see how many people, young and old, love the show, many recounting stories of their own Quickfire exercises at home with their loved ones. In my book, whatever gets kids thinking about their food and involved in the preparation of it will make them healthier eaters for life. And let's face it, anything that gets the family to break bread together is also a good thing. So, while no one at home will ask you to make a dessert using Diet Dr Pepper, isn't it nice to have the recipe just in case?

INTRODUCTION

A line of young chefs files into a spotless kitchen. The mood is somber, the music ominous, as if they are readying for a police lineup—or maybe a firing squad. It is early morning: some of the chefs are sleepy-eyed, and all are buzzing with anticipation because the glamorous Indian woman standing before them in sexy jeans is about to announce the next Quickfire Challenge.

What will it be today? Create an entrée using ingredients from a gas station mini-mart? Cook a trout dish for star chef Eric Ripert using a camping stove? Participate in a chaotic team relay race to see who can supreme oranges and whisk mayonnaise the fastest? Whatever the contest, you know it will be an all-out, tongs-to-the-wall show of grit, skill, and creativity.

Sure, anyone can turn out fine food with adequate time and top ingredients. But only someone with exquisite talent, speed, and nerves can fashion something not only edible but delicious under the tightest of constraints. One of the things that sets *Top Chef* and other Bravo shows apart from most of the genre of reality television is their focus on creative talent. The chef'testants who jump through extraordinary hoops every week have incredible skills, and we get to watch in amazement as an ever-dwindling number show us what they can do under intense pressure. For every Quickfire Challenge, host Padma Lakshmi explains the drill, calls time, and the chef'testants take off running. If they don't make it to the fridge fast enough, all the proteins will be gone. If they don't hustle, they might end up without a burner. If they didn't hear every detail of the challenge, they're toast.

Most of us will never have to nail a squirming eel to a chopping block and peel off its skin with our bare hands (though we now know how) or whip up a one-pot wonder in 45 minutes before the legendary Martha Stewart. But there is something exhilarating about watching professionals in top form compete with each other—pulling out all the stops of skill and resourcefulness to create something beautiful and original. It's like watching elite athletes, and we wait breathlessly to hear the judges as they bestow criticism and hard-earned praise.

Perhaps we home cooks will never perform at that level, but we all want to be inspired to try new things, to take chances, and to challenge ourselves in the kitchen. Such are the ideas behind *Top Chef: The Quickfire Cookbook*. Every once in a while, we need something to shock us, to knock us out of our cooking comfort zone.

THE QUICKFIRE CHALLENGE—WITH ITS CURVEBALL INGREDIENTS AND SADISTIC TIME LIMITS—IS NOT ONLY THE MOST AUDACIOUS AND HAIR-RAISING PART OF THE SHOW, IT IS ALSO THE MOST INSPIRING.

The challenges have given us fresh ideas and introduced us to unusual and exotic ingredients. And they have resulted in the many extraordinary recipes collected in these pages, from Casey's Foie Gras with Strawberry Gin Rickey (page 21) to Spike's Sensual Beef Salad (page 91).

If these chefs can turn canned Spam into a delicious entrée in fifteen minutes, we can push ourselves to do better and more interesting things in the kitchen, too. We will cook without recipes and bust out of old routines. Instead of merely sautéing the same old chicken breasts, we will flambé them! We will forego the safety of salmon in favor of octopus or monkfish. We will move beyond salt and pepper and try adding za'atar or vadouvan to a weeknight soup.

This book is not only a collection of Quickfire recipes, techniques, and memorabilia. It is a call to arms: Pick up that oyster knife and pry that little mollusk open as if your life depended on it! No meat mallet handy? Whack your chicken paillards with a cast-iron skillet! Week after week, the chef'testants have shocked and amazed us, from Ken dipping a finger in the sauce at Hubert Keller's Fleur de Lys to Ilan serving Chef Eric Ripert a chocolate bonbon stuffed with chicken liver. So we ask you: Will you continue to play it safe in the kitchen, or will you channel your creative energies and push yourself to enjoy further culinary adventures in the enduring spirit of the Quickfire?

Executive producer Dan Cutforth says of the Quickfires: "We've always joked about the day when we jump a motorbike over a tank of sharks, then force the chef'testants to cook them." Until then, it is surprising, unnerving at times, and never, ever boring to watch what happens.

BEHIND THE SCENES WITH *TOP CHEF* PRODUCERS

Top Chef executive producer Shauna Minoprio describes what happens behind the scenes:

I get in ridiculously early, drink too much coffee, eat too many bagels, and deal with a series of crises. I check the look of any food displays we are using for the Quickfire. Then, when Padma and the guest judge arrive and are ready from wardrobe and makeup, I walk them through the setup of the challenge.

During the challenge, I watch on video along with our director. We lay bets on who is going to do well, and I am almost always wrong. When the time is up, and Padma and the judge go into the kitchen to taste the dishes, our field producers grab the camera plates (a second plate of each dish is made exclusively to be photographed) and take them to our "food porn" area to be shot.

After Padma and the guest judge have gone around and tasted everything, I meet them in the food porn area, where the guest judge uses the plates laid out there as a memory aid to work out who their top and bottom picks are. (If any of the dishes have already been shot, we eat them!) Then Padma and the judge go back out to announce the winners and losers.

Executive producer Dan Cutforth is involved in every element of the show, including developing challenges, casting, and determining the look and feel of *Top Chef* every season. He describes the show's production and development:

Top Chef has exposed me to many wonderful culinary experiences that I would never otherwise have had. In terms of cooking I can truly say that I do incorporate techniques and ideas I learned on *Top Chef* into my own cooking. I am now a lot less satisfied by average cooking than I once was.

Who came up with the Quickfire Challenge concept, and what was the inspiration?

I have a tendency to think I invented everything, but in this case I think I did at least name the challenge "Quickfire," inspired by quickfire rounds at the end of game shows where everything happens at great speed. From the earliest stages of development of this show it seemed clear that every episode would need more than one challenge, but one of them would need to be fast-paced and more simple. The concept developed from there, and they were originally referred to internally as the "mini-challenge" until "Quickfire" began to stick. In the first season they were supposed to be quick tests of the various skills you needed to be a good chef, like knife skills (the fruit plate), calmness under pressure (being thrown onto the line in Hubert Keller's kitchen), a good palate (the Blind Taste Test). In subsequent seasons they have become much more involved and varied.

What would viewers be surprised to know about what happens behind the scenes?

You have to have stamina to work on *Top Chef*. The show often tapes into the early hours of the morning. The pace at which things happen is always a surprise to visitors—it's mostly very slow, and we also shoot a lot of stuff that ends up on the cutting room floor.

What do you look for when casting chef'testants? Do they usually conform to your expectations once they are on the show, or do they surprise you?

We look for excellent chefs, who are creative, versatile, and who have compelling personalities. There are always surprises when the chefs come together and each cast seems to have its own personality, and you never know what that's going to be like. One thing that is consistent is that they all get very frustrated with the production, until the show starts to air and everyone is treating them like rock stars and offering them jobs and sex.

TOP: EXECUTIVE PRODUCER DAN CUTFORTH WITH THE SEASON 4 CAST AND CREW. **BOTTOM LEFT:** KEY GRIPS ON SET IN THE *TOP CHEF* KITCHEN
BOTTOM RIGHT: CAMERA CREW ON SET FOR THE OPENING QUICKFIRE OF SEASON 5 ON GOVERNORS ISLAND, NEW YORK

FROM FRIES TO FOIE GRAS: INGREDIENT CHALLENGE

RECIPES

ILAN'S CORNER-STORE DEVILED EGGS

1 ounce salami, cut into thin strips

8 Original, Chile Picante, or Caliente flavor CornNuts

2 large or jumbo eggs

1 tablespoon olive oil

¼ teaspoon paprika, plus more for garnish

Salt

¼ cup chopped fresh oregano

Grated zest and juice of ¼ lime

MAKES 4

15 MINUTES

SEASON 2, EPISODE 4

QUICKFIRE CHALLENGE:

Create an amuse-bouche with $10 of ingredients from a vending machine.

1. In a small sauté pan over medium-high heat, fry the salami until crisp. Cover and set aside.
2. Put the CornNuts in a resealable plastic bag and crush with a mallet or heavy pan.
3. Have ready an ice-water bath. Bring a small saucepan with 6 inches of cold water to a boil over high heat. Use a spoon to carefully lower the eggs into the boiling water. Reduce heat to medium-high and boil the eggs for 8 minutes. Transfer the eggs to the ice-water bath and allow to cool.
4. Once cool, peel the eggs and cut them in half. Pop the yolks into a small bowl, reserving the whites. To the yolks, add the olive oil, crushed CornNuts, the ¼ teaspoon paprika, salt to taste, oregano, and the lime juice. Divide the mixture among the egg-white "cups," mounding the filling. Garnish with lime zest, paprika, and fried salami.

FUN FACT

WHAT MAKES A FOOD "DEVILED?"

Traditionally, "deviled" food equaled hot and spicy—kicked up with cayenne, paprika, or hot mustard. Deviled eggs are often spiced with paprika, but their name dates to the Underwood Company, which began selling canned foods with the Red Devil logo in the mid-nineteenth century. Foods mashed and seasoned in this way since then have been called "deviled," including the iconic eggs, which became a chic appetizer in the 1950s and '60s.

SARA N.'S APPLE, BACON, AND RICOTTA TARTLETS

One 9-inch frozen piecrust, thawed

½ cup (4 ounces) ricotta cheese

¼ cup (2 ounces) soft goat cheese, such as chèvre

1 tablespoon minced fresh chives

3 slices bacon, diced

1 Granny Smith apple, peeled, cored, and diced

2 tablespoons grated Parmigiano-Reggiano

1 HOUR

SEASON 3, EPISODE 5

QUICKFIRE CHALLENGE:

Create a dish using a premade piecrust.

TECHNIQUE

DOUGH!

The easiest way to make dough is using a food processor. The most important element is cold, so keep your hot little hands off the butter and chill your water with an ice cube before measuring it. Some cooks like to keep butter in the freezer for a few hours before using, especially in a hot climate. Place the dry ingredients (flour and salt or sugar, depending on your recipe) in the bowl of the processor and pulse. Add the butter and pulse again until the mixture resembles small peas. Add the water a little at a time until the dough just pulls together. Don't get excited and overmix. Pop the mixture out onto a floured work surface, preferably marble (again, cold), form into disks, wrap in plastic, and chill for about 30 minutes before using. Double the recipe and freeze the extra dough to make your work even easier the next time.

MAKES 4

1. Preheat an oven to 375°F.
2. On a floured work surface, roll out the piecrust and cut into 4-inch rounds. Press the rounds into tartlet pans or muffin cups, forming 4 individual tart shells. Prick the bottom of each tart shell with a fork, then fill with dried beans or pie weights. Bake until pale gold, 10 to 15 minutes. Cool on a wire rack, then remove the weights. Reduce the oven temperature to 350°F.
3. Meanwhile, in a medium bowl, combine the ricotta, goat cheese, and chives. Set aside.
4. In a medium skillet over medium heat, cook the bacon until crisp. Use a slotted spoon to transfer the bacon to a bowl, and add the apple to the skillet, cooking until soft. Remove the apple from the pan, mix with the bacon, cover, and set aside.
5. Form the grated cheese into a thin round on a Silpat- or parchment-lined baking sheet. Bake at 350° F until golden brown, about 5 minutes. Cool slightly, then cut the round into 4 triangles.
6. To assemble, put a spoonful of goat cheese mixture into each tart shell. Top with the apple-bacon mixture, dividing it evenly among the shells. Return to the oven until heated through, about 5 minutes. Serve each tartlet topped with a cheese triangle.

ILAN'S SPINACH, ARTICHOKE, AND WHITE BEAN SALAD

One 15-ounce can whole new potatoes, rinsed, drained, and quartered

2 teaspoons olive oil plus 3 tablespoons

Salt and pepper

Three 15-ounce cans chopped spinach, rinsed and drained

½ teaspoon red pepper flakes

One 15-ounce can cannellini beans, rinsed and drained

One 15-ounce can artichoke hearts, rinsed and drained

One 4-ounce can Vienna sausages, drained

SERVES 2–4

1. Preheat an oven to 425° F.
2. In a large baking dish, combine the potatoes, the 2 teaspoons olive oil, and salt and pepper to taste. Roast until crisp, about 20 minutes.
3. In a large sauté pan over medium heat, warm 1 tablespoon of the olive oil. Add the spinach and pepper flakes and stir to warm. Add the beans, stir to combine, and reduce heat to medium-low.
4. In another saucepan, heat the remaining 2 tablespoons olive oil over medium-high heat. Fry the artichokes, turning, until crispy, 2 to 3 minutes. Remove from the pan and pour off any excess oil. Add the sausages to the pan, increase heat to high, and sear until lightly browned on all sides, 1 to 2 minutes.
5. To serve, stir together the potatoes, spinach mixture, and artichokes. Mound on a platter and top with the sausages.

"EVERY TIME YOU CHANGE THE FORMAT IN WHICH YOU'RE DOING YOUR CRAFT YOU LEARN MORE."

ILAN

45 MINUTES

SEASON 2, EPISODE 6

QUICKFIRE CHALLENGE:

Make something delicious using three different canned foods.

ABOUT AN INGREDIENT

VIENNA SAUSAGES

The God's honest truth is, if you're not starving and therefore in need of a protein that costs under 50 cents a can, you probably have no business eating Vienna sausages. These little bundles are composed of ground chicken, pork, and beef, packed tightly together in a can with a little chicken broth. They were first sold in the late nineteenth century when Arthur and Charles Libby started a Chicago-based company selling beef packed in brine. And if, like Ilan, you can brown them and make them taste moderately good in some kind of flavorful preparation, perhaps you truly are worthy of being a Top Chef.

SARA M.'S GOAT CHEESE, FETA, AND FIG TARTS

One 9-inch frozen piecrust, thawed

1 tablespoon sherry vinegar

1/4 cup olive oil

1/4 cup finely diced shallots

Salt and pepper

1 cup heavy cream

4 egg yolks

1/2 cup (4 ounces) soft goat cheese

1/2 cup (4 ounces) crumbled feta cheese

10 dried figs, cut into rounds

4 stalks asparagus, trimmed and cut into 1-inch pieces

1 1/2 cups (6 ounces) Brie, cut into 1-inch pieces

1 small bunch of chives, cut into 1-inch pieces

MAKES 4

1. Preheat an oven to 375° F.
2. On a floured surface, roll out the piecrust and cut into 4-inch rounds. Place the rounds into 4 tartlet shells. Fill shells with pie weights and bake until pale gold, 10 to 15 minutes. Remove from the oven and remove the pie weights. (Leave the oven on.)
3. Meanwhile, make the vinaigrette. Combine the vinegar, olive oil, and shallots in a small bowl. Season with salt and pepper.
4. In a medium bowl, whisk together the cream and yolks. Add the goat cheese and feta, mashing with a wooden spoon to combine. Season with salt and pepper. Divide the mixture among the tartlet shells and top with figs. Bake until a knife inserted in the center of a tart comes out clean, 30 to 40 minutes.
5. Toss together the asparagus, Brie, and chives. Toss with the vinaigrette.
6. To serve, position each tart on a plate and top with salad.

35 MINUTES PREP, 40 MINUTES BAKING

SEASON 3, EPISODE 5

QUICKFIRE CHALLENGE:

Create a dish using a premade piecrust.

ABOUT AN INGREDIENT

WHAT IS WHITE ASPARAGUS?

Albino asparagus? This unusual vegetable has become popular in the United States, despite its high price and short growing season. White asparagus is grown in abundant dirt that does not permit light to reach the plant, thus depriving it of chlorophyll and the color green. White asparagus has a milder flavor than regular asparagus, and is better for braising or other long preparations—but some cooks simply find it bland, as did Leah from Season 5, who, nonetheless, made a good enough soup from it to win the Quickfire Challenge (see recipe, page 163).

> "I DIDN'T OVERANALYZE THINGS LIKE SOME PEOPLE. I DIDN'T KNOW WHAT TO EXPECT, SO I JUST PREPARED MYSELF TO TAKE IT DAY BY DAY AND DO THE BEST I COULD."

SARA M.

CASEY'S FOIE GRAS WITH STRAWBERRY GIN RICKEY

Foie Gras

1 teaspoon butter plus 4 tablespoons

1 small shallot, minced

2 cups fresh blackberries, rinsed and dried

2 cups fresh raspberries, rinsed and dried

1/4 cup water

1/4 cup sugar plus 1 tablespoon

1 sprig thyme

2 eggs, beaten

2 cups heavy cream

6 thin slices baguette

2 tablespoons brandy

1 vanilla bean

1/2 cup packed brown sugar

1 banana, peeled and cut into 1/4-inch slices

1 pound foie gras, cut into 6 even pieces

Salt and pepper

1/2 cup pecans, chopped

1 tablespoon grapeseed oil

1 pound arugula, stemmed

1 teaspoon balsamic vinegar

Juice of 1 lemon (about 3 tablespoons)

Strawberry Gin Rickey

6 fresh strawberries, diced

1 1/2 ounces Bombay Sapphire gin

3/4 ounce fresh lime juice

1/4 ounce good-quality aged balsamic vinegar

1 ounce simple syrup (see page 22)

Drizzle of strawberry liqueur (optional)

Crushed ice

Lime twist for garnishing

45 MINUTES FOR FOIE GRAS, 5 MINUTES FOR DRINK

SEASON 3, EPISODE 4

QUICKFIRE CHALLENGE:

Create an appetizer to pair with a Bombay Sapphire cocktail.

SERVES 6; MAKES 1 DRINK

FOR FOIE GRAS

1. Preheat an oven to 350° F.

2. Heat a medium sauté pan over medium heat. Add the 1 teaspoon butter. When the butter is melted, add 1 teaspoon of the shallot and cook, stirring, until softened. Add the berries, water, the 1/4 cup sugar, and the thyme. Reduce heat to low, and simmer until the berries break up and the juices thicken. Remove from heat and cool to room temperature. Strain the juice and set aside.

3. Whisk together the eggs, cream, and the remaining 1 tablespoon sugar. Add the baguette slices and soak until moist.

4. Heat a medium sauté pan over medium heat. Add 1 tablespoon of the butter. When the butter is melted, add the prepared bread slices. Brown on one side, then flip so both sides are golden and toasted. Remove from heat, cover, and set aside.

5. In a medium sauté pan melt 2 tablespoons of the butter over high heat. Add the brandy and cook to reduce to about 2 teaspoons. Split the vanilla bean and scrape out the seeds. Reduce heat to medium-low, add the vanilla seeds and brown sugar, and cook slowly until thickened. Add the banana slices and coat in syrup. Set aside.

6. Season the foie gras with salt and pepper. Pour the pecans into a shallow bowl. Press each piece of foie gras into the pecans, rolling to coat.

CONTINUED

ABOUT AN INGREDIENT
FOIE GRAS

Beloved by many chefs, demonized by animal-rights activists—perhaps no other food has elicited so much passion in the recent past as foie gras, the fattened liver of a duck or goose. Though some American cities and counties have flirted with a ban on foie gras, chefs and diners have proven unready to give up the goose.

Foie gras can be served in several ways: American chefs often sear it in a pan and serve with some kind of sweet, fruity reduction or jam; in France it is traditionally made into a rich terrine or pâté and served cold on toast points, accompanied by Sauternes or other dessert wine. With its silky taste and texture, and pricey cachet, it's not unusual for chefs to employ foie gras whenever they want to evoke luxury and decadence.

TECHNIQUE
SIMPLE SYRUP

Simple syrup is essentially sugar syrup. Use 2 parts water to 1 part sugar for a medium-density syrup, bring the mixture to a boil over medium heat, and simmer for a few minutes. Keep covered in a refrigerator indefinitely.

7. Heat the grapeseed oil in a large oven-safe sauté pan over medium-high heat. Sear the foie gras on one side until browned, about 3 minutes. Flip to cook the other side. When browned, slide the pan into the oven for 2 minutes to finish.
8. While the foie gras is finishing, move the toast to the oven to warm.
9. In a large sauté pan over medium heat, melt the remaining 1 tablespoon butter and add the remaining shallot. Cook until soft, about 2 minutes. Add the arugula; stir to wilt and coat with the butter and shallot. Add the vinegar and lemon juice and season with salt and pepper. Remove from heat.
10. To assemble, mound the arugula in the center of each warmed plate. Lean French toast against the pile of arugula. Place a spoonful of bananas beside the toast. Place a piece of foie gras on top of the toast, and drizzle the foie gras and salad with the berry sauce.

FOR STRAWBERRY GIN RICKEY

1. In a cocktail shaker, muddle the strawberries. Add the gin, lime juice, vinegar, simple syrup, strawberry liquor (if using), and ice. Shake vigorously. Strain into a rocks glass filled with crushed ice. Balance the lime twist on the rim for garnish.

"I HAVE TO SAY, I DON'T KNOW IF IT'S ONLY ME, BUT I'M TIRED OF EATING FOIE GRAS."

CHEF ROCCO DISPIRITO

BOMBAY SAPPHIRE MIXOLOGIST AND GUEST JUDGE JAMIE WALKER

ANTONIA'S RICE SALAD WITH SEARED SKIRT STEAK

One 1½-pound skirt, flap, or hanger steak (see sidebar)

Olive oil for coating

Salt and pepper

2 egg yolks, beaten

About 3 tablespoons rice vinegar

½ teaspoon sugar

About ⅓ cup canola oil

1 bunch cilantro, leaves and stems chopped

4 handfuls mâche

4 handfuls arugula

2 cups cherry tomatoes, halved

1 small red onion, thinly sliced

One 8.8-ounce bag Uncle Ben's Ready Rice, Garden Vegetable, prepared according to package instructions, or 2 cups cooked white or brown rice, at room temperature

SERVES 3–4

1. Coat the steak with olive oil and season with salt and pepper. Grill or broil to the desired degree of doneness, 3 to 4 minutes per side for medium-rare. Set aside to cool, then slice into thin strips.

2. Whisk the yolks, vinegar, ¼ teaspoon salt, and sugar with the canola oil. Stir in the cilantro. Taste and add a little more vinegar or canola oil if desired.

3. In a large bowl, toss the mâche, arugula, tomatoes, and onion. Mix in the prepared rice and toss with the vinaigrette.

4. To serve, mound the salad in the center of each plate and top with the steak strips.

"THIS IS SOMETHING I'VE EATEN AS A CHILD AND MY MOM LOVES TO MAKE. . . . IT'S A STRANGE COMBINATION BUT ONE THAT I ACTUALLY LOVE."

ANTONIA

25 MINUTES

SEASON 4, EPISODE 8

QUICKFIRE CHALLENGE:

Create a healthful entrée using Uncle Ben's microwavable rice.

ABOUT AN INGREDIENT
STEAK

Steaks come in many varieties and price ranges. When a steer is butchered, it is cut precisely into traditional cuts, which range from the lean and tender (sirloin $$$) to the tough but tasty (flank and skirt steak $$), or the marbled and in need of *mucho* cooking time (brisket, short ribs $). "Flat" steaks such as flank, skirt, and hanger are popular for their low price and great flavor, especially when seasoned and grilled, though if not cooked properly, these cuts can be very tough. They are all large, flat pieces of meat that come from the same region of the cow's side, or flank, between the ribs and hips.

BLIND TASTE TEST

The Blind Taste Test, the culinary equivalent of a spelling bee, has been a humbling and intense Quickfire Challenge since Season 1. Think you've got the palate of a Top Chef? It's time to put your taste buds to the test!

RULES

1. The host and judge assigns each person a secret ingredient. Participants should bring a lower-end and a higher-end version of their ingredient.
2. Each taster is blindfolded. One by one (and separately!), participants taste both versions of each secret ingredient.
3. The player who can correctly identify the most "gourmet" versions of the ingredients wins.

BLIND TASTE TEST GOURMET EDITION

SUGGESTED SECRET INGREDIENTS:

OLIVE OIL: Store brand versus imported extra-virgin, first-press oil

WINE: Wine in a box versus a bottle in the $15 to $20 range (pinot noir, pinot grigio, and pinot gris are fine tasting wines)

CRAB: Imitation versus lump

CHEESE: Velveeta versus imported Cheddar or Gouda

CHOCOLATE: Hershey's Special Dark bar versus a dark chocolate with 65 percent or higher cacao content, such as Scharffen Berger, Lindt, or Guittard

ICE CREAM: Store-brand vanilla versus Häagen-Dazs vanilla

WHIPPED CREAM: Cool Whip versus homemade whipped cream

MAPLE SYRUP: Mrs. Butterworth's versus Vermont Grade B maple syrup

BUTTER: I Can't Believe It's Not Butter! versus European butter

"FOR ALL THE FAT PEOPLE ACROSS AMERICA—YOU TOO CAN BE A GRAND MASTER OF SNACKS!"

MIGUEL "CHUNK LE FUNK" MORALES

BLIND TASTE TEST
JUNK FOOD EDITION

In Season 1, the chef'testants held an impromptu junk food taste test, and Miguel dominated. In this junk food taste test, whoever identifies the most ingredients wins.

SUGGESTED SECRET INGREDIENTS:

JELLY DONUTS: Different fillings (lemon, raspberry, strawberry)

JELLY BEANS: Different flavors

ICE CREAM: Different flavors and fun mix-ins such as candy and fruit

CEREAL: Lucky Charms, Cap'n Crunch, Froot Loops, Frosted Flakes, Cocoa Puffs, Apple Jacks, Cheerios

CANDY: Nerds, Pop Rocks, Whoppers, Red Vines versus Twizzlers

CHIPS: Funyuns, Flamin' Hot Cheetos, Doritos, Combos, CornNuts

SNACK CAKES: Ho Hos, Ding Dongs, Twinkies, Zingers

SODA CHALLENGE: Regular versus diet (Dr Pepper, cola)

STEPHANIE'S WHITE ALE–ORANGE JUICE MUSSELS

4 oranges

1 egg yolk at room temperature

¼ cup white wine vinegar

1 teaspoon Dijon mustard

1½ cups canola oil

½ jalapeño chile, seeded and chopped

1 teaspoon honey

Salt and pepper

2 tablespoons butter

1 fennel bulb, trimmed and thinly sliced

½ small onion, thinly sliced

3 cloves garlic, minced

1 pound mussels, cleaned and debearded (see sidebar)

½ cup white ale, such as Hoegaarden

Fresh cilantro leaves for garnishing

Chopped scallions for garnishing

1 loaf crusty bread, warmed in oven or sliced and grilled, for serving

SERVES 2 AS AN APPETIZER

1. Juice 3 of the oranges, and cut the other into supremes (see Glossary).
2. Combine the yolk, vinegar, and mustard in a blender or food processor. Pour the oil through the feed tube in a slow drizzle until the mixture thickens and becomes creamy. Transfer the mixture to a small bowl and add the chile and honey. Season with salt and pepper.
3. In a medium saucepan over medium heat, melt the butter. Add the fennel, onion, and garlic and sweat for a few minutes (see page 93). Increase heat to medium-high and add the orange juice and mussels. Cover and steam just until the mussels open and the meats loosen from their shells, 3 to 5 minutes. Add the ale and season with salt and pepper.
4. To serve, pour the mussels and broth into a large bowl. Drizzle with the vinaigrette and garnish with the orange segments and some cilantro leaves and scallions. Serve with the warmed bread.

45 MINUTES

SEASON 4, EPISODE 6

QUICKFIRE CHALLENGE:

Create a simple dish to pair with a beer for Chef Koren Grieveson.

TECHNIQUE

DEBEARDING MUSSELS

Soak the mussels in cold water for about 30 minutes. One by one, pull out the tough little mass of fibers the mussel uses to attach to rocks by grasping the "beard" tightly and yanking it swiftly toward the hinge end of the mussel. If you buy farmed mussels, which are increasingly common, you can skip the soaking and debearding and go straight to the cooking and eating.

C.J.'S GRILLED SQUID WITH AVOCADO, WATERMELON, AND ENDIVE

10 pieces cleaned, drained baby squid (tentacles only)

1 clove garlic, grated

1 tablespoon minced fresh flat-leaf parsley

7 tablespoons olive oil

Grated zest and juice of 1 lemon

Salt and pepper

½ teaspoon cayenne pepper

3 tablespoons diced watermelon rind

1 avocado

One 1-inch-thick round of watermelon, quartered, rind removed and reserved

1 head endive, thinly sliced

Minced fresh chives

SERVES 2

1. Combine the squid, garlic, parsley, 3 tablespoons of the olive oil, a drop of lemon juice, and salt and pepper to taste and marinate for 30 minutes. Refrigerate until ready to use.

2. In a small bowl, combine the lemon zest, remaining lemon juice, cayenne, and watermelon rind. Macerate for 20 minutes, then add 2 tablespoons of the olive oil.

3. Place a grill pan over high heat and let the pan get very hot. Cut the avocado lengthwise and twist the halves apart. Tap a knife blade against the pit until it dislodges, then twist the pit to pull it out. Rub each half with ½ tablespoon olive oil and sprinkle with salt and pepper. Carefully place the pieces flesh side down on the grill pan, and after both halves are marked with grill lines, about 2 minutes, remove from the grill. Use a spoon to scoop the avocado flesh free of the skins. Keep each half intact.

4. Heat the remaining 1 tablespoon olive oil in a medium sauté pan over medium heat. Season the watermelon quarters with salt and pepper. Lay the pieces flat in the pan to caramelize both sides, about 3 minutes per side.

5. Place a grill pan over high heat and let the pan get very hot. Add the squid and grill, stirring, until opaque, about 2 minutes.

6. Toss the endive with the vinaigrette, reserving some vinaigrette for drizzling. Place 2 pieces of watermelon on each plate. Top the watermelon with endive, chives, and avocado. Spoon watermelon rind into the indentation in the avocados, and top with grilled squid. Drizzle with the remaining vinaigrette.

1 HOUR, 30 MINUTES, INCLUDING MARINATING

SEASON 3, EPISODE 4

QUICKFIRE CHALLENGE:

Create an appetizer to pair with a Bombay Sapphire cocktail.

TECHNIQUE

CLEANING SQUID

To clean a squid, first gently pull off the head, which should also pull out most of the innards. Reach inside the body cavity with your finger and pull out any innards left behind as well as the hard little piece called the "quill." Discard the innards, quill, and head (unless you plan to harvest the ink for another cooking purpose or for printing). Depending on your preference, peel off the dark filmy skin from the squid (some cooks like to leave it on), and prepare the squid according to the recipe instructions.

JEFF'S OAT-FRIED CHICKEN AND GRITS

2 cups old-fashioned rolled oats (see sidebar below)

1½ tablespoons olive oil

Salt and pepper

2 skinless, boneless whole chicken breasts (about 1 pound), butterflied (see sidebar, page 31)

2¼ cups buttermilk

Leaves from 5 sprigs fresh thyme

2 tablespoons butter, melted, plus 3½ tablespoons

2 cups chicken stock

¾ cup stone-ground or instant grits

1½ cups heavy cream

2 cups fresh or frozen corn kernels

1½ cups all-purpose flour

3 eggs, beaten

1 zucchini, cut into 1-inch-thick sticks

Canola oil for frying

1 HOUR PREP, PLUS 3 HOURS FOR SOAKING

SEASON 5, EPISODE 10

QUICKFIRE CHALLENGE:

Create a dish using oats.

ABOUT AN INGREDIENT
ROLLED OATS

Old-fashioned rolled oats are made by steaming groats and flattening them with a roller. Instant oats are usually packaged with salt and sugar.

SERVES 4

1. Preheat an oven to 300°F. Toss the oats, ½ tablespoon olive oil, and salt and pepper to taste to combine. Spread the mixture on a baking sheet and toast for about 25 minutes, stirring once or twice. Transfer the baking sheet to a wire rack to cool. Grind finely in a blender or food processor and set aside.

2. Combine the chicken and 2 cups of the buttermilk. Soak for at least 3 hours, refrigerated.

3. Remove the chicken from the buttermilk and pat dry. Season with thyme and salt and pepper. Lay one piece of chicken between two sheets of clear plastic wrap. Using a meat mallet and working from the center of the chicken out toward the edges, pound gently until the chicken is very thin. Brush each breast with the melted butter and roll in 2 tablespoons of the oat mixture.

4. In a large sauté pan, heat 2 tablespoons of the butter over medium-high heat until it begins to foam. Add the chicken breasts and sauté until golden, about 1 minute. Flip the breasts and cook until the other side is golden, about 1 minute. Transfer to the oven to keep warm until serving.

5. In a large saucepan over medium-high heat, bring the chicken stock to a simmer. Whisk in the grits and stir well. Bring back to a boil. Cook the grits for 5 minutes, stirring constantly. Lower heat to medium, add the cream and cook for 15 minutes. Season with salt and pepper, then stir in the remaining 1½ tablespoons butter. Reduce heat to low and cook the grits, stirring occasionally, until they become creamy, about 15 minutes.

6. In a separate sauté pan, heat the remaining 1 tablespoon olive oil over medium heat. Add the corn and cook until tender, 2 to 3 minutes. Season with salt and pepper, and then stir into the grits along with ¼ cup of the ground toasted oats.

7. Season the flour with salt and pepper. Mix in the remaining ground oats. Whisk the remaining 1/4 cup buttermilk into the eggs.
8. Dip the zucchini sticks first into the egg mixture, and then roll in the oat mixture to coat. Lay the coated sticks on a wire rack as you work.
9. Heat 1 inch of canola oil to 350° F in a large sauté pan over medium-high heat. Working in batches, fry the zucchini until golden, 2 to 3 minutes. Drain on paper towels and sprinkle with salt.
10. To serve, mound a scoop of grits on each plate. Top with a chicken breast and a couple of zucchini sticks.

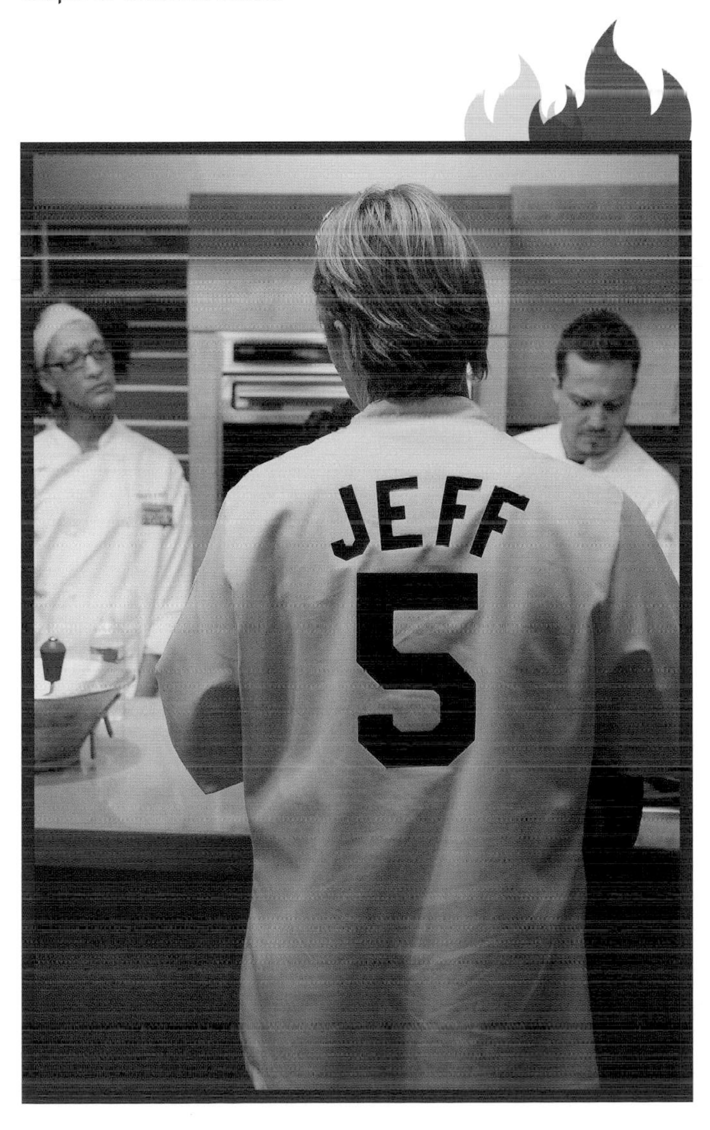

TECHNIQUE

BUTTERFLIED CHICKEN BREASTS

To butterfly chicken breasts, slice the breasts in half lengthwise and approximately three quarters of the way through, leaving a seam on one end. The opened breast will resemble butterfly wings. Plan ahead to make this recipe, as you'll need to soak the chicken breasts for a few hours.

"WE GO SHOPPING, THE SOCCER MOMS ARE THERE BUYING THEIR GROCERIES, AND FIFTEEN ADRENALINE-FILLED YOUNG CHEFS COME RUNNING IN WITH OUR BASKETS. IT LOOKS LIKE WE'RE ROBBING THE PLACE."

JEFF

IDENTIFY THAT INGREDIENT

Every season, the chef'testants eagerly anticipate the Taste Test Quickfire, where their mastery of ingredients is put to the ultimate test. See if you're fit to join the ranks of champions by identifying the ingredients below.

1

2

3

4

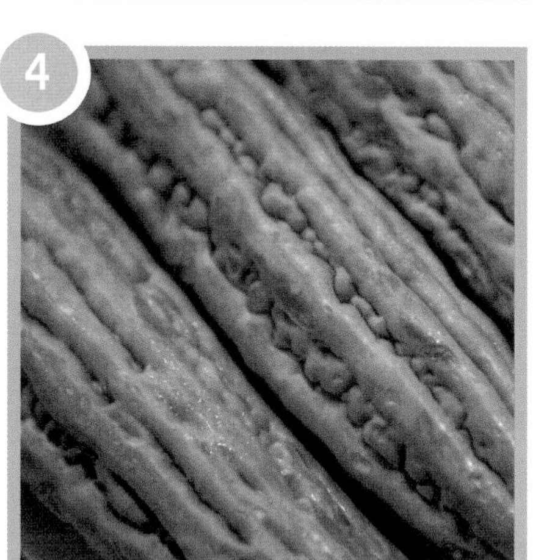

5

6

TASTE TEST HALL OF FAME

ANDREA	SEASON 1 BLIND TASTE TEST CHAMPION
MIGUEL	SEASON 1 JUNK FOOD BLIND TASTE TEST CHAMPION
CASEY	SEASON 3 TASTE TEST CHAMPION
ANTONIA	SEASON 4 BLIND TASTE TEST CHAMPION
HOSEA	SEASON 5 IDENTIFY THAT INGREDIENT CHAMPION

ANSWER KEY: 1) QUAIL EGGS, 2) EGGPLANT, 3) FARRO, 4) BITTER MELON, 5) ANCHOVIES, 6) TAPIOCA

SAM'S TEMPURA SHRIMP AND PEACH SANDWICH

- 1/2 cup Italian dressing
- 1 ripe peach, peeled, pitted, and sliced
- 1 tablespoon mayonnaise
- 1 tablespoon bottled barbecue sauce
- 3/4 cup all-purpose flour
- 1 tablespoon baking powder
- 1/2 cup water
- 1 teaspoon Sriracha sauce
- 1 egg, lightly beaten
- 3 medium shrimp, peeled, deveined (see page 122), and split lengthwise
- 2 tablespoons butter
- 2 thick slices white bread
- 4 slices red or yellow tomato
- 6 leaves baby arugula

MAKES 1

1. In a small bowl, combine the Italian dressing and peach slices. Refrigerate for at least 20 minutes to pickle.
2. Meanwhile, in another bowl, combine the mayonnaise and barbecue sauce. Set aside.
3. In a large bowl, whisk together the flour, baking powder, water, Sriracha, and egg until smooth. Dip the shrimp halves into the batter, coating evenly.
4. In a large sauté pan, melt the butter over medium-high heat. When the butter is bubbling, add the shrimp to the pan and fry until golden, 2 to 3 minutes per side.
5. Meanwhile, toast the bread to the desired doneness.
6. To assemble, spread the warm toast pieces with the mayonnaise mixture. Layer the tomato slices, arugula, and peaches on a piece of toast. Top with the shrimp, then finish with the other toast slice. Cut the sandwich in half and serve.

30 MINUTES PREP, PLUS 20 MINUTES FOR PICKLING

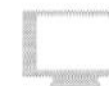

SEASON 2, EPISODE 10

QUICKFIRE CHALLENGE:

Create a snack using three Kraft Foods products.

"I WENT TO CULINARY SCHOOL, MY PARENTS SPENT TONS OF MONEY, AND I CAME OUT WITH A DEEP LOVE OF MAYONNAISE."

SAM

LIA'S PORK TENDERLOIN WITH ARTICHOKE TART

3 shallots, minced

Juice of 5 Meyer lemons (about 1 cup)

1 tablespoon butter, plus more for sautéing (optional)

2 cloves garlic, minced

Leaves from 1 bunch thyme

One 1¼ pound pork tenderloin

4 tablespoons olive oil

Salt and pepper

One 9-inch frozen piecrust, thawed

One 15-ounce can artichoke hearts, rinsed and sliced

½ cup dry white wine

2 tablespoons sour cream

1 tablespoon minced Spanish chorizo

2 tablespoons diced piquillo pepper

1 tablespoon sherry vinegar

1 ounce aged Gruyère cheese

1 pear, peeled, cored, and julienned

1 bulb fennel, trimmed and cut into thin slices

SERVES 4

1. Preheat an oven to 375° F.
2. In a large saucepan over medium heat, cook one-third of the shallots in a little butter or white wine until golden, about 10 minutes. Add the lemon juice, increase heat to medium-high, and cook to reduce until the liquid is syrupy, about 20 minutes. Reduce heat to low, add the butter, and whisk to thicken. Keep the reduction warm while you prepare the pork.
3. Combine the garlic, half the remaining shallots, and the thyme. Crush with a knife. Coat the tenderloin with 2 tablespoons of the olive oil, season with salt and pepper, and then rub the pork with the garlic-thyme mixture. Set aside.
4. Fill the piecrust with pie weights and bake until firm, about 15 minutes. Remove the pie weights and continue baking until golden, another 10 minutes. Remove from the oven and set aside. Increase the oven temperature to 450° F.
5. Meanwhile, in a large saucepan, combine the sliced artichokes and wine, and add water to cover. Cook over medium-high heat until the liquid is reduced to a syrup, about 10 minutes. Season with 1 teaspoon salt, add the sour cream, and pour into the tart shell.
6. In a large sauté pan or Dutch oven, heat 1 tablespoon olive oil over medium-high heat. Add the remaining shallots and the chorizo and sweat for 2 minutes (see page 93). Add the tenderloin and brown on all sides, about 5 minutes. Place in the oven and roast to the desired doneness, 15 to 20 minutes.
7. Combine the piquillo pepper, vinegar, and the remaining 1 tablespoon olive oil. Season to taste. Using a peeler, shave a few thin slices of the Gruyère. Mix the cheese with the pear, fennel, and dressing.
8. To assemble, divide the tart among 4 plates and top with salad. Slice the tenderloin, overlap on the plates, and drizzle with the warm lemon reduction.

1 HOUR, 30 MINUTES

SEASON 3, EPISODE 5

QUICKFIRE CHALLENGE:

Create a dish using a premade piecrust.

TOP CHEF ROYALTY

Some chef'testants returned to the same ingredients over and over. In commemoration, we hereby bestow the following titles:

VEGETARIAN GODDESS:
ANDREA, SEASON 1

ULTIMATE SOMMELIER:
STEPHEN, SEASON 1

FOAM MASTER:
MARCEL, SEASON 2

SEAFOOD KING:
BRIAN, SEASON 3

PASTA PRINCESS:
NIKKI, SEASON 4

SCALLOPS QUEEN:
JAMIE, SEASON 5

STEFAN'S BANANA MOUSSE WITH OATMEAL-ALMOND CRISP

8 ounces white chocolate, chopped

4 egg yolks

$^{1}/_{3}$ cup sugar

1 tablespoon dark rum

$1^{1}/_{2}$ cups heavy cream

$^{1}/_{3}$ cup diced banana

$^{2}/_{3}$ cup old-fashioned rolled oats (see page 30)

$^{1}/_{2}$ cup sliced almonds

30 MINUTES PREP, PLUS COOLING

SEASON 5, EPISODE 10

QUICKFIRE CHALLENGE:

Create a dish using oats.

TEAM EURO

Stefan and Fabio made up Season 5's "Team Euro," often stealing the show with their unforgettable quips and chummy camaraderie (not to mention their mad cooking skills).

"We're always together."—Fabio

"Stefan can't be apart from Fabio for more than ten minutes at a time." —Jamie

"I love the guy. He is so much fun and so good to be around with." —Stefan

SERVES 4–6

1. Melt $3^{1}/_{2}$ ounces of the chocolate in a double boiler or metal bowl set over, but not touching, simmering water in a saucepan. Cool to room temperature.
2. Meanwhile, in a large bowl, whisk the egg yolks with the sugar until fluffy. Whisk in the melted chocolate and rum.
3. With a balloon whisk or in a stand mixer, whisk the cream until it holds soft peaks. Fold the whipped cream into the chocolate mixture. Fold in the diced banana. Cover with plastic wrap and refrigerate for at least 3 hours.
4. Preheat an oven to 350° F.
5. Combine the oats and almonds in a bowl. Spread on a baking sheet and toast until golden, about 10 minutes. Remove from the oven, transfer the pan to a wire rack, and cool.
6. In a double boiler, melt the remaining $4^{1}/_{2}$ ounces chocolate. Break the almond mixture into chunks and gently mix with the chocolate until it holds together. Form into 4 or 6 even pieces and refrigerate for at least 30 minutes.
7. To serve, spoon or pipe the banana mousse into goblets or small bowls. Garnish with an almond crisp.

DALE L.'S STRAWBERRY SAFFRON FREE-FORM TART

One 9-inch frozen piecrust, thawed

2 cups fresh strawberries, quartered

1 pear, peeled, cored, and thinly sliced

1/2 cup sugar

1/3 cup dry white wine

Pinch of saffron threads

1/2 cup soft goat cheese, such as chèvre

1/2 cup cream cheese

Seeds of 1 vanilla bean

1 HOUR PREP, PLUS COOLING

SEASON 3, EPISODE 5

QUICKFIRE CHALLENGE:

Create a dish using a premade piecrust.

SERVES 6–8

1. Preheat an oven to 350° F.
2. On a lightly floured surface, roll the dough out to a 10-inch round. Prick all over with a fork, transfer to a baking sheet, and bake until golden, about 15 minutes. Transfer to a wire rack to cool.
3. Meanwhile, in a large saucepan over medium heat, sauté the strawberries, pear, 1/4 cup of the sugar, the wine, and saffron until the fruit is softened and the wine is reduced to 2 tablespoons, about 45 minutes.
4. With an electric mixer, beat the goat cheese, cream cheese, the remaining 1/4 cup sugar, and the vanilla bean seeds until thoroughly combined.
5. Spread or pipe the goat cheese onto the crust and top with the strawberry filling. Refrigerate for at least 30 minutes before serving.

"A TRUE CHEF WILL EXTEND HIMSELF AND TAKE RISKS."

DALE L.

TRE'S APPLE-FENNEL TARTE TATIN

1 cup milk

1 cup heavy cream

1 cup sugar plus 2 tablespoons

1 vanilla bean, split lengthwise

4 egg yolks, beaten

2 cups packed fresh mint leaves

2 cups dried cherries

3 cups port wine

1 cinnamon stick

1/4 cup toasted walnut pieces

1 tablespoon unsalted butter

1 bulb fennel, trimmed and diced, trimmings reserved

3 Granny Smith apples, peeled, pitted, and diced, peels reserved

One 9-inch frozen piecrust, rolled out and cut into four 2 1/2-inch rounds

Fresh mint sprigs for garnishing

Special equipment: four 4-ounce ramekins

1 HOUR, 30 MINUTES

SEASON 3, EPISODE 5

QUICKFIRE CHALLENGE:

Create a dish using a premade piecrust.

TECHNIQUE

TARTE TATIN

How many times have we heard the chef'testants complain that they are "*not* pastry chefs"? Desserts always have been a challenge on the show, but Tre impressed the judges with his unusual take on this classic French treat. What is tarte Tatin? Caramelize apple slices in sugar, cook them in a cast-iron skillet, add a pastry crust on top, and turn the whole thing upside-down—you've got tarte Tatin!

MAKES 4

1. Prepare an ice bath. Combine the milk, cream, 1/2 cup of the sugar, and the vanilla bean in small saucepan over medium heat. Bring just to a bubble. Whisk a few tablespoons of hot liquid into the yolks, then pour the mixture back into the pot. Whisk to combine. Cook, stirring constantly, another 3 or 4 minutes, until the mixture thickens enough to coat a spoon. Remove from heat, add the mint, and steep for 10 minutes.

2. Strain the mixture through a fine-mesh sieve and chill in a bowl set in the ice bath until you're ready to serve the dessert.

3. Combine the cherries, port, the 2 tablespoons sugar, and the cinnamon stick in a medium saucepan over medium-high heat. Cook to reduce until evaporated to a syrup, about 45 minutes. Cool for at least 10 minutes, then remove the cinnamon stick and stir in the walnuts. Set aside.

4. Preheat an oven to 375° F.

5. In a medium sauté pan over medium heat, melt the butter. Add the fennel and cook, stirring, about 6 minutes. Mix in the diced apples and cook until soft, about 5 minutes.

6. In a small saucepan over medium-high heat, caramelize the remaining 1/2 cup sugar.

7. Carefully pour the caramel to cover the bottom of the ramekins. Fill each ramekin three-quarters full with the apple mixture. Top each ramekin with a dough round. Bake until the crust is golden brown, about 12 minutes. Cool for 2 minutes, then invert each ramekin onto a baking sheet topped with a wire rack.

8. To assemble, pool some vanilla-mint sauce on each plate. Top with a tart. Top with the cherries and garnish with a mint sprig.

WHO COOKED IT?

Identify the chef'testants that cooked each of the following dishes.

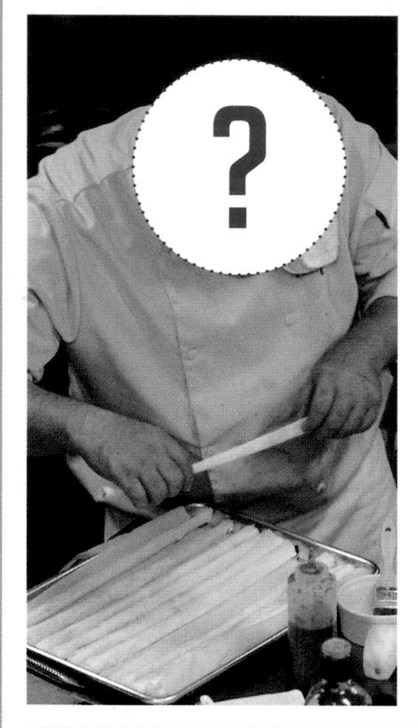

1. STEAK AND EGGS WITH GRAND MARNIER SHAKE
SEASON 3

Padma doesn't typically like steak, but this chef'testant's dish was her favorite breakfast.

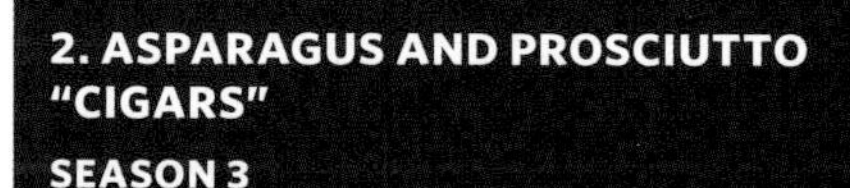

2. ASPARAGUS AND PROSCIUTTO "CIGARS"
SEASON 3

This chef'testant's greasy dish was the losing one on the yacht.

3. BAKED LOBSTER WITH ASPARAGUS AND HOLLANDAISE SAUCE
SEASON 5

Eric Ripert liked this chef'testant's imitation of a Le Bernardin dish the best.

4. ORANGE TURNED-ON ASPARAGUS
SEASON 4

The judges called this duo's dish "phallic."

6. ROASTED DUO OF BABY LAMB
SEASON 5

This chef'testant admittedly didn't know how to butcher lamb.

5. VIETNAMESE FISH SPRING ROLLS
SEASON 4

This duo was inspired by the movie *Good Morning, Vietnam.*

ANSWER KEY: 1) HUNG, 2) HOWIE, 3) STEFAN, 4) JENNIFER AND STEPHANIE, 5) SPIKE AND MANUEL, 6) ARIANE

HUNG'S CHOCOLATE PIE WITH BANANAS

6 ounces dark chocolate, finely chopped

1 cup heavy cream, whipped to stiff peaks

½ cup water

1 cup plus 2 tablespoons sugar

3 eggs

One 9-inch frozen piecrust, thawed

1 tablespoon butter

4 bananas, peeled and cut into ½-inch slices

½ cup dark rum, such as Myers's

½ cup packed brown sugar

1 tablespoon vanilla extract

½ teaspoon ground cinnamon

SERVES 8

1. Melt the chocolate in a double boiler or metal bowl set over over, but not touching, simmering water in a saucepan. Remove from heat and fold in the whipped cream.
2. Stir the water and granulated sugar together in a medium saucepan over medium-high heat. Cook, without stirring, until the mixture registers 325° F on a candy thermometer and is syrupy, about 8 minutes.
3. In a stand mixer, beat the eggs on high speed until frothy. Add the sugar mixture in a gradual stream while the machine is running. Whip until stiff. Fold the egg mixture into the chocolate mixture, cover, and refrigerate for 1 hour.
4. Preheat an oven to 375° F. Bake the piecrust until golden brown, about 15 minutes. Cool to room temperature on a wire rack.
5. Melt the butter in a medium saucepan over medium heat. Stir in the bananas, rum, brown sugar, vanilla, and cinnamon. Cook until the bananas soften and the liquid thickens, about 5 minutes.
6. To assemble, fill the cooled piecrust with the chocolate mixture. Top with the bananas, spreading them evenly to the edges of the pie. Refrigerate for at least 1 hour before serving.

1 HOUR PREP, PLUS COOLING

SEASON 3, EPISODE 5

QUICKFIRE CHALLENGE:

Create a dish using a premade piecrust for Chef Maria Frumkin.

ABOUT AN INGREDIENT

BUYING CHOCOLATE

When baking with chocolate, always buy the best quality you can find. The percentage of cocoa is not necessarily the best indicator of how chocolate will taste when baked. Look for the following indicators of quality: glossy sheen to surface, a crisp snap when broken, and a rich cocoa fragrance. Find a brand you trust.

CLIFF'S HOLIDAY BAILEYTINI AND STEAK TAPAS

Baileytini

3 mini marshmallows

1/4 vanilla bean, split lengthwise

Crushed ice

1 shot (1 1/2 ounces) Baileys Original

2 ounces vodka

2 ounces rum

1 cinnamon stick

Tapas

One 8-ounce filet mignon, halved

Sea salt (preferably fleur de sel), and pepper

1 tablespoon olive oil

1 cup crème fraîche or sour cream

2 canned plum tomatoes

1 shallot, coarsely chopped

16 sesame crackers

Whole nutmeg

1 ounce 70 percent dark chocolate

MAKES 1 DRINK; SERVES 8

FOR BAILEYTINI

1. Thread the marshmallows on the tines of a fork and hold over a gas burner or match and toast until golden.

2. Scrape the seeds from the vanilla bean and deposit in a cocktail shaker. Fill the shaker with ice, and add the Baileys, vodka, and rum. Shake vigorously and strain into a martini glass. Garnish with marshmallows and grate cinnamon on top.

FOR TAPAS

1. Season the filet mignon with sea salt and pepper. In a sauté or grill pan over medium-high heat, warm the olive oil, add the meat, and sear to the desired doneness. Allow the meat to rest for 10 minutes, then thinly slice.

2. In a blender or food processor, combine the crème fraîche, tomatoes, and shallot. Season to taste.

3. To assemble, place 2 sesame crackers on each plate. Place a slice of beef on top of each cracker. Grate nutmeg to taste over the beef, then add a grating of chocolate, a dollop of tomato crème fraîche, and a sprinkle of sea salt.

"BE TRUE TO THE PRODUCT."

CLIFF

5 MINUTES FOR COCKTAIL, 35 MINUTES FOR TAPAS

SEASON 2, EPISODE 8

QUICKFIRE CHALLENGE:

Create a drink using Baileys Irish Cream and an accompanying dish.

FUN FACT

VANILLA

Vanilla beans are the fruit of an orchid plant native to Mexico. Because it could only be pollinated by a bee native to Mexico, it wasn't until a 12-year-old slave named Edward Albius on the French-colonial island of Réunion figured out a method for hand-pollinating that the plant was able to grow elsewhere. This same labor-intensive method is still used today, making vanilla the second-most-expensive spice after saffron.

Now, the majority of the world's vanilla is grown on the island of Madagascar, though it also comes from Tahiti and Indonesia; only a tiny amount still comes from Mexico.

RADHIKA'S PEACH-LAVENDER BREAD PUDDING

1 tablespoon minced crystallized ginger
1 pint vanilla frozen yogurt, softened
1 loaf challah or brioche bread
3 eggs
1 cup heavy cream
1 teaspoon ground cinnamon
1 teaspoon ground ginger
1 teaspoon butter
½ teaspoon grated, peeled fresh ginger
2 peaches, peeled, pitted, and sliced
1 teaspoon fresh lavender or rosemary
4 tablespoons honey (or Diet Dr Pepper)
½ cup chopped cashews
About 2 cups fresh blueberries

45 MINUTES PREP, PLUS SOAKING

SEASON 5, EPISODE 7

QUICKFIRE CHALLENGE:

Create a delicious no-sugar dessert.

SERVES 6–8

1. Stir the minced ginger into the softened frozen yogurt. Return the mixture to the freezer.
2. Preheat an oven to 350° F.
3. Cut the loaf of bread into 1-inch cubes. Transfer the cubes to an 8-by-8-inch baking dish with 2-inch sides.
4. Whisk together the eggs, cream, and dry spices. Pour the mixture over the bread cubes and soak for 30 minutes.
5. Transfer the baking dish to a larger pan with at least 1-inch sides. Fill the larger pan with hot water until it reaches halfway up the sides of the smaller pan. Carefully transfer the pan to the oven and bake until browned on top, about 30 minutes.
6. Meanwhile, melt the butter in a large sauté pan over medium heat. Add the grated ginger and stir. Add the peaches, lavender or rosemary, and honey. Sauté until softened, about 8 minutes.
7. Slice the bread pudding into squares, or spoon into bowls. Top the bread pudding with a spoonful of warm peaches and a scoop of frozen yogurt. Sprinkle with cashews and blueberries.

TEST YOUR FOODIE IQ

1 WHICH OF THE FOLLOWING ARE NOT LEGUMES?

a. black-eyed peas
b. peanuts
c. alfalfa
d. walnuts

2 WHAT IS TEMBLEQUE?

a. the first four-star restaurant ever to open in Paris
b. a Puerto Rican coconut pudding
c. an egg-poaching technique
d. a lemon pastry cream

3 ABOUT HOW MANY POUNDS OF MILK DOES IT TAKE TO MAKE 1 POUND OF CHEESE?

a. 20
b. 10
c. 5
d. 2

4 PESTO SAUCE IS ORIGINALLY FROM WHAT PART OF ITALY?

a. Parma
b. Turin
c. Rome
d. Genoa

5 INJERA BREAD IS A STAPLE OF WHICH CULTURE'S CUISINE?

a. Ethiopian
b. Navajo
c. Mongolian
d. Korean

6 WHICH OF THE FOLLOWING IS NOT CLASSIFIED AS A CRUCIFEROUS VEGETABLE?

a. broccoli
b. rutabaga
c. asparagus
d. bok choy

7 WHAT WAS THE FIRST MEAL EATEN IN SPACE?

a. beef stroganoff
b. puréed applesauce
c. carrot cake
d. vanilla pudding

8 CASSEROLE MEANS WHAT IN FRENCH?

a. seasonal mix
b. combination
c. dish of dreams
d. saucepan

9 VEAL SALTIMBOCCA TYPICALLY USES WHAT INGREDIENT?

a. prosciutto
b. Brie
c. shrimp
d. clams

10 WHAT IS THE IDEAL TEMPERATURE FOR STORING WINE?

a. 35°F
b. 45°F
c. 55°F
d. 65°F

11 "DEVILS ON HORSEBACK" TYPICALLY MEANS:

a. prunes or dates wrapped in bacon
b. eggs topped with chile peppers
c. fried jalapeños
d. cherry and cheese tartlets

12 WHAT IS THE CULINARY TERM FOR KANGAROO MEAT?

a. Kangasaurus
b. Australus
c. Kangarly
d. S'moroo

ANSWER KEY: 1) D, 2) B, 3) B, 4) D, 5) A, 6) C, 7) B, 8) D, 9) A, 10) C, 11) A, 12) B

CHEF BIOS: SEASON 1

HAROLD DIETERLE

Highly competitive and also a likable team player—that was the M.O. for this season's talented winner. Harold started off strong by winning the first Elimination Challenge, and went on to clinch the 'wichcraft Sandwich and the Reimagined Junk Food Quickfires, staying consistent throughout the season. He is now chef-owner of his own successful restaurant, Perilla, in New York City.

"I'M A COOK. THAT'S WHAT I DO: I COOK."

HAROLD

TIFFANI FAISON

Competitive, confident, and dead set on being Top Chef, the villain of Season 1 rubbed many of her cohorts the wrong way, to put it mildly. But no one contested the fact that this woman can cook. She won several Eliminations and Quickfires to make it to the finals in Las Vegas. Tiffani was last seen working at Todd English's new restaurant, Riche, in New Orleans.

DAVE MARTIN

Watching Dave fret, stress, and—yes—cry several times during the season, few would have predicted he would have the nerves to make it all the way to the finals. But this former technology professional had gump-tion, humor, and a great way with complex, flavorful comfort food. Since *Top Chef*, Dave has moved to New York, where he owns and operates As You Wish Catering.

LEE ANNE WONG

This talented and amiable chef performed well all season. She may not quite have made it to the finale, but she so impressed the producers at *Top Chef* with her technique and attention to detail that she was hired as supervising culinary producer of the show. Not a bad consolation prize!

STEPHEN ASPRINIO

This sommelier and chef boasted and bombasted his way through Season 1. It's only when you consider that Stephen was a mere 24 years old, and already well on his way to success, that you can cut him some slack. Now you will find Stephen in Palm Beach, Florida, educating and, one hopes, feeding his customers at Forté di Asprinio.

MIGUEL MORALES

This young Jewish-Puerto Rican chef from New York made an indelible impression on *Top Chef* audiences. With the nickname "Chunk le Funk" and his jaunty beret, Miguel fulfilled the role of jokester. But he also cooked extremely well. Last we checked, he was cooking his heart out at the Mandarin Oriental in New York City.

ANDREA BEAMAN

The first chef to be eliminated twice from the show, Andrea seemed a little out of her league at first, but turned out to be a fan favorite for her down-to-earth persona and emphasis on healthy home cooking and lifestyle. Andrea keeps busy as an author, consultant, and TV personality specializing in healthful nutrition and living.

LISA PARKS

This former lawyer and mother of three brought a can-do optimism to *Top Chef*'s first season, despite having less restaurant experience than some of the others. Lisa loves cooking and it comes through in her food: she is still teaching, cooking, and catering in Los Angeles.

CANDICE KUMAI

This former model and culinary student struggled to keep up with some of her more experienced competitors, but charmed the judges and the audience. Candice attended Le Cordon Bleu's California School of Culinary Arts and is now cooking, consulting, and expanding her TV appearances.

BRIAN HILL

This personal chef to the stars from Los Angeles made himself known as a personality on the show before being told to pack his knives for serving disappointing monkfish to a class full of kids. Brian is busy working with Match BBQ & Grill in Pomona, California, among other enterprises.

CYNTHIA SESTITO

Cynthia seemed to be enjoying herself on the show, despite being one of the few chef'testants over 30. Unfortunately, Cynthia chose to leave the competition early due to her father's illness. She now runs Cynful Catering in the Hamptons, New York.

KEN LEE

The first chef'testant to be told to pack his knives and go was also the first to stick a finger in Chef Hubert Keller's sauce, and probably the first ever to talk back to Keller in such an obnoxious way. The volatile Irishman was sent home for cooking bland hotel banquet-style food, but he probably would not have lasted long in any case.

UTENSILS DOWN HANDS UP: TIME CHALLENGE

RECIPES

LEAH'S GRILLED BREAD WITH BACON AND EGG

- 1 shallot, minced
- 1 tablespoon olive oil
- ½ teaspoon red pepper flakes
- ¼ cup canned whole tomatoes, mashed
- One 1-inch-thick slice crusty bread
- 1 clove garlic, halved
- 2 slices bacon
- 1 quail or chicken egg
- Parmesan cheese for grating
- Salt and pepper
- 1 fresh sage leaf, fried

Special equipment:
1-inch cookie or biscuit cutter

30 MINUTES

SEASON 5, EPISODE 4

QUICKFIRE CHALLENGE:

Create a breakfast amuse-bouche in 30 minutes.

SERVES 1

1. In a medium saucepan over medium heat, sweat the shallots in olive oil until tender, about 5 minutes (see page 93). Add the pepper flakes, stir, and add the mashed tomatoes. Increase heat to medium-high and cook until the tomatoes thicken, about 8 minutes.
2. Grill the bread on both sides to the desired doneness and rub with garlic.
3. In a small sauté pan over medium heat, cook the bacon until the fat is rendered and the strips are crisp. Remove from the pan, then increase heat to medium-high and add the toasted bread to the pan. Heat until crunchy on both sides, about 1 minute per side. Remove the bread from the pan. Reduce heat to low, crack the egg into the pan, and cook until the white is set and the yolk is sunny-side up, about 2 minutes.
4. To serve, use the round cutter to punch a 1-inch round in the bread. Spread with 1 teaspoon of the tomato mixture. Cut the bacon to fit the bread and lay the bacon on top of the tomato. Cut out the same shape from the egg and position that on top of the bacon. Grate a bit of Parmesan on top, and sprinkle with salt and pepper. Top with the sage leaf.

"NOTE TO CHEFS: IF YOU WANT TO MAKE PEOPLE HAPPY, GIVE THEM BACON."

JUDGE TED ALLEN

ABOUT AN INGREDIENT
ALTERNATIVE EGGS

As home chefs become more adventurous, specialty eggs such as quail, duck, and ostrich are gaining in popularity and availability. Tiny quail eggs, which are a lovely speckled blue when in the shell, are similar in taste to chicken eggs, but slightly richer. Their diminutive size makes them hard to work with but adorable when fried and placed on top of toast, bacon, and other treats. Duck eggs are higher in cholesterol as well as some nutrients, as they have a larger yolk than chicken eggs, and they are popular with some bakers. Ostrich eggs are enormous and gooey—one ostrich egg is the equivalent of about two dozen chicken eggs.

MICAH'S TUSCAN SUSHI REVISITED

- 1 slice prosciutto
- 1 dried fig, sliced
- 1 tablespoon fig jam
- 1 fresh basil leaf, cut into chiffonade
- ¼ cup Gorgonzola cheese, crumbled
- 2 pecans, halved
- Balsamic vinegar for drizzling
- Fresh lemon juice for drizzling

10 MINUTES

SEASON 3, EPISODE 1

QUICKFIRE CHALLENGE:

Create an amuse-bouche from buffet appetizer ingredients in 10 minutes.

MAKES 1

1. Lay the prosciutto flat on a clean work surface.
2. Place the fig, jam, basil, cheese, and pecan slices on the prosciutto and roll the meat tightly around the filling, sushi style.
3. Drizzle balsamic vinegar and a couple drops of fresh lemon juice on the roll.

ABOUT AN INGREDIENT
BASALMIC VINEGAR

Most of what sells under the name "balsamic vinegar" in the supermarket is actually everyday wine vinegar with a little caramel color and sweetener added. The real thing is an entirely different animal, a thick, syrupy liquid of complex, concentrated flavor, created slowly and lovingly by food artisans on the plains of northern Italy. Genuine balsamic vinegar, labeled *Aceto Balsamico Tradizionale* and sold in special hourglass-shaped bottles, comes from two towns in the region of Emilia-Romagna: Modena and Reggio Emilia. It begins with pure grape must (*mosto cotto*, or cooked grape juice), which is then aged in a succession of small wooden barrels for up to 35 years. In the process, the liquid evaporates and the vinegar ages like a fine wine, taking on the flavors of the various wood barrels, from cherry to oak.

FIVE-MINUTE CHALLENGES

Practice your basic and not-so-basic kitchen skills against the clock.

WARM-UPS

- Whip five egg whites by hand with a whisk until soft peaks form. (Leftover tip: Use in Hubert Keller's Berry Verrine with Mousse and Swan, page 145.)
- Make a mirepoix: finely dice 1 onion, 1 carrot, and 1 celery stalk. Off the clock, you can add herbs, sauté the mixture in butter, and use it to season soups and sauces. (Use in Danny's Leek, Ham, and Egg Soup, page 166.)
- Make a perfect fried egg. (See page 89.)

ELIMINATION ROUNDS

- Separate a dozen eggs, making sure not to break the yolks. (Use in Carla's Green Eggs and Ham, page 161.)
- Blanch and peel 1 pound of pearl onions. (Store in a brine or vinegar mixture and use in cocktails.)
- Cut the peel off a pineapple and cut the flesh into bite-sized chunks with no "eyes." (Use to make Spike's Sensual Beef Salad, page 91.)
- Remove the heads, peel, and devein 1 pound of shrimp. (Make Jennifer's Shrimp and Scallop Beignets, page 122.)

TIEBREAKERS

- Create a rockin' guacamole. (Eat with tortilla chips or jícama wedges.)
- Using a sharp knife or kitchen shears, remove the back and neatly cut a chicken into ten serving pieces. (Use for Harold's Pan-Roasted Chicken with Potato Gnocchi, page 107.)
- French a rack of lamb: trim the fat neatly down to the meat and scrape the bones clean with your knife. (Roast and pair with Ryan's Lamb Patties with Pipérade, page 133.)

CLIFF'S SPOT PRAWN AND DAIKON SUSHI

3/4 cup soy sauce

1/2 cup mirin

1 bunch fresh shiso or mint leaves, cut into chiffonade

2 fresh oysters

1/2 cup finely diced mango

1 jalapeño chile, seeded and diced

1/2 cup julienned daikon radish

2 tablespoons brown sugar

1 tablespoon ginger juice, such as Ginger People brand

3 ounces hamachi, cut into thin slices

1 spot prawn, head removed, tail cut lengthwise in two equal parts

SERVES 2

1. In a small bowl, combine the soy sauce, mirin, and shiso.
2. Over another small bowl, shuck the oysters, reserving the meat and juice in the bowl, and reserving the bottom shells separately. Add the mango, jalapeño, and half of the soy mixture to the bowl. Let stand at least 5 minutes.
3. In another bowl, combine the daikon, sugar, ginger juice, and the remaining soy mixture. Let stand for at least 5 minutes.
4. To serve, spoon the oyster meat and some mango mixture back into the reserved shells. Form a cross with a thin slice of hamachi and one-half of the prawn tail. Drain the daikon salad and place a mound on the plate.

30 MINUTES, INCLUDING MARINATING

SEASON 2, EPISODE 2

QUICKFIRE CHALLENGE:

Create a sushi dish in 30 minutes.

"IT'S FOOD, IT'S NOT A ROCKET SCIENCE. IT'S REALLY JUST INNATE; WHATEVER I FEEL IS WHAT I DO."

CLIFF

CASEY'S THAI VEGETABLE ROLL WITH SMOKED SALMON

- 8 sheets rice paper
- 16 fresh Thai basil leaves
- 16 fresh mint leaves
- 16 fresh chives
- 1 large carrot, shredded coarsely
- 1 English cucumber, julienned
- 4 ounces rice vermicelli, softened in hot water and drained
- 1/2 cup chopped cashews
- 8 ounces smoked salmon, cut into 8 thin slices
- 1/4 cup basil pesto for dipping

10 MINUTES

SEASON 3, EPISODE 1

QUICKFIRE CHALLENGE:

Create an amuse-bouche from buffet appetizer ingredients in 10 minutes.

MAKE YOUR BETS

Organize a *Top Chef* viewing and place bets with your friends about the results of the episode. The person with the least points owes everyone else a homemade dinner.

SCORING:

Correctly guess Quickfire Challenge winner: 5 points

Correctly guess Elimination Challenge winner: 5 points

Correctly guess eliminated chef'testant: 10 points

MAKES 8

1. Set out all the ingredients before you begin, and prepare a shallow bowl of hot water.
2. Moisten the rice paper sheets by carefully dipping them, one by one, into the hot water. Place one moistened sheet of rice paper on a flat, clean surface or a bamboo mat.
3. Layer the ingredients on the paper, leaving one-third of the sheet from the bottom uncovered (the area closest to you). For each roll, add 2 basil leaves, 2 mint leaves, 2 chives, a large pinch of shredded carrot, a large pinch of cucumber, and a large pinch of vermicelli in a row on the paper. Sprinkle with cashews.
4. Starting at the bottom of the roll, begin rolling the paper tightly around the filling until you have a thin tube. Wrap the outside of the tube with a slice of smoked salmon. Repeat to make 8 rolls. Serve with the pesto for dipping.

"COOKING TECHNIQUE UNDER PRESSURE IS A TRUE GIFT. IF YOU'RE PRACTICED AND EXPERIENCED, THEN IT WILL GO EASILY FOR YOU."

JUDGE TED ALLEN

BRIAN M.'S TRES RIOS

1 pound mixed shellfish, such as mussels, clams, scallops, crayfish, shrimp, and lobster, cleaned (see page 69)

2 cups dry white wine

2 red bell peppers, seeded and cut into thin strips

Juice of 1 lemon (about 3 tablespoons)

2 tablespoons finely diced shallots

2 tablespoons unsalted butter

2 tablespoons Pernod

Salt and pepper

6 oysters (such as Kumamoto)

1 tablespoon red wine vinegar

½ baguette, sliced on the bias

3 farmed conch, cleaned and chopped

1 clove garlic

½ cup olive oil

1 handful fresh flat-leaf parsley, chopped

SERVES 6

1. Preheat an oven to 375° F.

2. Place a large pot over medium-high heat and add the shellfish, wine, bell peppers, lemon juice, 1 tablespoon of the shallots, the butter, Pernod, and salt and pepper to taste. Cover and steam until the shellfish open, about 5 minutes. Discard any unopened shells.

3. Shuck the oysters and arrange on a platter or clean work surface. Combine the remaining 1 tablespoon shallots, the vinegar, and ½ teaspoon black pepper and divide evenly over each oyster.

4. Spread the baguette slices on a baking sheet and toast until golden brown, about 10 minutes. Meanwhile, combine the conch, garlic, olive oil, parsley, and salt and pepper to taste in a blender or food processor and pulse until smooth. Spread on the toast slices and return to the oven for about 5 minutes. Remove from the oven and serve with the poached shellfish and oysters.

30 MINUTES

SEASON 3, EPISODE 3

QUICKFIRE CHALLENGE:

Catch and cook a shellfish dish in 30 minutes.

ABOUT AN INGREDIENT

SCALLOPS: WHAT'S THE DEAL?

If any ingredient has been the favorite of chef'testants it is the humble scallop. Why so popular? They are quick and simple to cook, and beautiful to present on the plate. One of the most common preparations is to sear scallops so they have a nice, brown caramelization on the outside and are still medium-rare on the inside. These perfect little bundles have a mild and pleasing, un-fishy flavor that complements both sweet and acidic flavors such as butter, lemon, and white wine.

QUICKFIRE HALL OF SHAME

Even the greatest chefs make mistakes. The Quickfire Challenge has seen plenty of these. While striking out in the Quickfire doesn't mean instant elimination, the results can tip the competition. Here's a look back at some of the not-so-glorious moments in Quickfire history.

SEASON 1

The first ever Quickfire was not an easy one: Work the line at Chef Hubert Keller's Fleur de Lys restaurant for 30 minutes. Under Chef Keller's watchful eye, Ken did the unmentionable—sticking his finger into a sauce to taste it. After Keller threw him out (along with the sauce), Ken made matters worse by talking back to the chef, only to get lip from a rightfully angry Tom Colicchio.

SEASON 2

For the Ice Cream Quickfire, Marcel mixed things up with his Bacon and Avocado Ice Cream. At Redondo Beach, people were confused by the odd combination, and several children spat it out. By popular vote, Marcel's ice cream was named worst out of the bunch. Though he gets a point for creativity, anyone who ruins ice cream *or* bacon deserves a slap on the wrist.

SEASON 3

In the Grocery Aisle Quickfire, Howie attempted to make a banana mousse with mandarin orange sauce. In the last few seconds, he dumped out the mousse from his cocktail glass and placed the empty glass on his plate, explaining that he didn't want to serve food he wasn't happy with. Needless to say, everyone was disappointed. As C.J. put it: "Make something; you're a chef. You can't give up. You don't ever give up."

SEASON 4

The rules for the Farmers' Market Quickfire were simple enough: Use no more than five ingredients (other than salt, pepper, sugar, and oil). Somewhere between the chopping and the cooking, Andrew slipped in an extra ingredient to his list: lamb chops, peaches, onions, mint, potatoes, and balsamic vinegar. When confronted by Padma, he argued that he'd thought balsamic vinegar was a given ingredient. Rules are rules!

SEASON 5

Filleting fish under the clock and in front of Chef Eric Ripert is no easy feat. Leah held her own in the first round, but by the time they got to the Arctic char, she started to lose steam. After making a mistake while removing the bones, she stopped working with just one fillet on her tray because she was "just over it." Like C.J. before him, Hosea sagely noted that you can't give up—especially in front of Eric Ripert.

HOWIE'S SHELLFISH CEVICHE WITH AVOCADO AND CRISPY PLANTAINS

6 mussels, rinsed and debearded (see page 27)

6 crayfish or Gulf shrimp, rinsed (shells on)

8 scallops, rinsed and trimmed

4 small conch, cleaned and shelled (optional)

6 Kumamoto oysters, shelled

Juice of 2 limes (1/4 cup)

Juice of 2 lemons (1/4 cup)

Juice of 1 tangerine (about 1/4 cup)

1 ripe avocado, peeled, pitted, and chopped

4 cups canola oil for frying

1 green plantain, very thinly sliced

6 grape tomatoes, halved

1 red bell pepper, seeded and finely diced

1 jalapeño chile, seeded and diced

4 scallions, white part only, thinly sliced

2 tablespoons olive oil

Salt and pepper

SERVES 4

1. Prepare an ice-water bath. Bring a large pot of salted water to a boil over high heat. Add the mussels and crayfish and poach for 1 minute. Remove with a slotted spoon and transfer to the ice-water bath to cool. Remove the shells from the mussels and crayfish or shrimp. Combine all the seafood in a bowl. Add the citrus juices and refrigerate for at least 20 minutes.

2. Smash the avocado with a fork and season with salt to taste. Set aside.

3. In a large sauté pan over high heat, heat the canola oil to 350° F. Add the plantain slices to the hot oil in batches and fry until crispy, about 1 minute. Remove the plantain slices from the oil with a slotted spoon, drain on paper towels, and season with salt.

4. Drain the seafood, pouring off the juices. Toss the seafood with the tomatoes, bell pepper, chile, scallions, and olive oil. Season with salt and pepper.

5. To serve, spoon the shellfish into martini glasses. Spoon the avocado on top. Garnish with the fried plantains.

45 MINUTES, PLUS MARINATING

SEASON 3, EPISODE 3

QUICKFIRE CHALLENGE:

Catch and cook a shellfish dish in 30 minutes for Chef Alfred Portale.

TECHNIQUE

CEVICHE: "COOKING" WITHOUT THE HEAT

There's the raw, there's the cooked, and then there's ceviche. Unlike sushi, in which fish is unabashedly raw, this Latin American dish uses the acid in citrus juice to "cook," or denature the proteins in fish, changing the chemical makeup and turning it opaque, a process much like curing or pickling. Cooks love ceviche because it is one of those flexible dishes that allows them to really show their stuff, combining unusual flavors and textures.

STEPHANIE'S SHRIMP, PORK, AND BANANA FRITTERS

- 1/4 pound medium shrimp, peeled, deveined (see page 122), and chopped
- 1/4 pound ground pork
- 1 banana, peeled and smashed
- 1 yellow plantain, peeled and diced
- 1 clove garlic, minced
- Salt and pepper
- 1/2 cup (1 stick) unsalted butter
- Juice of 2 limes (1/4 cup)
- 1 tablespoon sugar
- 1 1/2 cups peanut oil for frying
- 1/4 cup panko (breadcrumbs)
- 2 tablespoons fresh basil, cut into chiffonade (see sidebar)

45 MINUTES

SEASON 4, EPISODE 13

QUICKFIRE CHALLENGE:

Create two *frituras* (fried beach snacks), both with plantains, in 20 minutes.

TECHNIQUE

CHIFFONADE

Chiffonade is just a fancy French way of saying "very thinly sliced," and it is used almost exclusively for basil, mint, or other leafy herbs. The technique goes like this: neatly stack the herb leaves on top of each other, roll them lengthwise into a little bundle, and slice the bundle crosswise into thin strips—voilà!

MAKES ABOUT 2 DOZEN

1. In a small bowl, mix together the shrimp, pork, banana, plantain, and garlic. Season the batter with salt and pepper. Set aside.
2. Melt the butter in a heavy-bottomed saucepan over medium heat until browned, about 5 minutes. Pour the butter through a fine-mesh sieve to remove solids. Wash the pan and return it to very low heat. Strain the clarified butter back into the pan. Add the lime juice, sugar, and a pinch of salt. Heat, stirring occasionally, until the sugar is dissolved.
3. Heat the peanut oil to 350° F in a large sauté pan over medium-high heat. Fry 1/4 teaspoon of fritter batter until golden on all sides. Taste to check the seasoning and add more salt or pepper if necessary.
4. Spread the panko on a plate. Form the fritter batter into teaspoon-size balls; roll in panko to coat. Fry in batches without crowding the pan until golden brown, adding more oil if necessary. Remove the fritters from the pan with a slotted spoon and drain on paper towels.
5. Serve the fritters immediately, drizzled with the sauce and sprinkled with the basil.

"IT'S REFRESHING TO FIND A CHEF LIKE STEPHANIE WHO SURPRISES US."

JUDGE GAIL SIMMONS

FANTASY QUICKFIRE

What would the chef'testants do if they had a chance to put others through the gauntlet?

RICHARD, SEASON 4

"Have the chef'testants run through a hardware store to find a 'nontraditional' cooking utensil."

HOSEA, SEASON 5

"Have them cook something without being able to use a knife."

SPIKE, SEASON 4

"Make them go spear fishing in scuba diving gear to catch their meal, or maybe hunting. Or cook a death-row inmate's last meal."

RYAN, SEASON 4

"Throw the judges in there! I want to see them make an amuse-bouche from a vending machine in ten minutes."

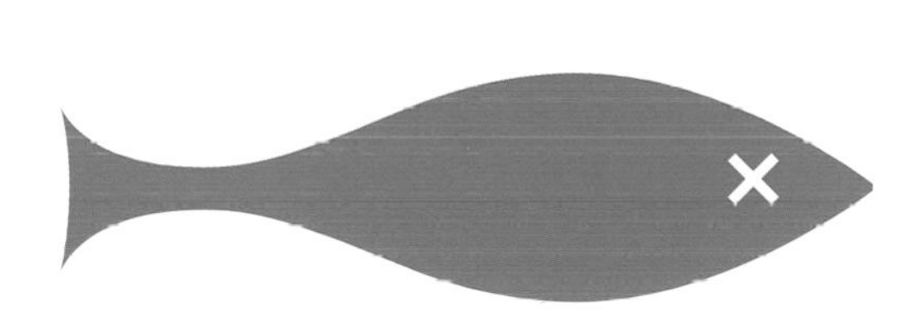

DALE, SEASON 4

"Have them do a live fish demo. It takes nerve to grab something and humanely kill it. Chefs spend most of their time preparing fish to be cooked, not cooking it."

JAMIE, SEASON 5

"Why not make the chef'testants cook something in fifteen minutes or less while being suspended from a rooftop of a building that is on fire? Let's see how well they do with that one!"

C.J.'S SEAFOOD-CAULIFLOWER-PEPPER SALAD

1 small head cauliflower

5 mussels, rinsed and debearded (see page 27)

5 clams, scrubbed

5 crayfish, rinsed, or large shrimp, peeled and deveined (see page 122)

5 scallops, rinsed and trimmed

Salt and pepper

1 teaspoon olive oil plus 3 tablespoons

2 sprigs thyme

1 teaspoon paprika, preferably Spanish

1 teaspoon saffron threads

1 teaspoon fresh lemon juice

1 teaspoon red wine vinegar

2 piquillo peppers, torn into strips

2 slices prosciutto, preferably prosciutto di Parma, cut into thin strips

2 tablespoons black oil-cured olives

30 MINUTES

SEASON 3, EPISODE 3

QUICKFIRE CHALLENGE:

Catch and cook a shellfish dish in 30 minutes.

SERVES 2 AS A MAIN COURSE

1. Use a mandoline or serrated knife to shave the cauliflower into very thin slices (see page 84).
2. Season the seafood with salt and pepper.
3. Heat a large cast-iron pan over high heat. Add the 1 teaspoon olive oil and the shellfish and thyme to the pan and cover. Cook, shaking the pan occasionally. Remove the shellfish in stages as the shells open and the shrimp (if using) and scallops become opaque, about 5 minutes total. Set aside to cool slightly. Then remove the crayfish meat from their shells.
4. In a small dry sauté pan over medium-low heat, toast the paprika and saffron. Once fragrant, transfer the spices to a medium bowl.
5. In a small bowl, whisk together the lemon juice, vinegar, and the 3 tablespoons olive oil. Add to the toasted spices. Season with salt and pepper.
6. Toss together the cauliflower, pepper strips, and prosciutto and then toss with some of the vinaigrette. Mound on plates, topping each serving with shellfish, olives, and additional dressing.

ABOUT AN INGREDIENT

CHOOSING SHELLFISH

Obviously the most important thing to look for when choosing shellfish is freshness. Try to get to know and trust a local fishmonger so you will always get the lowdown on the day's catch. If you are buying lobster, that's easy: They should be alive and kicking. Crab are also best if purchased live, but that is not always an option. When buying clams or mussels, look for a label on the bag indicating that they were harvested according to FDA guidelines. Shrimp should have a shiny appearance, either grayish or pinkish depending on the variety, and a mild, pleasant smell. Do not buy shellfish that has chipped or cracked shells.

ANTONIA'S POACHED EGG SALAD

- 4 strips bacon, cut into ¼-inch-wide pieces
- ½ pound wild mushrooms such as shiitake, cremini, oyster, or chanterelle, sliced
- ½ pound sunchokes, peeled and diced
- About ¼ cup olive oil
- Salt and pepper
- 10 squash blossoms
- 2 teaspoons sherry vinegar
- 4 large eggs
- 1 teaspoon Dijon mustard
- 1 small bunch chervil, chopped
- 1 small bunch chives, chopped
- 4 cups mâche

1 HOUR

SEASON 4, EPISODE 10

QUICKFIRE CHALLENGE:

Bring the sexy back to salad in 45 minutes for guest judge "Sexy" Sam Talbot.

SERVES 2–4

1. In a large sauté pan over medium-low heat, slowly crisp the bacon. Using a slotted spoon, transfer to paper towels to drain. Pour off and reserve the bacon fat, leaving enough in the pan to sauté the mushrooms. Add the mushrooms to the pan and increase heat to medium. Sauté the mushrooms until soft, 2 to 3 minutes. Transfer to a medium bowl.

2. Add the sunchokes to the pan, adding some olive oil if necessary. Sauté over medium heat until tender, about 5 minutes. Add to the mushrooms, and season with salt and pepper.

3. Heat about a teaspoon of the olive oil in the same pan over medium heat. Add the squash blossoms and stir until wilted, about 3 minutes. Remove from heat.

4. Meanwhile, poach the eggs. Bring a quart of water and 1 teaspoon vinegar to a slow boil over high heat in a large saucepan. Have ready an ice-water bath. Crack the eggs into a cup with a spout. When the water boils, reduce heat to simmer. Stir the water counterclockwise until it swirls, then carefully add 1 egg, then another, to the moving water. Simmer gently until the whites are firm, 1 to 2 minutes, then remove with a slotted spoon and deposit the poached egg in the ice-water bath. Repeat to poach the remaining eggs.

5. Combine the oil and the remaining 1 teaspoon vinegar in a small bowl. Add the Dijon mustard and reserved bacon fat and mix well.

6. To assemble, mix the mushrooms and sunchokes in a medium bowl with the chervil, chives, and crisped bacon. In another medium bowl, mix the mâche with the squash blossoms, then toss with bacon vinaigrette to taste. Heap the salad on plates, topping each with a poached egg and additional bacon vinaigrette, if desired.

JAMIE'S CHICKPEA SOUP

1 jalapeño or serrano chile, cut into very thin rounds

Grated zest and juice of 1½ lemons

1 tablespoon sugar

¼ cup olive oil

6 cloves garlic

1 yellow onion, sliced

Salt and pepper

4 tablespoons vadouvan or Madras curry

1 Fresno chile, seeded and minced

Two 15-ounce cans salt-free garbanzo beans, drained and rinsed

32 ounces low-sodium chicken broth

½ cup heavy cream

1 cup Greek yogurt

½ bunch cilantro, stemmed and chopped

½ bunch mint, stemmed and chopped

SERVES 3–5

1. In a small bowl, stir together the chile rounds, two-thirds of the lemon zest and juice, and the sugar, and macerate for at least 30 minutes.
2. In a large pot, heat the olive oil over low heat. Add the garlic, stir for 15 seconds, then add the onion and cook until softened. Season with salt. Increase heat to medium; add the vadouvan and heat, stirring, until the spice becomes fragrant, about 2 minutes. Add the Fresno chile and stir for 1 minute. Add the beans and broth and bring the liquid to a simmer. Simmer until the onion is tender, 15 to 20 minutes. Transfer the liquid to a blender or food processor and purée until smooth. Thin with cream, season to taste, and return to the pan over medium-low heat.
3. Combine the yogurt, cilantro, mint, and the remaining lemon zest and juice in a small bowl and mix well.
4. To serve, pour the soup into warmed bowls. Garnish with the cilantro yogurt and a sprinkle of pickled chile rounds.

30 MINUTES PREP, PLUS MACERATING

SEASON 5, EPISODE 3

QUICKFIRE CHALLENGE:

Make a dish from *Top Chef: The Cookbook*. Wait!—and turn it into a soup in 1 hour.

TOP CHEF: THE COOKBOOK

This Quickfire was based on the *New York Times* best-seller *Top Chef: The Cookbook*. For more information, go to www.chroniclebooks.com.

"WATCH OUT, GUYS. I'M SOMEBODY TO RECKON WITH HERE."

JAMIE

SEASON 6 REUBEN BENEDICT WITH THOUSAND-ISLAND HOLLANDAISE

- 1/2 cup white wine
- 1 bay leaf
- 1/2 teaspoon chopped fresh thyme
- 3 egg yolks, plus 4 large eggs
- 1 cup (2 sticks) unsalted butter, cut into 1-inch pieces
- 2 tablespoons ketchup
- 1 tablespoon sweet relish
- 1/4 teaspoon salt
- 1/4 teaspoon white or black pepper
- 1/2 teaspoon vegetable or olive oil
- Four 1/8-inch thick slices rye bread
- 8 ounces thinly sliced corned beef or pastrami
- 1 cup drained sauerkraut, at room temperature
- 4 slices Swiss cheese
- 1/2 teaspoon whole or ground caraway seed

30 MINUTES

SEASON 6, EPISODE 11

QUICKFIRE CHALLENGE:

Make breakfast for dinner in 30 minutes.

SERVES 2–4

1. In a large, heavy-bottomed saucepan over high heat, bring the wine to a boil. Add the bay leaf and thyme, and reduce to 1/4 cup, about 5 minutes. Remove from the heat and strain out bay leaf and thyme.
2. Pour 1 inch of water into the bottom of a double boiler set over medium-high heat. Bring to a simmer, then turn the heat to medium-low.
3. In the bowl of the double boiler, whisk the egg yolks. Place over the simmering water and continue to whisk the yolks until thick and pale, about 2 minutes. Whisk in the wine reduction, then add the butter, a few pieces at a time, whisking constantly until incorporated before adding more. Continue whisking until the mixture thickens enough to coat a spoon. Stir in the ketchup, relish, salt, and pepper, then turn off the heat but leave the sauce in the pan to keep warm.
4. Crack the 4 eggs into a cup with a spout. Heat a large, nonstick frying pan over medium-low heat. Add the oil to the pan, tilting the pan to coat. Add the eggs to the pan and cook until the whites are opaque throughout, about 2 minutes. Cover the eggs and allow them to cook for another 2 minutes, until the yolks are set, but still runny. Alternatively, poach the eggs (see page 71).
5. Meanwhile, toast the bread.
6. To finish, divide the toast between plates. Top each slice with corned beef, sauerkraut, and Swiss cheese. Top each Reuben with a fried egg, then drizzle with sauce and sprinkle with the ground caraway. Serve immediately.

"BREAKFAST HAS ALWAYS BEEN MY FAVORITE MEAL OF THE DAY."

CHEF WYLIE DUFRESNE

DALE T.'S GRILLED SCALLOPS WITH XO-PINEAPPLE FRIED RICE

3 teaspoons peanut oil

2 eggs, beaten

½ small pineapple, peeled, cored, and chopped

½ small onion, diced

3 tablespoons chopped scallion

3 cloves garlic, minced

1 tablespoon minced, peeled fresh ginger

One 8.8 ounce bag Uncle Ben's Ready Rice Whole Grain Brown Rice, prepared according to package instructions, or 2 cups cooked brown rice

1 tablespoon soy sauce

1 teaspoon XO sauce, such as Lee Kum Kee brand

1 teaspoon fish sauce

1 teaspoon chili oil plus 1 tablespoon

12 scallops, trimmed

½ teaspoon salt

10 long beans, blanched, trimmed, and cut into 1-inch pieces

1 tablespoon sweet soy sauce

2 tablespoons chopped fresh cilantro

30 MINUTES

SEASON 4, EPISODE 8

QUICKFIRE CHALLENGE:

Create a healthful entrée using Uncle Ben's microwavable rice in 15 minutes.

SERVES 4

1. Heat 1 teaspoon of the peanut oil in a medium sauté pan over medium heat. Add the eggs and cook, stirring, until firm. Remove the eggs from the pan and set aside.
2. Add another 1 teaspoon peanut oil to the pan. Add the pineapple, onion, scallion, garlic, and ginger and sauté for 2 minutes. Add the rice and stir to coat and heat the rice, 2 to 4 minutes.
3. Season with the soy sauce, XO sauce, fish sauce, and the 1 teaspoon chili oil, then add the eggs to the rice.
4. Season the scallops with ¼ teaspoon of the salt.
5. Heat a medium sauté pan or grill over medium-high heat. Add the 1 tablespoon chili oil, then the scallops and cook to medium, turning once, about 2 minutes.
6. Heat the remaining 1 teaspoon peanut oil in a sauté pan over medium-high heat. Add the long beans and cook, stirring, until crisp-tender, about 3 minutes. Season with sweet soy sauce and the remaining ¼ teaspoon salt.
7. To assemble, pile the rice on a large plate. Top with the long beans, then the scallops. Sprinkle with cilantro and serve.

SAM'S LIME GINGERSNAP CRUMBLE SUNDAE

2½ quarts heavy cream

4 cups milk

Grated zest and juice of 3 limes

2 cups sugar plus 3 tablespoons

24 egg yolks

26 gingersnaps

Grated zest of 1 lemon

Special equipment:
ice cream maker

15 MINUTES PREP, PLUS CHILLING

SEASON 2, EPISODE 3

QUICKFIRE CHALLENGE:

Create an original ice cream flavor in 2 hours, 45 minutes and serve it to passersby along the Redondo Beach Boardwalk.

HOME QUICKFIRE: DESSERT ISLAND

Put a wide variety of foods into a hat and have each person choose two. Make them decide which of the two they would rather eat for all eternity. Then ask them to bring a dish made with that food to your next *Top Chef* party.

SERVES 10

1. In a heavy saucepan over medium heat, bring the cream and milk to a simmer. Add the lime juice and the 2 cups sugar.

2. In a large bowl, whisk the yolks. Add one-third of the warm cream mixture, stirring constantly. Pour the mixture into the saucepan, whisking constantly until it thickens and begins to bubble. Remove from heat. Cool for 15 minutes at room temperature, then refrigerate until cool, at least 1 hour.

3. Grind 16 of the gingersnaps in a food processor until they are the consistency of fine breadcrumbs. Stir the ground cookies into the cooled ice cream mixture and pour into an ice cream maker. Freeze according to the manufacturer's instructions.

4. Combine the 10 remaining gingersnaps, the lime and lemon zest, and the remaining 3 tablespoons sugar in a food processor. Pulse until finely ground.

5. To serve, scoop ice cream into bowls and sprinkle with the gingersnap crumble.

"I WAS LIKE, OH MY GOD. TALL, DARK, AND HANDSOME, *AND* HE CAN COOK. SIGN ME UP."

ANTONIA, ON SEASON 4, EPISODE 10
QUICKFIRE JUDGE "SEXY" SAM

LAST SUPPER

Have you ever wondered what you'd want to eat if your next supper were your last?

JAMIE SEASON 5
"Coconut-fish curry and dosa on the beach in Southern India while drinking champagne straight from the bottle."

HOSEA SEASON 5
"A really delicious BLT."

RYAN SEASON 4
"Potluck with friends and family, where everyone brings one dish. With beer and vodka, too."

SPIKE SEASON 4
"A nice roast chicken and lemon-oregano potatoes, something very comforting, very simple. Something that reminds me of sitting around the table with family."

DALE SEASON 4
"Double burger with blue cheese and bacon, prime rib of beef, whole-roasted pig with crispy skin (Filipino), my mom's oxtail stew with peanuts, real Japanese ramen noodles, a couple pieces of toro sashimi, gyro from The Works in Chicago, deep-dish Pequod Sicilian-style pizza, and a bánh mì sandwich."

NIKKI SEASON 4
"Apple cider-braised pork with spicy apples, sweet potato, and kale or Brussels sprouts. Most of my fans would be surprised that the answer to that question is not 'pasta.'"

Season 5, Episode 12 was devoted to the last-supper wishes of famous chefs and culinary luminaries. Here's what they chose.

CHEF JACQUES PÉPIN

requested Squab and Peas. "When I was a kid, the squab was one of the special treats that we would get occasionally."

CHEF MARCUS SAMUELSSON

wanted Seared Salmon with Twice-Cooked Spinach, Roasted Potatoes, and Dill Sauce.

CHEF LIDIA BASTIANICH

wanted Chicken Roasted with Lemon and Herbs, and Roasted Potatoes.

SUSAN UNGARO,

president of the James Beard Foundation, requested Shrimp Scampi with Tomato Provençal.

CHEF WYLIE DUFRESNE

desired Eggs Benedict. "Proper egg cookery is the sign of a good chef."

CHEF BIOS: SEASON 2

ILAN HALL

The winner of Season 2 knows how to cook up controversy as well as a good pan of paella. Audiences came away with a sense of this Long Island native of Scottish and Israeli descent as a shrewd young cook with a snide attitude. Regardless of your feelings toward Ilan, he proved he could cook over and over again on the show. After traveling the world, Ilan is planning to open his own restaurant in downtown Los Angeles.

"GO IN WITH AN OPEN MIND, BECAUSE THAT'S HOW YOU'LL WIN. THAT'S HOW I WON!"

ILAN

MARCEL VIGNERON

Marcel studied at the Culinary Institute of America and worked with star chef Joël Robuchon to hone his craft before coming to *Top Chef*. There was something about this talented young chef with a passion for foams and a flamelike hairdo that pissed off the people around him during Season 2. His silly rap and other tics can be forgiven, however, given that Marcel shows true creativity and flair in his cooking. He is now behind the stoves at The Bazaar in Beverly Hills.

SAM TALBOT

This North Carolina native worked his way up to executive chef in some of New York City's best kitchens before becoming a chef'testant on *Top Chef*, and his finesse and experience showed in his consistently strong performance. "Sexy" Sam was probably a favorite to win. He ended up winning Fan Favorite and now cooks at the picturesque Surf Lodge in Montauk, New York.

ELIA ABOUMRAD

This young Mexican-born chef of Lebanese descent honed her competitive edge as a swimmer in her teens and early twenties before switching her passion and drive to cooking. She displayed her talent during many of this season's challenges. She was the only woman this season to make it to the finals in Hawaii, going a little crazy and shaving her head along the way.

CLIFF CROOKS

This tall, charismatic executive chef showed plenty of talent during Season 2, but was sent home after a misguided attempt to shave Marcel's head, against Marcel's will, on camera. This was one of the most shocking episodes in *Top Chef* history. Cliff aced several Quickfires, including the Sushi and Ice Cream Challenges. He was last seen cooking at the Blue Water Grill in New York City.

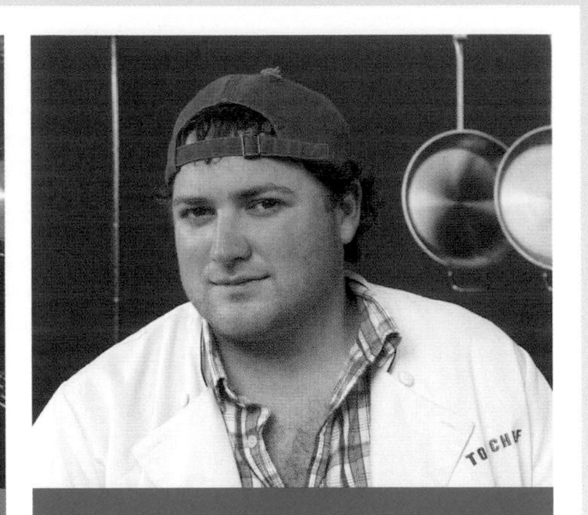

MICHAEL MIDGLEY

This line cook from Lodi, California, was a long shot, to say the least, but he lasted all the way to the final six. He made his mark as the first chef'testant to win a Quickfire and Elimination Challenge in one episode. Michael has since moved to Stockton, California, where he caters and teaches cooking.

BETTY FRASER

This former actress and Los Angeles chef and restaurateur brought a sense of fun and mischief to the *Top Chef* proceedings. She had the stamina to cook her heart out in every episode until the judges decried her contribution to the Seven Sins dinner—a trio of vegetable soups—as a bit too slothful.

MIA GAINES-ALT

The judges loved her fried frog legs, and audiences appreciated her cowgirl hat and down-home appeal. In the end, Mia chose to quit rather than see Elia sent home as team leader of the *Los Angeles* magazine Catering Challenge. Since shooting, Mia has relocated to Hawaii, where she is chef at the Hotel Molokai.

FRANK TERZOLI

"Frankie the Bull" seemed like a nice enough guy, aside from threatening Marcel with bodily harm a couple of times. This San Diego chef had the experience and talent to make it on *Top Chef*, but perhaps not the personality to live comfortably with a bunch of strangers for six weeks.

CARLOS FERNANDEZ

Carlos shined in the infamous Amuse-Bouche Vending Machine Quickfire but was later voted off for his contribution to the Thanksgiving meal. But don't fret: he of the great smile and warm demeanor is still cooking at the Hi-Life Café in Fort Lauderdale, where he is chef and partner.

JOSIE SMITH-MALAVE

Everyone had high expectations for this talented and vivacious Miami native, but like her friend Marisa, Josie was sent home early for a misguided trio at the six-course lunch for actress Jennifer Coolidge. These days, you'll find Josie cooking and consulting for some of New York's hippest eateries.

MARISA CHURCHILL

Marisa and Josie became fast friends on the show, but their ill-fated collaboration on a fruity palate cleanser as part of a six-course tasting menu got them both sent home. Formerly pastry chef at the acclaimed Ame in San Francisco, Marisa now works as a restaurant consultant.

EMILY SPRISSLER

In the short time she was on the show, Emily seemed to have a bit of a prickly personality. Before we got to know her better, however, she was sent home when the firemen at a South Pasadena fire station were unimpressed with her reinvention of a childhood classic. Now she has her hands full as a private chef and new mom.

OTTO BORSICH

Otto seemed out of his element in the first two episodes, perhaps regretting his decision to participate in a reality cooking contest with a bunch of chefs half his age. He quickly decided to call it quits right after the "Lychee-gate" incident. Last seen, Otto was executive chef at Mahi Mah's in Virginia Beach.

SUYAI STEINHAUER

This New Yorker didn't seem that surprised when she was sent home in the first episode for serving the judges a plate of so-so escargots with cheese sauce. Suyai is now back in the Big Apple, cooking for New York Fork meal delivery service.

OUTSIDE THE BREAD BOX: CREATIVITY CHALLENGE

RECIPES

MICHAEL'S CARROT CHIPS

Peanut or canola oil for frying

8 carrots, peeled and thinly sliced (see sidebar)

1 tablespoon cornstarch

Salt

15 MINUTES

SEASON 2, EPISODE 9

QUICKFIRE CHALLENGE:

Create a dish based on a color.

SERVES 4

1. Fill a large, heavy saucepan halfway with oil. Over high heat, bring the oil to 350° F to 365° F.
2. Toss the carrot slices with the cornstarch. Working in batches, fry until golden and crisp, about 3 minutes. Remove with a slotted spoon to drain on paper towels.
3. Sprinkle liberally with salt and serve.

EQUIPMENT HEADS-UP

USING A MANDOLINE

Chefs love tools that make fine cooking easier, and one good example is the mandoline. When you need to get potato slices uniform and paper-thin for making chips or if you are slicing beef fillet or zucchini for carpaccio, a mandoline does the work for you. Simply place a piece of food on the flat surface of the mandoline and slide it over the flat blade, much like a wide razor. Some models can also make julienne and crinkle cuts. What was once the province of restaurant chefs is now widely available to home cooks!

"IT'S LIKE BETTY CROCKER AND CHARLES MANSON HAD A LOVE CHILD, AND HE'S COOKING FOR ME."

JUDGE ANTHONY BOURDAIN

DALE T.'S SEXY SALAD WITH POACHED CHICKEN

2 whole chicken breasts (about 3 pounds)

1/4 cup salt

1/4 cup chicken stock

8 cups water

2 cups sake

1 cup mirin

10 nori sheets, torn into pieces

2 tablespoons soy sauce

1 cup rice wine vinegar

1/2 cup chili bean sauce, such as Lee Kum Kee Toban Djan

1/2 cup sugar

1/8 cup coriander seeds

1/8 cup fish sauce

1 cup thinly sliced Brussels sprouts

1/2 head napa cabbage, cored and thinly sliced

10 pearl onions

SERVES 4

1. Combine the chicken breasts, salt, chicken stock, and water in a large saucepan over medium-high heat. Bring the liquid to a boil, then turn off the heat, cover the pot, and let the breasts poach for 20 minutes. Remove the chicken from the liquid and slice into thin strips.

2. Combine the sake and mirin in a large saucepan over medium heat. Bring the liquid to a simmer, cook to reduce by one-third, then add the nori. Simmer to soften the nori, about 5 minutes. Add the soy sauce and carefully transfer the hot liquid to a blender or food processor. Blend into a smooth paste and set aside.

3. Combine the vinegar, chili bean sauce, sugar, coriander, and fish sauce in a large saucepan over medium-high heat and bring to a boil. Remove from heat. Add the Brussels sprouts, cabbage, and pearl onions to the liquid, stir to combine, and let the vegetables marinate for 10 minutes.

4. To assemble, place a spoonful of nori paste on each plate. Artfully arrange chicken strips and pickled vegetables on top of and around paste.

1 HOUR PREP, PLUS POACHING

SEASON 4, EPISODE 10

QUICKFIRE CHALLENGE:

Bring the sexy back to salad.

TECHNIQUE

POACHING CHICKEN

Though not as popular as sautéing and pan frying, poaching chicken is a perfect technique to use for something like chicken salad, where you want mild, succulent meat without added fat. Here's another way to poach chicken breasts: Fill a pot with water, chicken broth, and any aromatics that you like, such as chopped onion, celery, and carrot, which will impart subtle flavor to the meat. Add the chicken so that it fits in one layer, bring to a simmer, and cook on a low simmer until the chicken is opaque throughout, about 20 minutes.

MIA'S BEAN SALAD

1 tablespoon Dijon mustard

½ teaspoon sugar

1 teaspoon capers

1 tablespoon fresh mint, julienned

2 teaspoons ground black pepper

½ teaspoon salt

½ cup olive oil

½ cup fresh, blanched, or canned green beans, drained

½ cup canned garbanzo beans, drained

½ cup canned kidney beans, drained

2 fresh cooked or canned beets, drained, cut into thin slivers

¼ cup canned artichoke hearts, drained, cut into thin slivers

4 cups baby mixed greens

SERVES 4

1. Whisk together the mustard, sugar, capers, mint, black pepper, salt, and olive oil in a small bowl. Set aside.
2. Toss the beans, beets, and artichoke hearts in a medium bowl. Dress with some of the vinaigrette.
3. To serve, heap a mound of greens on a platter. With a slotted spoon, remove the bean mixture from the vinaigrette and spoon it over the greens. Then drizzle the remaining vinaigrette over the greens and serve.

15 MINUTES

SEASON 2, EPISODE 6

QUICKFIRE CHALLENGE:

Make something delicious using three different canned foods.

RECIPE SCRAPBOOK

Hosting your own home Quickfire Challenge or *Top Chef* viewing party? Ask each guest to bring a favorite recipe to your party, with enough copies for everyone. Purchase binders before the party and create a cover sheet as a memento of the night. Then collect all the recipes and give them as party favors. Challenge each guest to cook one recipe (not their own) for your *Top Chef* parties in coming weeks.

"I'M VERY SURPRISED WHEN I LOOK OVER MY SHOULDER AND SEE THAT THERE ARE PEOPLE THAT ARE BRAVE ENOUGH TO ATTEMPT HOT FOOD IN FIFTEEN MINUTES. I'M THINKING IMMEDIATELY SALAD."

MIA

THROWING A HOME QUICKFIRE PARTY

It's a party, so all the usual etiquette applies: Let guests know what they're in for. Send invitations by e-mail or phone (or mail!), so guests have time to reply at least one week in advance of your shindig. For a home Quickfire Challenge party, you'll want to invite people who love the Quickfire Challenge as much as you do. The people with whom you often watch *Top Chef* are the perfect guests!

PLAN IT

- ☐ You'll need ingredients and other materials based on the number of guests and what they will be doing.
- ☐ Are you farming out responsibilities to your guests? Let them know at least five days in advance what they should bring.
- ☐ Invite enough guests to allow for some healthy competition. But only invite as many guests as you have counter space to fit.
- ☐ Consider *Top Chef* props: orange plastic cutting boards, white chef coats, a gray and orange *Top Chef* color scheme.
- ☐ Choose your theme(s): Bombay Nights (everyone uses Indian spices), Junk Food Challenge (cooking from a vending machine), Mystery Ingredient (what is that surprising texture or flavor?), Who Chops Fastest?

DO IT

- ☐ Make sure the rules are clearly expressed and understood by all.
- ☐ Have a separate workstation for each guest.
- ☐ Let your guests choose their own teams or split them up by choosing straws or drawing team numbers from a hat. (For the true *Top Chef* touch, have your guests draw numbered knives from a block.)
- ☐ Have enough tools to go around. You can't compete in timed challenges if someone lacks a knife.
- ☐ Make sure someone takes photos (the non-cook in your crowd is a good choice).
- ☐ A good Home Quickfire Challenge party should be documented for the ages! You can share your photos with other fans on the Bravo Web site (www.bravotv.com/top-chef).
- ☐ Let the booze flow! Make sure you have wine openers and a well-stocked bar. Rule of thumb is to buy twice as much alcohol as you think you'll need. (And, of course, agree ahead of time with everyone on safe ways for getting home after the bash.)

STEPHEN'S BRUNCHWICH OF EGG, MANGO, AND MANCHEGO

1/2 bulb fennel, trimmed

1 small plantain, peeled and diced

1/2 mango, peeled, cut from pit, and diced

2 slices prosciutto

1 tablespoon butter

1 egg, cracked into a ramekin

2 slices brioche bread

1 thin slice manchego cheese

MAKES 1

1. Use a serrated knife or mandoline to shave the fennel into thin slices.
2. In a medium sauté pan over medium heat, cook the plantain and mango until caramelized, about 20 minutes. Remove from the pan and mix with the fennel. Set aside. Add the prosciutto to the pan and cook until the edges are crisped. Set aside.
3. Wipe any excess fat from the pan. Increase heat to medium-high and add the butter. Pour the egg into the pan and immediately reduce heat to medium-low. Cover the pan and let sit for 5 minutes, until the egg edges are opaque and the yolk is set.
4. Meanwhile, toast the brioche slices to the desired doneness. Place the egg on one slice of brioche. Top with the cheese, then with the fennel salad, prosciutto, and the other slice of brioche.

30 MINUTES

SEASON 1, EPISODE 7

QUICKFIRE CHALLENGE:

Create a signature sandwich to be featured at Tom Colicchio's 'wichcraft.

TECHNIQUE

THE PERFECT FRIED EGG

Here are a few tricks for achieving the perfect fried egg:

The right pan: A small nonstick pan is the way to go, not only for cooking but also for easy cleaning.

The right heat: Relatively low heat on the stove top will keep the egg from getting too browned and crackly around the edges.

The right fat: A small amount of butter works best.

Putting it all together: Crack the egg into a small bowl, being careful not to break the yolk. Heat the butter over medium-low heat and lower the egg into the pan. Cook until the white is set, about 4 minutes. Cover the pan, reduce the heat to low, and cook until the yolk is barely set, about 1 minute more.

SPIKE'S SENSUAL BEEF SALAD

1 pineapple, peeled, cored, and diced

1 bunch radishes, cut into thin rounds

1 cucumber, cut into thin rounds

1 bunch cilantro, stemmed and chopped, a few sprigs reserved for garnish

Leaves from 5 sprigs mint, a few leaves reserved for garnish

Salt and pepper

1 cup Sprite

½ cup fish sauce

1 cup fresh lime juice

4 jalapeño chiles, cut into thin rounds

1 tablespoon distilled white vinegar

One 1½-pound skirt, flap, or hanger steak (see page 23)

1 tablespoon olive oil

Pinch of red pepper flakes

SERVES 4

1. Combine the pineapple, radishes, and cucumber in a large bowl. Toss with the chopped cilantro and mint leaves. Season with salt and pepper.

2. Combine the Sprite, fish sauce, lime juice, and chiles in a large saucepan over high heat and bring to a boil. Cook to reduce to ½ cup, about 15 minutes, then add the vinegar.

3. Season the steak with olive oil, salt, pepper, and red pepper flakes. Grill or broil to the desired degree of doneness, 3 to 4 minutes per side for medium-rare. When cool, cut into thin strips, slicing against the grain of the meat.

4. Toss together the salad and three-quarters of the dressing. Arrange the salad on 4 plates, topping each with one-quarter of the sliced steak. Drizzle each plate with dressing and garnish with cilantro and mint.

"MY FAVORITE CHALLENGE WAS THE TOMAHAWK CHOP, NOT JUST BECAUSE I WON, BUT BECAUSE THAT WAS A REAL CHEF-Y CHALLENGE, FROM BUTCHERING THE CHOPS THROUGH TO MAKING A COOKED DISH."

SPIKE

45 MINUTES

SEASON 4, EPISODE 10

QUICKFIRE CHALLENGE:

Bring the sexy back to salad.

ABOUT AN INGREDIENT

PINEAPPLE

Though many of us grew up associating pineapples with Hawaii, they are, in fact, native to Latin America. Christopher Columbus found these odd-looking fruits on the island of Guadeloupe and took them back to Europe. In turn, it was European voyagers who brought them to the Philippines and Hawaii, where Mr. Dole and Mr. Del Monte eventually created quite a monopoly. Pineapples have come full circle, and are now mainly grown in Costa Rica, Mexico, and elsewhere: only about ten percent of the world's pineapples are now grown in Hawaii.

JEFF'S APPLE-FENNEL SOUP WITH BLUE CHEESE TOASTS

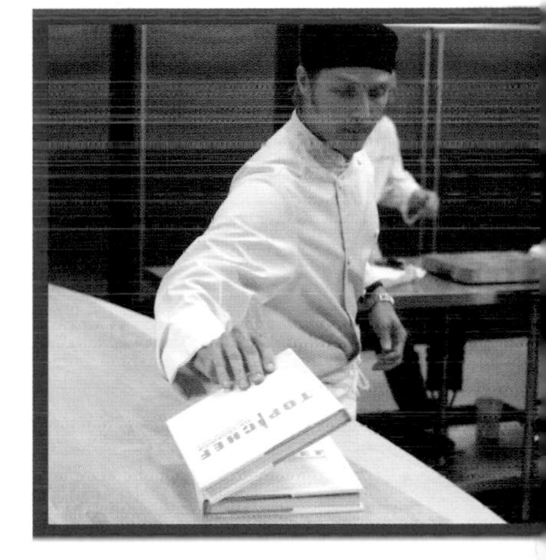

- 3 Granny Smith apples, peeled, cored, and diced
- 1 cup chopped fennel (fronds reserved for garnish)
- 2 large shallots, minced
- 2 cloves garlic
- 1/2 cup dry white wine
- 2 cups chicken stock
- 1 cup heavy cream
- Leaves from 3 sprigs thyme
- Leaves from 3 sprigs sage
- 3 fresh mint leaves
- 1/4 small baguette, cut into 1/2-inch slices
- 1/2 cup crumbled blue cheese
- 5 dried figs, chopped

SERVES 4

1. Preheat an oven to 350° F.
2. In a large saucepan over medium-low heat, sweat the apples, fennel, shallots, and garlic, reserving a few tablespoons of the apples. When the vegetables are softened, add the wine and cook to reduce by half. Add the chicken stock, reduce heat to low, and cook to reduce the liquid by half. Add the cream and herbs and simmer until bubbles form on the surface. Transfer to a blender or food processor and purée until smooth.
3. Meanwhile, place the baguette slices on a baking sheet and toast until golden, about 10 minutes. Spread each toast with a bit of blue cheese and top with a few slices of fig.
4. To serve, fill each soup bowl three-quarters full. Float several toasts in the soup and sprinkle with the reserved chopped apple and a few chopped fennel fronds.

1 HOUR

SEASON 5, EPISODE 3

QUICKFIRE CHALLENGE:

Make a dish from *Top Chef: The Cookbook*. Wait!—and turn it into a soup.

TECHNIQUE

"SWEATING" VEGETABLES

Perhaps not the most appetizing culinary term, "sweating" vegetables really means sautéing them in minimal fat at a low heat, usually in a covered pan, so they essentially poach in their own liquid. The result is a lot better than it sounds. Some good candidates for this kind of cooking are apples, leeks, onions, and other fruits and vegetables that contain lots of water. Long, slow cooking over low heat allows vegetables to become tender and well cooked without caramelizing or browning.

"IT'S LIKE A MOSH PIT—HANDS AND ELBOWS FLYING EVERYWHERE."

JEFF, ON WHAT HAPPENS WHEN PADMA SAYS "GO"

ELIA'S AHI TUNA WITH SPINACH SALAD AND MINI ONIONS

2 teaspoons honey

3 teaspoons toasted sesame oil

2 teaspoons canola oil

2 teaspoons sesame seeds

4 teaspoons fresh lime juice

1 teaspoon rice vinegar

2 teaspoons ginger juice

Salt and pepper

4 teaspoons olive oil

2 teaspoons whole-grain mustard

1 teaspoon soy sauce

1 teaspoon honey

One 6- to 8-ounce ahi tuna steak

2 cups baby spinach

2 tablespoons minced scallion, white part only

SERVES 2

1. Combine the honey, 2 teaspoons of the sesame oil, the canola oil, sesame seeds, 2 teaspoons of the lime juice, the rice vinegar, and 1 teaspoon of the ginger juice in a small bowl. Season with salt and pepper.

2. In another bowl, whisk together the olive oil, the remaining 2 teaspoons lime juice, the whole-grain mustard, the remaining 1 teaspoon ginger juice, the soy sauce, and honey. Season with salt and pepper.

3. Rub the tuna with the remaining 1 teaspoon sesame oil. Season with salt and pepper. Heat a medium skillet over medium-high heat. Add the tuna and sear on both sides, about 1 minute per side (see sidebar). Remove the pan from heat and transfer the fish to a cutting board. Slice the tuna into thin strips across the grain.

4. To serve, drizzle the tuna with the sesame-ginger sauce. Toss the spinach with the mustard vinaigrette. Divide the tuna and spinach between 2 plates and sprinkle with the minced scallion.

25 MINUTES

SEASON 2, EPISODE 7

QUICKFIRE CHALLENGE:

Create an entrée using ingredients from Redondo Beach farmers' market.

TECHNIQUE

SEARING

Searing is a fairly simple technique for cooking meat, poultry, and fish to create a crispy crust and moist and tender interior. And it's simple: you need a very hot, heavy pan and dry protein.

Add a bit of oil to a heavy sauté pan and heat to medium-high or high (about 300° F). Pat your protein dry with a paper towel, add salt and pepper, and place it in the pan. After a few minutes, when the edges of the meat have browned, reduce the heat to medium and flip. Cook to your desired doneness.

"THERE HAS TO BE DEFINITION IN YOUR DISH, A DEFINITE TASTE THAT COMES OUT. THE OTHERS JUST SUPPLEMENT IT."

ELIA

COCKTAILS GALORE

Occasionally, chef'testants are challenged to become bar chefs. Over five seasons of Quickfires, they've been challenged to mix cocktails with Baileys Irish Cream and Bombay Sapphire gin, or to conjure a new mixed drink from scratch. Some chef'testants thought a liquid concoction would be the perfect complement to their winning dish. Here are a few of the best liquid concoctions to emerge from those challenges.

CARLA'S CRANBERRY-GINGER SPRITZER

SEASON 5, EPISODE 13
SERVES 4

2 TABLESPOONS SUGAR FOR RIMMING
1 CUP CRANBERRY JUICE
1 CUP WATER
GRATED ZEST AND JUICE OF 1 LIME, PLUS 4 LIME SLICES FOR GARNISH
1 INCH FRESH GINGER, PEELED AND CUT INTO 1/8-INCH-THICK SLICES
1 TABLESPOON BROWN SUGAR
ICE CUBES
12 OUNCES LIME SODA

Pour the sugar onto a small plate. Moisten the rim of 4 tall glasses. Dip each glass into the sugar to rim.

In a small pitcher, combine the cranberry juice, water, lime zest and juice, ginger, and brown sugar. Stir until the sugar dissolves. Fill the glasses with ice. Strain the cranberry mixture into the glasses, filling each three-quarters full and topping with lime soda. Garnish with a lime slice.

SANDEE'S MOJITO

SEASON 3, EPISODE 4
SERVES 1

2 KEY LIMES OR 1/2 PERSIAN LIME, QUARTERED
8 FRESH MINT LEAVES
1 TABLESPOON SUGAR
1 OUNCE LIGHT RUM, SUCH AS BACARDI
CRUSHED ICE
CLUB SODA
EDIBLE FLOWER OR FRESH MINT SPRIGS FOR GARNISH

Combine the lime, mint, and sugar in an old-fashioned glass and bruise with a muddler or wooden spoon. Stir in the rum. Add crushed ice, then top off with club soda as desired. Stir to combine. Garnish with a flower or mint.

HOME QUICKFIRE CHALLENGE

During a *Top Chef* commercial break, race into your kitchen (or home bar area) and create a new cocktail. You have until the show comes back on, about three minutes, and then it's martini shaker down, hands up! You must use ingredients you have in your cupboard already (be creative and improvise).

BETTY'S CHILLED CHRISTMAS COCKTAIL

SEASON 2, EPISODE 8
SERVES 1

3 TABLESPOONS COCONUT RUM, SUCH AS PARROT BAY

JUICE OF 1 LIME

1 TABLESPOON SUGAR

CRUSHED ICE

2 TABLESPOONS CARAMEL OR REGULAR BAILEYS

1 TABLESPOON HEAVY CREAM

DASH OF GROUND CINNAMON

Chill a martini glass for at least 15 minutes. In a cocktail shaker, combine the rum, lime juice, and sugar with crushed ice and shake well. Strain into the chilled martini glass.

Mix the Baileys, cream, and cinnamon and serve in a shot glass. Chase sips of the martini with sips of the creamy Baileys mixture.

HOSEA'S GRAND MARNIER HURRICANE WITH POMEGRANATE AND BLOOD ORANGE

SEASON 5, EPISODE 13
SERVES 4

1 CUP BLOOD ORANGE JUICE (FROM ABOUT 5 BLOOD ORANGES)

1 CUP GRAPEFRUIT JUICE (FROM ABOUT 2 GRAPEFRUITS)

1/4 CUP POMEGRANATE JUICE

1 CUP LIGHT RUM

1 CUP DARK RUM

2 TABLESPOONS GRAND MARNIER

1/2 CUP SIMPLE SYRUP (SEE PAGE 22)

ICE CUBES

Combine juices and liquors. Add the simple syrup to taste. Pour over ice in 4 tall glasses. Serve immediately.

ARIANE'S GROUND CHICKEN AND BACON SAUSAGE

1 pound ground chicken

1/4 cup sauternes or other semisweet wine, such as sherry or Madeira, plus more if needed

1/4 cup diced bacon, preferably applewood smoked

2 tablespoons ground celery seeds

Salt

1 tablespoon ground fennel seeds

1/2 tablespoon garlic powder or 1 tablespoon minced garlic

Ground white pepper

Casings

1 tablespoon olive oil

1 tablespoon butter

2 yellow onions, diced

1 medium-hot chile, seeded and minced

3/4 cup amber beer

1/2 cup sweet relish or chopped sweet pickles

Whole-wheat hot dog buns

MAKES 8–12

1. Preheat an oven to 350°F.
2. In a large bowl, mix together the chicken, 1/4 cup wine, bacon, celery seeds, 1 tablespoon salt, fennel, garlic powder, and 1 tablespoon white pepper until the ingredients are evenly distributed. Scoop some of the mixture into a piping bag with a 1-inch tip. Tie off one end of the casing with a knot. Holding the casing in one hand and the piping bag in the other, pipe the mixture to fill the casing. Tie off every 4 inches. (Alternatively, make meatballs or patties by forming small handfuls of meat, see sidebar).
3. Heat a large sauté pan over medium-high heat. Working in batches, sear the sausages on both sides, about 2 minutes per side (see page 95). Arrange the seared sausages in a baking pan, and finish in the oven until cooked through, about 5 minutes.
4. Heat a large sauté pan over medium heat. Add the oil and butter, then the onions and chile. Cook, stirring, until soft, about 15 minutes. Increase heat to high, and add the beer to the pan in 3 additions, each time bringing to a boil and then adding more. Stir in the relish and remove from heat. Season with salt and pepper.
5. Toast the buns in the oven. Lay a sausage in each and top with the relish mixture.

1 HOUR

SEASON 5, EPISODE 2

QUICKFIRE CHALLENGE:

Create a signature hot dog.

TECHNIQUE

MAKING YOUR OWN SAUSAGE

You could go to town with a meat grinder, an extruder, and natural casings to make your own sausages like a pro. A far simpler proposition is to make loose sausage or sausage patties with already-ground meat and your own spice mixtures.

Choose your spice mixture and toast for best effect. Blend all ingredients thoroughly. Form the mixture into the shape you desire (patties, thin cigars, thick logs, or kebabs, patted onto a skewer). Cook immediately or chill for up to one week.

RADHIKA'S KEBAB SAUSAGE WITH TOMATO JAM

- 1 pound ground lamb
- ½ pound ground pork
- ½ pound ground chuck
- 1 tablespoon minced garlic
- 2 tablespoons tandoori masala
- Salt and pepper
- 1 tablespoon canola oil
- 1 large red onion, halved, then cut into thin slices
- 1 large cucumber, seeded, peeled, and diced
- ½ cup chopped fresh cilantro
- 1½ tablespoons white wine vinegar
- 1 teaspoon capers
- 3 tablespoons tomato paste
- ½ teaspoon ground cumin
- 1 cup mayonnaise
- ½ cup heavy cream
- 4 soft rolls, split

SERVES 4–6

1. Preheat an oven to 350° F.
2. In a large bowl, combine the lamb, pork, chuck, garlic, tandoori masala, 1 teaspoon salt, and ½ teaspoon pepper. Divide the mixture in two, and place one half in a food processor. Blend until very finely ground, then return to the bowl and mix with the rest of the meat. Form the sausage into logs the size and shape of bratwurst.
3. Heat a large sauté pan over medium-high heat. Working with a few sausages at a time, brown the meat, turning once. You will have to work in batches. Transfer the sausages to a baking sheet and bake in the oven for 10 minutes.
4. Meanwhile, in a large sauté pan over medium heat, heat the oil and cook the onion, stirring, until brown, about 20 minutes.
5. Combine the cucumber, cilantro, vinegar, capers, and salt and pepper to taste in a bowl and let macerate for 20 minutes.
6. Whisk the tomato paste, cumin, mayonnaise, and heavy cream together in a small bowl.
7. To assemble, toast the rolls. Spread both sides of each roll with tomato jam. Place 2 sausages in each roll and top with pickled cucumber and caramelized onion.

 **1 HOUR**

 **SEASON 5, EPISODE 2**

 **QUICKFIRE CHALLENGE:**
Create a signature hot dog.

"THE HARDEST PART WAS GOING INTO THE QUICKFIRE AND NOT KNOWING WHAT YOU WERE DOING. FOR ALL WE KNEW, THERE WERE TEN DANCING MONKEYS BEHIND THAT WALL THAT WE HAD TO SLAUGHTER AND BARBECUE."

RADHIKA

FABIO'S MEDITERRANEAN HOT DOG

$1^{1}/_{4}$ cups (4 ounces) sun-dried tomatoes

$^{1}/_{2}$ pound andouille sausage

$^{1}/_{2}$ pound ground pork

One 7-ounce jar roasted red peppers, drained and minced

4 ounces soft goat cheese, such as chèvre

Salt and pepper

4 tablespoons olive oil, plus more for brushing

1 small zucchini, cut into thin slices lengthwise

About 10 fresh basil leaves, chopped

1 clove garlic, minced

1 baguette

1 HOUR

SEASON 5, EPISODE 2

QUICKFIRE CHALLENGE:

Create a signature hot dog.

MAKES 4

1. Preheat an oven to 375° F.
2. Cover the sun-dried tomatoes with hot water in a bowl and soak for 30 minutes.
3. Slice the casing of the andouille and empty the contents into a food processor. Add the ground pork, peppers, and goat cheese. Season with salt and pepper, and pulse until combined.
4. Form the meat into patties, meatballs, or links. At this point, if you have a smoker, you may want to smoke the meat at 225° F for about 10 minutes.
5. In a large skillet, heat 2 tablespoons of the olive oil over medium-high heat. Brown the sausages, cooking for about 4 minutes per side. Transfer to a baking sheet and bake in the oven for about 10 minutes.
6. Meanwhile, heat 1 tablespoon of the olive oil in a large skillet over medium-high heat. Add the zucchini, season with salt and pepper, and sauté until golden, about 7 minutes.
7. Drain the sun-dried tomatoes and toss with the remaining 1 tablespoon olive oil, the basil, garlic, and season with salt and pepper.
8. Cut the baguette lengthwise and place on a baking sheet, cut sides up. Brush with olive oil and sprinkle with salt. Transfer to the oven to toast, about 5 minutes. Remove from the oven. Build the sandwich by piling sausage, then zucchini, then sun-dried tomato mixture on the baguette. Return to the oven until hot and the bread is crispy, about 10 minutes. Cut into 4-inch portions and serve.

TECHNIQUE

ROASTING PEPPERS

Here's one for Home Culinary Training 101: Preheat a broiler. Line a baking sheet with aluminum foil. Place the bell peppers on the sheet about 4 inches from the heat source and broil, keeping a close watch and turning several times, until the peppers are blackened all over but not on fire, about 15 minutes. Place the roasted peppers in a paper bag or in a bowl covered with plastic wrap for 10 minutes. Then, peel off the skin with your fingers (this will be a little messy), and use as desired in your recipe.

MAD LIBBIN' QUICKFIRE

Fill in the blanks to reveal your Quickfire Challenge.

NEIGHBORHOOD THROWDOWN

Things are heating up in the neighborhood. Pay a friendly visit to ______________ (name of a neighbor), and challenge them to a Quickfire Challenge. You'll each get to choose ______________ (number from 1-10) ingredients from the other person's fridge to work with. Your recipe must include ______________ (-ing verb for a cooking technique), and your neighbor's recipe should incorporate ______________ (-ing verb for another cooking technique). ______________ (name of another neighbor) will be the judge of this challenge.

COOKIN' UP THE PAST

Reimagine ______________ (favorite childhood dish) by incorporating new ingredients: ______________ (favorite hot dog condiment) and ______________ (ingredient you can't live without). Be inspired by the flavors of ______________ (name of the foreign country you last visited or would like to visit).

QUICKFIRE EXTREME

Use ______________ (ingredient you don't like), ______________ (ingredient you've never used), and ______________ (number from 5-10) other ingredients to create a delicious dinner dish. Incorporate ______________ (your favorite color) and ______________ (your least favorite color) into the dish. You cannot use ______________ (the two tools you use most in the kitchen). You'll get ______________ (number from 30-60) minutes to create your dish. Go! Invite at least ______________ (number from 2-10) people over to sample the results of this culinary adventure.

CLIFF'S SNAPPER WITH BLACKBERRY-BEET COMPOTE

1/4 teaspoon ground coriander

1/2 cup diced pickled beets

1/2 cup sliced fresh blackberries

1 tablespoon champagne vinegar

2 tablespoons grapeseed oil plus 1/4 cup

4 tablespoons olive oil, plus more for cooking

1 eggplant, peeled and cut into 1/2-inch cubes

1 tablespoon soy sauce

1 clove garlic, chopped

1 teaspoon five-spice powder

1/2 teaspoon paprika

Salt and pepper

Two 8-ounce snapper fillets

One 15-ounce can hominy, rinsed and drained

1/2 cup chopped pitted kalamata olives

3 tablespoons butter

1 HOUR

SEASON 2, EPISODE 9

QUICKFIRE CHALLENGE:

Create a dish based on a color.

SERVES 2-4

1. Preheat an oven to 400° F.

2. In a small dry saucepan over medium-low heat, toast the coriander until fragrant. Transfer to a small bowl. Add the beets, blackberries, vinegar, the 2 tablespoons grapeseed oil, and 2 tablespoons of the olive oil. Toss well to coat.

3. Toss the eggplant with 2 tablespoons of the olive oil, the remaining 1/4 cup grapeseed oil, the soy sauce, garlic, five-spice powder, paprika, and salt and pepper to taste. Spread the cubes evenly on a baking sheet. Roast until tender, stirring occasionally, about 20 minutes.

4. Fifteen minutes into roasting, cook the fish. Heat 1 to 2 tablespoons olive oil in a large sauté pan over medium heat. Add the fish fillets and brown on each side, about 2 minutes per side.

5. Reduce the oven temperature to 200° F. Transfer the fish to serving plates and keep warm in the oven while you prepare the hominy butter.

6. In the same sauté pan, heat the hominy and olives over medium-low heat, stirring the brown bits into the mixture. Transfer to a blender or food processor, add the butter, and process until smooth.

7. To serve, top each fillet with a heaping spoonful of hominy butter and the berry-beet compote. Mound roasted eggplant alongside.

MARK'S SIRLOIN STEAK, TURNIPS, MUSHROOMS, AND PEACH BUTTER

1 ripe peach, peeled, pitted, and sliced

3 1/2 tablespoons butter, at room temperature

Salt

2 small turnips, peeled and diced

1/4 pound hen-of-the-woods or oyster mushrooms, sectioned into bite sized pieces

One 1 1/2-pound boneless top sirloin steak

30 MINUTES

SEASON 4, EPISODE 2

QUICKFIRE CHALLENGE:

Create a dish using no more than five ingredients from Chicago's Green City Market.

SERVES 1-2

1. In a blender or food processor, purée the peach. Add 2 tablespoons of the butter, a pinch of salt, and pulse until combined. Set aside.
2. Bring a small pot of salted water to a boil. Add the turnips and boil until tender. Drain and then purée the turnips with 1 tablespoon of the butter.
3. In a medium sauté pan over medium heat, melt the remaining 1/2 tablespoon butter. Add the mushrooms and sauté until golden brown, 2 to 3 minutes. Set aside. In the same pan, sear the sirloin to the desired doneness, about 8 minutes per side for medium-rare (see page 95). Remove the steak from the pan, let rest for 5 minutes, then cut into thin strips.
4. To assemble, pool the turnip purée on a platter. Top with the sliced steak, then with the peach butter. Serve the mushrooms on the side.

"YOU WALK IN AND YOU SEE MING TSAI OR DANIEL BOULUD. YOU'RE TRYING TO PUT SOMETHING ON A PLATE, AND YOU'RE SHAKING LIKE CRAZY FOR THEM."

MARK

HAROLD'S PAN-ROASTED CHICKEN WITH POTATO GNOCCHI

2 tablespoons olive oil

1 bone-in, skin-on chicken breast half (about 10 ounces)

4 ounces fresh potato gnocchi

3 ounces (about 1 cup) hen-of-the-woods mushrooms, separated

1 cup frozen petite peas

½ cup chicken stock

1 tablespoon minced fresh thyme

Salt and pepper

SERVES 1

1. Preheat an oven to 425° F.
2. Heat 1 tablespoon of the olive oil in a medium oven-safe pan over medium-high heat. Add the chicken, skin side down, and brown for about 6 minutes. Flip the chicken and transfer the pan to the oven.
3. While the chicken is roasting (about 20 minutes), add the remaining 1 tablespoon olive oil to a large sauté pan over medium-high heat. Add the gnocchi and mushrooms and cook, stirring frquently, until the mushrooms are softened and the gnocchi is browning, about 7 minutes. Add the peas, then the chicken stock and thyme. Cook until thickened. Season with salt and pepper. Stir in any juices from the roasted chicken.
4. Place the chicken on the plate and surround with the mushroom, pea, and gnocchi mixture. Pour the juices over the chicken and vegetables.

45 MINUTES

SEASON 1, EPISODE 10

QUICKFIRE CHALLENGE:

Serve three different plates of high-protein, high-carb, low-fat food for the cast of Cirque du Soleil's *KA*.

FUN FACT

PERILLA

Harold's West Village restaurant is proof that *Top Chef* winners go on to create big things. Well, maybe not physically big—Perilla has eighteen tables and ten seats at the bar—but big in neighborhood popularity and big in flavor. Specializing in seasonal American cuisine, Perilla opened in 2007 and already has a list of regulars and a line of customers waiting to try unlikely dishes: Japanese-inspired meatballs made of duck meat and mountain yam? Clearly an inspired Top Chef.

"I DON'T REALLY HAVE A STRATEGY. I THINK MY FOOD SPEAKS FOR ITSELF. I JUST TAKE IT BY THE HORNS WHEN IT COMES TO ME."

HAROLD

IT DOESN'T GET ANY WACKIER THAN THIS

On the Season 3 *Top Chef* reunion show, the prize for "wackiest dish" went hands down to Hung's infamous Smurf Village.

The Quickfire Challenge for Season 3, Episode 10 was to create a dish using only $10 worth of products from a supermarket aisle. Hung's assignment: the cereal, coffee, and canned milk aisle. "I'm not excited or thrilled about this aisle," Hung said, "but I have to make use of what I have. I want to do something that I did when I was a kid."

As soon as Hung got into the kitchen, things started to look up.

"THIS CHALLENGE IS DEFINITELY THE MOST FUN I'VE HAD BECAUSE I MISS PLAYING WITH FOOD—BASICALLY, LIKE A CHILD."

In Season 2, Episode 3, Frank left the judges speechless with his *Alice in Wonderland*–inspired Mushroom Fantasy Salad. The Elimination Challenge was to create a dish updating a childhood classic. After viewing Frank's concoction in the kitchen, Judge Tom Colicchio asked: "Is this a childhood memory or a drug experience?"

"I ALWAYS LOVE FRANK'S PLATES. FRANK'S PLATES ARE CRAZY."

ILAN

CASEY'S GINGERSNAP AND PUDDING PARFAIT

- 2 cups gingersnaps
- 1/4 cup all-purpose flour
- 1/2 cup granulated sugar
- 1 egg
- 2 cups heavy cream
- 1 vanilla bean, split lengthwise, seeds removed and reserved
- Pinch of salt
- One 8-ounce jar mango preserves
- Juice of 1 lemon (about 3 tablespoons)
- 1 tablespoon confectioners' sugar

30 MINUTES PREP, PLUS CHILLING

SEASON 3, EPISODE 10

QUICKFIRE CHALLENGE:

Create a dish using $10 of ingredients from one aisle of a supermarket.

SERVES 4

1. In a blender or food processor, grind the gingersnaps to a fine crumb. Set aside.
2. In a large, heavy saucepan, whisk the flour, granulated sugar, and egg until pale yellow. Add 1 cup of the cream, whisking until combined. Add the vanilla bean and seeds and the salt. Set the pan over medium heat and bring the mixture to a bubble, whisking constantly, until it thickens to pudding consistency, about 10 minutes. Remove from heat, strain through a fine-mesh sieve, cool to room temperature, then cool in the refrigerator for 1 hour.
3. In a medium saucepan over medium heat, warm the mango preserves until they are runny. Add the lemon juice, remove from heat, and cool.
4. Whip the remaining 1 cup cream and sweeten with the confectioners' sugar.
5. To assemble, layer components in 4 wine goblets or parfait glasses. Begin with gingersnaps and top with pudding, then jam, then more gingersnaps. Finish with a dollop of the whipped cream.

"THE TOUGHEST THING ABOUT QUICKFIRES IS, OF COURSE, THE TIME. IT'S A MUCH QUICKER PACE THAN AT A RESTAURANT. BUT, LIKE RUNNING, IT'S A HUGE RUSH."

RICHARD B.

FRANK'S CREAMY FRUIT SALAD

1 cup heavy cream
1/2 cup sugar
1 teaspoon plain yogurt
1/4 cup blueberry pie filling
1/2 cup mascarpone
1/2 cup chopped, drained, canned peaches
1/2 cup fruit cocktail, drained
1/2 apple, thinly sliced

SERVES 4

1. Whip together 1/2 cup of the heavy cream, 1/4 cup of the sugar, and the yogurt on high speed in an electric mixer until stiff peaks form. Fold in the blueberry pie filling until well combined.
2. Whip together the mascarpone, the remaining 1/2 cup heavy cream, and the remaining 1/4 cup sugar on high speed until stiff peaks form. Fold in the chopped peaches until well combined.
3. To serve, put 2 tablespoons fruit cocktail in each of 4 chilled glasses. Layer one-quarter of the mascarpone peach cream on top. Top with one-quarter of the blueberry cream. Garnish with an apple slice.

15 MINUTES

SEASON 2, EPISODE 6

QUICKFIRE CHALLENGE:

Make something delicious using three different canned foods.

"TRYING TO MAKE TASTY FOOD WITH CANNED FOODS IS ASKING A LOT, AND I'M THINKING THAT IT'S GOING TO BE AN EXTREMELY DIFFICULT CHALLENGE."

MARCEL

FABIO'S BRÛLÉED BANANA WITH ESPRESSO ZABAIONE

4 thin slices brioche bread
8 large eggs
½ cup sugar plus 1 tablespoon
2 tablespoons all-purpose flour
1 teaspoon ground cinnamon
1 cup coffee beans
1 cup milk
1 teaspoon vanilla extract
½ cup honey
2 tablespoons butter
1 banana, peeled and thinly sliced
Heavy cream for drizzling

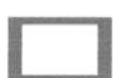

45 MINUTES

SEASON 5, EPISODE 4

QUICKFIRE CHALLENGE:

Create a breakfast amuse-bouche.

SERVES 4

1. Prepare an ice bath.
2. With a 2-inch biscuit or cookie cutter, cut a round from each slice of brioche. Set aside.
3. In a large bowl or stand mixer with a whisk attachment, whisk 6 of the eggs with the ½ cup sugar until pale and thick. Stir in the flour and cinnamon.
4. In a heavy medium saucepan over medium heat, combine the coffee beans and milk. Heat until bubbles begin to appear at the edges of the pan.
5. Whisk one-quarter of the milk mixture into the eggs, stirring constantly so the eggs don't curdle. Add the egg mixture to the milk mixture in the saucepan, whisking constantly over medium heat until the mixture thickens enough to coat a spoon, about 5 minutes.
6. Strain through a fine-mesh sieve into a metal bowl. Stir in the vanilla. Place bowl in the ice bath until cooled.
7. In a small bowl, whisk the remaining 2 eggs and the honey until smooth. Dip each brioche round into this mixture, coating evenly on both sides.
8. Melt 1 tablespoon of the butter in a large sauté pan over medium-high heat. Brown the brioche circles, turning when golden, about 12 minutes per side, and remove.
9. In the same saucepan, melt the remaining 1 tablespoon butter over medium-high heat. Add the banana slices, sprinkle with the 1 tablespoon sugar and sauté until golden and caramelized, about 5 minutes. Alternatively, caramelize the banana with a hand-held blowtorch (see note, page 142).
10. To serve, place a round of toast on a plate, top with a few slices of banana, and drizzle with cream. Serve additional coffee cream alongside in a shot glass.

"I'M A PROFESSIONAL CHEF; THERE'S NOTHING THAT CAN STRESS ME OUT. IF THEY'RE GONNA GIVE ME A MONKEY ASS TO FILL WITH FRIED BANANA, I'LL COME UP WITH SOMETHING ANYWAY. IT'S NOT A PROBLEM."

FABIO

CHEF BIOS: SEASON 3

HUNG HUYNH

Hung announced himself on the very first episode as a CPA: Certified Professional Asshole. He wasn't the meanest villain the show has ever seen, but certainly one of the cockiest. Hung's technical virtuosity, top-line restaurant experience (he worked at Guy Savoy in Las Vegas, and elsewhere), and sheer demonic speed propelled him to the finale in Aspen, where he edged out Dale for the win. He put a lot of himself and his Vietnamese flavors into the dishes, which led him to ultimate victory.

"I WORKED SO HARD TO GET HERE AND TO PROVE MYSELF."

HUNG

DALE LEVITSKI

The self-dubbed "Big Gay Chef" of Season 3 had not cooked for a year and a half before starting the competition, but during the season, Dale found redemption as well as his "inner chef." His last dish of the season, Rack of Colorado Lamb with a Deconstructed Ratatouille, simply took the judges' breath away. Dale is opening his own restaurant, Town & Country, in Chicago.

CASEY THOMPSON

This self-taught chef with a sunny disposition had the mojo to go all the way to the finale in Aspen. Casey proved herself as a hard worker and a team player. She may have missed out on being Top Chef by preparing a disappointing pork belly at the finals, but she did go on to win Fan Favorite. In 2008, Casey left her longtime position as executive chef at Dallas's Shinsei restaurant, and is now living and cooking in the San Francisco Bay Area.

BRIAN MALARKEY

Brian started out the season bold with a combo plate of snake and eel, and cooked with gusto to the end. An executive chef at a seafood restaurant, he had a likable persona and easygoing vibe that helped him on the show: he met his Waterloo in the form of too much gusto at the Aspen Rodeo Challenge. Brian is still leading the kitchen at Oceanaire in San Diego.

SARA MAIR

Sara M, a cheesemaker who had worked as sous chef in fine restaurants, proved she could really cook. Her adorable accent, pleasant personality, and interesting flavor combinations made her an asset to the show, despite winning no Quickfires and just one Elimination. She hopes to start her own goat farm and dairy back in Jamaica.

C.J. JACOBSEN

A testicular cancer survivor, C.J. won over audiences with his amiable disposition. He beat out the competition in the Daniel Boulud Burger Quickfire, but his shambolic attempt to make a tasty airplane meal left him vulnerable to elimination. Since leaving the show, C.J. moved from California to Chicago to help castmate Dale open Town & Country restaurant.

HOWIE KLEINBERG

This prickly self-taught Miami chef, nick-named "The Bulldog," was up and down during the season: he won two Elimination Challenges, and yet he also served the judges a plate of nothing—twice. The nick-name seems to be working out for Howie: he is now chef and owner of Bulldog BBQ in Miami.

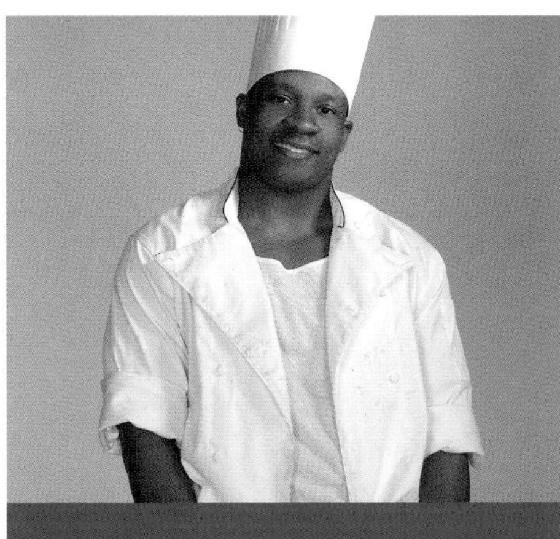

TRE WILCOX

One of the most talented and experienced chefs of the season met his downfall the time-honored way, by assuming leadership of the losing team in Restaurant Wars. Tre's quiet assuredness in the kitchen and classical training were honed as chef de cuisine at award-winning Abacus restaurant in Dallas. He is now working as a private chef and consultant.

SARA NGUYEN

Sara N. managed to keep a pretty low profile, hanging out in the middle of the pack, until she was faulted for serving sloppy sliders—in high heels—to a bunch of drunken party-goers. After leaving the show, Sara gave former castmate Dale a call and is now working with him in Chicago.

JOEY PAULINO

A quintessential New Yorker with a tough competitive streak, Joey was out to win, but he seemed to have made some friends despite himself. You will still find Joey channeling his intensity behind the stoves as executive chef at Café des Artistes in New York City.

LIA BARDEEN

Everyone was surprised and disappointed to see this talented sous chef from New York City's Jean Georges leave so soon, especially as she had just won the Trio Tasting Menu Challenge with her elegant, complex oil-poached shrimp. She is now cooking for Jean Georges' J+G Steakhouse in Mexico City.

CAMILLE BECERRA

Camille is a single mother and experienced chef-owner of the popular Paloma restaurant in Brooklyn. Unfortunately, her pineapple upside-down cake was deemed a disaster by the judges in Episode 4, but she did make a kind of triumphant return as one of the All Stars during the Top Chef Bowl of Season 5.

MICAH EDELSTEIN

A Floridian, by way of South Africa and Italy, Micah started off with a bang by winning the Amuse-Bouche Quickfire in the first round. However she fizzled fairly quickly when her reinvented meatloaf was judged to be a dry imitation of the original. Micah is still running her own catering company in South Florida.

SANDEE BIRDSONG

One of two chef'testants with mohawks this season, Sandee did not have far to go to take part in *Top Chef*, as she was already living in Miami and working as chef at Tantra. She was ousted in the second episode, during the Upscale Barbecue round, for her faulty grilled vanilla lobster.

CLAY BOWEN

This Mississippi native had a sweet personality and charming drawl, but his misfired amuse-bouche in the first Quickfire was but a preview of his disappointing entrée made of wild game and scorpion fish served in the Elimination Challenge. Clay did not take his early departure too much to heart, as he is still behind the stoves at a private club in Mississippi.

SHOW YOUR MAD SKILLS: TECHNIQUE CHALLENGE

RECIPES

RICHARD B.'S VEGETARIAN TACOS

One 1½-pound jícama, peeled

2 avocados, halved and pitted

1 papaya, peeled, seeded, and diced

1 tomato, seeded and diced

Juice of 6 Key limes (about ¾ cup)

Juice of 2 Persian limes (3 to 4 tablespoons)

1 teaspoon ground coriander

1 teaspoon ground cinnamon

2 tablespoons olive oil

Salt and pepper

½ cup chopped fresh cilantro, stems included

MAKES 16–20 TACOS

1. Use a mandoline or a very sharp knife to cut the jícama into approximately ⅛-inch-thick rounds. Alternatively, cut the jícama into wedges.
2. Mash together the avocados, papaya, and tomato in a bowl. Mix the lime juices, coriander, cinnamon, olive oil, and salt and pepper to taste into the avocado mixture. Reserve 2 tablespoons of the cilantro, then add the rest to the mixture.
3. Place 2 tablespoons of the avocado mixture in the center of a jícama slice. Roll into a faux taco, enclosing the filling within the jícama. Repeat until you run out of filling. (Alternatively, arrange the avocado mixture in the center of the platter. Surround with the jícama wedges, and sprinkle with cilantro. Use the wedges to scoop the avocado, like chips and guacamole.) Place the jícama tacos on a platter, sprinkle with the remaining cilantro, and eat up.

15 MINUTES

SEASON 4, EPISODE 3

QUICKFIRE CHALLENGE:
Create an upscale taco.

ABOUT AN INGREDIENT

CHOOSING AN AVOCADO

Avocados come in many varieties. The best-tasting and most widely available is the Hass, which has a thick, bumpy skin and rich, creamy flavor. Look for avocados that are dark green to black in color and that yield slightly to pressure. If the avocado is green and hard, it is underripe; if it is brownish-black and mushy, it is overripe. If only underripe avocados are available, you can hasten the ripening by placing them in a paper bag with an apple for 2 to 3 days. If only overripe avocados are available, buy something else!

"A TOUGH LESSON TO LEARN ON *TOP CHEF* IS THAT THE GOAL IS TO BEND EACH CHALLENGE TO YOUR STRENGTH, NOT TO TOTALLY CHANGE YOUR STYLE TO THE CHALLENGE."

RICHARD B.

JENNIFER'S SHRIMP AND SCALLOP BEIGNETS

One 4-ounce jar roasted peppers

1 clove garlic, minced

1 teaspoon sherry vinegar

2 tablespoons olive oil

Salt

1/4 teaspoon ground pepper

2 ripe avocados, peeled, pitted, and mashed

Juice of 1 lime (1 to 2 tablespoons)

1 teaspoon cayenne pepper

3 cups all-purpose flour

1/2 cup cornstarch

1 teaspoon baking powder

2 eggs, separated

3/4-inch piece fresh ginger, peeled and grated

Two 12-ounce bottles lager, such as Land Shark

1/2 pound shrimp, peeled, deveined (see sidebar), and chopped

10 scallops, trimmed and chopped

1/4 teaspoon Sriracha sauce

Juice of 1/2 grapefruit (about 1/3 cup)

2 cups canola oil for frying

2 tablespoons chopped fennel tops for garnishing

45 MINUTES

SEASON 4, EPISODE 6

QUICKFIRE CHALLENGE:

Create a simple dish to pair with a beer.

SERVES 6

1. In a blender or food processor, combine the peppers, garlic, vinegar, olive oil, 1 1/2 teaspoons salt, and the ground pepper. Purée until smooth. Set aside.
2. Combine the avocado, 1 tablespoon salt, lime juice, and cayenne.
3. In a large bowl, combine the flour, cornstarch, baking powder, and 1 teaspoon salt. Stir with a whisk to blend. Form a well, then add the yolks to the well. Add two-thirds of the ginger and mix until just combined, using as few strokes as possible. Whisk in the beer until just combined. The mixture should be thicker than pancake batter. In a large, clean bowl, beat the egg whites to stiff peaks. Fold into the batter. Season the seafood with salt to taste, Sriracha, grapefruit juice, and the remaining grated ginger. Stir into the batter.
4. In a large saucepan or deep-fryer, heat the oil to 375°F. Working in batches, drop golf ball-sized balls of batter into the hot oil and fry until golden brown, about 1 minute. Carefully flip and brown the other sides. Transfer with a slotted spoon to paper towels and sprinkle with salt. Place in a warm oven while you finish frying the remaining beignets.
5. To serve, spoon some pepper sauce on a plate. Pile each with 3 beignets. Garnish with fennel and serve with avocado.

TECHNIQUE

DEVEINING SHRIMP

Before cooking with shrimp, you need to remove the vein, the black threadlike bit that runs along the back. First, gently remove the shell with your fingers, leaving the tail intact (you can reserve the shells for making stock, or discard). Using a paring knife, cut an indent along the back of the shrimp, along the line of the vein. Then, use the tip of the knife or your finger to pull out the vein.

WHO COOKED IT BEST?

EGGS

CARLA'S GREEN EGGS AND HAM, SEASON 5 (PAGE 161)

JILL'S OSTRICH EGG QUICHE, SEASON 5

SARA M.'S EGGS IN A HOLE, SEASON 3

CEVICHE

JEFF'S ROCK SHRIMP CEVICHE, SEASON 5

HOWIE'S SHELLFISH CEVICHE, SEASON 3 (PAGE 65)

WINNER!

ANDREW'S SQUID CEVICHE, SEASON 4

GUMBO

WINNER!

HOSEA'S CHICKEN, DUCK, AND ANDOUILLE GUMBO, SEASON 5

CARLA'S CRAYFISH AND ANDOUILLE GUMBO, SEASON 5

STEFAN'S DUCK AND RABBIT GUMBO, SEASON 5

JOEY'S SCALLOPS WITH JASMINE RICE RISOTTO

1 cup jasmine rice

2 cups coconut milk

8 shiitake mushrooms, stemmed and sliced

1 mango, peeled, pitted, and diced

1 tablespoon thinly sliced fresh mint, plus more for garnishing

1 tablespoon thinly sliced fresh basil, plus more for garnishing

8 sea scallops

Salt

Cayenne pepper

1 teaspoon peanut oil for frying

SERVES 2-4

1. In a medium dry saucepan over medium heat, lightly toast the rice. Add the coconut milk, stir, and cover. Bring to a slow boil, then reduce heat to medium-low. Cook the rice until creamy, about 15 minutes. Remove from heat, then stir in the mushrooms, mango, the 1 tablespoon mint, and the 1 tablespoon basil.
2. Season the scallops with salt and cayenne pepper. In a large sauté pan over medium-high heat, heat the peanut oil. Add the scallops and sear for 1½ to 2 minutes per side.
3. To serve, place several spoonfuls of rice on each plate and top with scallops. Garnish with mint and basil.

35 MINUTES

SEASON 3, EPISODE 4

QUICKFIRE CHALLENGE:

Create an appetizer to pair with a Bombay Sapphire cocktail for Bombay mixologist Jamie Walker.

TECHNIQUE

RISOTTO

Risotto is one of those versatile dishes for which the actual recipe may vary but the technique remains the same. A couple of tips: Always stir the rice in the sautéed onions for a minute or two before adding wine. Also, contrary to what you may have learned, you do not actually need to stir risotto the entire time it cooks: stir vigorously when you add a new batch of hot liquid and then give your arm a break while the liquid is absorbed, stirring occasionally and making sure the rice doesn't scorch or stick to the bottom of the pot. Add flavor to your risotto by using stock spiked with dried porcini mushrooms, using chopped pancetta in the base, or adding a mild soft cheese, such as stracchino or mascarpone, toward the end of cooking, for a creamier texture as well.

"THIS IS *TOP CHEF*; IT'S NOT *TOP SCALLOP*!"

FABIO, ON JAMIE'S REPEATED USE OF SCALLOPS IN CHALLENGES

RICHARD B.'S VEGETARIAN "SASHIMI"

A few drops olive oil
4 cremini mushrooms, thinly sliced
Salt and pepper
A few drops yuzu juice
A few drops truffle oil
1 English cucumber, sliced lengthwise
A few drops kecap manis
Juice of 2 limes (1/4 cup)
1 teaspoon distilled white vinegar
1 small red beet, peeled
1 small yellow beet, peeled
4 slices radish
1 scallion, very thinly sliced
Honey for drizzling

30 MINUTES

SEASON 4, EPISODE 4

QUICKFIRE CHALLENGE:
Create a vegetable platter that showcases three techniques for Chef Daniel Boulud.

SERVES 2–4

1. In a small saucepan heat the oil over medium-high heat. Add the mushrooms and sear for about 5 minutes. Season with salt, pepper, yuzu, and truffle oil.
2. Cut the cucumber slices into very thin, long strips that resemble capellini noodles. Toss the strips with the kecap manis and 1/2 of the lime juice.
3. Bring a small pot of salted water to a boil. Add the vinegar. Boil the beets until just tender, about 15 minutes. Rinse immediately with cold water.
4. Dress the radish with the scallion, remaining lime juice, and salt and pepper to taste.
5. Position all vegetables on plate as a sashimi tasting. Cut the beets into sashimi shapes and drizzle with honey. Present the cucumber strips as a "pasta" in a spoon.

"RICHARD IMPRESSED ME ALL SEASON LONG WITH HIS AVANT-GARDE TECHNIQUE, WHIMSICAL STYLE, CREATIVITY, AND GRACE UNDER PRESSURE."

JUDGE GAIL SIMMONS

HOWIE'S VANILLA BUTTER–POACHED LOBSTER WITH WATERCRESS-CITRUS SALAD

1 navel orange

1 blood orange

1 ruby red grapefruit

1 tangerine

1 lime

1 pound (4 sticks) unsalted butter at room temperature

3 vanilla beans, preferably Tahitian, split lengthwise

Leaves from 2 sprigs thyme

Two 6-ounce lobster tails, in shell

1/8 teaspoon minced habanero or Scotch bonnet chile

4 tablespoons canola or olive oil, plus more for drizzling

Sea salt (preferably fleur de sel), pepper, and sugar (optional) for seasoning

4 cups stemmed watercress

4 grape tomatoes, cut into thin rounds

1 kumquat, cut into thin rounds

SERVES 2

1. Supreme the citrus fruits: with a sharp knife, cut off the tops and bottoms of the citrus fruits. Slicing downward from top to bottom, cut off the peels and white pith. Working over a bowl, cut between the membranes to dislodge the segments of fruit. Reserve the citrus juice.
2. In a small saucepan over low heat, melt the butter. Add the vanilla beans and thyme and heat for 1 minute. Add the lobster tails and cook until the flesh is opaque, 12 to 15 minutes. Remove the lobster and set aside to cool.
3. In a small bowl, combine the reserved citrus juice and the chile. Whisk in the 4 tablespoons canola oil in a thin stream. Whisking constantly, add 2 tablespoons of the vanilla-lobster butter. Season with sea salt, pepper, and sugar (if using).
4. Gently combine the watercress and half of the citrus supremes in a bowl. (Save the remaining supremes for another use.) Toss with a few tablespoons of the vinaigrette.
5. To assemble, crack open the lobster tails and carefully remove the meat. Toss with a couple tablespoons of the vinaigrette.
6. Divide the salad between 2 plates. Top with lobster. Garnish with tomatoes, kumquat, sea salt, and a drizzle of canola oil.

45 MINUTES

SEASON 3, EPISODE 2

QUICKFIRE CHALLENGE:

Create a dish featuring Florida citrus.

HOME-CHEF TIP

POACHING IN BUTTER

Have you ever wondered why food served in fine restaurants may taste better than what you make at home? One answer is butter. In a single dish, chefs will blithely throw in as much butter as you might use in a month! They want you to enjoy yourself and come back; you can bet they're not thinking about your waistline and your cholesterol level. Poaching in butter is a method of cooking a food slowly while it is submerged in a large amount of simmering butter plus water or stock. The technique is used mostly with mild fish and shellfish, particularly lobster, to bring out the natural sweetness and a silky, buttery (yes!) texture.

MISE-EN-PLACE TIPS AND TRICKS

The Mise-en-Place Relay Race Quickfire has been an exciting feature in most seasons. Mise-en-place ("put in place" in French) is all the prep work that has to happen before a chef can begin cooking, such as chopping, peeling, blanching, and breaking down large cuts of meat into individual portions. The key is taking care of mise-en-place quickly and efficiently, and good technique is developed by practice, practice, practice. Here's the way to mise like a Top Chef!

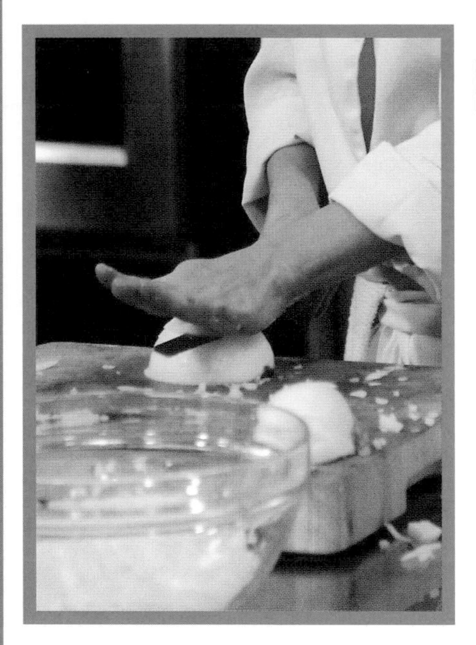

DICING AN ONION

1. On a flat, clean surface, hold the onion with your fingertips curled under, using the knuckles as a guide for the knife. Cut the onion in half through the root. Peel off the skin.
2. Place the onion flat side down, and cut vertical slices from one end to the other. Do not cut through the root end; it will hold the onion together.
3. Turn the onion 90 degrees, hold your knife blade parallel to the counter, and make 3 or 4 horizontal cuts in the onion.
4. Now cut across the onion. Voilà! Diced onion. If your dice isn't fine enough, chop the pieces some more.

In the Season 3 Mise-en-Place Relay Race Quickfire, Casey's knife apparently wasn't sharp enough, and she took extra long to dice her onions. By the time Hung finished quartering his chickens, Casey was still dicing!

TRIMMING ARTICHOKES

1. Rinse the artichoke in cold water, making sure to clean between the leaves.
2. Cut off the top inch or two of the artichoke. Cut the sharp thistles from the end of all the leaves (kitchen shears are great for this). Cut the stem off at the base.
3. Cut the artichokes in half lengthwise. Scoop out the hairy choke with a spoon, or cut it out with a small knife. As you trim the artichokes, place them in a bowl of water mixed with a squeeze of lemon juice, as they discolor quickly.

When artichokes were suggested for the Season 4 Mise-en-Place Relay Race Quickfire, judge Tom Colicchio thought the technique might take too long for the segment. After testing out some artichokes in the production offices, he decided to feature the ingredient.

SHUCKING OYSTERS

1. Buy oysters the day you plan to eat them, and rinse and scrub them before shucking. After they are opened, set them on ice.
2. Hold the oyster with the hinge facing you. Wriggle a very sharp knife into the shell at the hinge until you hit a firm muscle.
3. Give your knife a little twist, and the shell should pop open. Hold it carefully so the tasty juice doesn't slush out.
4. Slide the knife beneath the oyster to separate it from the shell.

You never know what skills will come in handy on *Top Chef*. Brian M., who had previously competed in oyster-shucking competitions, blew away the other team with his assembly-line method.

BREAKING DOWN CHICKENS

1. Lay the chicken, breast side down, on a clean surface. With a very sharp knife, find the leg joint. There's no need to cut through bone; simply cut along the outside of the leg bone to sever the leg. Repeat on the other side.
2. Flip the chicken over. Position your knife slightly to the left or right of the breastbone. With your blade, find the ribcage, and run the blade along the ribcage, working close to the bone to sever the breast. Repeat on the other side.
3. Cut off both wings by finding the wing joints. Wriggle your knife into the point of no resistance and cut off both wings.

Hung, king of the time crunch, quartered his chickens in about two minutes. According to Hung, "Not only do you need to be smart, creative, and have a great palate, but you also need the speed."

"HUNG, I JUST WANT TO SAY ONE OTHER THING: YOU HAVE INCREDIBLE KNIFE SKILLS, AND YOU'RE VERY FAST. JUST ONE WORD OF ADVICE: BE VERY CAREFUL WITH YOUR KNIFE BECAUSE YOU ALMOST CUT CASEY. PLEASE BE CAREFUL."

JUDGE TOM COLICCHIO

HOST YOUR OWN MISE-EN-PLACE RELAY RACE WITH ANY OF THESE TECHNIQUES:

- Separate 2 eggs and beat the whites until they are stiff enough to stand in an upside-down bowl without spilling out.
- Peel 3 apples using only a paring knife.
- Cut 4 oranges into supremes (cut off pith and peel, then slice orange flesh from the membranes into small, delicious segments).
- Fillet a monkfish.
- Crack a lobster by hand.
- Peel, seed, and dice a watermelon.

HOSEA'S PAELLA

- 2 skinless, boneless chicken breasts (about 1 pound), cut into medium dice
- Salt and pepper
- 1/2 cup olive oil
- 2 Spanish chorizo sausages, sliced into rings
- 1 red bell pepper, seeded and cut into strips
- 1 green bell pepper, seeded and cut into strips
- 1 yellow bell pepper, seeded and cut into strips
- 1 yellow onion, diced
- 4 cloves garlic, minced
- 1/2 teaspoon saffron threads
- 1 1/2 cups parboiled or Spanish short-grain rice
- 3 cups chicken stock, plus extra if needed
- 12 extra-large shrimp (16–20 count), peeled and deveined (see page 122)
- 1 teaspoon chopped fresh flat-leaf parsley
- 1 teaspoon minced fresh thyme
- 1 teaspoon minced fresh oregano
- 1 teaspoon minced fresh chives
- 3 teaspoons thinly sliced scallions

 1 HOUR

 SEASON 5, EPISODE 6

 QUICKFIRE CHALLENGE:

Create a one-pot holiday meal.

SERVES 4

1. Season the chicken with salt and pepper. In a paella pan or large sauté pan, heat the oil over medium-high heat and brown the chicken and chorizo on both sides, about 3 minutes per side.
2. Reduce heat to medium. Add the bell pepper strips and onion and cook until softened, about 5 minutes.
3. Stir in the garlic, saffron, rice, and stock. Reduce heat to low, cover, and simmer until the rice is al dente, about 20 minutes.
4. Add the shrimp, herbs, and scallions. Cook until the shrimp are pink and opaque, and the liquid is absorbed, about 5 minutes more.
5. Season with salt and pepper and serve immediately.

RYAN'S LAMB PATTIES WITH PIPÉRADE

2 tablespoons olive oil

1 tablespoon butter

1 large red onion, thinly sliced

1 red bell pepper, seeded and julienned

1 yellow bell pepper, seeded and julienned

Salt

1½ teaspoons Espelette pepper

1 small bunch cilantro, stemmed and chopped, with 1 teaspoon reserved and minced

1 tablespoon pine nuts, toasted

6 ounces ground lamb

1 teaspoon Dijon mustard

1 teaspoon ground ginger

1 teaspoon red pepper flakes

1 teaspoon black pepper

½ teaspoon sherry vinegar

¼ cup lager beer, such as Beck's

Grilled baguette slices for serving (optional)

SERVES 2-3

1. Heat 1 tablespoon of the olive oil and the butter in a large saucepan over medium-low heat. When the butter is melted, add the onion and bell peppers and sweat until softened, about 30 minutes. Remove from heat, stir in 1 teaspoon salt, 1 teaspoon Espelette pepper, and the 1 teaspoon minced cilantro. Set aside.

2. Toast the pine nuts in a small, dry skillet over medium heat, stirring occasionally, until they start to sizzle and brown. Remove from heat and let cool. Meanwhile, blanch the cilantro in a medium pot of boiling water for about 30 seconds. Drain and squeeze dry. Combine the cilantro, pine nuts, 1 tablespoon of the olive oil, and a pinch of salt in a blender or food processor and purée until smooth. Let cool.

3. Combine the ground lamb, mustard, ginger, red pepper flakes, 1 teaspoon salt, the black pepper, the remaining ½ teaspoon Espelette pepper, the vinegar, and lager in a large bowl. Use your hands to mix all ingredients thoroughly. Divide the mixture in half and form two patties, each about 4 inches in diameter.

4. Grill the patties to the desired doneness over medium-high heat on a grill or fry in a large sauté pan, 3 to 4 minutes per side for medium-rare.

5. To serve, pool the pipérade on each plate and top with a lamb patty. Drizzle with the cilantro purée. Serve with slices of grilled baguette, if desired.

 45 MINUTES

 SEASON 4, EPISODE 6

 QUICKFIRE CHALLENGE:

Create a simple dish to pair with a beer.

ABOUT AN INGREDIENT
ESPELETTE PEPPER

If you've ever meandered the French village of Espelette in late summer you might have wondered about the hundreds of bright red bundles of peppers hanging from the walls and balconies of the village houses. These are Espelette peppers (*piments d'espelette*), for which the village is famous. The peppers are dried in the sun and then used whole or ground into powder for cooking, or in the production of Bayonne ham. South American in origin, they have become a staple in Basque cooking, adding a spicy (yet not overly so) bite to pâtes, sausages, rolls, and even chocolate.

STEFAN'S GOULASH

2 tablespoons olive oil

1/2 pound veal shoulder, cut into 1/2-inch cubes

1 potato, cut into 1/2-inch cubes

1 carrot, peeled and coarsely chopped

1 parsnip, peeled and coarsely chopped

1/4 small head cabbage, cored and chopped

2 strips bacon, chopped

2 cups chicken stock

1 cup dry white wine

Leaves from 2 sprigs thyme

Leaves from 1 sprig rosemary

1 bay leaf

Salt and pepper

Chopped fresh flat-leaf parsley for garnishing

SERVES 4

1. In a large pot or a Dutch oven, heat the oil over medium-high heat. Sauté the veal in the oil until browned, about 10 minutes. Add the potato, carrot, parsnip, cabbage, bacon, stock, wine, thyme, rosemary, bay leaf, and salt and pepper to taste. Reduce the heat and simmer for about 1 hour, until the veal is tender. Taste for seasoning and adjust if necessary. Remove the bay leaf.

2. To serve, spoon into bowls and sprinkle with parsley.

1 HOUR, 30 MINUTES

SEASON 5, EPISODE 6

QUICKFIRE CHALLENGE:
Create a one-pot holiday meal.

FUN FACT
GOULASH

Known as the national dish of Hungary, goulash (*gulyás*) is a soup or stew made with beef or veal, vegetables, and, like many Hungarian dishes, frequently a healthy dose of paprika. The word *goulash* means "herdsman," as the dish was originally made by cattle herders, who could obtain the freshest ingredients for its contents. But the stew has become highly popular around the world for its hearty consistency, satisfying warmth, and unique flavor.

"I'M THE ONLY COCK IN THE STALL, AND I LOVE IT."

STEFAN

ARIANE'S FILET MIGNON AND CAULIFLOWER

- 1 head cauliflower, trimmed and cut into 1-inch pieces
- 2 small Yukon gold potatoes, peeled and cut into 1-inch dice
- 2 cups heavy cream, plus more as needed
- 1/4 cup chopped fresh flat-leaf parsley
- 1 clove garlic, chopped
- 3 sprigs thyme
- Salt and white pepper
- Black pepper
- 4 petite filet mignons (about 1 1/2 pounds total)
- 2 tablespoons butter
- 1 lemon, cut into wedges

SERVES 4

1. Combine the cauliflower, potatoes, 2 cups cream, parsley, garlic, thyme, and salt and white pepper to taste in a medium saucepan over medium-high heat. Bring to a boil, then reduce heat to low to simmer. Simmer until the cauliflower is fork-tender, about 20 minutes. Remove the thyme and discard.
2. With a slotted spoon, transfer the solids to a food processor. Blend until smooth. With the machine running, add more cream as needed to smooth the mixture. Season with salt and white pepper, then return to the pot and keep warm.
3. Preheat a broiler.
4. Season the beef generously with salt and black pepper. Place in a broiler pan and broil to the desired doneness, 8 minutes per side for medium-rare. Remove from the broiler and transfer to a separate plate. Top each filet with 1/2 tablespoon butter and let rest for 10 minutes before serving.
5. To serve, place a large spoonful of cauliflower purée in the center of each plate. Top with a filet. Squeeze lemon juice over the meat and serve with lemon wedges.

45 MINUTES

SEASON 5, EPISODE 6

QUICKFIRE CHALLENGE:

Create a one-pot holiday meal.

GET POT-LUCKY

Invite partygoers to bring a home-cooked dish. At the end of the night, ask them to vote for their favorite. Give the winner the title of Top Chef.

> "I'VE LEARNED JUST TO DO MY FOOD AND KEEP IT SIMPLE. SIMPLE IS GOOD."
>
> ARIANE

THE MAGIC OF

MOLECULAR GASTRONOMY

Molecular gastronomy, the practice of using physical and chemical processes in cooking, has taken the culinary world by storm. Several chef'testants have shown off with the many innovative techniques, including sous vide and cutting-edge spherification.

RECIPES TO TRY AT HOME

ANDREW'S FAUX CAVIAR

SEASON 4, EPISODE 4

1. Bring 3 quarts water to a boil.
2. Add 1 box tapioca pearls and cook for 17 to 20 minutes, stirring every 5 minutes to avoid clumping.
3. Once tapioca is three-quarters cooked (centers are still white), drain in a chinoise and rinse with cold water for 5 minutes.
4. Combine 2 cups mushroom soy sauce, 1 cup white soy sauce, and 1/4 cup balsamic vinegar, and submerge the tapioca in the mixture.
5. Float tapioca with 1/2 cup olive oil and allow to sit for at least 1 hour.

FABIO'S SPHERICAL KALAMATA OLIVES

SEASON 5, EPISODE 2

1. Mix 14 ounces olive purée, 0.6 ounce calcic acid, and 0.5 ounce xanthan gum in a blender or food processor to form a thick paste. Let the mixture rest at room temperature for 3 hours.
2. Add 1.5 quarts water to a food processor or large bowl. With the machine running, or while stirring to form a vortex in the water, slowly sprinkle .01 ounce sodium alginate (or algin) into the water and let rest for 3 hours.
3. Scoop olive mixture into a squeeze bottle. Squeeze the mixture out into small ovals into a tablespoon. Dip the spoon in the water mixture, release the ball, and let it soak in the mixture until a skin forms on the "olive," about 3 minutes. Gently scoop out the olive. Use immediately.

ANDREW'S YUZU-MINT GELÉE "GLACIER"

SEASON 4, EPISODE 2

1. Bring 8 cups water to a simmer. Add 1 1/2 cups honey or corn syrup and 2 cups yuzu juice, and then turn off heat.
2. Add 1 bunch mint, let stand for 3 minutes, and remove from the stove.
3. Add 1 cup agar-agar powder and blend until the agar is fully dissolved.
4. Strain and place in a waxed paper–lined chinoise. Let set for 3 hours.
5. Turn the chinoise over and carefully peel the waxed paper away.

MARCEL'S CHERRY FOAM

SEASON 2, EPISODE 9

1. In a medium bowl, combine 2 cups cherry juice with 1 teaspoon lecithin.
2. Using an immersion blender, foam the mixture until thick.

MOLECULAR GASTRONOMY SUPERSTARS

WYLIE DUFRESNE

Wylie Dufresne—chef and owner of wd~50 restaurant in New York, and chef'testant on *Top Chef Masters*—is a leading American proponent of molecular gastronomy. When Wylie was a guest judge on Seasons 4 and 5, the chef'testants battled it out to impress him with their inventive techniques.

HUNG

SEASON 3, FINALE

In the Season 3 finale, Hung blew away the judges and clinched the Top Chef win with his Sous-Vide Duck with Mushroom Ragout and Truffle Sauce. Guest judge Todd English called it "three-star Michelin," and co-judge Michelle Bernstein raved: "It's perfect—I'm a little jealous."

RICHARD B.

SEASON 4, EPISODE 4

Inspired by the movie *Willy Wonka & the Chocolate Factory*, Richard led his teammates, Dale T. and Andrew, to victory with their Smoked Salmon with Tapioca Pearls and Wasabi White Chocolate Sauce. Richard's trusty mini-smoker and several burning pieces of wood were involved.

STEFAN

SEASON 5, EPISODE 12

Stefan whipped up a clever take on the egg with two components: Poached Egg on Brioche with Ham and Béarnaise, and Panna Cotta with Mango Purée and Sweet Béarnaise (resembling a poached egg) with egg yolk in the panna cotta and egg white in the mango purée.

FABIO

SEASON 5, EPISODE 12

Fabio also played on the concept of the egg with Lychee Soup with Melon Yolk (served in an eggshell and resembling a raw egg) and Coconut Milk Panna Cotta with Mango Purée. Sodium alginate thickened the mango purée into a bright jelly "egg."

ANDREW

SEASON 4, EPISODE 2

Andrew used a thickening agent to create a "flavored glacier jelly mold concept" and won over the judges with his Squid Ceviche with Soy-Balsamic Tapioca (faux caviar) and Yuzu-Mint Gelée "Glacier."

DALE L.'S PEACH COBBLER CHÈVRE ICE CREAM

3/4 cup (1 1/2 sticks) butter, at room temperature, plus 1 tablespoon

1/4 cup brown sugar plus 6 teaspoons

1/2 cup granulated sugar plus 6 teaspoons

1/2 cup old-fashioned rolled oats (see page 30)

1/4 cup all-purpose flour

1/4 cup chopped pecans, plus 1 cup pecan halves

Pinch of salt

2 large peaches, peeled, pitted, and diced

3 tablespoons Grand Marnier or Cointreau

3/4 cup (6 ounces) chèvre or other soft goat cheese

2 tablespoons candied ginger, chopped

1/2 gallon vanilla ice cream, softened

SERVES 4–6

1. Preheat an oven to 350° F.
2. Line a baking sheet with a Silpat or parchment paper. In a stand mixer, cream the 3/4 cup butter on medium speed. Decrease the speed to low and add the 1/4 cup brown sugar, 1/4 cup of the granulated sugar, the oats, flour, chopped pecans, and salt, all at once, until they are evenly distributed and the mixture forms moist clumps. Spread evenly on the baking sheet and bake until crisp, about 30 minutes. Let cool, then break into bite-sized chunks.
3. In a large sauté pan over medium heat, melt the remaining 1 tablespoon butter. Add the peaches and cook, stirring, 1 minute. Add 1/4 cup granulated sugar and stir until dissolved, about 3 minutes. Add the Grand Marnier and carefully set afire. Allow the flame to burn out and remove the pan from heat. Strain the peaches, reserving the syrup separately.
4. Toss the halved pecans with the remaining 6 teaspoons *each* granulated and brown sugars until evenly coated. In a large saucepan over medium heat, brown the nuts, shaking the pan constantly, until golden and the sugar is melted, 7 to 10 minutes. Cool.
5. Bake the goat cheese at 350° F until softened, about 5 minutes. In a large bowl, fold the peaches, warm goat cheese, and candied ginger into the ice cream. Place in the freezer to harden, if desired. Spoon the ice cream mixture into bowls. Top each with cobbler, candied pecans, and peach syrup.

1 HOUR, PLUS CHILLING

SEASON 3, EPISODE 7

QUICKFIRE CHALLENGE:

Create an original mix-in for Cold Stone Creamery ice cream.

"I IMMEDIATELY THINK: PEACH COBBLER. I'M LOOKING FOR TEXTURE, I'M LOOKING FOR FLAVOR. FOR ME THAT'S KIND OF A NO-BRAINER."

DALE L.

HOWIE'S BERRIES AND ICE CREAM

2 cups each fresh strawberries, blueberries, raspberries, and blackberries, quartered

1/2 cup balsamic vinegar

1/2 cup sugar

2 sprigs thyme

1 quart vanilla ice cream, softened

Fresh-cracked black pepper for sprinkling

Sea salt (preferably fleur de sel) for sprinkling

20 MINUTES PREP, PLUS MACERATING AND CHILLING

SEASON 3, EPISODE 7

QUICKFIRE CHALLENGE:

Create an original mix-in for Cold Stone Creamery ice cream.

SUNDAE SPECIAL HOME QUICKFIRE PARTY

Create a make-your-own-sundae bar and invite the troops. You provide the ice cream and have each guest bring their own topping. Partygoers vote for the best (or most creative) topping. The winner gets to take home all the leftovers.

SERVES 6–8

1. In a large bowl, combine the berries with 1/4 cup of the vinegar and 1/4 cup of the sugar. Macerate for 15 minutes.
2. In a medium saucepan over medium-high heat, cook to reduce the remaining 1/4 cup vinegar and 1/4 cup sugar along with the thyme until the liquid is syrupy. Strain through a fine-mesh sieve, return to the pan, and scorch using a hand-held blowtorch (see note).
3. To assemble, stir the berries into the ice cream. Refreeze if desired. Scoop into bowls and top with the balsamic syrup, a pinch of black pepper, and a pinch of sea salt.

NOTE

Use caution and follow the manufacturer's directions when using a blowtorch. Sear one dish at a time. Hold the blowtorch about 4 inches from the top of the dish, moving the torch constantly so that the sugar browns evenly. Alternatively, use a broiler, placing the dishes about 4 inches from the heat source. Watch carefully and turn the baking sheets holding the dishes if necessary to brown evenly.

"IT WAS INTERESTING HOW YOU INCORPORATED THE TEXTURES WITH THE SALT. IT WORKED OUT REALLY WELL."

GUEST JUDGE GOVIND ARMSTRONG

MARCEL'S CARAMELIZED BANANA AND AVOCADO TOWER WITH RUM COCO

1 cup canola oil

About 10 small corn tortillas, cut into 2-inch circles

1 tablespoon granulated sugar, plus more for seasoning

Salt

3 cups coconut milk

1/4 cup water

2 tablespoons banana schnapps

1/2 tablespoon soy lecithin granules

1 avocado, peeled, pitted, and diced

1 teaspoon fresh lime juice

4 tablespoons butter

1/4 cup packed brown sugar plus 4 teaspoons

2 bananas, peeled and sliced

About 1 cup dark rum

2 cups (1 pint) vanilla ice cream

SERVES 4

1. In a large skillet over medium-high heat, heat the oil to 350° F. Working in batches, fry the tortillas for 15 seconds on each side, until crisp. Using a slotted spoon, transfer to paper towels to drain. Season the chips with granulated sugar and salt.
2. In a large saucepan combine 2 cups of the coconut milk, the water, and schnapps and heat to 180° F over medium-high heat. Add the lecithin granules and stir until dissolved. Remove from heat and foam the liquid with an immersion blender.
3. Combine the avocado, lime juice, a pinch of salt, the 1 tablespoon granulated sugar, and 1/2 cup of the coconut milk in a blender or food processor and purée on high speed. Add as much of the remaining coconut milk as needed to achieve a smooth purée. Set aside.
4. In a large sauté pan, melt the butter over medium-high heat. Add the 1/4 cup brown sugar and cook until melted. Add the bananas and saute for 30 seconds. Add 3/4 cup of the dark rum and carefully ignite with a match. Continue cooking until the flame extinguishes and the sauce reduces to caramel consistency, about 2 minutes.
5. To assemble, place 2 tablespoons rum and 1 teaspoon brown sugar on each of 4 large spoons. Fill 4 glasses with coconut milk foam. Set a spoon on top of each glass but do not tip the rum into the foam. Spoon avocado purée onto each plate and spread into a circle. Place corn chips on top of the purée. Arrange alternate layers of bananas and avocado purée. Scoop ice cream on top. To finish, ignite the rum on each spoon and pour into the foam.

1 HOUR, 30 MINUTES

SEASON 3, EPISODE 7

QUICKFIRE CHALLENGE:

Create an original mix-in for Cold Stone Creamery ice cream.

"GASTRO BOY HAS BALLS!"

JUDGE ANTHONY BOURDAIN

TOP CHEF

HUBERT KELLER'S BERRY VERRINE WITH MOUSSE AND SWAN

Meringue Swan

6 egg whites

1 cup granulated sugar

1 teaspoon fresh lemon juice

Orange Sabayon

6 egg yolks

¾ cup granulated sugar

2 cups fresh orange juice

Chocolate Mousse

1 cup whipped cream

3 tablespoons unsweetened cocoa powder

1½ tablespoons confectioners' sugar

Raspberry Sauce

1½ cups fresh or thawed frozen raspberries

1 teaspoon granulated sugar

Verrine

15 fresh blueberries

5 fresh strawberries

15 fresh raspberries

6 crumbled Girl Scout Thin Mints cookies or other mint chocolate wafers, plus two whole cookies for garnish

Final Presentation

6 ounces dark chocolate, chopped

2 fresh strawberries

2 HOUR PREP, PLUS CHILLING

SEASON 1, EPISODE 11
TOP CHEF MASTERS

QUICKFIRE CHALLENGE:

Create a dessert for people who really know something about sweets—Girl Scouts.

SERVES 2

FOR MERINGUE SWAN

1. Preheat an oven to 275° F.
2. In a large bowl, beat the egg whites with an electric mixer on low speed with 2 tablespoons of the sugar and the lemon juice for 4 to 5 minutes.
3. Add 2 tablespoons of the sugar and beat for 10 minutes more.
4. Gradually add the remaining sugar, increase the speed to medium-high, and beat for 3 minutes.
5. Using a fluted pastry tip, pipe S-curve swan necks and spirals to form the two wing sides of the body onto a Silpat- or parchment-lined baking sheet.
6. Bake for 30 minutes, until firm. Remove from the oven and cool.

FOR ORANGE SABAYON

Combine the yolks and sugar in a bowl and beat until frothy. Add the orange juice and place in a stainless-steel bowl over, but not touching, simmering water in a saucepan. Whisk constantly until thickened. Set aside and keep warm.

CONTINUED

TOP CHEF MASTERS

This recipe—whimsical, over-the-top, and quintessentially chef-y—comes out of Bravo's new show *Top Chef Masters*, which pits 24 of the biggest stars of the culinary world against each other in gritty competition for the title of Top Chef Master. A whole new set of stringent standards and tastes are in play, as the judges panel features former *New York* magazine critic Gael Greene; editor-in-chief of a food magazine James Oseland; and British journalist Jay Rayner. Same format, same stress: Just like the original, each episode has a Quickfire Challenge and an Elimination Challenge. Only, this time, it's the top dogs of the culinary world—including Wylie Dufresne, Rick Bayless, Wilo Benet, Roy Yamaguchi, Art Smith, and John Besh (all former *Top Chef* guest judges)—who are jumping through the hoops!

This winning recipe comes from superchef Hubert Keller of Fleur de Lys restaurant in San Francisco and Burger Bar in Las Vegas. Chef Keller was a guest judge on *Top Chef* Seasons 1, 2, and 5.

> "SINCE I WAS A GUEST JUDGE ON *TOP CHEF*, I FELT LIKE IT'S FAIR TO SAY: I WANT TO FEEL WHAT IT FEELS LIKE ON THE OTHER SIDE."
>
> HUBERT KELLER

FOR CHOCOLATE MOUSSE

1. Combine the cream and cocoa powder in a bowl and beat until firm peaks form. Fold in the confectioners' sugar.

2. Use the chocolate mousse to fill the swan and hold the parts together, and to make the little mouse (see photo on page 144 as a guide).

FOR RASPBERRY SAUCE

Purée the raspberries and sugar in a food processor or blender. Strain through a fine-mesh sieve.

FOR VERRINE

Layer the fresh berries, raspberry sauce, sabayon, and crumbled cookies in a short glass so all the ingredients are visible. Garnish with the whole cookies.

FOR FINAL PRESENTATION

1. Melt the chocolate in a double boiler or metal bowl set over, but not touching, barely simmering water in a saucepan.

2. Dip 2 strawberries in the chocolate. On a large plate, arrange meringue swan, mousse mouse, verrine, and strawberries. Use excess chocolate to drizzle on plate.

LEFT TO RIGHT: CHEFS MARK PEEL, JOHN BESH, DOUGLAS RODRIGUEZ, AND ANITA LO GET READY TO FACE OFF IN THE *TOP CHEF MASTERS* KITCHEN.

LISA F.'S CHOCOLATE-BERRY WONTONS

2 cups strawberries, sliced, plus 4 whole strawberries with stems

1/2 cup raspberry purée

2 tablespoons granulated sugar plus 1/4 cup

Grated zest and juice of 1/2 orange

8 fresh mint leaves, cut into chiffonade

4 fresh basil leaves, cut into chiffonade

1 cup heavy cream plus 1/4 cup

1 tablespoon confectioners' sugar

1/2 cup plain Greek yogurt

1 cup balsamic vinegar

4 ounces dark chocolate, plus 1 tablespoon grated

1/2 cup Kahlúa

1 teaspoon ground cinnamon

Canola oil for frying

4 square wonton wrappers

SERVES 4

1. In a large bowl, mix together the sliced strawberries, raspberry purée, the 2 tablespoons granulated sugar, the orange zest, orange juice, mint, and basil. Let sit for at least 30 minutes.
2. Whip the 1 cup heavy cream and the confectioners' sugar until stiff peaks form. Fold in the Greek yogurt. Set aside.
3. In a saucepan over medium-high heat, reduce the vinegar until it is thick and syrupy, about 10 minutes.
4. Melt the 4 ounces of chocolate in a double boiler or metal bowl set over, but not touching, barely simmering water in a saucepan, stirring until smooth. Dip the whole strawberries in the chocolate and cool on waxed paper or parchment until the chocolate is firm.
5. Combine the Kahlúa, the remaining 1/4 cup cream, and the grated chocolate in a small saucepan. Melt over low heat and stir until smooth. Divide among 4 shot glasses.
6. Combine the 1/4 cup granulated sugar and the cinnamon. Heat the oil in a large sauté pan over medium-high heat until shimmering. Fry the wonton squares until golden brown, about 5 minutes. Using a slotted spoon, transfer to paper towels. Toss with the sugar mixture.
7. To serve, divide the cream and berries among the 4 wontons, topping each wonton with a layer of cream and then berries. Drizzle the balsalmic reduction on top. Serve each with a shot glass of Kahlúa and a chocolate-covered strawberry.

1 HOUR, 30 MINUTES PREP, PLUS MACERATING

SEASON 4, EPISODE 7

QUICKFIRE CHALLENGE: Create an innovative dessert.

TOP CHEF DRINKING GAME

- Take a sip every time there is a kitchen mishap (someone drops something, cheftestants run into each other, the oven is set to the wrong temperature, etcetera).
- Take a drink every time a timer or clock is shown.
- Take two drinks every time Padma says "hands down."

PLATE LIKE A MASTER

In the *Top Chef* kitchen, it's not only about getting the food on the plate before the timer goes off. Because we all taste with our eyes before we bite, presentation of a dish is key, whether you're competing in a Quickfire Challenge or serving it up at home.

Always strive for your dishes to come across as simple, clean, and effortless.

Do not put too many contrasting elements on the plate.

Make sure there is enough visual and textural variety.

Follow these tips, and you'll plate like a master!

PLATING

Consider what you are serving food on or in. If you plan to pool sauce on a plate, warm the plate first so the sauce doesn't congeal. If you are serving a salad, do not pile it on a warm plate, as it will wilt. If you are serving soup, consider serving it in mugs or deep bowls; a wide, shallow bowl exposes a lot of warm soup, causing it to cool quickly and form an unappetizing film.

Use complementary items on the plate. Do not pair a chocolate tart with a sliced tomato for garnish.

COLOR

Use contrasting colors, and your food will pop.

Add color with sauces and garnishes.

Useful tool: a hand-held blowtorch (see page 142) to add caramelized color.

TEXTURE

Choose complementary textures, both for appearance and taste. If you're serving custard, garnish with something crunchy. If you're serving something substantial, like a steak, try an oozy blue cheese or a dollop of horseradish cream. Ingredients such as toasted breadcrumbs add visual appeal and a contrasting texture.

Add height to your dishes by stacking foods or including vertical elements on the plate, such as a shot glass.

SYMMETRY

Eye a plate and visualize an imaginary axis. Balance the size and shape of the elements on the plate on either side of this line.

SHAPES

Cut foods into interesting shapes. Use a small tourné (bird's beak) knife for carving vegetable shapes.

Food can be round, square, rectangular, peaked, or flat. You may want to contrast these shapes, or use several shapes of the same size to create a repeating theme.

SPACING

A bit of distance between each element will help each one stand out.

There should be one focal element on the plate. For example, the brightest, the largest, the meat. The focal element should stand out most.

SAUCES

Add visual interest with drizzles and dashes.

Consider pooling sauce beneath the food.

Useful tools: eye droppers, squeeze bottles, toothpicks for drawing lines through sauces, a piping bag for piping mousse and creams, a paintbrush for applying sauces or oil on dough that can be cut and browned.

GARNISHES

Garnishes add an extra splash of color and texture to the plate. Sprinkle them on the rim of a plate or arrange alongside the main elements.

Dried herbs: paprika, cracked pepper, sea salt, colored salts, tarragon, rosemary, parsley, chives.

Minced or sliced vegetables: red bell pepper, yellow tomato, red or purple cabbage.

Edible flowers and fresh herbs: nasturtiums, borage, chamomile, chive flowers, broccoli florets, lavender, rosemary, chervil.

CHEF BIOS: SEASON 4

STEPHANIE IZARD

This Chicagoan won over the audiences as well as the judges with her infectious smile, calm demeanor, and soulful cooking style. Both first female *Top Chef* winner and Season 4 Fan Favorite, she is living proof that you can be a champion in the kitchen without sacrificing personality or sense of humor. Stephanie is opening The Drunken Goat restaurant in Chicago.

"WINNING *TOP CHEF* IS JUST A REAFFIRMATION THAT THIS IS WHAT I'M MEANT TO DO IN LIFE."

STEPHANIE

LISA FERNANDES

Lisa had her ups and downs during the season. Her constant scowl and prickly personality rubbed many fellow chefs the wrong way, and got her voted Top Reality Villain by the editors of *TV Guide*. But it was skill, not personality, that landed this experienced New York chef in the Final Three: at the last minute she nearly walked away with the whole thing, as the judges marveled at her distinctive pan-Asian menu.

RICHARD BLAIS

Richard cooked like a rock star throughout the competition, in fact, he was a safe bet to win the season, until he second-guessed himself during the last match in Puerto Rico. A veteran of culinary school and training with the likes of Daniel Boulud and Thomas Keller, Richard stood out with his molecular gastronomy and envelope-pushing creations. Now you'll find him spending time with his family at his upscale burger joint, Flip Burger Boutique, in Atlanta.

ANTONIA LOFASO

This talented chef and single mom from Los Angeles made it all the way to the Final Four in Puerto Rico when some undercooked pigeon peas proved to be her downfall. She rocked the Uncle Ben's Rice Quickfire in Episode 8 (see recipe, page 23) and stayed cool under pressure on the egg line at Lou Mitchell's. Antonia did time in the Spago kitchen before becoming executive chef at Foxtail restaurant in Los Angeles.

SPIKE MENDELSOHN

The hat-wearing provocateur of Season 4 who coined the term "culinary boner" proved that he could cook—and butcher tomahawk chops like nobody's business. Spike learned classic technique at the CIA before making his bones working at Mai House in New York City. When he's not mugging for the cameras, you'll find Spike in Washington D.C. at his gourmet burger joint, Good Stuff Eatery.

DALE TALDE

Dale had the skills to go all the way. His time at Morimoto and Buddakan in New York informed his dishes, but his fiery temperament seemed to get the better of him at times. He was ousted for a too-sweet dish of butter-scotch scallops in the Restaurant Wars Challenge. Dale is now consulting with restaurants in Chicago and checking out spaces for his casual noodle bar concept.

ANDREW D'AMBROSI

Feisty and funny Andrew made it all the way to Episode 10 before being told to pack his knives for serving quirky sushi to a roomful of hungry Chicago policemen. Originally from Fort Lauderdale, Andrew moved to New York to make it in the culinary big leagues. Andrew's good-natured and sometimes kooky antics on and off the set made an indelible impression.

NIKKI CASCONE

This strong-minded native New Yorker had worked with some of the biggest chefs in the land before finding acclaim as chef and partner of 24 Prince in New York City. Nikki was sent home for her lack of team leader-ship in the Wedding Wars competition; she is now planning her own wedding, as well as targeting a second restaurant opening in the near future.

MARK SIMMONS

The freewheelin' Kiwi who grew up on a sheep farm always looked like he was having fun in the kitchen—and in the apartment. He peaked early, then coasted along in the middle until he was ousted for a misguided vegetable curry in the Common Threads-Cooking with Kids Challenge. Kudos to Mark for making the judges eat Marmite in his very first challenge of the show.

JENNIFER BIESTY

Jennifer, the other half of this season's lesbian power couple with Zoi, has kitchen bona fides, classical training, and an execu-tive chef position at the popular restaurant Coco500 in San Francisco under her belt. She was sent packing for a phallic asparagus salad that she and Stephanie constructed. (Zoi and Jennifer have since split up.)

RYAN SCOTT

This outgoing San Franciscan may have been eliminated halfway through the competition, but he impressed the judges with his simple, heartfelt cooking. Ryan has been able to parlay his celebrity into fundraising for the American Heart Association, March of Dimes, and other worthy causes. Ryan now runs a catering business in San Francisco and is planning to open his own restaurant.

ZOI ANTONITSAS

It was a bit awkward when Zoi and Jennifer announced to the other chef'testants on the first day of filming that they were in fact a couple (they are currently split)—little did the two know they would become lesbian icons. Zoi focuses on Italian and Greek food in her cooking and has worked with some of the biggest chefs on the West Coast.

MANUEL TREVINO

It was a surprise when this native Texan with Mexican roots and classical French training got knocked out of the competition early on. His collaboration with Spike to incorporate the film *Good Morning, Vietnam* into a summer roll with sea bass and Swiss chard seemed like a winner, but didn't make the cut with guest judge Daniel Boulud.

ERIK HOPFINGER

Erik grew up in restaurant kitchens in New York and moved to San Francisco to cook as executive chef for restaurants including Circa. Despite his obvious chops in the kitchen and his self-deprecating humor at the judges' table, Erik didn't make it past Episode 3 when he was sent home for some soggy corn dogs at the Block Party Elimination.

VALERIE BOLON

Valerie, who works as a personal chef in Chicago, met her culinary demise early in Season 4 after the daunting Zoo Food Challenge. Valerie took one for Team Gorilla and went home for a bad plate of blinis.

NIMMA OSMAN

Nimma came all the way from Georgia, only to be eliminated in the first episode of Season 4 for her lackluster interpretation of shrimp scampi. This soft spoken young chef seemed to lack the killer competitive instinct necessary to survive in the *Top Chef* kitchen. But never fear, Nimma is still cooking her heart out in Atlanta.

FACE THE FIRING SQUAD: JUDGES' CHALLENGE

RECIPES

HAROLD'S ONION RINGS

About 4 cups canola oil for frying

One 12-ounce bottle beer

1 egg

$1^1/_2$ cups buttermilk

1 cup all-purpose flour

Salt and pepper

1 large yellow onion, cut into $^1/_4$-inch rings

 **30 MINUTES**

 **SEASON 1, EPISODE 10**

 QUICKFIRE CHALLENGE:

Make a food platter with four different snacks for high-stakes poker players, including Phil Hellmuth.

SERVES 4

1. Preheat an oven to 200° F.
2. Fill a large saucepan with 2 inches of oil. Over medium-high heat, heat the oil to 350° F.
3. In a large bowl, whisk together the beer, egg, and buttermilk. Gradually whisk in the flour, season with salt and pepper, and whisk until smooth.
4. Have ready a baking sheet covered in paper towels. Working in batches, dip the onion rings in the batter, then lower into the hot oil. Fry until golden brown and crispy, about 2 minutes. Using a slotted spoon or mesh strainer, transfer the onion rings to paper towels to drain (remove excess batter from the pan). Keep warm in the oven until you're finished frying. Season with salt and pepper.

TECHNIQUE

FRY, FRY BABY

Most of the deep-frying that takes place on *Top Chef* happens in a professional deep fryer. You can buy a home version, a Fry Daddy, Fry Baby, or other brand, but you do not need any special equipment to fry at home. A deep, heavy, medium saucepan will do the trick just fine. Fill the pan with about 2 inches of a neutral-tasting, high-smoking-point oil such as canola or peanut. Heat the oil over medium-high heat until it reaches about 350° F. (You can measure the temperature on a deep-frying or candy thermometer, or else test it by placing a small amount of food in the oil and seeing if it makes a nice bubbling and hissing commotion.)

Make sure the food you are frying is dry, then fry away, baby!

Judging a home Quickfire Challenge? Use this handy scoreboard and the buzzwords when critiquing your friends' dishes, and you'll sound just like the pros.

CATEGORY	DESCRIPTION	POSSIBLE POINTS	POINTS AWARDED
TASTE	GOOD FLAVOR COMBINATION APPROPRIATE SEASONING APPROPRIATE TEXTURE RIGHT TEMPERATURE EVENLY COOKED CHEWABLE	10	
PLATING AND PRESENTATION	GOOD USE OF COLOR GOOD USE OF SHAPES GOOD USE OF SPACING GOOD USE OF SYMMETRY APPROPRIATE GARNISH APPEALING AROMA	5	
ORIGINALITY	CREATIVE USE OF INGREDIENTS CREATIVE TECHNIQUE	5	
FOLLOWED RULES OF CHALLENGE	REQUIREMENTS OF THE CHALLENGE ARE MET	5	
BONUS POINTS	DIFFICULT RECIPE OR TECHNIQUE	2	
	TOTAL POINTS	27	

COMMENTS

OTTO'S QUICKFIRE SUSHI

Spicy Tuna Roll

Sriracha sauce, as needed

1 tablespoon mayonnaise

¼ pound sushi-grade tuna, diced

2 sheets nori

½ cup cooked sushi rice

1 avocado, peeled, pitted, and cut into long strips

Crab and Scallion Roll with Ginger

1 tablespoon mayonnaise

2 tablespoons chopped scallion

2 teaspoons grated, peeled fresh ginger

⅓ cup (about 3 ounces) fresh lump crabmeat

Salt

2 sheets nori

½ cup cooked sushi rice

Wasabi, soy sauce, and pickled ginger for serving

MAKES 4

FOR SPICY TUNA ROLL

1. In a small bowl, add Sriracha to the mayonnaise until it is to your taste. Stir in the diced tuna.
2. Lay a sheet of nori flat on a clean surface. Spoon half of the rice onto the nori and pat it evenly over the surface of the nori to cover the sheet. Arrange several slices of avocado at one end of the sheet.
3. On the opposite end of the sheet, place half of the tuna, forming a thin strip. Gently roll the tuna and nori upward toward the avocado, tucking the roll around the fillings. Slice the roll into 6 to 8 pieces. Repeat to make a second roll.
4. Serve immediately with wasabi, soy sauce, and pickled ginger.

FOR CRAB AND SCALLION ROLL WITH GINGER

1. Combine the mayonnaise, scallion, and ginger in a small bowl. Stir in the crab. Season with salt.
2. Lay a sheet of nori flat on a clean surface. Spoon half of the rice onto the nori and pat it evenly over the surface of the nori to cover the sheet. Place half of the crab mixture in a single line along the short end of the nori. Gently roll, tucking the roll around the fillings. Slice the roll into 6 to 8 pieces. Repeat to make a second roll.
3. Serve immediately with wasabi, soy sauce, and pickled ginger.

30 MINUTES

SEASON 2, EPISODE 2

QUICKFIRE CHALLENGE:

Create a quick sushi dish for sushi chef Hiroshi Shima.

TECHNIQUE

ROLLING SUSHI

Make sure all of your ingredients are prepped in advance and readily available when you begin your roll. Have a small bowl of warm water ready to moisten your fingers.

Lay a rolling mat on a flat, clean surface. Lay your base layer (nori or whatever you're using) on the mat. Wet your fingers and spread a thin layer of rice (or whatever your first layer is) on top of the base layer. Make sure it evenly covers the base layer, and press it down firmly. Place your additional filling in thin, even layers on top of the rice.

Carefully lift up the side of the mat nearest you, tucking the edges of the roll under. Use your fingers to compress and shape the roll.

CARLA'S GREEN EGGS AND HAM

- 1 cup chopped fresh flat-leaf parsley
- 1 jalapeño chile, roasted and peeled, plus 1 jalapeño, seeded and diced
- 1 cup olive oil
- Salt and pepper
- 1 avocado, peeled, pitted, and diced
- 1 green tomato, seeded and diced
- 1/2 teaspoon minced garlic
- 1 small bunch fresh chives, minced
- Canola oil for cooking
- 1/2 pound country ham, diced
- 1/2 cup spinach leaves, cut into chiffonade
- 4 egg whites, divided into two bowls
- 4 quail egg yolks
- 4 slices brioche or challah bread
- Butter for toast

SERVES 4

1. Blend the parsley, roasted jalapeño, and olive oil in a food processor until smooth. Season with salt and pepper.
2. Combine the avocado, tomato, diced jalapeño, garlic, and 1/2 teaspoon chives. Season with salt and pepper.
3. In a large sauté pan, heat canola oil over medium heat, and cook the ham until crisp. Transfer to paper towels to drain. Wipe the oil from the skillet. Toss the ham with the chives.
4. Combine the spinach and 2 egg whites in a bowl. Whisk in the remaining 2 egg whites until combined.
5. Heat canola oil in the pan used to cook the ham. Pour the spinach mixture into the pan. Gently spoon the quail yolks into the whites, spacing them as the pan will allow. Cover and cook until the whites are set, about 4 minutes.
6. Meanwhile, toast the bread slices and spread with butter.
7. To serve, top each slice of toast with a wedge of "green" egg. Top the egg with the ham and salsa. Drizzle with parsley-jalapeño oil.

QUICKFIRE CHALLENGE:

Create an egg dish for Chef Wylie Dufresne.

"I FEEL LIKE I AM JUST THIS TORTOISE, AND I'VE BEEN PICKING UP SPEED! I'LL SEE YA, SEE YA!"

CARLA

STEFAN'S AMUSE-BOUCHE EGG

1 egg

1 teaspoon shredded Cheddar cheese

3 tablespoons heavy cream

Salt and pepper

1 teaspoon prepared salsa

1 sprig fresh cilantro

 **15 MINUTES**

 **SEASON 5, EPISODE 4**

 QUICKFIRE CHALLENGE:

Create a breakfast amuse-bouche for Chef Rocco DiSpirito.

EQUIPMENT HEADS-UP

EGG TOPPER

Even star chef Rocco DiSpirito was impressed with Stefan's neatly decapitated egg. You can do the same show-off move at home with a nifty little gadget called an egg topper or egg cutter. You'll find several different kinds available online and at specialty stores; most of them feature a round blade that neatly cuts through the eggshell to form a perfect little cup for serving.

MAKES 1

1. Add 2 inches of water to a small, deep saucepan. Bring to a slow boil over medium-high heat. Bring a teakettle of water to boil over high heat.

2. With a sharp knife or an egg cutter, carefully remove the pointed top quarter of the eggshell (see sidebar). Discard the eggshell top. Separate the egg yolk from the white. Return the yolk to the shell and balance upright in an egg cup or ramekin.

3. In a small bowl, combine the egg white, cheese, heavy cream, and a sprinkling of salt and pepper. Carefully fill the shell to the top with this mixture. Reduce heat to medium-low and carefully set the egg in its cup in the water bath. Pour additional boiling water from the kettle into the saucepan until the water comes up to the top rim of the egg cup. Cover the pan and poach the egg in the simmering water bath until it is set but soft, about 5 minutes.

4. Serve in the egg cup, topped with salsa and a sprig of cilantro.

LEAH'S ASPARAGUS SOUP WITH OLIVE-TUNA TOASTS

2 1/2 tablespoons butter

2 bunches white or green asparagus, cut into 1-inch pieces (about 4 cups)

2 shallots, chopped

2 1/2 cups chicken stock

1/2 cup heavy cream

1 teaspoon salt

1/8 teaspoon cayenne pepper, plus more for garnish

1 1/2 teaspoons fresh lime juice, plus more for finishing soup

1/2 cup oil-cured black olives, pitted and chopped

3 tablespoons minced fresh chives

Grated zest and juice of 1/2 lemon

3 tablespoons olive oil

4 thick slices bread, preferably brioche

1 clove garlic, minced

Leaves from 1 sprig thyme, chopped

1/4 cup oil-packed tuna, flaked

Microgreens, such as baby arugula or radish sprouts, for garnishing

SERVES 4

1. In a large saucepan melt the 2 tablespoons butter over medium heat. Add the asparagus and shallots. Sweat until tender, 8 to 10 minutes (see page 93). Add the chicken stock.

2. Transfer to a blender or food processor and blend until smooth. Strain through a fine-mesh sieve. Stir in the heavy cream and season with the salt, 1/8 teaspoon cayenne, and the 1 1/2 teaspoons lime juice. Let cool, cover, and refrigerate.

3. Combine the olives, chives, lemon zest and juice, and 2 tablespoons of the olive oil in a small bowl and set aside.

4. Using a 2-inch cookie or biscuit cutter, punch out 4 rounds of bread. Heat the remaining 1 tablespoon oil and the 1/2 tablespoon butter in a medium sauté pan over medium heat. Stir in the garlic and thyme. Toast the bread in this butter and oil mixture.

5. To serve, ladle the soup into chilled bowls. Working on a clean surface, spread a dollop of tapenade on each toast round. Top with 1 tablespoon tuna, then another dot of tapenade. Float a bread round in each serving of soup, top with microgreens, and sprinkle with additional lime juice and cayenne.

45 MINUTES

SEASON 5, EPISODE 3

QUICKFIRE CHALLENGE:

Make a dish from *Top Chef: The Cookbook*. Wait!—and turn it into a soup for Chef Grant Achatz.

COOKBOOK CHALLENGE

Turn to a random recipe (the first one that's not a salad) in this book, and make a salad inspired by the ingredients and flavors in the recipe.

NAME THAT JUDGE

The guest judges of the Quickfire Challenges are like a who's who of the culinary world. And Top Chefs not only know their food, they know their people. Think you have what it takes to keep up? Match the photo of the guest judge to the name and description on the next page.

SCORING

0-3 PLEASE PACK YOUR KNIVES AND GO.

4-6 WAS SOMEONE SNOOZING DURING CULINARY SCHOOL?

7-8 YOU'RE A FOODIE FORCE TO BE RECKONED WITH.

9 CONGRATULATIONS. YOU'RE ON YOUR WAY TO BECOMING TOP CHEF!

DONATELLA ARPAIA

STEPHEN STARR

RICK BAYLESS

JOHNNY IUZZINI

NORMAN VAN AKEN

MING TSAI

ALFRED PORTALE

SIRIO MACCIONI

DANIEL BOULUD

A. Raised on a farm outside of Lyon, this French chef has restaurants in New York City, Palm Beach, Miami, and Las Vegas.

B. This Italian chef is the illustrious owner of Le Cirque restaurant in New York City.

C. This Jean Georges pastry chef was identified by Forbes.com in 2007 as one of the ten most influential chefs working in America today.

D. Born into a family of barbecue restaurateurs, this American chef is known for modern interpretations of traditional Mexican cuisine.

E. This chef and restaurateur hosts two cooking shows and is known for Asian inspired American fusion cuisine.

F. This highly successful restaurateur is the owner of restaurants in Philadelphia, New York City, and Atlantic City, and of the ________ Restaurant Organization.

G. An important force in the New American Cuisine movement, this chef is the owner of Gotham Bar and Grill in New York City.

H. This Floridian chef founded New World Cuisine, melding the influences of Latin America, the Caribbean, the southern United States, and Asia.

I. This attorney-turned-restaurateur has opened a string of highly acclaimed restaurants in New York City, and is a recognized food expert and authority on entertaining.

ANSWER KEY: 1) Ming Tsai / E, 2) Norman Van Aken / H, 3) Rick Bayless / D, 4) Alfred Portale / G, 5) Donatella Arpaia / I, 6) Johnny Iuzzini / C, 7) Sirio Maccioni / B, 8) Daniel Boulud / A, 9) Stephen Starr / F

DANNY'S LEEK, HAM, AND EGG SOUP

4 cups any kind day-old bread

8 tablespoons unsalted butter, 4 melted, 4 cubed

2 tablespoons olive oil

1 cup button mushrooms, cut into 1/8-inch-thick pieces

1/2 cup chopped fresh tomatoes

1/2 cup diced canned peeled tomatoes

1/2 cup chopped leek

4 cups chicken stock

1 cup fresh or frozen corn kernels

Salt and pepper

3/4 pound ham, preferably Black Forest, cut into thin strips (about 1 cup)

1 teaspoon white vinegar

4 eggs

Special equipment:
juicer, blender, or food processor

1 HOUR

SEASON 5, EPISODE 3

QUICKFIRE CHALLENGE:

Make a dish from *Top Chef: The Cookbook* and turn it into a soup for Chef Grant Achatz.

EQUIPMENT HEADS-UP

VITA-MIX

Meet the Ferrari of blenders. Used by both professionals and home cooks, the Vita-Mix is a juicer/blender with a turbo-charged motor that cuts through the fibers of fruit and vegetables and pulverizes them into a smooth and creamy consistency. The blades move so fast they produce enough heat to cook food, and the dry blade can turn whole grains into flour. If you already have a Viking range, a Sub-Zero fridge, and the must-have accoutrements, the Vita-Mix could well become your next obsession. Home models cost between $300 and $500.

SERVES 4

1. Preheat an oven to 375°F.
2. Toss the bread with the melted butter in a bowl, spread on a baking sheet, and bake until crisp, 25 minutes.
3. In a large pot heat 1 tablespoon of the olive oil over medium heat. Sweat the mushrooms, chopped tomatoes, diced canned tomatoes, and leek until soft, about 10 minutes (see page 93). Add the stock and corn and simmer for 5 minutes. Add the cubed butter and remove from heat.
4. Carefully transfer the soup base to a blender or food processor. Blend until smooth, then strain through a fine-mesh sieve. Season with salt and pepper.
5. In a skillet, heat the remaining 1 tablespoon olive oil over medium heat. Crisp the ham. Set aside.
6. Fill a saucepan to a depth of 2 inches with salted water and bring to a low simmer over medium-high heat. Add the vinegar, stir, then crack the eggs into the simmering water. Poach the eggs until set but still soft, about 3 minutes. Remove with a slotted spoon.
7. To serve, divide the bread among warmed soup bowls. Ladle the soup over and garnish with the ham strips and a poached egg.

DAVE'S GRAPE APE SANDWICH

4 strips bacon, preferably applewood smoked

1 cup (about 4 ounces) shiitake mushrooms, stemmed and diced

1 cup (about 4 ounces) cremini mushrooms, diced

1 small shallot, diced

½ roasted red pepper, diced

Salt and pepper

2 tablespoons butter at room temperature

2 slices sourdough bread

8 thin slices ham, preferably Black Forest

4 slices pecorino cheese, preferably Toscano or Sardo

2 to 3 tablespoons grape jam or jelly

1 cup red seedless grapes for serving

MAKES 1

1. In a large sauté pan over medium heat, cook the bacon until crisp. Remove from the pan and chop. Without cleaning the pan, add the mushrooms and shallot and cook until soft, about 5 minutes. Remove from heat, mix with the bacon and diced red pepper. Season with salt and pepper and set aside.

2. Butter the bread and toast it in the same pan over medium heat. When toasted, still working in the pan over medium heat, pile ham and cheese on one slice of bread. Spread the other slice with jam and the mushroom-bacon mixture. Cover the pan and heat on medium heat until the cheese is melted through, 2 to 3 minutes. Remove from heat, carefully press the halves together, cut the sandwich on the diagonal, and serve with grapes alongside.

"I FOCUS ON FLAVORS AND FOOD THAT'S FUNCTIONAL."

HAROLD

20 MINUTES

SEASON 1, EPISODE 7

QUICKFIRE CHALLENGE:

Create a signature sandwich to be featured at Tom Colicchio's 'wichcraft restaurant.

TOP CHEF TITLES

Several *Top Chef* spinoffs have given former chef'testants another chance to score a Top Chef title.

4 STAR ALL STAR
Season 1 vs. Season 2

Chef'testants:
Stephen, Harold, Dave, Tiffani (Season 1); Ilan, Elia, Marcel, Sam (Season 2)

Winner: Season 1

HOLIDAY SPECIAL
Chef'testants:
Tiffani, Stephen (Season 1);
Josie, Marcel, Betty (Season 2);
C.J., Tre, Sandee (Season 3).

Winner: Tiffani

HAROLD'S MORTADELLA WITH WILTED DANDELION GREENS

- 2 slices sourdough bread
- ½ tablespoon olive oil
- 1 ounce oyster or stemmed shiitake mushrooms
- Salt and pepper
- 1 handful dandelion or other bitter green
- 1½ tablespoons mayonnaise
- 1½ tablespoons tapenade
- ¼ cup sliced red grapes
- 3 ounces sliced mortadella
- ½ teaspoon chopped fresh flat-leaf parsley

MAKES 1

1. Toast the bread to the desired doneness.
2. In a medium skillet over medium heat, sauté the mushrooms in the olive oil until tender, 2 to 3 minutes. Season with salt and pepper to taste. Set aside.
3. In the same pan, over medium heat, wilt the dandelion greens, stirring, about 3 minutes.
4. Mix together the mayonnaise and tapenade in a small bowl.
5. To serve, spread tapenade mixture on both slices of toasted sourdough. Top one slice with the mushrooms, greens, grapes, mortadella, and parsley. Top with the second piece of toast, cut the sandwich on a diagonal, and serve.

"SO LONG AS I'M NOT GOING TO ANOTHER GAS STATION OR COOKING FOR FIVE-YEAR-OLDS, I'M FINE."

HAROLD

25 MINUTES

SEASON 1, EPISODE 7

QUICKFIRE CHALLENGE:

Create a signature sandwich to be featured at Tom Colicchio's 'wichcraft restaurant.

ABOUT AN INGREDIENT

YOU SAY BOLOGNA, I SAY MORTADELLA

Bologna is the name used in America for an industrial version of mortadella, a coldcut that has its origins in the Italian city of Bologna (boh-lone-ya). Mortadella is made from finely ground pork that is extruded through a large cylinder. The pork is often mixed with peppercorns, pieces of pork fat, and sometimes pistachios. The version popularized in the United States by Oscar Mayer has helped give this delicacy a reputation as a common lunch meat, but artisanal versions can be as good as a fine salame.

ANDREW'S DUCK TACOS WITH PLANTAIN JAM

- 1 teaspoon canola oil
- ¼ cup diced white onion
- 1 cup diced ripe plantain
- ⅓ cup honey
- 1 teaspoon sherry vinegar
- ½ teaspoon diced dried chipotle chile
- 2 tablespoons water
- Salt and pepper
- 2 duck breasts
- ½ teaspoon chili powder
- 2 tablespoons grapeseed oil
- 4 small flour tortillas

MAKES 12

1. Heat the canola oil in a large saucepan over medium heat. Sauté the onion until translucent, about 5 minutes. Add the plantain and sauté for 2 minutes. Add the honey, vinegar, and chipotle. When the honey starts to bubble, add the water and remove from heat. Season with salt and pepper, then mash the mixture with a spoon.
2. Season the duck with salt and pepper and the chili powder. Set a large sauté pan over high heat. Add the grapeseed oil, and when it shimmers, place the duck, skin side down, in the pan. When the skin begins browning, reduce heat to medium. Cook slowly, allowing the fat to melt and occasionally basting the meat with the fat, about 15 minutes.
3. While the duck skin is crisping, grill the tortillas. Keep warm and set aside.
4. Once the duck breasts are ultra crispy, flip them over. Cook for 2 minutes, until medium-rare. Remove from heat and allow to rest for 4 to 5 minutes.
5. Slice the duck thinly and pile slices onto a tortilla. Top with plantain jam.

"I WOKE UP WITH A FIRE INSIDE MY STOMACH, LIKE, 'EITHER I'M GONNA STAB SOMEBODY OR I'M GONNA MAKE SOME AMAZING FOOD.'"

ANDREW

1 HOUR

SEASON 4, EPISODE 3

QUICKFIRE CHALLENGE:

Create an upscale taco for Chef Rick Bayless.

A T-SHIRT IS FOREVER

Many memorable things have been said throughout the seasons of *Top Chef*, and Bravo has captured the most unforgettable ones on T-shirts.

SEASON 1:
"I'm not your bitch, BITCH!"
(inspired by Dave)

SEASON 2:
"Oh, Big Time"
(inspired by C.J.)

SEASON 4:
"I Have a Culinary Boner"
(inspired by Andrew)

SEASON 5:
"I ♥ Padma"
(inspired by Hosea)

WYLIE DUFRESNE

"I appreciated your efforts with the whites. I thought that was a really smart technique, and I really liked the tempura. But I didn't see how ultimately they played off one another."

ON HOSEA'S EGG WHITE STICKY RICE, SEASON 5, EPISODE 12

ROCCO DISPIRITO

"When you cook prosciutto I think you have to be very careful how you do it. It takes on a really strong gamey flavor, and that's what I didn't like about your pizza."

ON STEPHANIE'S DEEP-DISH PIZZA, SEASON 4, EPISODE 1

ERIC RIPERT

"The salad was not seasoned; you did not put salt in it. So it was kind of bland."

ON BRIAN'S TROUT, SEASON 3, EPISODE 13

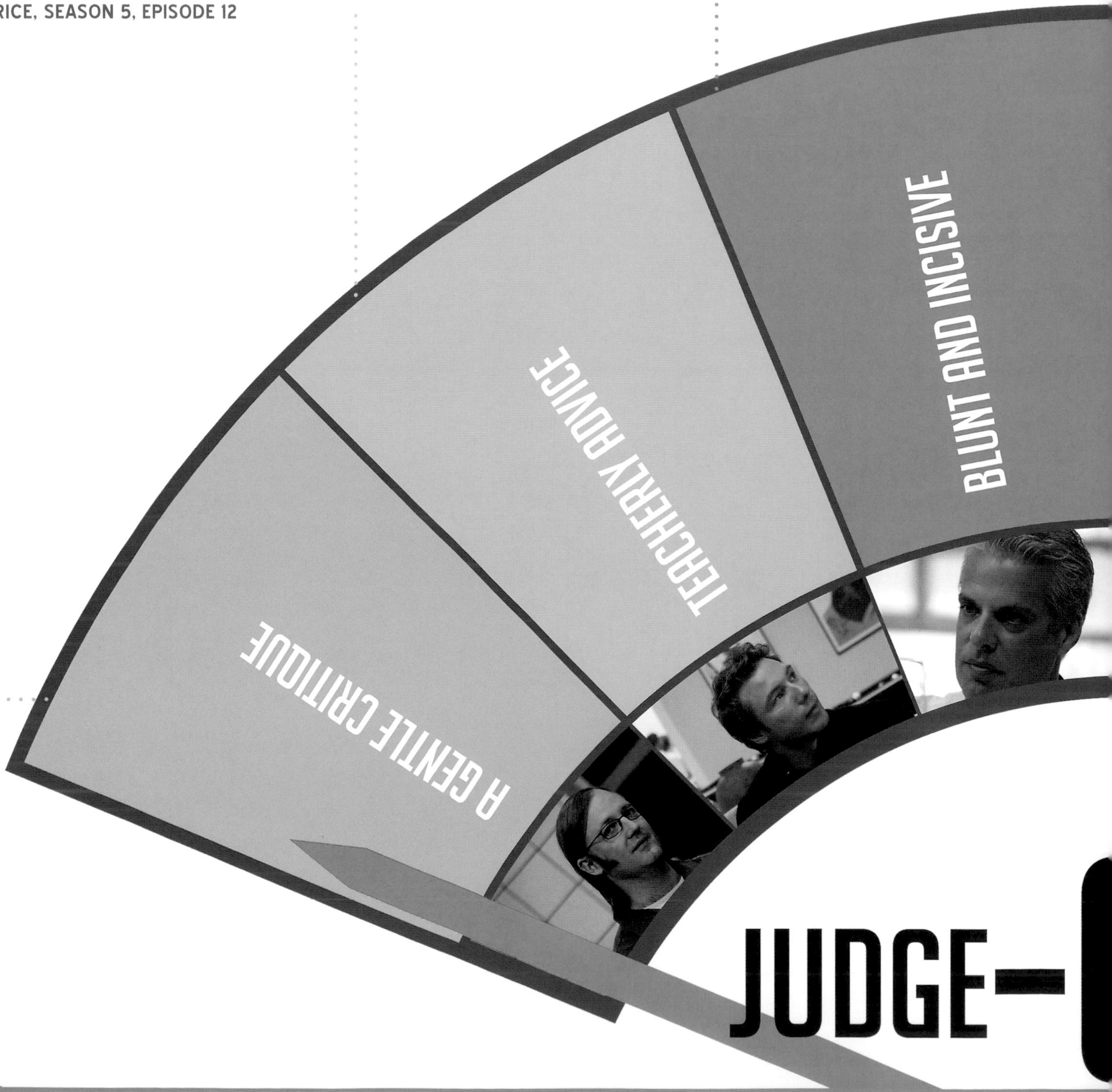

A GENTLE CRITIQUE

TEACHERLY ADVICE

BLUNT AND INCISIVE

JUDGE-

HUNG HUYNH

"That was the dish where I thought: You opened cans up and tossed a little bit of olive oil on a piece of toast. I wish you could have done a little more seasoning-wise."

ON JAMIE'S BRUSCHETTA, SEASON 5, EPISODE 8

MICHELLE BERNSTEIN

"Your kidneys tasted so much of kidney. No sauce was even attempted. You made something that I love to eat hard to eat."

ON ELIA'S SEARED KIDNEY WITH SWEETBREADS, SEASON 2, EPISODE 5

TED ALLEN

"Green is the color that I think would have been the easiest. I felt that your presentation was really kind of a mess. I mean it looked like something you raked up. Not to be unkind."

ON BETTY'S GREEN ZUCCHINI TAMALE, SEASON 2, EPISODE 9

OUTRIGHT DISMISSIVE

MAKING IT PERSONAL

UNKIND AND PROUD OF IT

-METER

ANTONIA'S DEEP-DISH PIZZA WITH PROSCIUTTO, BURRATA, AND ARUGULA

5 tablespoons olive oil, plus more for greasing the pan

3 cloves garlic, minced

1 pound arugula, stemmed and chopped

Salt and pepper

4 large ripe tomatoes in varied colors, quartered and seeded

Leaves from 1 bunch fresh thyme

1/2 pound low-moisture mozzarella cheese, shredded

1/4 pound fontina cheese, shredded

10 ounces pizza dough, thawed if frozen, and pressed or rolled into a 10-inch round

1 teaspoon aged balsamic vinegar

5 sprigs fresh marjoram, chopped

Canola oil for frying

1/4 pound thinly sliced proscuitto

1/2 pound burrata cheese (see sidebar)

One 4-ounce chunk Parmigiano-Reggiano, grated for serving

Special equipment: 14-inch deep-dish pizza pan

MAKES 1

1. Preheat an oven to 550° F.
2. In a large sauté pan, heat 1 tablespoon of the olive oil over medium heat. Add the garlic and three quarters of the arugula. Cook until wilted. Season with salt and pepper. Set aside.
3. Reserve 4 tomato slices in a small bowl. Place the rest of the tomatoes and the thyme on a baking sheet. Drizzle 2 tablespoons of the olive oil over the tomatoes and season with salt and pepper. Roast until softened, about 5 minutes (leave the oven on). Cool for 15 minutes, then peel and discard the tomato skins.
4. In a medium bowl, toss the mozzarella and fontina until blended.
5. Grease the pizza pan with olive oil. Place the dough in the pan and press it evenly on the bottom and halfway up the sides. Sprinkle one-half of the cheese mixture on the bottom. Top with a layer of sautéed arugula, then a layer of tomatoes. Repeat layering with the cheese, arugula, and tomatoes. Bake for about 20 minutes, rotating the pan every 5 minutes to ensure even baking.
6. Meanwhile, cut the reserved tomato slices into 1/4-inch pieces. Mix with the remaining 2 tablespoons olive oil, the vinegar, marjoram, and salt and pepper to taste. Marinate for 15 minutes.
7. Place a medium saucepan over high heat. Add canola oil to a depth of 1/2 inch and heat the oil to 350° F. Fry the remaining arugula until crispy and translucent. Using a slotted spoon, transfer to paper towels to drain.
8. When the pizza is done, allow to rest at least 10 minutes. Then sprinkle the prosciutto over the top. Using all of the burrata, divide 6 spoonfuls evenly on the pizza. Add mounds of marinated tomato over the burrata, and finish with a couple pieces of crispy arugula. Slice and serve with freshly grated Parmigiano-Reggiano.

1 HOUR

SEASON 4, EPISODE 1

QUICKFIRE CHALLENGE:

Create a signature deep-dish pizza for Chef Rocco DiSpirito.

ABOUT AN INGREDIENT

BURRATA

What is this mysteriously creamy cheese and where did it come from? The answer is Puglia, the heel of the southern Italian boot, and the cheese is actually less mysterious than you'd think: Burrata (the word means "buttered") is really just a hybrid of fresh mozzarella and cream. The mozzarella is molded into a little round with a cream filler that blends with the cheese. It's delicious on flatbread, pizza, or by itself with a side of olives. Burrata has a short shelf life, so it's a good thing that many domestic cheese artisans have begun producing it in the United States.

SPIKE'S PIZZA ALLA GREEK

2 tablespoons olive oil

1 onion, diced,
plus 1 onion, cut into thin rings

1 green bell pepper,
seeded and julienned

1 carrot, diced

Salt and pepper

One 28-ounce can peeled
Italian tomatoes

2 cloves garlic, crushed

1 teaspoon red pepper flakes

½ pound fennel or Italian sausage,
casing removed

1 small fennel bulb, trimmed,
halved, and thinly sliced

1 pound pizza dough

Pinch of dried oregano

1 cup shredded mozzarella

1 cup crumbled feta cheese

1 cup kalamata olives, pitted

6 button mushrooms, sliced

Asiago cheese for grating

Pinch of dried thyme

Special equipment:
8-inch deep-dish pizza pan

1 HOUR

SEASON 4, EPISODE 1

QUICKFIRE CHALLENGE:

Create a signature deep-dish pizza for Chef Rocco DiSpirito.

BEST BROS

A special bond formed among the Season 4 males:

 **SPIKE AND MARK** 

Oh, the infamous bubble bath. "If I want to get in a tub with Mark," Spike said, "I'm getting in a bubble bath with Mark."

ANDREW AND SPIKE

This dynamic duo was known for their "Vanilla Love."

RICHARD AND DALE

Richard openly admired Dale's "gorgeous nipples."

MAKES 1

1. In a large sauté pan over medium heat, heat 1 tablespoon of the oil. Add the diced onion, the bell pepper and carrot, and sauté until softened, about 5 minutes. Season with salt and pepper. Add the tomatoes, garlic, and half of the pepper flakes to the pan, breaking up the tomatoes with a spoon. Cook, stirring, until the sauce is thickened, about 15 minutes.

2. In another large sauté pan, heat the remaining 1 tablespoon oil. Add the sausage and brown, breaking up with a spoon. Using a slotted spoon, remove the sausage from the pan and pour off most of the grease. Add the onion rings and fennel to the pan; season with salt and pepper. Sauté until tender, strain, and cool.

3. Preheat an oven to 550°F. Oil an 8-inch deep-dish pizza pan.

4. Press the dough into the pan. Sprinkle with the remaining pepper flakes, salt to taste, and a little dried oregano. Spread the mozzarella on the crust. On top of the mozzarella, layer the sausage and the onion-fennel mixture. Ladle tomato sauce over vegetables, reserving a few tablespoons. Sprinkle the feta, kalamata olives, and mushrooms over the top. Dot the top of the pizza with a few more tablespoons of tomato sauce.

5. Bake until brown and bubbly, about 40 minutes. Grate the Asiago over the crust and sprinkle the pizza with dried oregano and thyme before serving.

TRE'S SURF-AND-TURF BURGER

Two 8-ounce beef tenderloin filets, halved lengthwise

$1\frac{1}{2}$ tablespoons kosher salt

1 tablespoon cracked black pepper

2 teaspoons chopped fresh rosemary

2 tablespoons olive oil, plus more for drizzling

1 cup (about 4 ounces) chopped cremini mushrooms

2 tablespoons chopped shallots

$\frac{1}{2}$ cup dry red wine

4 tablespoons unsalted butter at room temperature

8 large shrimp, peeled and deveined (see page 122)

3 tablespoons grapeseed oil

1 cup dry white wine

1 teaspoon chopped fresh thyme

Juice of $\frac{1}{2}$ lemon (about $1\frac{1}{2}$ tablespoons)

3 tablespoons sour cream

1 teaspoon prepared horseradish

4 ciabatta or crusty buns, toasted and halved

2 tomatoes, sliced

2 cups packed arugula leaves

4 slices provolone cheese

MAKES 4

1. Season the filets with 1 tablespoon of the kosher salt, $\frac{1}{2}$ tablespoon of the cracked black pepper, and the rosemary.
2. Heat the 2 tablespoons olive oil in a small sauté pan over medium-high heat. Sear each filet, 2 to 3 minutes per side. Transfer to a cutting board to rest. Drain the excess fat from the pan and return to heat, increasing to high. Add the mushrooms and sauté for 2 minutes, then add the shallots and cook, stirring, for another 2 minutes. Add the red wine and cook to reduce for 1 minute. Turn off the heat and whisk in 2 tablespoons of the butter. Keep warm until ready to serve.
3. Season the shrimp with the remaining $\frac{1}{2}$ tablespoon kosher salt. In a medium sauté pan over high heat, heat the grapeseed oil until it begins to smoke. Add the shrimp and sear, shaking the pan so the shrimp flips, 2 to 3 minutes. Add the white wine to the pan and cook to reduce the wine by half. Add the remaining 2 tablespoons butter, reduce the heat to medium-low, and stir in the thyme and lemon juice, then remove from heat. Split each shrimp in half lengthwise, then return to the pan. Keep warm over low heat.
4. To assemble, mix the sour cream and horseradish in a small bowl. Spread the mixture on the cut sides of the buns. Place a piece of tenderloin on a bottom bun. Top with mushrooms and shallots, then with 2 pieces of shrimp. Add sliced tomatoes, arugula, and cheese. Sprinkle with the remaining cracked black pepper, drizzle with olive oil, and cap with the top of the bun. Repeat to make the remaining sandwiches.

30 MINUTES

SEASON 3, EPISODE 8

QUICKFIRE CHALLENGE:

Create an adventurous burger for Chef Daniel Boulud.

FUN FACT
CIABATTA

This tasty Italian bread is called *ciabatta*, which means "slipper," because of its wide, flat, bedroom-footwear-like shape.

"THE SUN SHINES BRIGHTLY, BUT I THINK PADMA MIGHT EVEN SHINE BRIGHTER THAN THAT."
—C.J., SEASON 3

Indian model turned actress, cookbook author, and host extraordinaire, Padma has awed chef'testants and fans alike with her radiant smile, stunning outfits, and easy charm. Here's a tribute to the white-hotness that is Padma Lakshmi.

WE ♥ PADMA

"THE INNER QUEEN INSIDE ME IS SCREAMING TO KNOW: WHERE'S PADMA, AND WHAT IS SHE WEARING?"

RICHARD S., SEASON 5

"FOR ALL OF US CHEFS, SHE'S A PLEASURE WHEN SHE COMES IN THE KITCHEN . . . SHE COMES IN, AND IT BRIGHTENS OUR DAY."

SANDEE, SEASON 3

"HOW THE HELL CAN [THE CHEF'TESTANTS] COOK ANYTHING WITH THAT PADMA PARADING AROUND? CHRIST, I'D EAT ANYTHING OFF OF HER. I'D DRINK HER BATHWATER."

MARK, *TOP CHEF* FAN FROM SAN DIEGO

"IS PADMA THAT HOT? YES, ACTUALLY, HOTTER."

RICHARD B., SEASON 4

"I SEE PADMA—SMOKIN' HOT. I CAN'T EVEN COOK. I GET SWEATY ARMPITS JUST LOOKING AT HER."

STEFAN, SEASON 5

"PADMA BITING DOWN ON THAT BONE IN THE HOT DOG CHALLENGE WILL ALWAYS BE ETCHED IN MY MEMORY."

JAMIE, SEASON 5

RADHIKA'S SAMBAR-CRUSTED BUTTERFISH WITH CHORIZO AND CORN

1 tablespoon all-purpose flour

2 teaspoons sambar powder

1 pound butterfish fillets (escolar, walu, or black cod)

Salt and pepper

Olive or grapeseed oil for frying

1 cup Spanish chorizo, diced

1 tablespoon olive oil

3 cups chopped spinach leaves

1 cup fresh or frozen corn kernels

1½ cups dry white wine

1 shallot, minced

1½ cups heavy cream

5 saffron threads, crushed

1 cup (2 sticks) cold butter, cut into small pieces

SERVES 4

1. Preheat an oven to 325° F.
2. Combine the flour and sambar in a bowl. Season the fillets with salt and pepper, then coat on both sides with the sambar mixture.
3. Heat a large sauté pan over medium-high heat. Add the olive or grapeseed oil. Add the fish and sear until lightly browned, about 2 minutes on each side (see page 95). Transfer to the oven to keep warm.
4. Wipe out the pan and add the chorizo. Cook over medium heat until the sausage releases its oils, about 5 minutes. Remove the chorizo and set aside. Add the 1 tablespoon olive oil, spinach, and the corn to the pan. Cook, stirring, until the spinach is wilted and the corn is cooked, about 5 minutes. Season with salt and pepper. Keep warm until serving.
5. In a medium saucepan over high heat, cook the wine to reduce by half. Add the shallot and reduce heat to medium-high. Add the heavy cream and cook to reduce the liquid until thickened, about 10 minutes. Add the saffron and remove from heat. Gradually whisk in the butter, one tablespoon at a time, adding each tablespoon when the one before is incorporated. Season with salt and pepper.
6. To serve, divide the spinach among 4 plates. Top with the fish, then sauce. Sprinkle each plate with chorizo.

30 MINUTES

SEASON 5, EPISODE 9

QUICKFIRE CHALLENGE:

Create a tasting and pitch a new restaurant concept for restaurateur Stephen Starr.

TOP CHEF AWARDS

Critically acclaimed and immensely popular, *Top Chef* has received various recognitions:

- Won the award for Outstanding Editing in a Reality Series at the 60th Primetime Emmy Awards
- Nominated for Outstanding Cinematography for Reality Programming and Outstanding Reality-Competition Program at the 59th Primetime Emmy Awards
- Named one of *Time* magazine's Top 10 Returning Series of 2007

CASEY'S PAN-SEARED TROUT WITH SUMMER CORN, TOMATOES, AND GRAPES

3 tablespoons olive oil, plus more for drizzling

Four 6-to-8-ounce fillets lake or rainbow trout, sole, or tilapia, skin-on

3 tablespoons butter, cut into 1/2-inch cubes

Kernels cut from 1 ear sweet corn or 1 cup frozen corn

1/2 cup green grapes, halved

1 shallot, minced

Salt and pepper

1/2 cup dry white wine

1/4 cup chopped fresh tarragon, plus 1 tablespoon for sprinkling

Juice of 1/2 lemon

1 ripe red tomato, cut into 1/8-inch-thick rounds

2 cups watercress sprigs

SERVES 2

1. In a large cast-iron or stainless-steel skillet, heat 2 tablespoons of the olive oil over medium-high heat. Season the fillets with salt and pepper. Add the fillets, skin side down, and cook until the flesh side begins to turn opaque, about 4 minutes. Flip over and cook 1 minute more. Remove from the pan and keep warm.

2. Clean the skillet with paper towels and place back over medium-high heat. Add the remaining 1 tablespoon olive oil. Add 2 tablespoons butter to melt. Add the corn, grapes, and shallot and cook, stirring, until softened, about 5 minutes. Season with salt and pepper, then add the wine and simmer for 1 minute more. Add the 1/4 cup tarragon, lemon juice, and the remaining 1 tablespoon butter.

3. To assemble, season the tomato slices with salt and pepper. Drizzle the watercress with olive oil and season with salt. Lay a slice of tomato on each plate. Top with watercress, then fillets, then the grape-corn mixture. Drizzle with the pan sauce, then sprinkle tarragon on top.

"YOUR DISH HAS A SOUL, WHICH IS VERY IMPORTANT WHEN YOU ARE COOKING IF YOU HAVE THE SOUL OF SOMEONE BEHIND IT."

GUEST JUDGE ERIC RIPERT

25 MINUTES

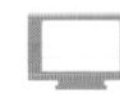
SEASON 3, EPISODE 13

QUICKFIRE CHALLENGE:

Create a fresh trout dish using a camp stove for Chef Eric Ripert.

FAN FAVORITES

Except for Stephanie, none of these chef'testants ended up with the *Top Chef* title, but they did win over viewers to snag the title of Fan Favorite.

SEASON 1: Not instituted

SEASON 2: Sam

SEASON 3: Casey

SEASON 4: Stephanie

SEASON 5: Fabio

TOP CHEF YEARBOOK 

CLASS CLOWN
Andrew, Season 4

BEST HAIR
Marcel, Season 2

MOST CHARMING
Fabio, Season 5

Delizioso

KING OF SPEED
Hung, Season 3

UNDERDOG
Carla, Season 5

hootie-hoo!

SEXIEST
Sam, Season 2

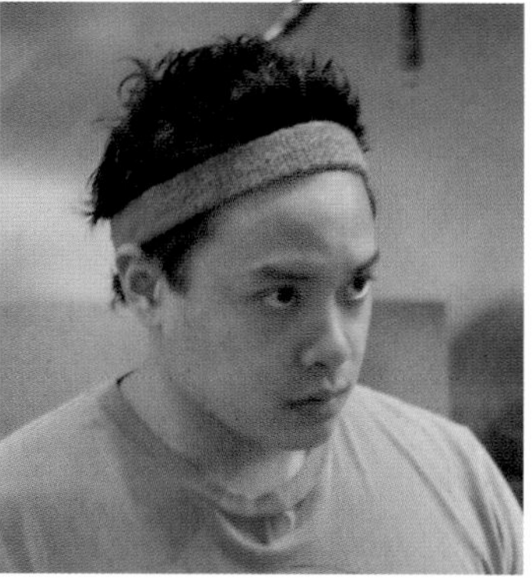

MOST COMPETITIVE
Dale, Season 4

SWEATIEST
Howie, Season 3

xoxo

BIGGEST FLIRT
Stefan, Season 5

Team Euro!

COUGAR IN THE KITCHEN
Ariane, Season 5

ROWR!

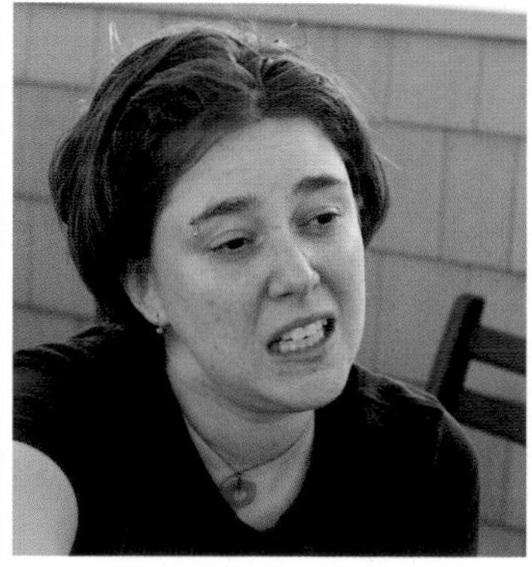

BIGGEST SCOWL
Lisa, Season 4

BEST PERSONALITY
C.J., Season 3

DAVE'S GRILLED KOBE TENDERLOIN

1 cup ruby port

1 cup chicken stock

4 dried figs, quartered

1 sprig fresh rosemary, plus more sprigs for garnish

1 teaspoon black peppercorns

1 teaspoon coarsely ground black pepper

Kosher salt

2 cups aged balsamic vinegar, preferably ten years old

1 cup packed brown sugar

Splash of cognac

Splash of Myers's rum

One 2-pound Kobe beef tenderloin

SERVES 4

1. In a large saucepan over medium-high heat, combine the port, stock, figs, 1 sprig rosemary, and peppercorns. Cook to reduce the liquid by half, 20 to 30 minutes, then remove the rosemary. Transfer to a blender or food processor and purée. Season with kosher salt.
2. Return the fig sauce to the saucepan over medium-high heat. Add the balsamic vinegar, brown sugar, cognac, and rum. Cook to reduce until thickened and syrupy, another 20 minutes.
3. Meanwhile, season the tenderloin with 2 teaspoons kosher salt and the ground black pepper. Grill over high heat until nicely browned on all sides, about 15 minutes per side for medium-rare. Alternatively, sear the meat on all sides in an oven-safe pan, then transfer to a preheated 400° F oven and roast to the desired doneness.
4. To serve, cut the tenderloin into 1-inch-thick slices. Fan the slices on a platter and pour the sauce over. Garnish with rosemary sprigs.

1 HOUR, 30 MINUTES

SEASON 1, EPISODE 10

QUICKFIRE CHALLENGE:

Serve three different plates of high protein, high carb, low-fat food for the cast of Cirque du Soleil's *KA*.

FUN FACT

KOBE BEEF

Here are some interesting morsels about the famous Kobe beef:

True beef: All Kobe beef is from the Hyōgo region of Japan; if it is made elsewhere, it can only be called "Kobe-style."

Nice life: Kobe beef cattle are massaged and fed some sake and beer, along with a steady diet of grain.

Chew on this: Kobe beef is so marbled with fat that if it were graded on the USDA system (Select, Choice, Prime), it would require its own grade.

STEPHANIE'S BITTERSWEET CHOCOLATE CAKE

8 ounces bittersweet chocolate, chopped, plus 3 ounces

4 tablespoons butter

2 egg whites

1 1/3 cups sugar, plus 1 tablespoon

7 egg yolks

1 cup fromage blanc

1 1/2 cups heavy cream

Zest and juice of 2 1/2 lemons, plus 1 cup fresh lemon juice

1 bunch basil, coarsely chopped, plus a handful of fresh basil leaves

Pinch of salt

2 cups pomegranate juice, such as POM

 **1 HOUR, 30 MINUTES**

 **SEASON 4, EPISODE 7**

 QUICKFIRE CHALLENGE:

Create an innovative dessert for Chef Johnny Iuzzini.

SERVES 6–8

1. Preheat an oven to 350° F. Line two 8-by-8-inch (or one 17-by-11-inch) rimmed baking sheets with parchment paper. Spray the parchment with nonstick cooking spray.

2. In a metal bowl set over, but not touching barely simmering water in a saucepan, melt the chopped chocolate and the butter, stirring occasionally. Remove from heat and allow to cool for about 5 minutes.

3. Meanwhile, whisk the egg whites to soft peaks. Gradually whisk in 1/2 cup of the sugar and whisk to stiff peaks.

4. Put the yolks in a large bowl. Pour the chocolate mixture through a fine-mesh sieve into the yolks. Whisk until smooth. Fold the egg whites into the chocolate mixture just until combined.

5. Divide the batter between the prepared pans and smooth the tops. Bake until a toothpick inserted into the center of a cake comes out clean, 16 to 18 minutes (or up to 35 minutes if baking as a single cake). Cool the cakes on a wire rack. After 10 minutes, carefully lift the cakes and parchment from the pans.

6. Whisk together the fromage blanc, 1/2 cup of the heavy cream, the zest and juice of 1 lemon, and the 1 tablespoon sugar until smooth.

7. Chop the 3 ounces of chocolate. Put the chopped chocolate in a large heatproof bowl. In a medium saucepan, heat the chopped basil and the remaining 1 cup heavy cream over medium heat until bubbles appear on the surface of the cream. Do not allow the cream to boil. Remove from heat and strain the mixture. Pour the hot cream over the chocolate and let sit for 1 minute, then stir until smooth. Stir in the salt. Refrigerate until ready to use.

8. Combine the 1 cup lemon juice, the zest of 1 lemon, and 1/3 cup sugar in a medium saucepan over medium-high heat. Cook to reduce until syrupy, about 10 minutes, then strain and cool.
9. Combine the pomegranate juice, remaining 1/2 cup sugar, zest and juice of 1/2 lemon, and basil leaves in a medium saucepan over medium-high heat. Reduce until syrupy, about 20 minutes, then strain and cool.
10. To assemble, spread filling on one cake. Top with the second cake. (Alternatively, cut the cake horizontally into two even pieces.) Frost the top with the ganache, spreading carefully so the cake does not tear. Cut the cake into servings. Drizzle or pool the lemon reduction and pomegranate reduction onto plates and position the servings on top of the syrups.

ABOUT AN INGREDIENT

CHOCOLATE PAIRINGS

Some flavors paired with chocolate are classics: cherry, hazelnut, orange, ginger, peanut butter. But many chocolatiers are testing the palates of experimental American chocolate consumers. Chocolate bars with no sugar (also known as 100 percent bars) are a connoisseur's favorite.

SURPRISING (-LY DELICIOUS!) CHOCOLATE COMBINATIONS:

Chocolate + tarragon
Chocolate + bacon
Chocolate + Szechuan pepper
Chocolate + Taleggio cheese
Chocolate + Pop Rocks

"BAKING IS DEFINITELY TECHNICAL. IT'S NOT LIKE COOKING WHERE YOU PUT A LITTLE BIT OF THIS, A LITTLE BIT OF THAT, YOU THROW THINGS IN. YOU HAVE TO MEASURE."

LISA F.

NIKKI'S YOGURT CAKES WITH TWO SAUCES

1 cup all-purpose flour

1 teaspoon baking powder

1/2 teaspoon salt

1/2 cup plain yogurt

2 tablespoons buttermilk

4 eggs

2 cups sugar

1 teaspooon vanilla extract

3/4 cup (1 1/2 sticks) unsalted butter, melted and cooled

1 cup *each* fresh strawberries, raspberries, and blackberries

1 cup water

2 cups Concord grapes or grape juice

Leaves from 1 sprig fresh rosemary, chopped

1/4 cup balsamic vinegar

1 cup heavy cream, whipped to soft peaks

MAKES 4

1. Preheat an oven to 350° F. Sift the flour, baking powder, and salt together into a bowl 3 times. Combine the yogurt and buttermilk in a small bowl.
2. Fill a medium saucepan halfway with water. Over high heat, bring the water to a boil, then reduce to a simmer. In a metal bowl that fits on top of the saucepan without touching the water, combine the eggs and 1 cup of the sugar. Whisk until smooth and thickened; the mixture should be at 120° F. Remove from heat and cool slightly. Transfer to a stand mixer and beat on high until doubled in volume, 7 to 8 minutes. Reduce heat to medium-low, return the mixture to the pan and drizzle in the vanilla and butter. Stir in the flour mixture, alternating with the yogurt mixture, ending with flour. Beat until combined between additions, scraping down the sides of the bowl as needed.
3. Spray four 1-cup cake molds or ramekins with nonstick cooking spray. Fill each mold almost to the top with cake batter. Bake until a toothpick inserted into the cake comes out clean, about 20 minutes. Remove from the oven and cool on a wire rack.
4. Mix the fruit with 1/2 cup of the sugar and water in a medium saucepan over medium heat. Cook, stirring, until the fruit thickens and coats the spoon, about 15 minutes. Purée in a blender or food processor until smooth.
5. In a medium saucepan over medium-high heat, cook the grapes with the remaining 1/2 cup sugar, the rosemary, and vinegar until the sauce thickens enough to coat a spoon, about 15 minutes. Purée and strain through a fine-mesh sieve.
6. Carefully unmold each cake by running a sharp knife around the edges of the mold and inverting the cake. Top each cake with both sauces and a dollop of whipped cream.

1 HOUR, 30 MINUTES

SEASON 4, EPISODE 7

QUICKFIRE CHALLENGE:

Create an innovative dessert for Chef Johnny Iuzzini.

TECHNIQUE

WHIP IT GOOD

No store-bought whipped cream can stand up to the rich and silky taste and texture of real stuff made by hand. You can whip cream with just a whisk, a large metal bowl, and some elbow grease. If you have an electric mixer, even better. Use chilled, fresh cream and put your whisk or whisk attachment and metal bowl into the freezer a few minutes before whipping. Pour a cup of heavy cream into the bowl and mix on medium speed until the cream begins to have soft peaks. Add a tablespoon of sugar and a teaspoon of vanilla extract before or during whipping for added flavor.

CHEF BIOS: SEASON 5

HOSEA ROSENBERG

This young executive chef from Boulder honed his skills by working with Wolfgang Puck and other top names. Although he achieved notoriety by getting cozy on the couch with castmate Leah, Hosea also made himself known this season by consistently coming up with strong dishes, with an emphasis on seafood. In a surprise upset worthy of March Madness, he out-cooked arch rival Stefan to claim the title of Top Chef.

STEFAN RICHTER

Love him or hate him, you have to admit that this pan-European smart-mouth has talent in the kitchen. He clinched multiple Quick-fires and Eliminations, and barely lost to Hosea in the finale in New Orleans. From a family of cooks, Stefan trained with some of Europe's greatest chefs, and at the Bellagio hotel in Las Vegas and other top locales before starting his own catering business in Los Angeles.

CARLA HALL

Carla tied with charmer Fabio for best quips during Season 5. Between meditating with her spirit guides and "sending the love" from the kitchen, this caterer and former model from D.C. proved she could dish it with the best of them. A Southern gal with classical training and a way with pastry can go far in this world, and on this show—in fact all the way to the final three. Carla, a hearty Hootie-Hoo to you!

FABIO VIVIANI

This native Florentine may not have won the title of Top Chef, but he is numero uno in our hearts. Audiences and judges alike couldn't get enough of Fabio's adorable accent, malapropisms, and authentic homemade pasta. He was charming and affable to the finish. He went down for an inauthentic Creole Maque Choux, but he did it with style and went home with the Fan Favorite title.

JEFF MCINNIS

This executive chef from Miami can cook, yet somehow his fine-dining execution and presentation never quite clicked with the judges. He was sent home for a lackluster ceviche in the Top Chef Bowl but returned for a surprise encore at the New Orleans finale, where he was able to show off his cooking prowess and knowledge of Southern cuisine.

LEAH COHEN

Leah sometimes came across as a little mopey on the show, but you don't make it as far as she did without some major skills. After her brief romance with castmate Hosea, and giving up in the Filet-O-Fish Quickfire, however, things went downhill. Leah was finally given the ax for her Eggs Benedict in the Last Supper Challenge. Leah is currently cooking at Centro Vinoteca in New York City.

JAMIE LAUREN

Jamie won over audiences and the judges with her down-to-earth personality and creative cooking style. People felt like they related to this very precise, determined person, who is executive chef at the popular Absinthe Brasserie & Bar in San Francisco. Obsessed with onions and seasonal produce, she is like the tattooed, teddy-bear-hugging sister you never had.

RADHIKA DESAI

Though classically trained in French cooking, Radhika often tinged her dishes with Indian and Middle Eastern flavors that won over the judges. She took two Quickfire wins before nerves got the better of her and she went down for lack of leadership in Restaurant Wars. She is executive chef at Between Boutique Café & Lounge in Chicago.

ARIANE DUARTE

Ariane was almost cut in the first two episodes, only to rise, phoenix-like, to win two Eliminations and a Quickfire. She impressed the fan base as a good sport and down-to-earth mama and a "Cougar in the Kitchen." After being sent home for a badly butchered lamb, she returned, undaunted, to her family and her restaurant, CulinAriane, in Montclair, New Jersey.

EUGENE VILLIATORA

Eugene got to show off his humor, fearless-ness in the kitchen, and abundant tattoos in Season 5 before being faulted for an overcooked fish during the Focus Group Challenge. We have not seen the last of this scrappy self-taught native Hawaiian; he has the talent, cojones, and sweet way with Pacific Rim flavors to go far.

MELISSA HARRISON

Melissa fought hard and struggled to emerge from the bottom three for much of the season. Unfortunately, her soulful Latin American cuisine did not strike a chord with the judges. Her somewhat pedestrian tuna tacos, made for the Focus Group Challenge, became objects of particular contempt from the judges, including newcomer Toby Young.

DANNY GAGNON

Was there ever a more quintessential New Yorker? Sadly, Danny's collaboration with Carla and Eugene on a plate of confusion for Gail's Bridal Shower resulted in a one-way ticket home. This CIA-trained chef is now sautéing and tournéeing his way to fame behind the stoves at the Babylon Carriage House in Long Island.

ALEX EUSEBIO

Born in Madrid, raised in the Dominican Republic and New York City, and currently cooking in Los Angeles, Alex is like a one-man melting pot. The pain of being sent home for a poorly planned crème brûlée during the *Today Show* Challenge was cushioned by his impending nuptials.

RICHARD SWEENEY

With classical training and a winning smile, affable Richard founded this season's "Team Rainbow" with Patrick and Jamie, and made quick friends with his fellow housemates. Unfortunately, the Foo Fighters weren't pleased with his Banana S'mores, and he was sent home to San Diego.

JILL SNYDER

An experienced chef from Baltimore, Jill's downfall was not so much lack of skill but lack of judgment. Her gigantic ostrich egg did not enhance what the judges considered a boring quiche served to diners at Tom Colicchio's Craft restaurant. Jill was sent packing before we could see what else she had up her sleeve.

PATRICK DUNLEA

Everyone agrees, Patrick is a sweetheart. That did not stop the judges from sending this culinary student to pack his knives and go back to school on the very first episode. His rookie effort at textbook Chinese—bland salmon with bok choy and soggy noodles—was not going to cut it with this New York crowd.

LAUREN HOPE

In the infamous first episode of Season 5, Lauren was kicked off Governors Island in the first-ever Quickfire Elimination Challenge. A recent culinary school grad, Lauren will now have plenty of time to refine her peeling and dicing technique.

EPISODE RECAPS: SEASON 1

EPISODE 1

QUICKFIRE: Work the line at Hubert Keller's restaurant Fleur de Lys.
WINNER: Lee Anne
GUEST JUDGE: Hubert Keller
TIME LIMIT: 30 minutes

It became clear in the very first challenge that those with the most restaurant experience would have an edge. But even Harold—future Top Chef of Season 1—couldn't conquer his nerves and was cut for his shaking hands. Tiffani made a strong early showing, and Ken mouthed off at Chef Keller.

EPISODE 2

QUICKFIRE: Prepare a beautiful fruit plate.
WINNER: Stephen
GUEST JUDGE: Elizabeth Falkner
TIME LIMIT: 30 minutes

Despite thinking that the challenge was unworthy of his talents, Stephen managed to show off his technical and presentation skills and win over guest judge Elizabeth Falkner of Citizen Cake with his "Deconstructed" Fruit Plate.

EPISODE 3

QUICKFIRE: Prepare an octopus dish.
WINNER: Tiffani
GUEST JUDGE: Laurent Manrique
TIME LIMIT: 1 hour

Again, Tiffani showed her talent early on, impressing chef Manrique and edging out the others for her clever fried octopus dish with basil, mint, and chives.

EPISODE 4

QUICKFIRE: Create a dish using $20 of ingredients from a gas station mini-mart.
WINNER: Lee Anne
GUEST JUDGE: Jefferson Hill
TIME LIMIT: 30 minutes

Some were excited by the challenge while others were barely able to hide their disgust as they struggled to make something edible from Spam and Cup-a-Soup. Chef Jefferson Hill gave the win to Lee Anne's Funyun and Oscar-Mayer Spiedini.

EPISODE 5

QUICKFIRE: Identify ingredients while blindfolded.
WINNER: Andrea
GUEST JUDGE: Mike Yakura

She may not have been the best cook, but we knew Andrea's crunchy-granola tastes would come in handy somehow. The Yogic One slayed the competition in this blindfolded taste test of obscure Asian ingredients that turned out to be all but impossible.

EPISODE 6

QUICKFIRE: Create an appetizer using $3 of ingredients.
WINNER: Stephen
GUEST JUDGE: Ted Allen
TIME LIMIT: 20 minutes

When guest judge Ted Allen entered the scene, the fun began. Apparently Andrea's idea of a sales pitch was telling Allen that her Carrot Salad would give him "a great bowel movement." Stephen made the most of his ingredients and turned out a pleasing plate of Poached Manila Clams.

EPISODE 7

QUICKFIRE: Create a signature sandwich to be featured at Tom Colicchio's 'wichcraft.
WINNER: Harold
GUEST JUDGE: Tom Colicchio
TIME LIMIT: 30 minutes

Dave and Stephen made strong showings, but Harold took it with a Mortadella with Wilted Dandelion Greens Sandwich (see recipe, page 169). Miguel dropped the ball with a Deconstructed Falafel that deconstructed in Tom's hands.

EPISODE 8

QUICKFIRE: Design and pitch a wedding menu to a real newlywed couple.
WINNER: Lee Anne
GUEST JUDGES: Betrothed couple Scott and Scott
TIME LIMIT: 1 hour, 30 minutes

Despite the fact that Dave had more catering experience, Lee Anne won over same-sex couple Scott and Scott with her over-the-top presentation—including origami swans and gallery-ready watercolors—and her inspired Pan-Asian menu.

EPISODE 9

QUICKFIRE: Reimagine a classic junk food dish as an upscale entrée.
WINNER: Harold
GUEST JUDGES: Tom Colicchio, Katie Lee Joel, and Gail Simmons
TIME LIMIT: 45 minutes

Tiffani professed her love for foods on a stick, and Dave had trouble with his nacho presentation. Harold confessed to being fed up with junk food challenges, but managed to claim victory with his Popcorn Cakes with Shrimp Ceviche, Calamari, and Clementines.

EPISODE 10

QUICKFIRE: Make a series of dishes in Las Vegas for high rollers, high-stakes poker players, and Cirque du Soleil performers.
WINNER: Harold
GUEST JUDGES: Former chef'testants Stephen, Miguel, Lee Anne, and others
TIME LIMIT: 30 minutes

All the chef'testants were feeling the heat during this grueling Quickfire in the Vegas finale. Though Dave was clearly about to lose it, he performed well until getting confused and serving two dishes instead of three for the final challenge.

EPISODE RECAPS: SEASON 2

EPISODE 1

QUICKFIRE: Create a flambé dish.
WINNER: Sam
GUEST JUDGE: Harold Dieterle
TIME LIMIT: 20 minutes

Elia made a weak early showing by unwisely choosing red wine as her flambé liquid: "Sexy" Sam walked away with the win, and Marcel managed to not light his hair on fire in the first show of a highly flammable season.

EPISODE 2

QUICKFIRE: Create a sushi dish.
WINNER: Cliff
GUEST JUDGE: Hiroshi Shima
TIME LIMIT: 30 minutes

After an early morning visit to a fishmonger, Cliff bested the others and showed off his sushi skills for chef Hiroshi Shima with a plate of Hama Hama Oysters with a Ginger-Soy Mango Salsa and Spot Prawn and Daikon Sushi (see recipe, page 59).

EPISODE 3

QUICKFIRE: Create an original ice cream flavor and serve it to passersby along Redondo Beach Boardwalk.
WINNER: Cliff
GUEST JUDGES: People at Redondo Beach
TIME LIMIT: 2 hours, 45 minutes

These chef'testants are not always best at serving the tastes of the Average Joe, which became clear when Marcel's avant-garde ice cream bombed, while Cliff's people-pleasing s'mores flavor was a hit with the kids. Cliff became the first chef'testants to win two Quickfires in a row.

EPISODE 4

QUICKFIRE: Create an amuse-bouche with $10 of ingredients from a vending machine.
WINNER: Carlos
GUEST JUDGE: Suzanne Goin
TIME LIMIT: 30 minutes

In this classic Quickfire, the chef'testants were given rolls of quarters and told to make an amuse from the sorry contents of some vending machines. Carlos edged out the win with his Sunflower Seed and Carrot Loaf with Cilantro, Sesame Seeds, and—believe it or not—Squirt.

EPISODE 5

QUICKFIRE: Create a dish using offal.
WINNER: Sam
GUEST JUDGE: Michelle Bernstein
TIME LIMIT: 2 hours

Most of the chef'testants seemed pretty comfortable working with innards. Sam won over guest judge and offal-lover Michelle Bernstein with his Sweetbread and Scallion Beignets, while Elia was scolded for her unblanched kidneys.

EPISODE 6

QUICKFIRE: Make something delicious using three different canned foods.
WINNERS: Sam, Mia, Ilan, Cliff, Frank
GUEST JUDGE: Tom Colicchio
TIME LIMIT: 15 minutes

Several of the chef'testants made strong showings in this one, despite being forced to work with canned foods. In an unusual twist, the lucky winners of this challenge got to skip the Elimination Challenge altogether.

EPISODE 7

QUICKFIRE: Create an entrée using ingredients from Redondo Beach farmers' market—no cooking allowed.
WINNER: Marcel
GUEST JUDGE: Raphael Lunetta
TIME LIMIT: 30 minutes

For this challenge, the chef'testants had 30 minutes to turn $20 of Farmers' Market food into a delicious uncooked dish. Marcel got props from guest judge Raphael Lunetta for his innovative Watermelon and Tomato Trio.

EPISODE 8

QUICKFIRE: Create a drink using Baileys Irish Cream and an accompanying dish.
WINNER: Cliff
GUEST JUDGE: Kristin Woodward
TIME LIMIT: 20 minutes

Dismayed by the challenge, Cliff told us he's not a bartender, he's a cook. Yet he cinched another Quickfire with his bold mix of flavors: a Baileys, Rum, Vodka, and Vanilla-Bean Cocktail served with Steak Tapas (see recipe, page 45).

EPISODE 9

QUICKFIRE: Create a dish based on a color.
WINNER: Michael
GUEST JUDGE: Ted Allen
TIME LIMIT: 30 minutes

Betty turned in what looked like an exploding green compost pile, and Cliff did an admirable job with purple considering he is colorblind. Underdog Michael, wiped out from having a tooth pulled, finally won one. He went on to win the Elimination Challenge in the same episode—a first on *Top Chef*.

EPISODE 10

QUICKFIRE: Create a snack using three Kraft Foods products.
WINNER: Sam
GUEST JUDGE: Mike Yakura
TIME LIMIT: 30 minutes

Sam confessed his love of mayonnaise and charmed the judges with the unusual flavor combo in his Southern Kraft Sandwich of Tempura Shrimp, Pickled Peaches, and Barbecue Aïoli (see recipe, page 33).

EPISODE 11

QUICKFIRE: Create a sensual dish using Nestlé Chocolatier products.
WINNER: Sam
GUEST JUDGE: Eric Ripert
TIME LIMIT: 1 hour, 30 minutes

In the last Quickfire of the season, Sam overcame his weak showing in the previous episode's Restaurant Wars to take another Quickfire win for his Shrimp and Banana with Chocolate Chipotle Sauce. Ilan got some well-deserved ribbing for serving Chef Ripert essentially a chocolate bonbon filled with chicken liver.

EPISODE RECAPS: SEASON 3

EPISODE 1

QUICKFIRE: Create an amuse-bouche from buffet appetizer ingredients.
WINNER: Micah
GUEST JUDGES: Tom Colicchio, Gail Simmons
TIME LIMIT: 10 minutes

Let the games begin! As the chef'testants scrambled to make something from the appetizer platters, Clay came up short with a misguided gazpacho in an apple, and Micah won for her "Tuscan Sushi Revisited" of prosciutto, fig jam, and balsamic vinegar (see recipe, page 56).

EPISODE 2

QUICKFIRE: Create a dish featuring Florida citrus.
WINNER: Hung
GUEST JUDGE: Norman Van Aken
TIME LIMIT: 30 minutes

C.J. was humbled to find he left citrus seeds in his dish, and Padma crunched on a shell remnant in Lia's crab salad. Hung showed his skills early in the game with an elegant Slow-Roasted Sea Bass with Citrus and Watercress Salad.

EPISODE 3

QUICKFIRE: Catch and cook a shellfish dish.
WINNER: Brian M.
GUEST JUDGE: Alfred Portale
TIME LIMIT: 30 minutes

After fishing the shellfish out of a tank, it was not a huge surprise when Brian, executive chef at a seafood restaurant, won with his "Tres Rios" of assorted shellfish (see recipe, page 61).

EPISODE 4

QUICKFIRE: Create an appetizer to pair with a Bombay Sapphire cocktail.
WINNER: Casey
GUEST JUDGE: Jamie Walker
TIME LIMIT: 30 minutes

More accustomed to wine than booze, many of the chefs expressed doubts about this one. The judge chose Casey's decadent French-Toast with Pecan-Crusted Foie Gras, paired with a Balsamic Strawberry Rickey (see recipe, page 21), as the winner.

EPISODE 5

QUICKFIRE: Create a dish using a premade piecrust.
WINNER: Joey
GUEST JUDGE: Maria Frumkin
TIME LIMIT: 1 hour, 30 minutes

Prickly Joey tried to pass himself off as a rookie, but later told the camera, "What the others don't know is that I actually have pastry experience." It showed in his well-made Trio of Tarts.

EPISODE 6

QUICKFIRE: Culinary Bee: identify ingredients either by taste or by sight.
WINNER: Casey
GUEST JUDGE: Rocco DiSpirito

Before this fun Quickfire began, Joey got things off to a sunny start by saying that things were beginning to get competitive, and if that meant throwing someone over the balcony, so be it. No violence ensued, as Casey snagged a second Quickfire win.

EPISODE 7

QUICKFIRE: Create an original mix-in for Cold Stone Creamery ice cream.
WINNER: Dale
GUEST JUDGE: Govind Armstrong
TIME LIMIT: 45 minutes

Hung and Casey both went (way) out on a limb for this one, mixing cauliflower and Sriracha, respectively, into sweet-cream ice cream. Chef Govind Armstrong preferred the more traditional flavors of Howie's berries, and for the win, Dale's Peach Cobbler Chèvre Ice Cream (see recipe, page 141).

EPISODE 8

QUICKFIRE: Create an adventurous burger.
WINNER: C.J.
GUEST JUDGE: Daniel Boulud
TIME LIMIT: 30 minutes

You can't accuse C.J. of playing it safe: His Scallop Mousse and Shrimp Burger with Tangerine was bold, and guest judge Daniel Boulud loved it.

EPISODE 9

QUICKFIRE: Mise-en-Place Relay Race.
WINNERS: Team of Dale, Howie, Hung, and Sara M.
GUEST JUDGE: Ted Allen
TIME LIMIT: 30 minutes

Divided into teams, the chef'testants raced to complete a series of technical tasks: shucking oysters, dicing onions, breaking down chickens, and separating and whipping eggs. Casey turned out to be the world's slowest, most methodical onion dicer, and Hung blew away the judges with his knife skills.

EPISODE 10

QUICKFIRE: Create a dish using $10 of ingredients from one aisle of a supermarket.
WINNER: Brian
GUEST JUDGE: Michael Schwartz
TIME LIMIT: 20 minutes

Brian won with his creative Spam, Corned Beef Hash, and Fried Egg with Onions and Balsamic Reduction, but perhaps the most memorable dish was Hung's Smurf Village made with cereal (see page 108). Howie turned in nothing.

EPISODE 11

QUICKFIRE: Make Padma breakfast using a Bunsen burner, a blender, and ingredients from the pantry.
WINNER: Hung
GUEST JUDGE: Padma Lakshmi
TIME LIMIT: 20 minutes

The chef'testants were surprised to be woken up by Padma, who told them they would be making her an impromptu breakfast right there in the apartment. Hung thrilled Padma with his Steak and Eggs, and his Papaya, Banana, and Grand Marnier Shake.

EPISODE 12

QUICKFIRE: Taste and then re-create a classic dish from Le Cirque restaurant.
WINNER: Hung
GUEST JUDGE: Sirio Maccioni
TIME LIMIT: 25 minutes

Cocky from several wins in a row (and just cocky in general), Hung didn't break a sweat as he strutted into the Le Cirque kitchen and turned out a replica of Sea Bass Wrapped in Thinly Sliced Potato over Braised Leeks.

EPISODE 13

QUICKFIRE: Create a fresh trout dish using a camp stove.
WINNER: Casey
GUEST JUDGE: Eric Ripert
TIME LIMIT: 20 minutes

Outdoorsy Aspen is a far cry from the Big Apple, but New York City's Le Bernardin Chef Eric Ripert was impressed with Casey's Pan-Seared Trout with Corn, Tomatoes, and Grapes (see recipe, page 183).

EPISODE RECAPS: SEASON 4

EPISODE 1

QUICKFIRE: Create a signature deep-dish pizza.
WINNERS: Antonia, Dale, Erik, Jennifer, Mark, Richard, Ryan, Spike
GUEST JUDGE: Rocco DiSpirito
TIME LIMIT: 1 hour, 30 minutes

Brand spanking new chef'testants tried to impress guest judge Rocco DiSpirito with an original Chicago-style deep-dish pizza. Many of the chef'testants struggled with the dough, while others used fillings that simply fell flat.

EPISODE 2

QUICKFIRE: Create a dish using no more than five ingredients from Chicago's Green City Market.
WINNER: Mark
GUEST JUDGE: Wylie Dufresne
TIME LIMIT: 30 minutes

After Padma called time, everyone scrambled to find their produce, including Richard, who made off with half a Eucalyptus tree. Despite being frazzled at the market, Mark pulled in a win for the Kiwis.

EPISODE 3

QUICKFIRE: Create an upscale taco.
WINNER: Richard
GUEST JUDGE: Rick Bayless
TIME LIMIT: 30 minutes

Latin-American cuisine star chef Rick Bayless was pleased with Richard's inventive jícama-wrapped Vegetarian Tacos (see recipe, page 121).

EPISODE 4

QUICKFIRE: Create a vegetable platter that showcases three techniques.
WINNER: Dale
GUEST JUDGE: Daniel Boulud

Getting back to basics, the chefs demonstrated three core skills as they sliced and diced a plate of veggies for Chef Boulud. Dale showed off his knife chops with a kick-ass sashimi platter.

EPISODE 5

QUICKFIRE: In a blind taste test, identify the better-quality ingredients.
WINNER: Antonia
GUEST JUDGE: Ming Tsai

Antonia showed off a discerning palate by identifying the higher-quality ingredient twelve out of fifteen times. Jennifer and Ryan also made strong showings.

EPISODE 6

QUICKFIRE: Create a simple dish to pair with a beer.
WINNER: Jennifer
GUEST JUDGE: Koren Grieveson
TIME LIMIT: 30 minutes

The *Top Chef* kitchen was decked out like a brew pub, as the chef'testants got to taste and choose a beer before creating a dish to pair with it. Poker-faced guest judge Koren Grieveson sampled the wares and chose Jennifer's Shrimp and Scallop Beignets (see recipe, page 122) as the winner.

EPISODE 7

QUICKFIRE: Create an innovative dessert.
WINNER: Richard
GUEST JUDGE: Johnny Iuzzini
TIME LIMIT: 30 minutes

Despite being ridiculously straightforward, this dessert challenge threw most of the chefs into a panic. Dale pulled out a favorite family recipe of shaved ice and coconut called Halo Halo. Richard sealed the win with his innovative Banana Scallops that became his signature dish.

EPISODE 8

QUICKFIRE: Create a healthful entrée using Uncle Ben's microwavable rice.
WINNER: Antonia
GUEST JUDGE: Art Smith
TIME LIMIT: 15 minutes

None other than Oprah's personal chef Art Smith joined the chef'testants as they were challenged to make a good-tasting dish using Uncle Ben's new brand of microwavable rice. Antonia led the pack with her Rice Salad with Seared Skirt Steak (see recipe, page 23).

EPISODE 9

QUICKFIRE: Mise-en-Place Relay Race.
WINNERS: Team of Richard, Antonia, Andrew, and Stephanie
GUEST JUDGE: Tom Colicchio

The chef'testants split into two teams of four and competed to see who could complete four tasks fastest. They supremed oranges, turned and cleaned artichokes, cleaned and filleted monkfish, and whipped mayonnaise by hand. Stephanie showed a strong whisking arm, and Dale didn't bother to hide his disappointment when his team lost.

EPISODE 10

QUICKFIRE: Bring the sexy back to salad.
WINNER: Spike
GUEST JUDGE: Sam Talbot
TIME LIMIT: 45 minutes

Top Chef brought back fan favorite, "Sexy" Sam Talbot from Season 2, who decided that Spike's Sensual Beef Salad (see recipe, page 91) warmed his cockles. Antonia's Egg and Bacon Salad was also a strong contender.

EPISODE 11

QUICKFIRE: Work as a short-order cook making eggs in a Chicago diner.
WINNER: Antonia
GUEST JUDGE: Heleen Thanas
TIME LIMIT: 30 minutes

Dale did an admirable job, but Antonia won this one by keeping her cool at the egg station of Lou Mitchell's, a wonderfully chaotic Chicago breakfast institution.

EPISODE 12

QUICKFIRE: Butcher tomahawk chops and then cook them perfectly.
WINNER: Spike
GUEST JUDGE: Rick Tramonto
TIME LIMIT: 30 minutes

Turned out Spike has butchery in the blood. While others toiled and troubled with big knives in the cold packing house, Spike didn't sweat it, turning out near-perfect chops. Antonia's steak was also cooked perfectly, but Spike earned the win here, only to be eliminated later in the same episode.

EPISODE 13

QUICKFIRE: Create two frituras (fried beach snacks), both with plantains.
WINNER: Stephanie
GUEST JUDGE: Wilo Benet
TIME LIMIT: 20 minutes

Down to the last four chef'testants in the first part of the finale, Stephanie won over Wilo Benet with her tasty Tostones, flattened fried plantains with seared tuna, and her Kosher Special: Shrimp, Pork, and Banana Fritters (see recipe, page 66).

EPISODE RECAPS: SEASON 5

EPISODE 1

QUICKFIRE: Beat the clock to peel fifteen apples without a peeler, then dice them into brunoise (fine dice), and finally make a dish using the apples.
WINNER: Stefan
GUEST JUDGE: Tom Colicchio
TIME LIMIT: 20 minutes for cooking round

Welcome to New York, now go home. In this first Quickfire that was also an Elimination, chef'testants fresh off the plane had to perform some basic tasks under the gun. Speedy Stefan finished first, and Radhika and Leah landed in the bottom.

EPISODE 2

QUICKFIRE: Create a signature hot dog.
WINNER: Radhika
GUEST JUDGE: Donatella Arpaia
TIME LIMIT: 45 minutes

The chef'testants had to create a hot dog that could compete against Angelina from Dominick's legendary hot dog truck in Queens. Jill was busted for not making her own sausage, and Radhika nailed it with her Kebab Sausage with Tomato Jam (see recipe, page 101).

EPISODE 3

QUICKFIRE: Make a dish from *Top Chef: The Cookbook*. Wait!—and turn it into a soup.
WINNER: Leah
GUEST JUDGE: Grant Achatz
TIME LIMIT: 1 hour

All of the chef'testants proved to be quick on their feet in this switcheroo challenge. Danny and Jamie turned in strong dishes, but Leah overcame her dislike of white asparagus to create the winning soup (see recipe, page 163), according to Chef Grant Achatz.

EPISODE 4

QUICKFIRE: Create a breakfast amuse-bouche.
WINNER: Leah
GUEST JUDGE: Rocco DiSpirito
TIME LIMIT: 30 minutes

Leah won her second Quickfire in a row with her Grilled Bread with Bacon and Egg (see recipe, page 55). Fabio misstepped by ignoring Rocco's plea for bacon and going with a sweet breakfast instead.

EPISODE 5

QUICKFIRE: Identify the most ingredients in a dish.
WINNER: Hosea
GUEST JUDGE: Padma Lakshmi

The chef'testants competed head-to-head in a kind of "Name That Tune" of the taste buds to identify ingredients in Thai Green Curry and other complex dishes. This was a little more of a win-by-not-losing for Hosea, but he proved he has a sharp palate.

EPISODE 6

QUICKFIRE: Create a one-pot holiday meal.
WINNER: Ariane
GUEST JUDGE: Martha Stewart
TIME LIMIT: 45 minutes

Cooking for none other than Martha Stewart, Hosea, Ariane, and Jamie impressed The Great One with their one-pot wonders (see recipes, pages 132 and 137), while Eugene got called out for thickening his stew with cornstarch.

EPISODE 7

QUICKFIRE: Create a delicious no-sugar dessert.
WINNER: Radhika
GUEST JUDGE: Jean-Christophe Novelli
TIME LIMIT: 45 minutes

Radhika won another one with her Peach-Lavender Bread Pudding (see recipe, page 46). Ariane turned in a weak dessert and was taken to task for her over-whipped cream.

EPISODE 8

QUICKFIRE: Create a dish using only assorted canned products.
WINNER: Stefan
GUEST JUDGE: Hung Huynh
TIME LIMIT: 15 minutes

Jeff showed off his cooking and presentation skills with a deep-fried conch and two sauces. Former *Top Chef* winner Hung Huynh appraised the dishes with a cold eye and awarded top place to Stefan's Baked-Bean Soup with Spam and Velveeta.

EPISODE 9

QUICKFIRE: Create a tasting and pitch a new restaurant concept.
WINNERS: Leah, Radhika
GUEST JUDGE: Stephen Starr
TIME LIMIT: 30 minutes

Successful restaurateur Stephen Starr chose Leah's Asian fusion concept and Radhika's Middle East-meets-India as successful ideas, setting them both up for failure as team leaders for the notorious Restaurant Wars Challenge.

EPISODE 10

QUICKFIRE: Create an entrée using oats.
WINNER: Stefan
GUEST JUDGE: Scott Conant
TIME LIMIT: 45 minutes

Carla revealed herself as an oat lover, while Jamie was less than excited to cook with the grain. Most of the chef'testants used the oats as a breading for pan-frying various proteins, but Stefan won it with his Banana Mousse with Oatmeal-Almond Crisp (see recipe, page 36).

EPISODE 11

QUICKFIRE: Fillet several types of fish.
WINNER: Stefan
GUEST JUDGE: Eric Ripert
TIME LIMIT: 5 minutes (sardines), 5 minutes (Arctic char), 10 minutes (freshwater eel)

In this unforgettable challenge, seafood chef Hosea was embarrassed by his lack of skill taking apart tiny sardines. Leah quit halfway through filleting an Arctic char, and Stefan couldn't contain his glee as he nailed a squirming eel to a plank and tore off its skin.

EPISODE 12

QUICKFIRE: Create an egg dish.
WINNER: Carla
GUEST JUDGE: Wylie Dufresne
TIME LIMIT: 1 hour

Many of the chef'testants overthought this one, using molecular gastronomy and clever plating to impress Wylie Dufresne. Though Stefan made a beautiful duo plate, Dufresne preferred the flavorful simplicity of Carla's Green Eggs and Ham (see recipe, page 161).

EPISODE 13

QUICKFIRE: Create a dish using crayfish.
WINNER: Jeff
GUEST JUDGE: Emeril Lagasse
TIME LIMIT: 1 hour

The final four arrived in New Orleans to a major surprise: the recently eliminated Leah, Jamie, and Jeff returned for a second chance. The three duked it out before celebrity chef Emeril Lagasse. Jeff won this Quickfire, but failed to clinch the Elimination win needed to secure him a spot at the finale.

GLOSSARY

aioli (ay-OH-lee or i-OH-lee): A thick, cold mayonnaise-like sauce that originated in Provence, in southern France, often served with simply cooked or steamed vegetables and fish.

al dente (al-DEN-tay): Literally "to the tooth" in Italian, "al dente" describes food—most commonly pasta or rice—that is cooked through but still firm, offering slight resistance when chewed, and not soft or mushy at all. When cooking pasta, be sure to start tasting pieces well before the recommended cooking time is up: as soon as it loses its raw taste and the center is no longer bright white, it's done.

amuse-bouche (ah-MOOZE-boosh): Literally "mouth amuser" in French. A small, one- or two-bite-sized dish served just before the beginning of a meal.

beignet (ben-YAY): Refers to a donut popular in southern Louisiana—a deep-fried, yeast-raised rectangle (no hole), dusted liberally with confectioners' sugar.

blanch: To cook, usually very briefly, in boiling water. Vegetables are often blanched to loosen their skins for peeling or to soften them before cooking them further by another method.

braise (BRAYZ): To brown in fat, then cook slowly in liquid at low heat.

brioche (bree-OSH): A pliable, eggy bread made from yeast, butter, and eggs. Brioche can be sweet or savory, and can be shaped into rolls, loaves, rings, or fanciful shapes such as crosses or braids.

brûlée (broo-LAY): In informal use, to caramelize the surface of a food (as in crème brûlée) with a hand-held blowtorch, salamander grill, or oven broiler.

butterfish: (a.k.a. skipjack or dollarfish) A small, bony species indigenous to the eastern United States that has a distinctively rich flavor and fatty texture. Many different kinds of fish are sold as "butterfish," however, including escolar, walu, and black cod.

caramelize: To heat until the sugars in a food liquefy and turn brown.

ceviche (seh-VEE-chay): A South American dish of raw fish or other seafood tossed with citrus juice and fresh herbs and vegetables such as hot chilis, tomatoes, scallions, and cilantro.

chiffonade (shiff-on-AHD): To cut leafy greens or herbs into very thin ribbons. To chiffonade basil, for example, stack the leaves, roll them into a tight cigar-like roll, and cut across the roll into thin strips.

chorizo (chor-EE-soh): There are several varieties of chorizo, but Spanish and Mexican are the prevalent varieties in U.S. markets. Spanish chorizo is usually a cured pork sausage seasoned with pimento, a smoked paprika, and garlic, spices, and herbs. Mexican chorizo is a fresh sausage seasoned with chili peppers and vinegar, usually cut from its casing and fried.

clarify: To remove impurities or sediment from a liquid.

daikon (DIE-con): A large, white-fleshed Asian radish with white or black skin.

deglaze: To add a liquid (often wine) to a hot pan in which food is being or has been sautéed, stirring to scrape up any of the browned bits of food in the bottom of the pan, which add flavor and body to the dish or sauce.

flambé (flahm-BAY): To ignite the alcohol in a mixture using a long lighter or kitchen match.

fold: To very gently incorporate one ingredient or mixture into another.

gelée (jeh-LAY): Loosely used, a jelled or jellylike sauce.

jícama (HEE-kah-mah): A round root vegetable with creamy white, crisp flesh and light brown skin. Jícama can be eaten raw or cooked and is common in Latin American dishes.

julienne (joo-lee-EHN): To cut into long, thin strips.

Kobe (KOH-bee): An extremely expensive, high-grade type of beef from cattle raised to exacting standards in Kobe, Japan. The animals are fed a special diet that includes beers, and are massaged regularly with sake.

mince: To cut into very small pieces—as small as possible without pureeing. To mince garlic or fresh herbs, first chop roughly with a chef's knife, then hold the tip of the knife on the cutting board with your palm while you rock the blade over the food, moving it back and forth, always keeping the knife in contact with the board.

pan roast: To roast in a pan. A meat or vegetable can be seared in an ovenproof pan on the stove top, then transferred to the oven to finish cooking. Informally, "pan roasting" can refer to cooking meat or vegetables (in very little fat) on the stove top over high heat.

pipérade (peep-er-odd): Basque in origin, pipérade is a stewed mixture of onions, garlic, tomatoes, sweet peppers, and Espelette pepper used as a side dish or seasoning for eggs.

piquillo (pi-KEE-yo): Small, mildly spicy pickled peppers from northern Spain that can be eaten alone, stuffed, or used to bring a unique tang to meat, poultry, fish, and egg dishes.

plantain: A starchy member of the banana family, plantains are widely consumed throughout the world. Unlike bananas, platains are cooked before they are eaten.

poach: To cook food (usually meat, fish, or eggs) in barely simmering water or another liquid.

purée: To chop or mash a food until it achieves a smooth, uniform consistency.

reduce: To cook a liquid at a brisk simmer or boil in order to evaporate it, thickening and concentrating its flavor.

saffron: The dried stigmas of the saffron flower, used as a spice. Saffron has a strong flavor and lends a distinctive yellow color to dishes such as paella, risotto, and bouillabaisse.

sambar: Sambar powder is a mix of dried, crushed spices including coriander, fenugreek, mustard, and chili. Sambar also refers to a soup of vegetables, spices, and pulses that is eaten on its own or used as a condiment.

sashimi: Fresh raw seafood. If it has rice on it, under it, in it, or around it, it's sushi. If not, it's sashimi.

sauté: To cook food in a small amount of fat in a shallow pan or skillet on the stove top, stirring frequently.

sear: To cook over high heat in order to quickly brown the exterior of a piece of food (usually meat or fish).

sodium alginate: Alginic acid, a chemical compound derived from seaweed with various commercial and scientific uses. In cooking, food-grade sodium alginate is used as a thickener and stabilizer.

sommelier (suh-mel-YAY): A trained and experienced wine specialist. Working at a restaurant, the sommelier is responsible for buying and storing wines for the restaurant's cellar, and will work with the chef to pair wines (and sometimes beers and spirits) with dishes on the menu.

sous vide (soo-VEED): Literally "under vacuum" in French. A method of cooking food in vacuum-sealed plastic bags submerged in a bath of water at a carefully maintained temperature.

supreme: In cooking, the term "supreme" has several meanings, including "Chicken Supreme" (a chicken breast with wing attached). When referring to citrus fruits, to supreme means to cut off the skin, pith, and membrane, and separate the wedges.

tarte Tatin (tart TA-tan): An upside-down tart with caramelized fruit and a pastry crust, traditionally made with apples.

tempura: Batter-dipped, deep-fried foods, such as vegetables and fish. Traditionally Japanese.

truffle: A variety of underground mushroom of the genus Tuber. "Truffle" can also refer to a confection made of chocolate ganache formed into balls and either left plain, dusted with cocoa, or encased in a chocolate shell.

vadouvan (VA-doo-vanh): A mild, citrusy play on Indian curry popularized by Parisian chefs, this spice paste includes caramelized shallots and garlic, fenugreek, chilis, and turmeric.

verrine: A French appetizer or confection in which components are layered in a glass to present a contrast of textures. Verrines can be sweet or savory. Sweet verrines often consist of layers of cream, crumbled cake or biscuit, and fruit. Savory verrines can contain meat or seafood, cheese, vegetables, and custard.

yuzu: A distinctively sour and aromatic Japanese citrus fruit.

zabaione (Za-bah-YO-nee): A rich, sweet Italian custard thickened with eggs.

RECIPES BY TYPE

BREAKFAST

APPETIZERS

SALADS

SOUPS

SANDWICHES

TACOS

PIZZA

SIDES

MAIN COURSES

DESSERTS

DRINKS

MENUS

TROPICAL TASTES

FRIDAY NIGHT FAMILY DINNER

LIGHT SUMMER SUPPER FOR ONE

SEAFOOD SPECTACULAR

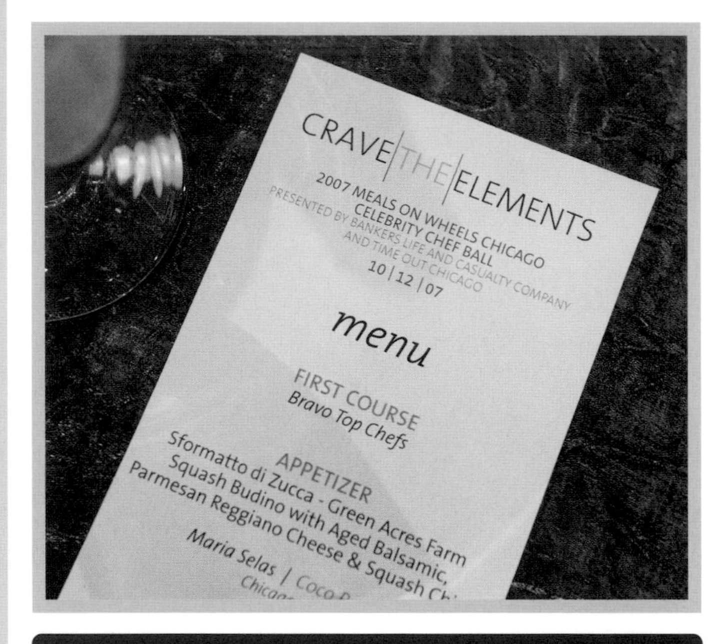

 SUNDAY DINNER

 I HEART SUSHI PLATTER

 SUPERBOWL MUNCHIES

 ELEGANT DINNER PARTY FOR SIX

ROMANTIC DINNER FOR TWO

BREAKFAST IN BED

COCKTAIL PARTY FINGER FOOD

FRY, DADDY, FRY

VEGETARIAN DELIGHT

INDEX

Note: Chef'testants are indexed by their first names.

ACKNOWLEDGMENTS

Grace under pressure. We all aspire to it, and we try, but don't always succeed, at exhibiting it ourselves. Bravo shows put talented people in extreme situations that move the ordinary into the realm of extraordinary. Whether it be food, fashion, beauty, design, or pop culture, Bravo series have captured what we are capable of when pushed to the edge: How fast, how resourceful, how creative can we be? How true do we remain to ourselves? The Quickfire segment of *Top Chef* has always been one of my favorite showcases of grace under pressure, and sometimes fire. This book—part cookbook and part fanzine of the show—is a celebration of what happens when innovative chefs are pushed to their limits in terms of time, ingredients, and creativity. I hope the recipes, tips, and Home Quickfire Challenges in these pages will inspire us all to push ourselves, get a little crazy, come back to our real selves, and, most of all, have fun in the kitchen.

—Lauren Zalaznick

President, NBC Universal Women & Lifestyle Entertainment Networks

Thanks to Tom Colicchio, Padma Lakshmi, and Gail Simmons.

Thanks to the team at Bravo Media: Frances Berwick, Cameron Blanchard, Victoria Brody, Andrew Cohen, Johanna Fuentes, Maile Marshall, Lauren McCollester, Kim Niemi, Kate Pappa, Dave Serwatka, Ellen Stone, Trez Thomas, Jennifer Turner, and Andrew Ulanoff.

Thanks to the team at Magical Elves: Doneen Arquines, Rich Buhrman, Liz Cook, Dan Cutforth, Bill Egle, Gayle Gawlowski, Chi Kephart, Jane Lipsitz, Shauna Minoprio, Molly O'Rourke, Erin Rott, Nan Strait, Andrew Wallace, and Webb Weiman.

Thanks to the chef'testants who participated in this book: Richard Blais, Nikki Cascone, Radhika Desai, Jamie Lauren, Spike Mendelsohn, Hosea Rosenberg, Ryan Scott, and Dale Talde.

Thanks also to Cynthia Arntzen, Sarah Billingsley, Mikyla Bruder, Kathy Chow, Vanessa Dina, Anne Donnard, Catherine Grishaver, David Hawk, Ben Kasman, Justin Kellis, Jennifer Kong, Laurel Leigh, Amy Nichols, Doug Ogan, Peter Perez, Molly Prentiss, and Ann Spradlin.

Thanks to all our sponsors, without whom *Top Chef* would not be possible.

TRADEMARKS

Apple Jacks, Froot Loops, and Frosted Flakes are registered trademarks of Kellog NA Co.

Bacardi and Bombay Sapphire are registered trademarks of Bacardi & Co. Limited.

Baileys Original and Baileys Irish Cream are registered trademarks of R&A Bailey and Co.

Beck's is a registered trademark of Import Brands Alliance.

Cap'n Crunch is a registered trademark of The Quaker Oats Company.

Cheerios, Cocoa Puffs, and Lucky Charms are registered trademarks of General Mills, Inc.

Cheetos, Doritos, and Funyuns are registered trademarks of Frito-Lay, Inc.

Cointreau is a registered trademark of Remy Cointreau Group.

Cold Stone is a registered trademark of Kahala Corporation.

Combos is a registered trademark of Mars, Incorporated.

Cool Whip, CornNuts, Nilla Wafers, Pop Rocks, and Velveeta are registered trademarks of Kraft Foods Global, Inc.

Cup-a-Soup and I Can't Believe It's Not Butter! are registered trademarks of Unilever.

Ding Dongs, Ho Hos, Zingers, and Twinkies are registered trademarks of Interstate Bakeries Corporation.

Dr Pepper is a registered trademark of Dr Pepper/Seven Up, Inc.

Frangelico is a registered trademark of C&C Group PLC.

Fry Baby and Fry Daddy are registered trademarks of National Presto Industries, Inc.

Ginger People is a registered trademark of The Ginger People.

Grand Marnier is a registered trademark of Société des Produits Marnier-Lapostolle.

Guittard is a registered trademark of Guittard Chocolate Company.

Häagen-Dazs is a registered trademark of HDIP, Inc.

Hama Hama Oyster is a registered trademark of Hama Hama Oyster Company.

Hershey's, Twizzlers, and Whoppers are registered trademarks of The Hershey Company.

Hoegaarden is a registered trademark of INBEV.

Kahlúa is a registered trademark of Malibu-Kahlúa Incorporated.

Land Shark Lager is a registered trademark of Margaritaville Brewing Co.

Lee Kum Kee is a registered trademark of Lee Kum Kee.

Lindt is a registered trademark of Lindt & Sprungli.

Mrs. Butterworth's is a registered trademark of Pinnacle Foods Group LLC.

Myers's Rum is a registered trademark of Diageo PLC.

Nerds and Nestlé Chocolatier are registered trademarks of Société des Produits Nestlé S.A.

Parmigiano-Reggiano is a registered trademark of Consorzio del Formaggio Parmigiano-Reggiano.

POM is a registered trademark of POM Wonderful LLC.

Red Vines is a registered trademark of American Licorice Company.

Scharffen-Berger is a registered trademark of SVS Chocolate LLC.

Spam is a registered trademark of Hormel Foods Corporation.

Squirt is a registered trademark of Cadbury Schweppes.

Sprite is a registered trademark of The Coca-Cola Company.

Sub-Zero is a registered trademark of Sub-Zero, Inc.

Thin Mints is a registered trademark of Girl Scouts of the United States of America.

Uncle Ben's Ready Rice is a registered trademark of Mars, Incorporated.

Viking is a registered trademark of Viking Range Corporation.

Vita-Mix is a registered trademark of Vita-Mix Corporation.

TABLE OF EQUIVALENTS

The exact equivalents in the following tables have been rounded for convenience.

LIQUID/DRY MEASURES

U.S.	METRIC
1/4 teaspoon	1.25 milliliters
1/2 teaspoon	2.5 milliliters
1 teaspoon	5 milliliters
1 tablespoon (3 teaspoons)	15 milliliters
1 fluid ounce (2 tablespoons)	30 milliliters
1/4 cup	60 milliliters
1/3 cup	80 milliliters
1/2 cup	120 milliliters
1 cup	240 milliliters
1 pint (2 cups)	480 milliliters
1 quart (4 cups, 32 ounces)	960 milliliters
1 gallon (4 quarts)	3.84 liters
1 ounce (by weight)	28 grams
1 pound	454 grams
2.2 pounds	1 kilogram

LENGTH

U.S.	METRIC
1/8 inch	3 millimeters
1/4 inch	6 millimeters
1/2 inch	12 millimeters
1 inch	2.5 centimeters

OVEN TEMPERATURE

FAHRENHEIT	CELSIUS	GAS
250	120	1/2
275	140	1
300	150	2
325	160	3
350	180	4
375	190	5
400	200	6
425	220	7
450	230	8
475	240	9
500	260	10